STREPTOCOCCI

THE SOCIETY FOR APPLIED BACTERIOLOGY
SYMPOSIUM SERIES NO. 7

STREPTOCOCCI

Edited by

F. A. SKINNER

AND

L. B. QUESNEL

1978

ACADEMIC PRESS

LONDON . NEW YORK . SAN FRANCISCO
A Subsidiary of Harcourt Brace Jovanovich, Publishers

ACADEMIC PRESS INC. (LONDON) LTD
24-28 OVAL ROAD
LONDON NW1 7DX

QR
82
S78
S76
1978

U.S. Edition published by
ACADEMIC PRESS INC.
111 FIFTH AVENUE
NEW YORK, NEW YORK 10003

Copyright © 1978 By the Society for Applied Bacteriology

ALL RIGHTS RESERVED

NO PART OF THIS BOOK MAY BE REPRODUCED IN ANY FORM BY PHOTOSTAT, MICROFILM, OR ANY OTHER MEANS, WITHOUT WRITTEN PERMISSION FROM THE PUBLISHERS

Library of Congress Catalog Card Number 78-54540
ISBN: 0-12-648035-4

Printed in Great Britain by
Whitstable Litho Ltd., Whitstable, Kent

Contributors

ELLA M. BARNES, *ARC Food Research Institute, Norwich, Norfolk NOR 70F, England*

P. G. H. BIJKER, *Department of the Science of Food of Animal Origin, Faculty of Veterinary Medicine, University of Utrecht, The Netherlands*

N. P. BURMAN, *Thames Water Authority, New River Head Laboratories, London EC1R 4TP, England*

W. A. COX, *Unigate Technical Centre, Bradford-on-Avon, Wiltshire BA15 1DH, England*

I. EELDERINK, *Department of the Science of Food of Animal Origin, Faculty of Veterinary Medicine, University of Utrecht, The Netherlands*

A. W. EVANS, *Thames Water Authority, New River Head Laboratories, London EC1R 4TD, England*

J. M. HARDIE, *Oral Microbiology Department and MRC Dental Epidemiology Unit, The London Hospital Medical College, London E1 2AD, England*

A. HURST, *Health Protection Branch, Health and Welfare Canada, Ottawa, Canada*

M. INGRAM, *ARC Meat Research Institute, Langford, Bristol, Avon BS18 7DY, England*

D. J. JAYNE-WILLIAMS, *National Institute for Research in Dairying, Shinfield, Reading, Berkshire RG2 9AT, England*

DOROTHY JONES, *Department of Microbiology, University of Leicester, Leicester, Leicestershire LE1 7RH, England*

M. J. LATHAM, *National Institute for Research in Dairying, Shinfield, Reading, Berkshire RG2 9AT, England*

B. A. LAW, *National Institute for Research in Dairying, Shinfield, Reading, Berkshire RG2 9AT, England*

J. E. LEWIS, *Unigate Food Ltd, Sturminster Marshall, Wimborne, Dorset BS21 4AX, England*

P. D. MARSH, *Oral Microbiology Department and MRC Dental Epidemiology Unit, The London Hospital Medical College, London E1 2AD, England*

W. R. MAXTED, *Central Public Health, Laboratory, London NW9 5HT, England*

G. C. MEAD, *ARC Food Research Institute, Norwich, Norfolk NOR 70F, England*

D. A. A. MOSSEL, *Department of the Science of Food of Animal Origin, Faculty of Veterinary Medicine, University of Utrecht, Utrecht, The Netherlands*

M. T. PARKER, *Central Public Health Laboratory, London NW9 5HT, England*

J. PAYNE, *ARC Food Research Institute, Norwich, Norfolk NOR 70F, England*

P. W. ROSS, *Department of Bacteriology, University of Edinburgh, University Medical School, Edinburgh EH8 9AG, Scotland*

G. F. H. SALT, *Central Veterinary Laboratory, Ministry of Agriculture, Fisheries and Food, Weybridge, Surrey KT15 3NB, England*

M. ELISABETH SHARPE, *National Institute for Research in Dairying, Shinfield, Reading, Berkshire RG2 9AT, England*

G. STANLEY, *Unigate Technical Centre, Bradford-on-Avon, Wiltshire BA15 1DH, England*

JANET K. STEVENS, *Thames Water Authority, New River Head Laboratories, London EC1R 4TP, England*

R. WHITTENBURY, *Department of Biological Sciences, University of Warwick, Coventry, Warwickshire CV4 7AL, England*

C. D. WILSON, *Central Veterinary Laboratory, Ministry of Agriculture, Fisheries and Food, Weybridge, Surrey KT15 3NB, England*

Preface

In 1977 the Society for Applied Bacteriology devoted its annual Symposium to the study of a single genus of micro-organisms, *Streptococcus*. This volume is a collection of the papers presented at that Symposium, held during the Summer Conference in the University of Manchester in July.

The genus *Streptococcus* was a most appropriate choice, as it comprises organisms whose impact on man range from the obviously inimical to the decidedly beneficial. The genus includes many disease-causing organisms such as *Streptococcus pyogenes* and *S. pneumoniae*, of obvious concern to the clinician; others, such as *S. agalactiae* and *S. equi* give rise to diseases in animals and fall within the province of the veterinarian, while many others are involved in the complex ecological interactions to be found in the oral cavity and have been implicated in tooth decay and plaque formation.

More beneficial streptococci are those used in the manufacture of dairy products such as cheeses and fermented milks of various kinds that are consumed in ever-increasing quantities and whose world-wide economic impact is enormous. Some of the important streptococci in these industries are *S. lactis*, *S. cremoris*, *S. thermophilus* and *S. diacetylactis*. In between these poles are species such as *S. faecalis*, a useful indicator of pollution of food and water supplies; and the streptococci involved in the digestive processes of ruminant animals. The significance of the presence of streptococci in various foods and the nature of cell damage and repair consequent upon food processing procedures are other topics of consideration.

These and other aspects of streptococci are reviewed in this volume, which includes papers, not only of immediate interest for the practical application of our knowledge of streptococci to medicine and industry, but also on fundamental, descriptive and research aspects of the study of these organisms. The final session of the Manchester Symposium was devoted to an exposition and discussion of the media used for routine and research purposes in the selection and identification of streptococci, and in the final section of this volume we have compiled a wide selection of media, presented succintly but in sufficient detail to be of real practical use to the practising bacteriologist.

F. A. SKINNER
Rothamsted Experimental Station
Harpenden AL5 2JQ
Hertfordshire
England

L. B. QUESNEL
Department of Bacteriology and Virology
University of Manchester
Manchester M13 9PT
England

Contents

LIST OF CONTRIBUTORS	v
PREFACE	vii

Composition and Differentiation of the Genus *Streptococcus*
DOROTHY JONES

Introduction	1
Early Classification	3
Serological Grouping	6
Current Classification	8
Concluding Remarks	37
References	40

Biochemical Characteristics of *Streptococcus* Species
R. WHITTENBURY

Introduction	51
The Aerobic Nature of Streptococci and Other Lactic Acid Bacteria	51
Soft Agar Culture	52
The Effect of Adding Haemin to Media	56
The Inclusion of Pediococci and Aerococci within the Genus *Streptococcus*	64
References	68

The Pattern of Streptococcal Disease in Man
M. T. PARKER

The Streptococci of Man	71
Streptococcal Disease in Man	77
Virulence of Streptococci for Man	85
Epidemiological Features of Streptococcal Diseases in Man	87
The General Picture	101
References	102

Group A Streptococci: Pathogenesis and Immunity
W. R. MAXTED

Introduction	107
The M Proteins	109
Heterogeneity Among Antigens of the M complex	113
M types Associated with Particular Streptococcal Diseases	118

The Pathogenesis of Post-streptococcal Glomerulonephritis and
Rheumatic Fever 120
References 123

Ecology of Group B Streptococci
P. W. ROSS
Historical 127
The Organism 128
Diseases Associated with Group B Streptococci 129
Carriage in Adults 132
Colonization and Infection in Humans 134
Distribution of Serotypes Among Streptococcal Isolates . . 135
Factors Involved in Human Infections by Group B Streptococci . 136
The Future 138
References 139

Streptococci in Animal Disease
C. D. WILSON AND G. F. H. SALT
Introduction 143
Diseases of Cattle 144
Streptococcal Diseases of Pigs 149
Streptococcal Diseases of Horses 151
Streptococcal Diseases of Sheep 152
Streptococcal Diseases of Poultry 153
References 153

Streptococci and the Human Oral Flora
J. M. HARDIE AND P. D. MARSH
Introduction 157
Streptococci Found in the Mouth 158
Some Aspects of the Biochemistry and Metabolism of Oral
Streptococci 175
Microbial Interactions Involving Oral Streptococci . . . 181
The Relationship of Oral Streptococci to Disease . . . 185
References 193

Streptococci in the Alimentary Tract of the Ruminant
M. J. LATHAM AND D. J. JAYNE-WILLIAMS
Introduction 207
Streptococci in the Alimentary Tract of the Healthy Ruminant . 210
Involvement of *Streptococcus bovis* in Alimentary Disorders . 219

Aspects of the Physiological and Biochemical Characteristics of
Streptococcus bovis and other Group D Streptococci of
Significance to Rumen Function. 224
Interrelationships Between *Streptococcus bovis* and other Rumen
Micro-organisms 230
Conclusions 232
References 233

Streptococci in the Intestinal Flora of Man and other Non-ruminant Animals
G. C. Mead
Introduction 245
Occurrence and Distribution in Different Regions of the Intestine. 246
Factors Affecting the Streptococcal Flora. 248
Development of the Streptococcal Flora 248
Influence of Diet 250
Species Distribution in Various Hosts 250
Some Possible Interactions between the Streptococcal Flora and
the Host 254
Summary 257
References 258

Streptococci in the Dairy Industry
B. A. Law and M. Elisabeth Sharpe
Introduction 263
Lactic Streptococci used as Starter Cultures in the Dairy Industry . 264
Utilization of Milk Proteins by Lactic Streptococci . . . 267
Lactose Utilization by Lactic Streptococci 268
Alternative Pathways of Carbohydrate Metabolism and their Effect
on the Properties of Short-life Fermented Products . . 270
The Influence of Lactic Streptococci on Cheese Ripening . . 271
The Formation of Toxic Amines in Cheese by Group D Streptococci 273
Causes of Slow Growth of Lactic Streptococci in Milk . . 274
Factors Stimulating the Growth of Lactic Streptococci . . 275
References 276

Starters: Purpose, Production and Problems
W. A. Cox, G. Stanley and J. E. Lewis
Introduction 279
The Composition and Purpose of Starter Cultures . . 279
Production of Starters in the Dairy Industry 281
Problems Associated With Production and Use of Starters . 291
References 293

Nisin: Its Preservative Effect and Function in the Growth Cycle of the Producer Organism
A. HURST
　General Introduction 297
　Nisin as a Food Preservative 298
　Function of Nisin in the Growth Cycle of the Producer Organism . 301
　References 311

Streptococci of Lancefield Groups A, B and D and Those of Buccal Origin in Foods: Their Public Health Significance, Monitoring and Control
D. A. A. MOSSEL, P. G. H. BIJKER AND I. EELDERINK
　Ecological Principles 315
　Monitoring of Foods for Streptococci 318
　Control of Contamination by Streptococci 328
　References 329

Streptococci as Indicators in Water Supplies
N. P. BURMAN, JANET K. STEVENS AND A. W. EVANS
　Introduction 335
　Routine Application of Faecal Streptococci Counts . . . 336
　The Use of Faecal Coliform: Faecal Streptococci Ratios to Indicate the Source of Faecal Pollution 340
　Acknowledgements 346
　References 346

Damage and Recovery in Streptococci
J. PAYNE
　Introduction 349
　Damage Caused by Exposure to Cold 351
　Damage Caused by Exposure to Heat 354
　Damage Caused by Other Stress Conditions 356
　Recovery from Cold-induced Damage 358
　Recovery from Heat-induced Damage 359
　General Comments 366
　References 368

Isolation Media for Streptococci: Proceedings of a Discussion Meeting 371
　Principles
　　Ella M. Barnes 372
　Streptococci of Human Diseases
　　P. W. Ross 375

CONTENTS

Streptococci of Animal Diseases
    C. D. Wilson 379
Oral Streptococci
    J. M. Hardie and P. D. Marsh 380
Streptococci of the Normal Intestinal Flora
    G. C. Mead 384
Dairy Streptococci
    M. Elisabeth Sharpe 386
Streptococci in Food
    D. A. A. Mossel 392
Streptococci in Water
    N. P. Burman and A. W. Evans 393
Summary of Discussion
    M. Ingram and Ella M. Barnes 394
Appendix 395

Selected Abstracts Presented at the Summer Conference . . . 397

SUBJECT INDEX 409

Composition and Differentiation of the Genus *Streptococcus*

DOROTHY JONES

*Department of Microbiology, University of Leicester,
Leicester, Leicestershire, England*

CONTENTS

1. Introduction 1
2. Early classification 3
3. Serological grouping 6
4. Current classification 8
    (a) Pyogenic streptococci 9
    (b) Pneumococci 17
    (c) Oral streptococci 19
    (d) Faecal streptococci 22
    (e) Lactic streptococci 29
    (f) Anaerobic streptococci 31
    (g) Other streptococci 34
5. Concluding remarks 37
6. References 40

1. Introduction

THE GENUS *Streptococcus* is now classified in the family Streptococcaceae together with the genera *Pediococcus, Leuconostoc, Aerococcus* and *Gemella* (Deibel & Seeley 1974). All these genera contain bacteria which are predominantly spherical or ovoid in shape, but rod forms do occur—especially on solid media. All have a Gram positive cell wall structure and, with the exception of *Gemella*, give a Gram positive staining reaction. None forms endospores. None synthesizes haem compounds. All are facultatively anaerobic, have complex nutritional requirements, and form mainly lactic acid or lactic, acetic and formic acids and ethanol and CO_2 from carbohydrates. The genera are distinguished from each other by two main features: the mode of cell division and the end products, formed from the metabolism of carbohydrates (see Table 1).

All anaerobic forms have been excluded from the family Streptococcaceae (Deibel & Seeley 1974) but Rogosa (1974) and Holdeman & Moore (1974) are of the opinion that Gram positive anaerobic coccoid bacteria which occur in pairs or chains and produce lactic acid as the major fermentation product should be included in the genus *Streptococcus*. Whether or not this is justifiable on our present knowledge of these anaerobic cocci, is debatable.

Equally debatable is the differentiation of five genera in the family Streptococcaceae on the basis of mode of cell division and the major products of

Table 1
Differential characters of the genera included in the family Streptococcaceae

Streptococcus	Cells divide in one plane. Glucose fermented to yield mainly dextrorotatory lactic acid (homofermentative)
Leuconostoc	Cells divide in one plane. Glucose fermented to yield levorotatory lactic acid, CO_2, ethanol and/or acetic acid (heterofermentative)
Pediococcus	Cells divide in two planes. Glucose fermented to yield inactive lactic acid (homofermentative)
Aerococcus	Cells divide in two planes. Glucose fermented to yield mainly dextrorotatory lactic acid (homofermentative)
Gemella	Cells divide in one plane. Glucose fermented (products of fermentation not known)

Deibel & Seeley (1974).

carbohydrate fermentation. Tetrad production is not always easy to detect in the genus *Pediococcus* and Orla-Jensen (1919) occasionally noted tetrad formation in *Streptococcus cremoris* and *Betacoccus (Leuconostoc)*. There are also many reports of inactive lactic acid production by streptococci and dextrorotatory lactic acid production by tetracocci—see Whittenbury (1965*a*) and London (1976) for references to literature.

There is a growing opinion amongst bacteriologists that taxonomic ranks above the species level are highly artificial and this is almost certainly true of the genera included in the family Streptococcaceae. However, the object of this communication is not to discuss the taxonomic status or interrelationships of all the bacterial taxa presently included in the family Streptococcaceae (Deibel & Seeley 1974), but to deal with those bacteria which are, or have been suggested to be, members of the genus *Streptococcus*. For this purpose, I propose to include in the genus the following groups: the 21 named species and seven groups listed '*species incertae sedis*' in the genus *Streptococcus* (Deibel & Seeley 1974); streptococci of serological groups T, U and V (de Moor 1963; Thal & Söderlind 1966; Jelínková & Kubín 1974); the group of oral streptococci referred to as *S. mutans*; the aerotolerant and strictly anaerobic species of Holdeman & Moore (1974) and Barnes *et al.* (1977). In thus delimiting the area to be discussed, I wish to make it quite clear that I am not necessarily implying that these groups form a good genus in the taxonomic sense. In fact, to borrow one of the terms used by Ravin (1963) to denote different kinds of bacterial species, they form what can be described as a *nomen genus*, that is a collection of groups of bacteria which bear the name *Streptococcus* with either a species epithet or a serological group designation.

2. Early Classification

The term 'streptococcos' was first used by Billroth (1874) to describe chain-forming, coccoid-shaped bacteria which he noted in wounds and discharges from the animal body. He also noted the presence of 'streptococcos' in about one-half of the cases of erysipelas which he examined. The terms 'streptococcos' and 'streptococcus' were subsequently used by various authors to designate a particular kind of cell congregation and were not used in the generic sense. The generic name *Streptococcus* was first used by Rosenbach (1884) to describe a coccus, growing in chains, which he had isolated from suppurative lesions in man. To this organism he gave the name *S. pyogenes*. Rosenbach (1884) was of the opinion that *S. pyogenes* was different from the organism isolated from cases of erysipelas (designated *S. erysipelatos*) mainly on the basis of the appearance of the two organisms in gelatin and agar cultures, but subsequent workers did not find these criteria, or additional ones, adequate for distinguishing between the two forms.

In the next twenty years the association between streptococci and a large number of diseases in man and animals was established, as was the importance of these organisms in the dairy industry. As a result, numerous bacteriological investigations were carried out in an attempt to classify and identify the various forms. These early classifications were based on pathogenicity, cultural appearances on gelatin and agar, cellular morphology, reactions on blood and in milk culture and growth temperature. The early literature is well documented by Sherman (1937) and Wilson & Miles (1975).

The first definitive classification of streptococci was that of Andrewes & Horder (1906). These workers used fermentation of various sugars, reduction of neutral red and growth characteristics in milk, together with morphological observations. An analysis of the results of these tests on some 1200 colonies isolated from human disease conditions, human mouth and intestine, air and milk, led Andrewes & Horder (1906) to distinguish eight groups of streptococci which they designated as follows: *S. pyogenes*, which they considered to be identical with *S. pyogenes* of Rosenbach (1884); *S. equinus*, a saprophytic group derived from the intestine of herbivores and found at that time to be the most common streptococcus in the London air and dust; *S. mitis*, an essentially saprophytic group occasionally associated with disease and found mainly in human saliva and faeces; *S. salivarius*, mainly isolated from saliva and the intestine; *S. anginosus*, a pathogenic, long-chained form of *S. salivarius* associated with sore throats and also found in the intestine; *S. faecalis*, isolated mainly from the human intestine; the pneumococci, a group distinguished by the possession of a capsule when growing in the animal body and in certain culture media. Although Andrewes & Horder (1906) did not give the pneumococci a species name in the genus, they stated "there seems no justification for removing the pneumococci from the genus streptococcus".

The next major contribution to streptococcal classification was made by Orla-Jensen (1919) as a result of a comprehensive study of lactic acid bacteria mainly from milk and dairy produce but also including isolates from fermented foods, bovine and human faeces and a few isolates from human and animal disease conditions. Orla-Jensen (1919) used a greater variety of tests including fermentation characteristics, tolerance to heat and sodium chloride, temperature limits of growth and cellular morphology. He paid particular attention to the composition of the culture media, a feature which has been notably lacking in more recent studies of this group. As a result, he recognized nine groups of streptococci which he designated: *S. lactis*, which he considered to be identical with *Bacterium lactis* (Lister); *S. cremoris*; *S. mastitidis*, which he noted was also called *S. agalactiae* (Lehmann & Neumann); *S. thermophilus*; *S. bovis*; *S. inulinaceus*; *S. faecium*; *S. glycerinaceus* and *S. liquefaciens* (Frankland & Frankland). In addition, he grouped together a number of pathogenic streptococci as *S. pyogenes* because he was of the opinion that he had too little experience of the pathogenic streptococci to be certain whether one or more species existed in the area. Orla-Jensen (1919) also noted the presence amongst his collection of saprophytic streptococci some which did not fit into any of the above species.

During this period and immediately afterwards a number of new tests were devised and old tests refined in attempts to improve the classification of streptococci. References to these tests may be found in the works of Orla-Jensen (1919), Sherman (1937) and Wilson & Miles (1975). Various classifications of the streptococci were produced and many new 'species' characterized or merely named. A list of many of these species may be found in *Index Bergeyana* (Buchanan *et al.* 1966).

In 1937, Sherman produced the first systematic classification of streptococci isolated from a number of different habitats. Sherman (1937) stated "A major factor which has rendered abortive attempts to classify the streptococci as a whole, has been the failure to segregate, first, in more or less homogeneous divisons, the members of the genus". He considered that this omission was particularly important in those classifications which were based largely on fermentation tests, and was of the opinion that, "Though valuable in their proper role of aiding in the differentiation of closely related types, the fermentation tests when applied to the streptococci as a whole, without previous subdivision by other methods, can lead only to utter confusion".

Sherman (1937) excluded from the genus *Streptococcus* all strictly anaerobic cocci. He also excluded the pneumococci, considered by Andrewes & Horder (1906) to form a species within the genus, on the basis of their extreme sensitivity to bile. The remainder of the streptococci he divided into four 'primary' divisions based on ability to grow at 10 and 45°C; ability to survive heating at 60°C for 30 min; ability to grow at pH 9.6 and in the presence of 0.1% methylene blue and strong reducing ability. These divisions he named

Table 2
Composition of Sherman's (1937) divisions of the streptococci

Pyogenic	Viridans	Lactic	Enterococcus
S. pyogenes	S. salivarius	S. lactis	S. faecalis
S. mastitidis	S. equinus	S. cremoris	S. liquefaciens
S. equi	S. bovis		S. zymogenes
Animal pyogenes	Varieties of S. bovis		S. durans
The human C*	S. thermophilus		
Minute haemolytic			
Group G* streptococci			
Group E* streptococci			
Group H* streptococci			

* C, G, etc refer to Lancefield serological groups.

'pyogenic', 'enterococcus', 'lactic' and 'viridans'. Table 2 lists the groups or species of streptococci allocated to these divisions by Sherman (1937).

The pyogenic division incorporated most of the species which were recognized at that time as being pathogenic for man and animals. The majority were usually β-haemolytic and contained a group polysaccharide which could be detected by the Lancefield (1933) precipitin technique. They did not grow at 10°C and usually not at 45°C; were killed by exposure to a temperature of 60°C for 30 min; possessed a weak reducing activity and were not tolerant to methylene blue, sodium chloride or alkali (i.e. did not grow at pH 9.6).

The viridans group differed from those in the pyogenic division by not being β-haemolytic; usually growing at 45°C and not producing ammonia from peptone. Some also survived heating at 60°C for 30 min. They resembled the pyogenic group in not growing at 10°C, possessing a weak reducing ability and not being tolerant to methylene blue, sodium chloride or alkali. Many members of this group produced greening (i.e. α-haemolysis) on blood, hence the name viridans. This was very much a name of convenience because some viridans species produced no effect on blood and some streptococci in the other 'physiological' divisions were α-haemolytic. The designation viridans was to have unfortunate consequences because it became commonly used either as a general term, or worse, a species epithet to describe a number of quite different streptococci which produced greening on blood media.

Sherman (1937) was aware of the fact that the designation of a division as lactic could be misleading, since all streptococci produce lactic acid. However, he pointed out that the ordinary milk souring organism had been known as the 'lactic acid streptococcus' for a very long time. He thought that this, coupled with the fact that these streptococci formed a homogeneous group, justified the designation of a lactic division. The lactic streptococci were differentiated from the pyogenic and viridans streptococci by their ability to grow at 10°C, their strong reducing action and their ability to tolerate methylene blue in a concentration of 0.1%. They were also characterized by the possession of a group-specific polysaccharide, Lancefield group N.

The streptococci of the enterococcus division differed markedly from the streptococci in other divisions by their combination of low minimum and high maximum temperatures of growth and by their greater tolerance to salt and alkali. They were also characterized by the possession of the Lancefield group D antigen. Sherman (1937) chose the name enterococcus for this division because, since its use by Thiercelin (1899), many workers had used the term in a loose sense to refer to faecal streptococci which closely resembled *S. faecalis*.

Within each of these physiological divisions other tests were used to differentiate species (see Sherman 1937) and for many years this classification remained the most generally accepted. Even today, the tolerance tests of Sherman (1937), on which he based his primary divisions, are still used in the classification of the streptococci, although the genus is no longer divided into his physiological groups (Deibel & Seeley 1974).

3. Serological Grouping

Serological grouping of the streptococci will be discussed separately because of the important role which it has played in the identification of these organisms. The soluble specific substance, upon which the grouping is based, was first noted by Hitchcock (1924) as a 'residue antigen' and was believed by him to be common to almost all haemolytic streptococci. This view was held until its group-specific nature was discovered by Lancefield (1933) as a result of a series of studies on the antigenic structure of haemolytic and non-haemolytic streptococci (see Sherman 1937, for early literature). When Lancefield (1933, 1934) recognized the presence of the group polysaccharides, the so-called 'C' substances which could be detected serologically by precipitin techniques, there appeared to be an excellent correlation between the serological groups designated by Lancefield (A to E and N) and *Streptococcus* species characterized on the basis of physiological and biochemical tests (Sherman 1937). This correlation between serological group and physiological type led to the equating of serological groups with certain species. This had two major consequences for the subsequent classification of streptococci.

Since serological grouping was relatively easy to perform using acid (Lancefield 1933) or formamide (Fuller 1938) extracts of whole organisms, streptococci were frequently identified merely as, for example, Group A (*S. pyogenes*) or Group B (*S. agalactiae*) and further tests were frequently not carried out (except serological typing mainly for epidemiological purposes). Thus, because in the cases cited the procedures worked well for many years, heavy reliance was placed on serological grouping in the identification of streptococci. This resulted in the designation of a number of serological groups (E, F, G, H, K, L, M, O, P, Q, R, S, T, U, V) of streptococci which had not been comprehensively studied by other criteria and the acceptance of these groups as homogeneous taxonomic entities, often as distinct species. Exceptions were serological group C, where distinction was always made between the various species or biotypes; group D which originally embraced *S. faecalis* and *S. durans* and group N which contained the species *S. cremoris* and *S. lactis*.

The other consequence was that much effort was put into detecting serological group substances amongst some well-studied streptococcal species in an attempt for better characterization. Thus, group D antigen was demonstrated in *S. bovis* and *S. equinus* (Shattock 1949; Smith and Shattock 1962). These two species are physiologically quite different from *S. faecalis,* enterococcus division of Sherman (1937), and argument then ensued as to whether *S. bovis* and *S. equinus* should be considered to be closely related to the species of the enterococcus division because all contained a serological group D antigen. This resulted in much confusion (see Hartman *et al.* 1966). The group D antigen was later (Jones 1959; Smith & Shattock 1964) also noted in strains designated group Q streptococci by Guthof (1955) which physiologically did resemble *S. faecalis*; in *S. suis* (Elliott 1966) and in the streptococci of serological groups R, S and T (Elliott 1966; Windsor & Elliott 1975; Wilson & Miles 1975). As further studies were carried out it became apparent that many of the other Lancefield serological group antigens, especially G, H, K, M and O are present in streptococci which are physiologically heterogeneous (see Williams 1956; Deibel & Seeley 1974; Wilson & Miles 1975).

Thus, although the Lancefield (1933) serological grouping technique has certainly been of value in providing a quick identification method for certain streptococci, e.g. *S. pyogenes*, *S. agalactiae*, heavy reliance on this method has not encouraged other fundamental biological and chemical studies of the streptococci, and has therefore hampered and is still hampering a better understanding of the relationships between these bacteria. Only recently Krause (1972) referring to the streptococci wrote "Efforts to classify bacteria primarily by metabolic behaviour rather than antigenic constitution is often a futile exercise". However, other bacteriologists (see Williams 1956; Colman 1968; Chapman 1972; Wilson & Miles 1975), while recognizing the value of serological techniques, advocate the investigation of other characters in attempts to resolve the classification and identification of the streptococci.

The chemical compositions of the group substances and type antigens and their cellular location have now been established for a number of streptococci. The group substances in serological groups A, B, C, E, F, G, H and K have been shown to be polysaccharides located in the cell wall, while those in groups D and N are teichoic acids located between the cell wall and membrane, and are referred to as intracellular (see Deibel & Seeley 1974, for list of chemical compositions).

4. Current Classification

Despite the heavy reliance placed on serological grouping, streptococci have been investigated by other techniques and in the last twenty years information has accumulated from a variety of different kinds of study.

These include the use of improved or additional biochemical tests, including detection of enzymes (deoxyribonuclease, ribonuclease, fibrinolytic enzymes, etc.), resistance to chemical compounds and antibiotics (see Deibel & Seeley 1974; Wilson & Miles 1975); further studies in serological grouping and typing (see Wannamaker & Matsen 1972; Wilson & Miles 1975); cell structure (Cole 1968); carbohydrate, teichoic acid and amino acid composition of the cell walls (Salton 1953; Cummins & Harris 1956; McCarty 1958; Jones & Shattock 1960; Roberts & Stewart 1961; Michel & Gooder 1962; Krause & McCarty 1962; Slade & Slamp 1962; Michel & Willers 1964; Wittner & Hayashi 1965; Colman & Williams 1965; Knox & Wicken 1973); studies of the peptidoglycan structure (Schleifer & Kandler 1967; Kandler *et al.* 1968; Ghuysen 1968); Mosser & Tomasz 1970; Slade & Slamp 1972; Schleifer & Kandler 1972); gel electrophoresis studies (Hess & Slade 1965; Lund 1965, 1967; Osborne *et al.* 1976); studies of biochemical pathways and electron transport systems (see Deibel 1964; Deibel & Seeley 1974; Whittenbury 1964; Bryan-Jones & Whittenbury 1969; Ritchey & Seeley 1976; London 1976); protein homology studies (London & Kline 1973; London *et al.* 1975; London 1976); G+C% base ratio studies (see Jones & Sneath 1970; Deibel & Seeley 1974; Holdeman & Moore 1974; Barnes *et al.* 1977; Coykendall 1977); nucleic acid homology studies (Weissman *et al.* 1966; Coykendall *et al.* 1971; Roop *et al.* 1974; Coykendall 1977); genetic studies, including plasmid transfer (see Jones & Sneath 1970; Courvalin *et al.* 1972; Tomura *et al.* 1973; Clewell & Franke 1974; Cords *et al.* 1974; Courvalin *et al.* 1974; Jacob & Hobbs 1974; Kayser 1974; Kozak *et al.* 1974; McKay & Baldwin 1974; Cleary *et al.* 1975; Clewell *et al.* 1974, 1975; Dunny & Clewell 1975; Fuchs *et al.* 1975; Jacob *et al.* 1975; Malke 1975; Malke *et al.* 1975; McKay & Baldwin 1975; Nakae *et al.* 1975; Cook 1976; Efstathiou & McKay 1976; Horodniceanu *et al.* 1976; Leblanc & Hassell 1976; McKay *et al.* 1976; Tagg *et al.* 1976; Tagg & Wannamaker 1976; Yagi & Clewell 1976; Anderson & McKay 1977; Miyamura *et al.* 1977); numerical taxonomic studies (Colobert & Blondeau 1962; Raj & Colwell 1966; Carlsson 1968; Colman 1968; Seyfried 1968; Drucker & Melville 1969; Jones *et al.* 1972; Wilkinson & Jones 1977). However, with one or two notable exceptions these studies have been confined to a few species or strains of streptococci.

Therefore, although some progress has been made, the relationships of streptococci to each other, to the strictly anaerobic and aerotolerant cocci which occur in chains and produce lactic acid, and to the other bacteria in the family Streptococcaceae, remain unclear.

In order to review the present state of streptococcal classification it is necessary to divide the bacteria to be discussed into groups of manageable size. The physiological divisions of Sherman (1937) undoubtedly provided a very useful primary sorting method at the time. However, as pointed out by Deibel & Seeley (1974), more recently recognized species often cut across the broad lines of Sherman's major divisions and as mentioned earlier, Sherman (1937) excluded the pneumococci and all anaerobic chain-forming, lactic acid-producing, cocci from the genus. I have therefore divided those bacteria which, for the purpose of this communication, I have decided to include in the genus *Streptococcus*, into seven groups (see Table 3). These groups are to some extent based on the physiological divisions of Sherman (1937) but with some modifications and additions to accommodate the more recently recognized species together with those species which, although not included in the genus by Sherman (1937), are now considered by many workers to be members of the genus *Streptococcus*. These groups are designated 'pyogenic', 'pneumococci', 'oral', 'faecal', 'lactic', 'anaerobic' and 'other streptococci' based on the method of presentation of Wilson & Miles (1975). I wish to stress that these groupings are used merely as a means of presenting the material. They are purely artificial and are largely based on such shared features as ability to cause disease, main habitat, oxygen tolerance, etc., and do not necessarily imply any close relationship between the streptococci included in any one group.

(a) *Pyogenic streptococci*

The streptococci included in this group are all animal parasites and most cause septicaemic or respiratory tract infections in man and animals but some are also associated with other clinical conditions. The majority are β-haemolytic. They can be divided into a number of Lancefield serological groups on the basis of cell wall polysaccharide antigens. Several of the species (or serological groups) appear to be fairly strictly adapted to certain animal hosts. None is especially resistant to heat or grows at extremes of temperature, pH or sodium chloride concentrations, nor exhibits strong reducing properties. The salient characters of the species (or serological groups) are listed in Table 4.

The species which has received most attention amongst this group of streptococci is *S. pyogenes* because of its importance as a disease-causing organism in man, although it has been isolated also from outbreaks of disease in animals (see Wilson & Miles 1975). The terms *S. pyogenes* and streptococci of serological group A are frequently used synonymously, but although all strains of

Table 3
Groupings of the streptococci

Groups	Species	Other description	Lancefield group	Haemolysis	Main Habitat
Pyogenic	S. pyogenes		A	β	Man
	S. equi		C	β	Horse
	S. zooepidemicus		C	β	Animals
	S. equisimilis		C	β	Man, animals
	S. dysgalactiae		C	α	Cattle
	—	Large colony group G	G	β (α or NH)	Man, animals
	S. agalactiae		B	β (α or NH)	Cattle, man
	—	S. infrequens	E, P, or U	β	Pig, cattle
	—	S. subacidus			
	—	S. lentus			
	—		L	β	Dog, pig
	—		V	β	Pig
	—	S. suis	R, S or T (±D)	β (α)	Pig, man
	—	Minute*	G	β	Man
	—		F	β	Man
	S. anginosus		—	α	Man
Pneumococci	S. pneumoniae	Diplococcus Pneumococcus			
	S. salivarius		K or —	NH or (α)	Human mouth, intestine
	S. milleri	Streptococcus MG	A, C, F, G, or — ('Ottens antigens')	NH	Mouth
	S. mitior	S. mitis	O, K, M or —	α	Throat, mouth
Oral	S. sanguis	S. viridans			
	S. mutans	Streptococcus SBE	H or —	α (β)	Mouth (soil?)
	S. rattus				
	S. cricetus	S. mutans	—	NH	Mouth, carious teeth
	S. sorbrinus				
	S. ferus				

Group	Species			Habitat
Faecal	S. faecalis	D	β or NH	Enterococci — Intestine of man and animals
	S. faecium	D	α, β or NH	Intestine of man and animals
	S. avium	D(Q)	α or NH	Birds
	S. faecium var. casseliflavus	D	α or NH	Plant material
	S. bovis	D	α or NH	Faeces of animals
	S. equinus	D	α or NH	Faeces of horses
Lactic	S. lactis	N	(α) or NH	Milk and dairy products
	S. lactis subsp. diacetylactis	N	(α) or NH	
	S. cremoris	N	(α) or NH	
Anaerobic	S. intermedius	—	NH	Peptostreptococcus Hare group VIa / Peptostreptococcus — Human clinical sources and faeces
	S. morbillorum	—	NH, or, wα	Diplococcus / Peptococcus — Human clinical sources
	S. constellatus	—	NH, or, wβ	Diplococcus / Peptococcus — Human clinical sources
	S. hansenii	—	NH	Human faeces
	S. pleomorphus	—	β or NH	Caeca of poultry
Other streptococci	S. uberis	(E) (C, D, P, U)	NH	Cows, soil
	S. acidominimus	(E)	NH	Cows
	S. thermophilus	—	NH	Milk

NH, non-haemolytic; (α), rarely α; wα or wβ, weak α- or β-haemolysis.

* Possibly related to S. milleri.

The G+C content of DNA of those Streptococcus species investigated is 33–46 moles % (see Deibel & Seeley 1974; Holdeman & Moore 1974; Barnes et al. 1977; Coykendall 1977).

Based on information from Deibel & Seeley (1974); Jelínková & Kubín (1974); Wilson & Miles (1975); Coykendall (1977).

Table 4
Some differential biochemical reactions of pyogenic streptococci and pneumococci*

Species	Growth on 40% bile	Hippurate hydrolysis	Aesculin hydrolysis	Acid production								
				Mannitol	Lactose	Sorbitol	Trehalose	Glycerol (aerobic)	Salicin	Raffinose	Inulin
S. pyogenes	−	−	V	−	+	−	+	−	+	−	−
S. equi	−	−	−	−	−	−	−	−	+	−	−
S. zooepidemicus	−	−	V	−	V	+	−	−	+	−	−
S. equisimilis	−	−	V	−	V	−	+	+	+	−	−
S. dysgalactiae	−	V	V	−	V	V	+	−	V	−	−
Large colony, group G	−	−	V	−	V	−	+	+	V	−	V
S. agalactiae	+	+	−	−	V	−	+	+	V	−	−
Groups† E, P, U	V	−	+	+	V	+	+	+	+	−	−
Group L	V	V	−	−	+	V	+	V	V	−	−
Group V	−	−	−	+	+	+	+	−	+	NI	−
Groups R, S, T	V	−	+	−	+	−	+	−	+	V	+
S. anginosus (Minute streptococci)	V	−	V	−	+	−	+	−	+	V	−
S. pneumoniae	−	−	−	V	+	−	+	W	−	+	+

+, 90% positive; −, 90% negative; V, some strains positive, some negative; W, weak positive; NI, no information.
* Information based on Deibel & Seeley (1974), Jélinková & Kubín (1974), Wilson & Miles (1975).
† Groups, Lancefield serological groups.

S. pyogenes contain the group A antigen, the converse is not always true. Streptococci which biochemically are not *S. pyogenes* have been demonstrated to contain the group A antigen (Chapman 1972) but the antigen is not widely found amongst streptococci other than *S. pyogenes* (see Wilson & Miles 1975). Information on the role of *S. pyogenes* in diseases of man may be found in the contributions of Parker and Maxted (this volume).

Streptococcus equisimilis, *S. zooepidemicus*, *S. equi* and *S. dysgalactiae* all contain the group C antigen. The first three species are all β-haemolytic and reasonably well characterized. *Streptococcus dysgalactiae* is non-haemolytic and less well studied. They are mainly distinguished from each other on the basis of fermentation of sorbitol and trehalose (see Table 4). The cell wall composition of *S. equisimilis*, *S. zooepidemicus* and *S. dysgalactiae* is identical (Colman & Williams 1965).

As with *S. pyogenes* heavy reliance has been placed on serological grouping reactions together with one or two biochemical tests in identifying streptococci of serological group C. Consequently, the relationships of the species to each other and to other pyogenic streptococci is unresolved. The nucleic acid homology studies of Weissman *et al.* (1966) shed little light on the problem because the species designation of the two serological group C strains used is not given. However, the serological relatedness among the fructose diphosphate aldolases (FDP aldolases) of various streptococci (using antiserum prepared against the enzyme of *S. faecalis*) indicates that the aldolases of *S. equi* and *S. dysgalactiae* share the same general properties and are similar to that of *S. pyogenes*. On the other hand, the aldolase of *S. equisimilis* appears to be quite different and *S. zooepidemicus* aldolase possesses properties intermediate between the *S. lactis* group and the *S. bovis* group (London & Kline 1973; London *et al.* 1975).

The large colony group G streptococci are only one group of streptococci which contain the group G antigen. There is no doubt that streptococci of this type cause infections in man and animals (see Wilson & Miles 1975). Physiologically they are quite different from other streptococci which contain the group G antigen. In many ways the large colony group G streptococci resemble *S. pyogenes*. At least three serological types have been noted, one of which cross-reacts with *S. equisimilis* and can give rise to confusion. The strains examined by Colman & Williams (1965) all had the same cell wall composition which was not found in any of the other streptococci except one strain which was serologically group L. The group is poorly studied and more than one physiological type may be present (Sherman 1937; Deibel & Seeley 1974). No species name has been suggested.

Streptococcus agalactiae contains the group B antigen and has long been recognized as an important cause of mastitis in cattle. It is frequently present in the human respiratory, genital and gastro-intestinal tract, and in the last decade has been recognized as an important pathogen of man, especially infants. Two distinct types of symptoms have been recognized amongst newborn infants: 'early onset' and 'late onset' neonatal disease (see Parker and Ross, this volume). Strains from human sources differ in a number of pathogenic characters from bovine strains (Parker, this volume), and it may well be that differences exist between the strains responsible for the different types of neonatal disease but not enough is yet known to be sure whether the human strains are sufficiently different from *S. agalactiae* to be regarded as a separate species. As with *S. pyogenes* in particular, and most of the pyogenic streptococci in general, undue emphasis has until recently always been placed on the serological grouping reactions. Thus the terms *S. agalactiae* and streptococci of serological group B are used synonymously.

London & Kline (1973) on the basis of serological relatedness of the FDP aldolases grouped *S. agalactiae* (ATCC 13813) with three strains of *S. salivarius* (ATCC 13419, 9222 and NIH 122). However, the numerical taxonomic study of Colman (1968) indicated that *S. agalactiae* and *S. salivarius* strains clustered as quite separate and distinct taxa.

There are some streptococci which contain the group B antigen which are clearly distinct from *S. agalactiae*. Robinson & Meyer (1966) isolated a non-haemolytic streptococcus from an outbreak of disease in captive tropical fish. Although serologically group B, the organism did not grow at 37°C, was not bile tolerant nor CAMP positive (Christie *et al*. 1944). Although these fish strains have been mentioned here there is no evidence that they cause clinical symptoms in man or higher animals, and they could equally well—and perhaps more appropriately—have been discussed with 'other streptococci'.

Streptococci of serological groups E, P and U have been isolated from milk and from cervical abscesses, pneumonia and septicaemic infections in pigs. The nomenclature of these bacteria is confused. Haemolytic streptococci corresponding to *S. infrequens* (Holman 1916) were isolated from human clinical cases by Brown *et al*. (1926), and subsequently designated serological group E by Lancefield (1933). Holman (1916) also used the term *S. subacidus* to describe another group of strains, also isolated from human pathogenic conditions, which differed from *S. infrequens* by not fermenting lactose, mannitol and salicin. Later, noting the similarity of their strains with *S. infrequens*, Frost *et al*. (1927) applied the name *S. infrequens* to the haemolytic streptococci of serological group E isolated from milk. Subsequently, Brown (1939) presented evidence that strains designated *S. infrequens* by Holman (1916) were probably group A streptococci, and he proposed the name *S. lentus* for the group E strains isolated from milk.

The association of serological group E streptococci with abscesses in the cervical lymph nodes of pigs was first noted by Newsom (1937). In the next twenty years there was a number of reports of the association of streptococci of this kind with clinical conditions in pigs (see Deibel *et al.* 1964 for references to early literature), but the first comparative study of the milk and pig strains was that of Moreira-Jacob (1956). Amongst strains studied by him, Moreira-Jacob (1956) detected two physiological groups. One group which hydrolysed aesculin and produced acid from mannitol and salicin he designated *S. infrequens* (Holman); the other, which failed to give positive reactions in these tests, he designated *S. subacidus* (Holman), on the grounds that the name *S. lentus* (Brown) merely produced confusion.

Recently, Colman (1970) has made a case for the designation of strains of serological group E as *S. lentus* (Brown) and this opinion is supported by Wilson & Miles (1975). However, while rejecting the names *S. infrequens* and *S. subacidus*, Deibel & Seeley (1974) make no reference to *S. lentus* and merely treat these streptococci as serological group E. The most recent comprehensive studies of these organisms are those of Deibel *et el.* (1964) and Yao *et al.* (1964).

Streptococci with similar biochemical properties to those of serological group E but showing no cross-reaction with group E antisera have been also isolated from clinical conditions in pigs. Moberg & Thal (1954) demonstrated the presence of the group P antigen in some of these strains while Thal & Söderlind (1966) demonstrated the presence of group U antigen in their strains. Noting the apparent physiological similarity of streptococci of serological groups E, P and U, de Moor & Thal (1968) examined strains representing all these serological groups and concluded that they belonged to a single species for which they used the name *S. infrequens.*

The serological relationships within the group are not clear. de Moor & Thal (1968) demonstrated that the group E polysaccharide is unrelated to that of groups P and U. The latter have a common antigenic factor which de Moor & Thal (1968) demonstrated to be present in formamide but not acid extracts of the whole cells. There is also no consensus of opinion on the number of serological types (see Moreira-Jacob 1956; Yao *et al.* 1964). The situation is complicated further by the demonstration by London *et al.* (1975) that there appears to be no very close relationship between group P and group U strains as measured by the serological relatedness of their FDP aldolases. However, only one strain of each group was used and no strains of group E were investigated by these authors. Colman & Williams (1965) noted quite different cell wall patterns in strains of group P and group E. They did not examine group U strains.

Streptococci designated serological group L were first described by Fry (see Hare & Fry 1938; Laughton 1948). They were isolated from dogs and have since been found to be associated with miscellaneous infections in these animals and in pigs (Jelínková & Kubín 1974). Whether they are indeed a separate species or merely a group of streptococci which contain a particular antigenic factor under the conditions of test, is not yet known. The aldolase studies of London & Kline (1973) placed the one strain of this group studied, in the same general class as *S. pyogenes.* Analysis of the cell walls of four different cultures of group L showed a different composition for each strain. Two were isolated from humans and two probably from animals (Colman & Williams 1965).

Streptococci of serological group V were proposed as a separate serological group by Jelínková & Kubín (1974). The strains were isolated from pig lymph nodes and found to be serologically distinct. For this reason, together with the biochemical properties of the strains, Jelínková & Kubín (1974) considered them to be distinct from other streptococci isolated from pig lymph nodes. Whether the strains of serological group V do form a separate taxonomic entity must await further study.

Streptococci of serological groups R, S and T have been isolated mainly from pigs where they cause septicaemia often with lesions in the bones and joints (Field *et al.* 1954; de Moor 1963; Elliott *et al.* 1966). Strains of serological group R have also been implicated in several cases of meningitis and other severe infections in man (Perch *et al.* 1968; Perch & Kjems 1971). Biochemically they are all very similar (see Table 4) but their serology is confused. de Moor (1963) divided them into three new serological groups R, S and T. Elliott (1966) investigated the serological relationships of streptococci isolated from outbreaks of disease in young pigs (2-6 weeks) in England, together with some of those isolated by de Moor (1963) in Holland. As a result of his studies he concluded that the strains isolated in England (designated by him PM streptococci) were identical with the serological group S strains of de Moor (1963). However, because Elliott (1966) noted that they all reacted with group D antisera he was of the opinion that "there would appear to be some justification for establishing within group D an additional subgroup with status equivalent to that of *S. faecalis* and *S. bovis*". For this subgroup he proposed the name *S. suis*. Elliott (1966) noted that biochemically the strains differed from *S. faecalis* and *S. bovis* and this has been my experience. A preliminary numerical taxonomic study of streptococci (Jones, unpublished) indicates that these streptococci form a group more closely related to *S. pyogenes* than to *S. faecalis* or *S. bovis*. The numerical taxonomic study of Colman (1968) also indicates that they are not closely related to *S. faecalis* or *S. bovis*.

More recently Windsor & Elliott (1975) investigated streptococci isolated from an outbreak of meningitis in weaned pigs (10-14 weeks old). These bacteria also cross-reacted with group D antiserum, but were not serologically identical with *S. suis* (Elliott 1966). Windsor & Elliott (1975) therefore designated the streptococci isolated from young piglets (2-6 weeks old) as *S. suis* capsular type 1, identical with serological group S of de Moor (1963), and those isolated from older piglets *S. suis* capsular type 2. This latter group they considered to be identical with serological group R of de Moor (1963). The serological situation is further confused because some of these strains also cross-react with group E and N antisera, probably due to shared type specific antigens (Elliott 1966) and some strains of serological group B also react with group R antiserum (see Wilson & Miles 1975). I have been unable to find any further information regarding the relationship of serological group T to either group R or S, except for the aldolase studies of London *et al.* (1975). These studies indicate a reasonably close relationship between the FDP aldolases of one strain of serological group S and one strain of group T.

Colman & Williams (1965) noted that the cell wall composition of two group R strains was the same as one group S strain (cell wall pattern 6). However, another group S strain possessed the same cell wall composition as two group T strains (cell wall pattern 11). Cell wall patterns 6 and 11 were the two most frequently found patterns amongst the streptococci examined by them.

The species *S. anginosus* was described by Andrewes & Horder (1906) to accommodate long-chained streptococci isolated from the human throat and gut of man. Deibel & Seeley (1974) consider that the 'minute' β-haemolytic streptococci of Long & Bliss (1934) and *Streptococcus* MG (Mirick *et al.* 1944) are members of this species and they do appear to form a homogeneous group on the basis of physiological and nutritional tests. However the species is serologically heterogeneous, some strains reacting with group F antiserum and some with group G (type 1). Wilson & Miles (1975) do not refer to the species *S. anginosus* as defined by Deibel & Seeley (1974), but they do point to a resemblance between the 'minute' β-haemolytic streptococci of Long & Bliss (1934) and *S. milleri* of Guthof (1956). Although 'minute' β-haemolytic streptococci have been frequently isolated from human clinical conditions, they remain a poorly studied group.

Other streptococci which generally conform to the description pyogenic have been isolated and characterized mainly on their serological grouping reactions. It is now generally agreed that H and K antigens do not serve to define any taxonomic group. These antigens together with the group O antigen (Boissard & Wormald 1950) are found irregularly distributed amongst non-haemolytic streptococci (Wilson & Miles 1975). The significance of the group M antigen is now also doubtful. Skadhauge & Perch (1959) isolated streptococci from human sources all of which reacted with group M antiserum but which could be divided into three biotypes. There were also serological differences between the biotypes and one biotype reacted with group K serum.

Differentiation of species amongst the pyogenic streptococci is not easy because so much emphasis has been placed on serological grouping reactions that little information has been accumulated on the broad biological characters of these bacteria. However, now that the limitations of the serological reactions are being accepted, at least by some workers, we hope other studies will be made.

(b) *Pneumococci*

As suggested by Andrewes & Horder (1906) the pneumococci are now regarded as a species, *S. pneumoniae*, within the genus *Streptococcus*. Detailed descriptions of the species are given by Deibel & Seeley (1974) and Wilson & Miles (1975). *Streptococcus pneumoniae* has received a great deal of attention because of its pathogenicity for man. The species is considered separately from the pyogenic group because it is not β-haemolytic but produces a strong greening reaction on blood, is extremely bile soluble, has a characteristic morphology and does not contain a group polysaccharide similar to those possessed by streptococci of the various Lancefield serological groups, although a species-specific group polysaccharide has been demonstrated. Chemical studies on the peptidoglycans of *S. pneumoniae* indicate a relationship to *S. pyogenes* (Mosser & Tomasz 1970).

Table 5
Some biochemical characters of oral streptococci

Species	Survive 60°C 30 min⁻¹	Growth at 10°C	Growth at 45°C	Growth in 4% NaCl	Growth at pH 9.6	Hippurate hydrolysis	Arginine hydrolysis	Aesculin hydrolysis	Acetoin from glucose	H₂O₂ production	Acid production from										
											Glycerol	Mannitol	Sorbitol	Raffinose	Salicin	Inulin	Sucrose	Trehalose	Lactose	Laevan from sucrose	Dextran from sucrose
S. salivarius	−	−	−	−	−	−	−	V	V	−	−	−	−	+	+	+	+	+	+	+	−
S. milleri	−	−	−	V	−	−	+	+	V	−	−	−	−	−	+	−	+	+	+	−	−
S. mitior (S. mitis)	−	−	−	+	−	−	+	−	+	+	−	−	−	V	−	−	V	−	+	−	V
S. sanguis	−	−	V	V	−	−	+	V	−	+	−	−	−	V	+	−	+	+	+	−	+
S. mutans*	−	−	V	+	−	−	−†	−	+	−‡	−	+	+	+	+	+	+	+	+	−	+

Symbols as for Table 4.
* For characters of S. mutans, S. rattus, S. cricetus, S. sorbrinus, S. ferus, as defined by Coykendall, see Coykendall (1977).
† Serotype b strains hydrolyse arginine.
‡ Serotype d–g strains usually produce H₂O₂.
Based mainly on information from Wilson & Miles (1975), Hardie & Bowden (1976).

On the other hand, studies on the FDP aldolases indicate that *S. pyogenes* and *S. pneumoniae* are not close on this criterion. The properties of the FDP aldolase of *S. pneumoniae* are more similar to those of *S. bovis* and *S. equinus* (London & Kline 1973). The cell wall composition of *S. pneumoniae* is distinct (Colman & Williams 1965). The main characters of the species are listed in Table 4.

(c) *Oral streptococci*

The streptococcal species listed in Table 5 are treated as a group only because together they comprise the predominant but not the exclusive streptococcal flora of the human mouth. Additionally most of them have been implicated to some degree in dental caries. Taxonomic studies of streptococci of the oral cavity have been reviewed by Hardie & Bowden (1976a, b). The differential characters of the species are listed in Table 5.

Streptococcus salivarius (Andrewes & Horder 1906) is characterized by the ability of most strains to produce laevan from sucrose and therefore give rise to large mucoid colonies on sucrose agar. Although many workers have demonstrated the presence of group K antigen (Hare 1935) in strains of this species, Williams (1956) showed that only slightly more than one-half of the laevan producing strains corresponding to *S. salivarius* contained this antigen, and further, that the group K antigen was not confined to streptococci with the cultural characteristics of *S. salivarius*. The serology of this species has not been fully resolved. There are two serotypes designated, type 1 and type 2, but other types have been reported. Type 1 strains cross-react with strains of *Streptococcus* MG but this latter group of streptococci are quite different physiologically from strains of *S. salivarius*. Montague & Knox (1968) demonstrated that only type 1 strains react with group K antiserum.

On the basis of a numerical taxonomic study Colman (1968) concluded that *S. salivarius* is a good species but the strains examined by Colman & Williams (1965) had different cell wall patterns. The molecular weight of the FDP aldolase of *S. salivarius* strains is more than twice that of other streptococci studied by London & Kline (1973). However the enzyme, surprisingly, has approximately the same electrophoretic mobility as the other smaller streptococcal enzymes and on serological relatedness criteria falls into the same group as *S. agalactiae*. A much more thorough study of this species including the relationship of strains containing the group K antigen to other slime-producing and non-slime-producing streptococci is required before the taxonomic status of *S. salivarius* can be resolved.

The species *S. milleri* is relatively unfamiliar except to dental bacteriologists (Hardie & Bowden 1976a, b). The species was named by Guthof (1956) to describe strains isolated from dental abscesses. Similar organisms have since been found in dental plaques although several workers have failed to isolate streptococci corresponding to *S. milleri* from the mouth (see Bowden *et al.* 1975). Colman & Williams (1965) drew attention to the similarity in cell wall composition between strains of *S. milleri* and non-haemolytic streptococci which reacted with group A, C, F and G antisera and those serologically ungroupable streptococci which possessed Ottens antigens (the so-called type antigens of serological group F, Ottens & Winkler 1962). There is now good general agreement on the biochemical and physiological characters of strains called *S. milleri*, but the serology is less clear (Hardie & Bowden 1976*a, b*). There is no general agreement over the use of the specific epithets *milleri* and *anginosus*. Deibel & Seeley (1974) include the 'minute' haemolytic streptococci (Long & Bliss 1934) and *Streptococcus* MG in the species *S. anginosus* Andrewes & Horder) but make no reference to *S. milleri*. Wilson & Miles (1975) treat the 'minute' haemolytic streptococci as a separate group in their pyogenic division but do point out a close physiological relationship between them and *S. milleri*. They also think *Streptococcus* MG is related to *S. milleri*. Cowan & Steel (1974) list both *S. anginosus* and *S. milleri* in their identification tables although there are few differences in the test reactions of the two species. Further work is required to establish the relationships between strains named *S. milleri, S. anginosus*, 'minute' haemolytic streptococci and *Streptococcus* MG. Hardie & Bowden (1976*a*) point out the necessity of carefully characterizing strains of *S. milleri* because they are resistant to sulphonomides and grow on some selective media used for the isolation of *S. mutans*.

Streptococcus mitior (Schottmüller 1903) was suggested by Colman & Williams (1972) as a name for α-haemolytic streptococci that do not hydrolyse arginine or aesculin, do not ferment mannitol or sorbitol, are inhibited by bile and do not possess rhamnose in the cell wall. In his numerical taxonomic study of oral streptococci, Carlsson (1968) noted that strains with these characteristics clustered in his group 1:A, together with strains described in the literature as *S. mitis* and *S. viridans*. Certain strains of *S. sanguis* serotype II also appear to be allied to *S. mitior* (Hardie & Bowden 1976*a*). Deibel & Seeley (1974) describe the species *S. mitis* (Andrewes & Horder) as containing α-haemolytic streptococci lacking a serological group antigen, and make no reference to *S. mitior*. Wilson & Miles (1975) "are in some doubt how to name the common α-haemolytic streptococcus of the throat that forms very long chains in broth culture" and named *S. mitior* by Schottmüller (1903) because "organisms of this general description have in the past been called *S. viridans* or *S. mitis*." It is also likely that strains assigned to serological groups O and M may be members of this group (see Wilson & Miles 1975). London & Kline (1973) and London *et al.* (1975) examined the properties of the FDP aldolases of strains labelled *S. mitis* and found that they fell into two distinct groups.

The original description of *S. sanguis* is that given to a streptococcus isolated by White & Niven (1946) from cases of subacute bacterial endocarditis and at that time known as 'Streptococcus SBE'. The organisms appeared to be very similar to a group of streptococci investigated serologically by Hare (1935) and designated by him 'group H streptococci'. In recent years strains conforming to the characteristics of *S. sanguis* as described by Colman & Williams (1972), have been isolated with increasing frequency from the human mouth. Serological studies have indicated that several different serotypes exist amongst these strains and only some of these serotypes possess the group H antigen (Colman & Williams 1972; Hardie & Bowden 1976a, b). The confusion over the identity of the group H antigen requires further clarification particularly since various commercial producers of streptococcal grouping sera use different strains to raise group H antibodies.

Colman & Williams (1965) noted several types of cell wall pattern amongst the *S. sanguis* strains examined by them but the aldolases of strains of *S. sanguis* examined by London *et al.* (1975) all appear to possess the same general properties.

Streptococcus mutans was originally described by Clarke (1924) to accommodate non-haemolytic streptococci isolated from carious teeth. After the report of Clarke (1924) little interest was shown in these organisms and the species is not listed by Deibel & Seeley (1974). However, in the last decade, this species has received much more attention because of its association with dental caries.

Streptococcus mutans strains produce acid from a wide variety of sugars and usually dextran from sucrose, but with the exception of strains of serotype b do not produce ammonia from arginine. A number of serotypes (Bratthall 1970; Perch *et al.* 1974) and biotypes (Facklam 1974; Hardie & Bowden 1976a) have been recognized, but Facklam (1974) thinks the biotypes merely represent atypical strains and that phenotypically, *S. mutans* strains form a well-defined taxon, a view supported by Hardie & Bowden (1976a). The latter authors do draw attention to the fact that strains of the d-g group show some physiological differences from the other serotypes. These strains less frequently hydrolyse aesculin or ferment sorbitol, raffinose and melibiose, while unlike the other serotypes they usually produce hydrogen peroxide.

Using serological methods London *et al.* (1975) distinguished two groups amongst the FDP aldolases of *S. mutans*. One group consisted of strains of serotypes a, b and c, the other of serotype d strains. These findings are in accord with the suggestion of Hardie & Bowden (1976a) that serotype d-g strains may be different.

Recently, Coykendall (1977), mainly on the basis of nucleic acid studies, has suggested that the *S. mutans* group contains five 'genospecies'; *S. mutans*, *S. rattus*, *S. cricetus*, *S. sorbrinus* and *S. ferus*. Whether these genospecies are in fact distinct taxonomic entities must await further study. At the present time it seems more sensible to retain the name *S. mutans* for the whole group.

(d) *Faecal streptococci*

Although these streptococci are not of great clinical importance, they are one of the most studied groups of streptococci. This is because of their distribution and importance in water and the food and dairy industries, where their presence can be correlated with faecal contamination from human and animal sources.

However, although these streptococci have been studied intensively and the individual species reasonably well characterized, there are still many problems of identification—especially of what may be called the intermediate forms, that is, those forms which cannot easily be assigned to one of the named species. Other problems involve the taxonomic status of the species as presently characterized, for example should haemolytic and/or gelatin liquefying strains of *S. faecalis* be given subspecies status; should *S. bovis* strains which produce acid from mannitol be assigned to a separate species? In addition to these problems there is confusion about the terms used collectively to describe these bacteria. The terms 'faecal streptococci', 'enterococci', 'group D streptococci' and sometimes even *S. faecalis* have been used interchangeably and by implication synonymously. A detailed account of the historical background to this situation is given by Hartman *et al.* (1966). Comprehensive reviews of early studies on these bacteria may be found in the papers of Orla-Jensen (1919); Dible (1921); Sherman (1937); Frost & Engelbrecht (1940); Skadhauge (1950); Seeleman (1954); Shattock (1962); Niven (1963); Deibel (1964) and Hartman *et al.* (1966).

All the species listed in Table 6 with the exception of *S. faecium* subsp. *casseliflavus* are found as part of the normal gut flora of animals and man. All contain the group D antigen. However, this antigen is not confined to these species but also occurs in streptococci designated serological groups R, S and T (see Elliott 1966; Windsor & Elliott 1975). On physiological and biochemical grounds *S. faecalis*, *S. faecium* and *S. avium* are quite distinct from *S. bovis* and *S. equinus* (see Table 6). The first three species have characteristics typical of the enterococcus division of Sherman (1937), although of course Sherman (1937) did not recognize *S. faecium* (Orla-Jensen) nor *S. avium* (Nowlan & Deibel 1967) as separate species. In contrast *S. bovis* and *S. equinus* were grouped in the viridans division of Sherman (1937). The grouping of these two last species with those of the enterococcus division of Sherman (1937) and the consequent confusion of terms to describe the whole group, stems from the demonstration of the group D antigen in *S. bovis* and *S. equinus* (Shattock 1949; Smith & Shattock 1962) and the undue emphasis placed on this one character in streptococcal classification.

Streptococcus faecalis was the name applied by Andrewes & Horder (1906) to the most common streptococcus isolated from human faeces. These streptococci were later shown by Dible (1921) to be almost certainly the same as those studied by Thiercelin (1899) and called enterococcus by him. Over the years this

Table 6
Some biochemical characters of faecal streptococci

Species	Survival 60°C 30 min⁻¹	Growth at 10°C	Growth at 45°C	Growth at 6.5% NaCl	Growth at pH 9.6	Growth on 40% bile agar	Growth and reduction 0.4% tellurite	Growth and reduction 0.1% tetrazolium	Arginine hydrolysis	Hippurate hydrolysis	Slime production from sucrose	Starch hydrolysis	Mannitol	Lactose	Sucrose	Arabinose	Sorbitol	Sorbose	Trehalose	Raffinose	Inulin	Glycerol (aerobic)	Glycerol (anaerobic)	Melibiose	Melezitose
S. faecalis	+	+	+	+	+	+	+	+	+	+	−	−	+	+	+	−	+	−	+	−	−	+	+	−	+
S. faecium	+	+	+	+	+	+	−	−	+	V	V	−	V	+	V	V	V	−	+	V	−	V	−	+	−
S. avium	+	V	V	W	+	+	−	V	−	−	−	−	+	+	+	+	+	+	+	−	V	+	+	V	+
S. faecium var. *casseliflavus*	V	+	V	+	+	+	V	V	V	V	−	−	+	+	+	+	V	NI	+	+	V	V	V	+	−
S. bovis	−	−	+	−	V	+	−	V	−	−	V	+	V	+	+	V	V	NI	V	+	+	−	V	+	−
S. equinus	−	−	+	−	−	+	−	−	−	−	−	V^a	−	−	+	−	−	NI	V	−	V	−	−	−	−

Symbols as for Table 4.
V^a, weak variable hydrolysis under aerobic conditions only.
Based on information mainly from Jones (1959), Shattock & Smith (1962), Mundt & Graham (1968), Facklam (1972), Deibel & Seeley (1974).

species has become progressively better characterized and now forms a reasonably well-defined taxon (Deibel 1964; Whittenbury 1965b; Jones et al. 1972; Deibel & Seeley 1974). There is however still some disagreement about the status of strains of *S. faecalis* which are haemolytic and/or possess the ability to liquefy gelatin. Deibel & Seeley (1974) divide the species into *S. faecalis* subsp. *faecalis*, *S. faecalis* subsp. *liquefaciens* (non-haemolytic, gelatin liquefied) and *S. faecalis* subsp. *zymogenes* (β-haemolytic, gelatin may or may not be liquefied). However, Deibel (1964) and Jones et al. (1972) have cast doubt upon the validity of these subspecies. In their opinion all the strains should be included in the species *S. faecalis*. It has been known for many years that both the ability to haemolyse blood and liquefy gelatin are not stable characters. Brock et al. (1963) showed that the haemolysin was also a bacteriocin, and more recent work by Jacob et al. (1975) has shown that the haemolytic ability of *S. faecalis* var. *zymogenes* is plasmid borne and they infer that plasmid transfer is by conjugation.

There are a number of reports in the literature of streptococci which cannot be allocated unequivocally to *S. faecalis* or *S. faecium*; these most frequently appear to be more closely related to *S. faecium*. Amongst atypical strains related to *S. faecalis* are those described by Mundt (1963a, b) and Martin & Mundt (1972). These authors reported that *S. faecalis* strains isolated from wild and cultivated plants and from wild animals and insects, differed markedly in their reaction in litmus milk from those strains of *S. faecalis* isolated from human sources. Another group of atypical *S. faecalis* strains are those isolated by Pette (1955) from Gouda cheese and named by him *S. faecalis* var. *malodoratus*. A taxonomic study by Jones et al. (1972) which included two strains of *S. faecalis* var. *malodoratus* did not resolve their taxonomic relationships. They were more closely related to *S. faecium* than to *S. faecalis*, but were not related closely enough to *S. faecium* to be included in that species.

Streptococcus faecium (Orla-Jensen) is now recognized as a well-defined taxon distinct from *S. faecalis* (see Deibel & Seeley 1974). It is also generally agreed that strains previously called *S. durans* (Sherman & Wing 1937) should be classified in this species (Deibel 1964; Jones et al. 1972). As mentioned above there are many streptococci which, while related to *S. faecium*, differ, sometimes quite markedly, from the typical reactions of the type strain. These streptococci have been isolated from animal and plant material but most frequently from vegetation and silage. Motile strains isolated from plant material by Langston et al. (1960) were named by them *S. faecium* var. *mobilis*. References to the early literature on motile strains of *S. faecalis* and *S. faecium* may be found in the papers of Graudal (1957), Hugh (1959) and Lund (1967). Graudal (1957) concluded that motile strains constituted a subgroup distinct from *S. faecalis* and called by him *S. glycerinaceus* (Orla-Jensen), *S. faecium* and *S. durans*. He also further divided his motile strains on the basis of whether or not they produced a yellow pigment. Lund (1967) studied a number of motile strains (including representatives of Graudal's pigmented and non-pigmented groups) and was of the opinion that physiologically they were very closely related to *S. faecium*, but gel electrophoresis of whole protein extracts and esterase enzymes indicated that, on these criteria, they were distinct from both *S. faecalis* and *S. faecium*.

Mundt & Graham (1968) proposed the name *S. faecium* var. *casseliflavus* for a group of motile, yellow-pigmented streptococci isolated from vegetation. Taylor *et al.* (1971) questioned the use of the single characteristic of pigmentation as a reliable basis for defining a new species but pigmentation was not the only basis on which Mundt & Graham (1968) proposed the subspecies. Mundt and his coworkers (Roop *et al.* 1974; Mundt 1975) have continued their studies on streptococci isolated from plants, and have amassed a large collection of atypical strains. These strains form a heterogeneous group but appear to fall into two subgroups. One group they term '*S. faecium*-like' and the second group 'other streptococci' (Mundt 1975). On the basis of DNA–DNA homology studies (Roop *et al.* 1974), they present evidence for the separateness of the subspecies *S. faecium* var. *casseliflavus* and this is in accord with the opinion of Amstein & Hartman (1973) based on the examination of the fatty acid composition of enterococci by gas chromatography.

There are numerous other reports of the isolation of atypical *S. faecalis* and *S. faecium* strains from diverse animal and plant sources (Colobert & Blondeau 1962; Hartman *et al.* 1966; Raj & Colwell 1966). Raj & Colwell (1966) suggested that certain of these strains deserved recognition as a distinct taxonomic entity at the species level and Colobert & Blondeau (1962) proposed the name *S. innominatus* for their unclassified strains. More comparative studies need to be undertaken before it can be established whether the intermediate forms are separate enough from *S. faecalis* and *S. faecium* to be regarded as separate species or whether they are merely variants of these species. If the latter is the case a re-definition of the species *S. faecalis* and *S. faecium* may be necessary.

The species name *S. avium* was proposed by Nowlan & Deibel (1967) to accommodate streptococci resembling *S. faecalis* and *S. faecium* first described by Guthof (1955) from human sources and designated by him streptococci of serological group Q. Subsequent studies demonstrated the presence of the group D antigen in these strains (Jones 1959; Smith & Shattock 1964); Nowlan & Deibel (1967) suspected that these strains might, at least partially, account for the occurrence of the 'intermediate strains' of *S. faecalis* and *S. faecium* referred to above. They therefore undertook a comparative study of representatives of strains isolated by Guthof (1955) together with apparently similar strains isolated from various animal and food sources and strains of *S. faecalis* and *S. faecium*. The results indicated that those streptococci previously referred to as serological group Q were distinct from *S. faecalis* and *S. faecium* and deserved species status. Nowlan & Deibel (1967) proposed the name *S. avium* for these strains because of their frequent occurrence in chicken faeces. The species is now recognized as a distinct taxon but there are few, if any, references in the literature to the isolation of streptococci with characters typical of *S. avium* from chicken or other animal sources.

The two remaining species of 'faecal streptococci', *S. bovis* and *S. equinus* are physiologically and biochemically quite distinct from the other streptococci discussed in this section (see Table 6).

The species *S. bovis* was first described by Orla-Jensen (1919) who found streptococci of this type to be particularly characteristic of cow dung. Orla-Jensen (1919) characterized *S. bovis* strains by their rather narrow growth temperature (25-45°C), ability to form capsules and ability to ferment arabinose and starch. He noted the ability of the strains to produce acid from inulin and raffinose when a good source of nitrogen was present whereas mannitol was never fermented. Closely related strains which grew at 22°C, did not form capsules but produced acid from raffinose and inulin even when the nitrogen source was poor, and frequently also fermented starch, mannitol and sorbitol, he placed in a separate species, *S. inulinaceus*. However, Orla-Jensen (1919) did stress that it was frequently very difficult to distinguish *S. bovis* from *S. inulinaceus* and further stated "I must admit that in the case of some intermediate forms it is difficult to say where they should be placed".

Following a number of investigations of strains of *S. bovis*, *S. inulinaceus* and the so-called 'Bargen streptococcus' frequently isolated from ulcerative colitis, Sherman (1937) concluded that there was no justification for considering *S. inulinaceus* as a species distinct from *S. bovis*, and that the Bargen streptococcus should also be included in the species *S. bovis*. However, Sherman (1937) did refer to variant strains of *S. bovis* (see Table 2). Orla-Jensen (1919) stressed the ability of *S. bovis* strains to form capsules and their inability to ferment mannitol, but Sherman (1937, 1938) reported that a substantial proportion of strains of *S. bovis* examined by him did ferment mannitol. Of course, included in the strains of *S. bovis* examined by Sherman (1938) were those which had previously been designated *S. inulinaceus* (Orla-Jensen 1919). Slime formation by *S. bovis* on sucrose agar was originally recognized by Niven *et al.* (1941). The slime was identified by Bailey & Oxford (1958) as a dextran. Mann *et al.* (1954) and Dain *et al.* (1956) noted that slime-forming strains of *S. bovis* were generally non-fermenters of mannitol while those strains which did produce acid from mannitol did not produce slime, but later work by Barnes *et al.* (1961) did not confirm this close correlation between ability to ferment mannitol and inability to produce slime from sucrose by *S. bovis*. However, as a result of a study of a number of *S. bovis* and related organisms isolated from cattle and sheep, Medrek & Barnes (1962) proposed the presence of two biotypes amongst strains of *S. bovis* on the basis of capsular slime production, non-fermentation of mannitol and production of greening on blood agar by one type (i.e. typical *S. bovis* strains) and fermentation of mannitol, no greening on blood agar and no capsular slime production by the other type. A numerical taxonomic investigation by Jones *et al.* (1972) which included mannitol and non-mannitol fermenting strains of *S. bovis* gave no indication of the presence of two distinguishable biotypes meriting separate species status amongst these strains. These authors further noted that amongst streptococci of serological group D investigated by them, some strains of *S. faecium* as well as *S. bovis* produced slime when grown on a sucrose gelatin medium. More recently, Kiel & Skadhauge (1973) examined mannitol-fermenting strains of *S. bovis* isolated mainly from human cases of bacteriaemia together with some mannitol-fermenting strains named *S. faecalis* var. *septicus* (Friedberg 1941), the one mannitol fermenting isolate of Medrek & Barnes (1962) and a strain of *S. bovis* (ATCC 9809). Kiel & Skadhauge (1973) concluded that the mannitol-fermenting strains which resemble *S. bovis* in many characters, form a distinct, rather homogeneous group showing resemblances to *S. faecalis* var. *septicus* (Friedberg 1941) and *S. inulinaceus* (Orla-Jensen 1919). The features they used to distinguish the group were the same as those listed by Medrek & Barnes (1962) with the additional feature of ability to grow at 22°C. Out of the 51 mannitol fermenting strains studied by Kiel & Skadhauge (1973) 42 grew at 22°C whereas not one of the eight non-mannitol fermenting strains of *S. bovis* grew at this temperature. Kiel & Skadhauge (1973) made a clear distinction between the possession of capsules and the ability to form slime on sucrose. They found that slime production on sucrose agar was characteristic of most of the mannitol-fermenting non-capsulated strains, and some of their non-mannitol-fermenting capsulated strains in the presence of air as well as in air with 10% CO_2. Medrek & Barnes (1972) noted slime production from sucrose only amongst their non-mannitol-fermenting strains and then only if 10% CO_2 was present. Facklam (1972) in an investigation of *S. bovis* strains from human clinical sources, divided his isolates into 'typical' *S. bovis* (91% produced acid from mannitol, 98% from raffinose, 100% hydrolysed starch, 84% produced slime on sucrose agar) and 'variant' *S. bovis* strains (which did not hydrolyse starch, produce acid from mannitol nor slime from sucrose agar, and only 29% of the isolates produced acid from raffinose). Further work is necessary on *S. bovis* strains from a variety of different sources. Since Orla-Jensen (1919) considered capsule production by *S. bovis* strains grown in milk to be an important character of the species, such future work should make a clear distinction between capsule production and extracellular slime production on sucrose agar in air with or without 10% CO_2.

The species *S. equinus* was described and named by Andrewes & Horder (1906) who found it to be the most common streptococcus isolated from the air of London. At that time the most common organic pollution of city air was horse dung and it is not surprising that later workers found strains of this species to be particularly characteristic of the intestine of horses, sometimes being isolated in pure culture (Sherman 1937). Although the only constant difference between the species *S. bovis* and *S. equinus* was the inability of the latter species to produce acid from lactose, Sherman (1937) thought that *S. equinus* merited separate species status because "the general pattern" of reactions given by *S. equinus*, aside from its inability to ferment lactose, was sufficiently distinct.

Although *S. bovis* and *S. equinus* are presently listed as distinct species (Deibel & Seeley 1974), there is no general agreement amongst bacteriologists on the status of the two species. Some workers (Seeley & Dain 1960) think they should be regarded as biotypes of the same species, while Smith & Shattock (1962) think the two species deserve separate status. The numerical taxonomic study of Jones *et al*. (1972) is in accord with the latter view. *S. equinus* is not a particularly well-studied species. It is rarely, if ever, isolated from human sources. Sherman (1937) reported the isolation of lactose negative streptococci from the human throat, but Facklam (1972) in a survey of a number of group D streptococci from human material did not isolate any strains of *S. equinus*. Further work is required to resolve the taxonomic status of *S. bovis*, *S. inulinaceus* and *S. equinus*.

There is no doubt that the species here treated as 'faecal' streptococci fall into two quite unrelated groups. I agree with Hartman *et al*. (1966) that any collective term to describe the group is misleading. *Streptococcus faecalis*, *S. faecium* and *S. avium* are phenotypically similar but distinct taxa, although many intermediate strains exist. DNA-DNA homology studies (Roop *et al*. 1974) indicate a close relationship between *S. faecalis* and *S. faecium*, and the FDP aldolases of *S. faecalis*, *S. faecium* and *S. avium* all possess similar properties (London & Kline 1973; London *et al*. 1975). However, the nucleic acid homology studies of Roop *et al*. (1974) based on one strain of *S. faecium* var. *casseliflavus* do not indicate a close relationship between this subspecies and either *S. faecalis* or *S. faecium*.

Streptococcus bovis and *S. equinus* are physiologically quite distinct from the above species and additionally the properties of their FDP aldolases are quite different (London & Kline 1973; London *et al*. 1975).

Kalina (1970) has argued that *S. faecalis* and *S. faecium* should be transferred to the genus *Enterococcus* (Thiercelin & Jouhaud 1903) as *E. faecalis* and *E. faecium* respectively. These streptococci do undoubtedly differ from most streptococcal species in many physiological characters and their FDP aldolases are also different (London & Kline 1973). Whittenbury (1965a) has drawn attention to the similarity between these species and the genus *Pediococcus* but until a comparative study of the whole lactic acid group has been conducted, the creation of another genus could only cause further confusion. What would happen to strains such as those isolated by Mundt (1975) which appear to have physiological characteristics resembling *S. faecalis* and *S. lactis*?

(e) *Lactic streptococci*

This group includes the species *S. lactis* and *S. cremoris* and a subspecies of *S. lactis*, *S. lactis* subsp. *diacetylactis*. The group is of particular importance in the dairy industry. All contain the serological group N antigen which is a glycerol teichoic acid containing galactose phosphate (Elliott 1963). Like the group D antigen it is not a cell wall constituent but occurs intracellularly (Smith & Shattock 1964). The group N teichoic acid cross-reacts with certain type-specific antipneumococcal sera (Heidelberger & Elliott 1966).

Streptococcus lactis was isolated from milk and described by Lister (1873) as *Bacterium lactis*. As the primary agent of the souring of raw milk, *S. lactis* has been recognized under a variety of names (see Sherman 1937). It is a common contaminant of milk and dairy products. Stark & Sherman (1935) isolated *S. lactis* very easily from certain plants and Sherman (1937) suggested that its natural habitat may well be vegetation. Of interest in this context is the isolation from plants of streptococci resembling *S. lactis* in certain characteristics and *S. faecalis* in others (Mundt 1975).

Streptococcus lactis is now a well recognized species (see Deibel & Seeley 1974) although early workers did not always find it easy to distinguish *S. lactis* from *S. faecalis* (see Sherman 1937). It does resemble *S. faecalis* in some respects, e.g. growth at $10°C$ and on 40% bile agar, and decolourization of litmus milk before clotting. However, it does not grow at $45°C$, in 6.5% NaCl or at pH 9.6. Some biochemical characters of *S. lactis* are listed in Table 7.

Some strains of *S. lactis* produce an antibiotic called nisin (Hirsch 1951) that inhibits many Gram positive organisms. There is good circumstantial evidence for plasmid involvement in nisin production. Nisin-producing strains of *S. lactis* spontaneously give up to 1% nisin non-producing cultures and at very high frequency after acridine, ethidium bromide or high-temperature treatment (Kozak *et al.* 1974). Certain of the nisin negative derivatives lacked the plasmid DNA present in their parent strains, but others did not (Fuchs *et al.* 1975). Thus it is very likely that autonomous plasmids are involved in nisin production. *Streptococcus lactis* strains have been shown by other workers to contain plasmids which are probably involved in the expression of other characters such as antibiotic resistance, lactose fermentation, proteinase activity, etc. (see McKay *et al.* 1973). Efstathiou & McKay (1976), in work which provided conclusive evidence that lactose metabolism and proteinase activity are plasmid linked in *S. lactis*, isolated five distinct plasmids from one strain.

Streptococcus lactis subsp. *diacetylactis* was described and named by Matuszewski *et al.* (1936). It has the same characteristics as *S. lactis* except that it metabolizes citrate in the presence of a fermentable carbohydrate to form acetoin, diacetyl and CO_2 (see Table 7). The taxonomic status of this subspecies is debatable. The establishment of a subspecies on the basis of one

Table 7
Some biochemical characters of lactic streptococci

Species	Growth at 10°C	Growth at 39.5°C	Growth at pH 9.2	Growth in 4% NaCl	Citrate utilized	Acetoin/diacetyl produced	Hippurate hydrolysis	Arginine hydrolysis	Aesculin hydrolysis	Acid production from												
										Glucose	Maltose	Lactose	Xylose	Arabinose	Sucrose	Trehalose	Mannitol	Salicin	Raffinose	Inulin	Glycerol	Sorbitol
S. lactis	+	+	+	+	–	–	v	+	v	+	+	+	v	v	v	v	v	v	–	–	–	–
S. lactis subsp. diacetylactis	+	+	+	+	+	+	v	+	v	+	+	+	v	v	–	v	v	v	–	–	–	–
S. cremoris	+	–	–	–	–*	–	–	–	v	+	–	+	–	–	–	v	–	v	–	–	–	–

Symbols as for Table 4.
* Some strains utilize citrate in presence of carbohydrate.
Based mainly on information from Deibel & Seeley (1974).

character is always questionable, as has already been discussed in the case of *S. faecalis* and its subspecies, and is especially so when genetic exchange involving that character has been demonstrated. Møller-Madsen & Jensen (1962) have reported that DNA extracted from *S. lactis* subsp. *diacetylactis* is capable of transforming the ability to ferment citrate and produce malty aroma to several *S. lactis* strains.

Although awarded full species status by Orla-Jensen (1919) the taxonomic position of the species *S. cremoris* is also doubtful. The species was created by Orla-Jensen (1919) to accommodate "the lactic acid bacteria first studied by Storch, which owing to its aroma formation has become generally used for souring cream in the manufacturing of butter". This description could equally well have been applied to *S. lactis* subsp. *diacetylactis*. Orla-Jensen (1919) also differentiated *S. cremoris* from *S. lactis* on the basis of longer chain formation by *S. cremoris*, its lower optimum growth temperature, its lower but more constant production of acid in milk, and its ability to form slime in milk. Although Sherman (1937) thought that the studies of Orla-Jensen (1919) indicated that *S. cremoris* was sufficiently distinct from *S. lactis* to merit separate species status, he pointed out that the differences noted by Orla-Jensen (1919) between the two species "are relative or quantitative ones rather than definitive".

The studies of Yawger & Sherman (1937), based in part on the observations of Ayers *et al.* (1924) that certain strains of *S. lactis*, designated by the latter authors *S. lactis* var. B, did not produce ammonia from peptone, resulted in the further differentiation of *S. cremoris* from *S. lactis*. The differential characters were, inability of *S. cremoris* to produce ammonia from peptone, to grow at 40°C in the presence of 4% NaCl or at pH 9.2 (see Table 7).

There are however many biochemical similarities between *S. lactis* and *S. cremoris*. Additionally both possess a peptidoglycan similar to that of *S. pyogenes* except for differences in cross-bridge compounds (Schleifer & Kandler 1967), although two types of cross-bridge occur in *S. cremoris*: one identical to that found in *S. lactis*, the other a dipeptide consisting of L-alanyl-threonine (Schleifer & Kandler 1967). It is highly probable that *S. cremoris* is a variety of *S. lactis* more adapted to a dairy environment. It has never to my knowledge been isolated from other sources (*S. lactis* and *S. lactis*-like streptococci have been isolated from plants). Further, preliminary DNA–DNA homology studies with one strain of each species indicate a relationship of 91% (Garvie 1978) and the properties of the FDP aldolases are very similar (London & Kline 1973).

(f) Anaerobic streptococci

The species included in this section are all aerotolerant to strictly anaerobic Gram positive cocci which occur in pairs and chains and produce lactic acid

Table 8
Some biochemical characters of anaerobic streptococci

Species	Aerotolerance	Aesculin hydrolysis	H$_2$S production	Cellobiose	Dextrin	Fructose	Galactose	Glucose	Lactose	Maltose	Mannitol	Salicin	Starch	Sucrose
				\multicolumn{11}{c}{Acid production from}
S. intermedius	+	+	−	+	V	+	+	+	+	+	+	+	V	+
S. morbillorum	+	−	−	−	−	V	V	W	−	W	W	−	−	W
S. constellatus	+	+	−	V	V	+	+	+	−	+	+	+	+	+
S. hansennii	−	V	+	−	−	−	+	+	+	−	−	−	−	−
S. pleomorphus	−	NI	W	−	−	+	−	+	−	−	V	−	−	−

Symbols as for Table 4.
Based mainly on information from Holdeman & Moore (1974), Barnes *et al.* (1977).

(with or without small amounts of acetic or formic acids, ethanol or CO_2), as the major product of fermentation. Until recently they would have been included in the genus *Peptostreptococcus* but because they produce lactic acid as a major fermentation product they are excluded from all the genera in the new family Peptococcaceae (Rogosa 1971). They are also excluded from the new family Streptococcaceae (Deibel & Seeley 1974) because they are not facultatively anaerobic. However, Rogosa (1974) suggested that because their major fermentation product is dextrorotatory lactic acid, they should be included in the genus *Streptococcus*, a view also held by Holdeman & Moore (1974). Therefore, although excluded from the eighth edition of *Bergey's Manual of Determinative Bacteriology*, these bacteria appear in the current literature as species of the genus *Streptococcus* and for this reason are included in this paper.

To date, five species have been described: *S. intermedius*, *S. constellatus* and *S. morbillorum* are all aerotolerant. *Streptococcus hansenni* and *S. pleomorphus* are strictly anaerobic.

The species *S. intermedius* was originally described and named by Prévot (1925). Transferred to the genus *Peptostreptococcus* (Smith 1957), streptococci of this kind have been more intensively studied and returned to the genus *Streptococcus* by Holdeman & Moore (1974). The source of the original isolates of Prévot is unknown but according to Holdeman & Moore (1974) strains of *S. intermedius* are isolated from many kinds of human clinical material and from faeces. According to Barnes *et al.* (1977) the anaerobic coccus group VIa strains of Hare (1967) and strains isolated from the caeca of poultry and designated group I peptostreptococci (Barnes & Impey 1970) are identical with *S. intermedius*.

Strains now included in the species *S. morbillorum* as defined by Holdeman & Moore (1974) have previously been called *Diplococcus morbillorum* (Prévot 1933), *D. robeblae* (Tunnicliffe 1933) and *Peptostreptococcus morbillorum* (Smith 1957). They are probably not so aerotolerant as *S. intermedius*. The strains have been isolated from human clinical material.

The species *S. constellatus* recently re-defined by Holdeman & Moore (1974) contains bacteria previously named *D. constellatus* (Prévot 1924) and *Peptococcus constellatus* (Douglas 1957). The anaerobic coccus group VIb of Hare (1967) is probably identical with this species. The strains have been isolated from human clinical material and from vaginal swabs.

Detailed descriptions of all these aerotolerant species are given by Holdeman & Moore (1974). The main characteristics of the species are listed in Table 8.

The two anaerobic species *S. hansennii* and *S. pleomorphus* are more recently described new species. *Streptococcus hansennii* (Holdeman & Moore 1974) contains strictly anaerobic organisms isolated from human faeces. *Streptococcus pleomorphus* (Barnes *et al.* 1977) contains anaerobic streptococci isolated mainly from poultry. Detailed descriptions of both these species may be found in the papers of Holdeman & Moore (1974) and Barnes *et al.* (1977). The main characteristics, based on the descriptions of these authors, are listed in Table 8.

The inclusion of these aerotolerant and anaerobic forms in the genus *Streptococcus* is at the moment a placement of convenience. More work is required before the relationships of these forms to the facultatively anaerobic streptococci is resolved.

(g) *Other streptococci*

In this section are included three of the most common or most studied of the named species of streptococci which have not been included in the other sections. They are mainly non-haemolytic streptococci which have been isolated from the body surfaces of animals and from dairy sources. They have been and still are frequently referred to as viridans streptococci. None of them can be serologically grouped by the criteria of Lancefield (1933). The main characters are listed in Table 9.

Streptococcus uberis was first described by Ayers & Mudge (1922) and named by Diernhofer (1932). It is found in cow's milk, on the skin and in the throat and faeces of cows (see Cullen 1966) and is responsible for a form of bovine mastitis which is particularly common in the winter. Detailed descriptions of the physiological characters of the strains may be found in the papers of Seeley (1951) and Roguinsky (1971). The serology of the species is confused. Some of the strains which have been considered to be *S. uberis* react with group E antiserum, but Stableforth (1959) was unable to obtain group E antibody by injection into rabbits of *S. uberis* strains which cross-reacted with group E antisera. In addition to reactions with group E antisera, cross-reactions have also been noted with group P and U antisera (Roguinsky 1971). Roguinsky (1969) has also reported cross-reactions with group C, G and D antisera. These reactions are probably due to common type antigens. *Streptococcus uberis* has not, to my knowledge, ever cross-reacted with group B antiserum but it can be confused with *S. agalactiae* because it hydrolyses hippurate and some strains are reported to be CAMP positive.

Deibel & Seeley (1974) draw attention to the similarity between *S. uberis* and *S. faecalis* and *S. faecium* because of the ability of *S. uberis* strains to grow at 10 and 45°C, frequently surviving heating at 60°C for 30 min, and because their folic acid requirement for growth is similar to that of *S. faecium*. However, the overall physiological pattern of *S. uberis* is quite different and Roguinsky (1971) reported that none of the 147 strains of *S. uberis* examined by him survived heating at 60°C for 30 min and only nine out of 175 strains grew at 45°C.

The species *S. acidominimus* was also first described by Ayers & Mudge (1922) and is commonly isolated from cow's milk and faeces. It has a fermenta-

Table 9
Some biochemical characters of other streptococci

Species	Growth at 10°C	Growth at 45°C	Survive 60°C 30 min⁻¹	Hippurate hydrolysis	Aesculin hydrolysis	Starch hydrolysis	Acid production												
							Glucose	Lactose	Sucrose	Maltose	Trehalose	Glycerol (aerobic)	Raffinose	Inulin	Arabinose	Xylose	Mannitol	Sorbitol	Salicin
S. uberis	+	V	V	+	+	−	+	+	+	+	+	+	−	+	−	−	+	+	+
S. acidominimus	−	−	−	+	−	−	+	+	+	+	+	−	−	−	−	−	NI	NI	NI
S. thermophilus	−	+	+	−	−	+	+	+	+	V	−	−	−	−	−	−	−	−	−

Symbols as for Table 4.
Based on information from Abd-el-Malek & Gibson (1948), Roguinsky (1971), Deibel & Seeley (1974), Wilson & Miles (1975).

tion pattern rather similar to *S. uberis* but produces smaller amounts of acid, the pH value of the carbohydrate-containing media rarely falling below 6.0 (see Deibel & Seeley 1974). Like *S. uberis* it can also be confused with *S. agalactiae* because it hydrolyses hippurate and all three species are associated with bovine environments. Unlike *S. uberis*, *S. acidominimus* does not produce acid from glycerol or hydrolyse arginine. Some strains react with group E antiserum. Wilson & Miles (1975) suggest that it may be a variant of *S. uberis*. This suggestion seems very reasonable at the present time but more studies are required to resolve the question.

The species *S. thermophilus* was described and named by Orla-Jensen (1919) during his studies on the bacteriology of lactic acid bacteria occurring in milk and dairy products. Orla-Jensen (1919) isolated streptococci of this kind mainly from heat-treated milk and dairy utensils, and Abd-el-Malek & Gibson (1948) found *S. thermophilus* together with *S. bovis* to be the predominant streptococci in bulked milk immediately after pasteurization. Not all the strains of *S. thermophilus* studied by these authors conformed exactly to the description of this species by Sherman (1937) but they appeared to be closely related. *Streptococcus thermophilus* grows very actively at 50°C and survives heating at 65°C for 30 min. On the other hand it also has a high minimum temperature of growth (20°C according to Sherman 1937; 18°C according to Abd-el-Malek & Gibson 1948). It has been reported that *S. thermophilus* not only does not grow at temperatures much below 20°C but is actually killed, but the experiments of Abd-el-Malek & Gibson (1948) do not agree with this observation. These workers inoculated the organisms into milk which was then held at 8 to 12°C for two to three weeks, and while no apparent growth occurred during that period. with very few exceptions, the dormant cultures of *S. thermophilus* showed prompt growth when incubated at 37°C.

Abd-el-Malek & Gibson (1948) failed to obtain growth of *S. thermophilus* strains on blood agar and this has also been my experience especially with fresh isolates. However, Sherman (1937) reported growth but no haemolysis on blood agar, while Deibel & Seeley (1974) report α-haemolysis. *Streptococcus thermophilus* grows weakly on media containing meat peptone compared with the growth obtained on casein. On first isolation strains demonstrate a marked preferential fermentation of the disaccharides, sucrose and lactose, rather than glucose, but this characteristic is not so apparent after several subcultures in the laboratory.

As pointed out by Sherman (1937), *S. thermophilus* strains produce acid from relatively few carbohydrates (see Table 9). However, while Sherman and his coworkers (see Sherman 1937) never isolated a strain which produced acid from maltose this has not been the experience of other workers. Orla-Jensen (1919) reported weak or late fermentation of maltose by certain strains of *S. thermophilus* and this was also the experience of subsequent workers including Abd-el-Malek & Gibson (1948). *Streptococcus thermophilus* is used as a 'starter culture' in the production of certain cheeses and yoghurt and is therefore of technical importance in the dairy industry. However, its natural habitat is not known. Like *S. cremoris* it has been isolated only from milk and dairy products and it appears to be isolated even more rarely from milk now that the method of pasteurization has been changed to a much higher temperature for a shorter time.

It is highly likely that the species *S. thermophilus* is a heat-resistant variant of another organism; whether or not this organism can now be recognized is not yet possible to answer. The peptidoglycan structure of *S. thermophilus* is identical with that of *S. faecalis* (Schleifer & Kandler 1967), but it would be unwise to place too much emphasis on this one character in the absence of other corroborative evidence. Especially so since on the properties of its FDP aldolase, London & Kline (1973) group *S. thermophilus* strains with *S. lactis* and *S. cremoris*.

Other organisms with the general characters of streptococci have been reported in the literature. They are not included here, because they have now been shown to belong to some of the named species discussed here or because they are not streptococci or because the descriptions are too meagre.

5. Concluding Remarks

It is apparent from the material presented and discussed here that the classification of the genus *Streptococcus* is unsatisfactory. Not only, with few exceptions, is the unequivocal characterization of the subdivisions (subgenera, species or serological groups, depending on the terminology used) within the genus difficult, but there is no consensus of opinion on the circumscription of the genus.

It is now over 100 years since streptococci were first recognized and over 70 years since Andrewes & Horder (1906) differentiated eight groups of closely related forms which they equated with distinct species. Since that time streptococci have been one of the most studied groups of bacteria. Why, therefore, is their taxonomy so underdeveloped?

Streptococci have complex nutritional requirements, they do not generally metabolize organic acids and they do not synthesize haem compounds. They therefore inhabit and thrive in environments where carbohydrate and protein materials abound. These environments include the tissues and intestinal tracts of animals and man, dairy products, food and vegetable material. In these situations some cause diseases, others produce favourable or unfavourable changes in milk, food, etc. They are therefore of great economic importance and have been studied intensively, but the clinical and veterinary bacteriologists have tended to concentrate their investigations on streptococci isolated from clinical conditions in man and animals, dairy bacteriologists on those isolated from milk and dairy produce, etc. This has resulted in very little comparative work on the streptococci as a whole.

The situation has been compounded by the heavy reliance placed by clinical and veterinary bacteriologists in particular, on one character, the so-called group antigen, in the identification of streptococci. Thus streptococci of clinical and veterinary importance have been poorly characterized by other criteria of potential value in establishing their taxonomic relatedness to other streptococci. Serological grouping has also been given undue emphasis in the classification of other streptococci. An example is the grouping together in serological group D of *S. faecalis* and *S. faecium* with the physiologically quite different species *S. bovis* and *S. equinus*.

On the whole more biochemical and physiological tests have been carried out on streptococci from sources other than human or clinical disease. These studies have, however, produced other problems for the better understanding of the relationships amongst the streptococci. For example, *S. faecalis* strains from the human gut are well characterized. Forms with the same characteristics can be isolated and re-isolated from this source. To make isolation easier, selective media have been developed for this purpose. These selective media are then used to isolate *S. faecalis* strains from food, water, vegetation, etc. The isolates are subjected to the battery of laboratory tests developed for the identification of *S. faecalis* strains from the human gut. The result is that in addition to forms corresponding to what are called typical *S. faecalis* strains, atypical strains are isolated. When this occurs, as it frequently does, the bacteriologist either discards the atypical strains, because they distort the tidiness of the results, or allocates them to a new species or subspecies or bemoans the state of streptococcal classification but does little about it. *Streptococcus faecalis* was chosen as an example but the same situation occurs with other streptococcal species.

All the early bacteriologists stressed the variation in biochemical characters amongst strains of similar streptococci examined by them in the laboratory (Andrewes & Horder 1906; Winslow & Winslow 1908; Orla-Jensen 1919). Orla-Jensen (1919) in his classic work on the lactic acid bacteria drew particular attention to the effect of the composition of the test medium on the results obtained. More recently, Whittenbury (1963) in an elegant study rarely referred to in the current literature and therefore probably ignored by most bacteriologists, showed the important effects of oxygen concentration, temperature and media composition on the results of sugar fermentations (see Whittenbury, this volume).

If one believes in evolution one must necessarily believe in variation. Without going into the detailed arguments as to whether streptococci represent high or low evolutionary forms of bacteria (see Mundt 1975; London 1976, for some theories and references to early literature) it is undoubtedly true that some streptococci are particularly highly adapted to certain environments. As new environments are created by man, e.g. prepacked foods, changes in dairy technology, etc., bacteria—including streptococci—will colonize these environments. Those bacteria which succeed will be those amongst the population best suited to thrive in these environments. What differences in the chromosomal genome, if any, exist between such forms we do not yet know.

It is now known that many phenotypic characters of importance in the identification of streptococci are, at least in some strains, plasmid linked, e.g. haemolysis in *S. faecalis*, lactose fermentation, proteinase activity and probably nisin production in *S. lactis*. These plasmids can be transferred to other strains where, if they are incorporated into the organism and their genetic information expressed, they may confer an advantage in a particular environment. If this occurs then it will be those strains which contain the relevant plasmid which will become the predominant bacterial flora of that particular environment. What effect plasmid-borne characters have on classification is not yet known, although preliminary work with the genus *Proteus* suggests not very much when the classification is based on many features (McKell 1977), but they can have a marked effect on identification if the diagnostic scheme used relies heavily on such characters.

All these factors—the tendency for streptococci from different sources to be studied independently; the undue emphasis given to serological grouping; the lack of attention given to the effect of laboratory culture conditions on the test reactions of streptococci; the effect of plasmid-borne characters (which can be lost or gained) in some diagnostic schemes—have contributed to the poor state of streptococcal classification.

The solution lies in more basic biological, chemical and genetic study of streptococci from different ecological sources. Some such studies have already been conducted (Colman & Williams 1965; Colman 1968, 1970; London & Kline 1973; London et al. 1975) but there is much still to be done. Such studies should also include other lactic acid bacteria. Orla-Jensen (1919) considered the lactic acid bacteria to be a closely related group. The protein homology studies of London et al. (1975) and London & Kline (1973) and the numerical taxonomic studies of Wilkinson & Jones (1977) agree with the opinion of Orla-Jensen (1919). Future studies should aim at producing a classification of the lactic acid bacteria which is based on morphological, physiological, biochemical, genetic and ecological data. Such a classification would indicate the broad relationships amongst these bacteria and would lead to a better circumscription of the main divisions (genera or subgenera) and a better differentiation of the subdivisions (species) within the main divisions. The species so obtained would be characterized on the basis of the possession of a general pattern of certain relatively stable characters and would accommodate strains which in 'general pattern' resembled the typical reactions of the type strain of that species, even if they did not correspond exactly.

A good and reliable identification scheme for the streptococci will be achieved only when the classification of these organisms is based on information obtained from broad biological and chemical studies. Only then will it be possible to define the composition of the genus *Streptococcus* and differentiate the subdivisions within it, with some confidence.

6. References

ABD-EL-MALEK, Y. & GIBSON, T. 1948 Studies in the bacteriology of milk. 1. The streptococci of milk. *Journal of Dairy Research* 15, 233-248.

ANDERSON, D. G. & McKAY, L. L. 1977 Plasmids, loss of lactose metabolism and appearance of partial and full lactose-fermenting revertants in *Streptococcus cremoris* Bl. *Journal of Bacteriology* 129, 367-377.

ANDREWES, F. W. & HORDER, J. 1906 A study of the streptococci pathogenic for man. *Lancet* 2, 708-713; 775-782; 852-855.

AMSTEIN, F. & HARTMAN, P. A. 1973 Differentiation of some enterococci by gas chromatography. *Journal of Bacteriology* 113, 38-41.

AYERS, S. H., JOHNSON, W. T. & MUDGE, C. S. 1924 Streptococci of souring milk with special reference to *Streptococcus lactis*. *Journal of Infectious Diseases* 34, 29-48.

AYERS, S. H. & MUDGE, C. S. 1922 The streptococci of the bovine udder. *Journal of Infectious Diseases* 31, 40-50.

BAILEY, R. W. & OXFORD, A. E. 1958 A quantitative study of the production of dextran from sucrose by rumen strains of *Streptococcus bovis*. *Journal of General Microbiology* 19, 130-145.

BARNES, E. M. & IMPEY, C. S. 1970 The isolation and properties of the predominant anaerobic bacteria from the caeca of turkeys and chickens. *British Poultry Science* 11, 467-481.

BARNES, E. M., IMPEY, C. S., STEVENS, B. J. H. & PEEL, J. L. 1977 *Streptococcus pleomorphus sp. nov*: an anaerobic streptococcus isolated mainly from the caeca of birds. *Journal of General Microbiology* 102, 45-53.

BARNES, I. J., SEELEY, H. W. & VAN DEMARK, P. J. 1961 Nutrition of *Streptococcus bovis* in relation to dextran formation. *Journal of Bacteriology* 82, 85-93.

BILLROTH, T. 1874 *Untersuchungen über die Vegetationsformen von Coccobacteria Septica*. Berlin: G. Reimer.

BOISSARD, J. M. & WORMALD, P. J. 1950 A new group of haemolytic streptococci for which the designation "group 0" is proposed. *Journal of Pathology and Bacteriology* 62, 37-41.

BOWDEN, G. H., HARDIE, J. M. & SLACK, G. L. 1975 Microbial variations in approximal dental plaque. *Caries Research* 9, 253-277.

BRATTHALL, D. 1970 Demonstration of five serological groups of streptococcal strains resembling *Streptococcus mutans*. *Odontologisk Revy* 21, 143-152.

BROCK, T. D., PEACHER, B. & PIERSON, D. 1963 Survey of the bacteriocines of enterococci. *Journal of Bacteriology* 86, 702-707.

BROWN, J. H. 1939 The taxonomy of streptococci. 3rd International Microbiological Congress, New York, p. 172.

BROWN, J. H., FROST, W. D. & SHAW, M. 1926 Hemolytic streptococci of the beta type in certified milk. *Journal of Infectious Diseases* 38, 381-388.

BRYAN-JONES, D. G. & WHITTENBURY, R. 1969 Haematin-dependent oxidative phosphorylation in *Streptococcus faecalis*. *Journal of General Microbiology* 58, 247-260.

BUCHANAN, R. E., HOLT, J. G. & LESSEL, E. F. 1966 *Index Bergeyana*. London: Livingstone.

CARLSSON, J. 1968 Numerical taxonomic study of human oral streptococci. *Odontologisk Revy* 19, 137-160.

CHAPMAN, S. S. 1972 Unusual Group A streptococci: Colonial appearance and haemolysis. In *Streptococci and Streptococcal Diseases,* eds Wannamaker, L. W. & Matsen, J. M. New York & London: Academic Press.
CHRISTIE, R., ATKINS, N. E. & MUNCH-PETERSEN, E. 1944 A note on a lytic phenomenon shown by Group B streptococci. *Australian Journal of Experimental Biology and Medical Science* **22,** 197-200.
CLARKE, J. K. 1924 On the bacterial factor in the aetiology of dental caries. *British Journal of Experimental Pathology* **5,** 141-147.
CLEARY, P. P., JOHNSON, Z. & WANNAMAKER, L. W. 1975 Genetic instability of M protein and serum opacity factor of group A streptococci: evidence suggesting extrachromosomal control. *Infection and Immunity* **12,** 109-118.
CLEWELL, D. B. & FRANKE, A. E. 1974 Characterisation of a plasmid determining resistance to erythromycin, lincomycin and vernamycin Bα in a strain of *Streptococcus pyogenes. Antimicrobial Agents and Chemotherapy* **5,** 534-537.
CLEWELL, D. B., YAGI, Y., DUNNY, G. M. & SCHULTZ, S. K. 1974 Characterization of three plasmid deoxyribonucleic acid molecules in a strain of *Streptococcus faecalis*: identification of a plasmid determining erythromycin resistance. *Journal of Bacteriology* **117,** 283-289.
CLEWELL, D. B., YAGI, Y. & BAVER, B. 1975 Plasmid-determined tetracycline resistance in *Streptococcus faecalis*: evidence for gene amplification during growth in the presence of tetracycline. *Proceedings of the National Academy of Sciences, Washington* **72,** 1720-1724.
COLE, R. M. 1968 Structure of the group A streptococcal cell and its L-form. In *Current Research on Group A Streptococcus,* ed. Caravano, R. Amsterdam: Excerpta Medica Foundation.
COLMAN, G. 1968 The application of computers to the classification of streptococci. *Journal of General Microbiology* **50,** 149-158.
COLMAN, G. 1970 The classification of streptococcal strains. Ph.D. thesis: University of London.
COLMAN, G. & WILLIAMS, R. E. O. 1965 The cell walls of streptococci. *Journal of General Microbiology* **41,** 375-387.
COLMAN, G. & WILLIAMS, R. E. O. 1972 Taxonomy of some human viridans streptococci. In *Streptococci and Streptococcal Diseases,* eds. Wannamaker, L. W. & Matsen, J. M. New York & London: Academic Press.
COLOBERT, L. & BLONDEAU, H. 1962 L'espèce *Streptococcus faecalis.* I. Étude de l'homogénéité par la méthode Adansonnienne. *Annales de l'Institut Pasteur* **103,** 345-362.
COOK, A. R. 1976 The elimination of urease activity in *Streptococcus faecium* as evidence for plasmid-coded urease. *Journal of General Microbiology* **92,** 49-58.
CORDS, B. R., McKAY, L. L. & GUERRY, P. 1974 Extrachromosomal elements in group N streptococci. *Journal of Bacteriology* **117,** 1149-1152.
COURVALIN, P. M., CARLIER, C. & CHABBERT, Y. A. 1972 Plasmid-linked tetracycline and erythromycin resistance in group D *Streptococcus. Annales de l'Institut Pasteur* **123,** 755-759.
COURVALIN, P. M., CARLIER, C., CROISSANT, O. & BLANGY, D. 1974 Identification of two plasmids determining resistance to tetracycline and to erythromycin in group D *Streptococcus. Molecular and General Genetics* **132,** 181-192.
COWAN, S. T. & STEEL, K. J. 1974 *Manual for the Identification of Medical Bacteria,* 2nd end. Cambridge: University Press.
COYKENDALL, A. L. 1977 Proposal to elevate the subspecies of *Streptococcus mutans* to species status, based on their molecular composition. *International Journal of Systematic Bacteriology* **27,** 26-30.
COYKENDALL, A. L., DAILY, O. P., KRAMER, M. J. & BEATH, M. E. 1971 DNA-DNA hybridization studies of *Streptococcus mutans. Journal of Dental Research* **50,** 1131-1139.
CULLEN, G. A. 1966 The ecology of *Streptococcus uberis. British Veterinary Journal* **122,** 333-339.

CUMMINS, C. S. & HARRIS, H. 1956 The chemical composition of the cell wall in some Gram-positive bacteria and its possible use as a taxonomic character. *Journal of General Microbiology* **14**, 583-600.

DAIN, J. A., NEAL, A. L. & SEELEY, H. W. 1956 The effect of carbon dioxide on polysaccharide production by *Streptococcus bovis*. *Journal of Bacteriology* **72**, 209-213.

DEIBEL, R. H. 1964 The group D streptococci. *Bacteriological Reviews* **28**, 330-366.

DEIBEL, R. H. & SEELEY, H. W. 1974 Streptococcaceae fam. nov. In *Bergey's Manual of Determinative Bacteriology*, 8th edn, eds Buchanan, R. E. & Gibbons, N. E. Baltimore: Williams & Wilkins.

DEIBEL, R. H., YAO, J., JACOBS, N. J. & NIVEN, C. F. 1964 Group E streptococci: 1. Physiological characterisation of strains isolated from swine cervical abscesses. *Journal of Infectious Diseases* **114**, 327-332.

DIBLE, J. H. 1921 The enterococcus and the faecal streptococci: their properties and relations. *Journal of Pathology and Bacteriology* **24**, 3-35.

DIERNHOFER, K. 1932 Aesculinbouillon als Hilfsmittel für die Differenzierung von Euten und Milchstreptokokken bei Massenuntersuchungen. *Milchwirtschaftliche Forschungen* **13**, 368-374.

DOUGLAS, H. C. 1957 Genus IV. *Peptococcus* Kluyver and Van Niel 1936. In *Bergey's Manual of Determinative Bacteriology*, 7th edn, eds Breed, R. S., Murray, E. G. D. & Smith, N. R. Baltimore: Williams & Wilkins.

DRUCKER, D. B. & MELVILLE, T. H. 1969 Computer classification of streptococci mostly of oral origin. *Nature* **221**, 664.

DUNNY, G. M. & CLEWELL, D. B. 1975 Transmissible toxin (hemolysin) plasmid in *Streptococcus faecalis* and its mobilization of a noninfectious drug resistance plasmid. *Journal of Bacteriology* **124**, 784-790.

EFSTATHIOU, J. D. & McKAY, L. L. 1976 Plasmids in *Streptococcus lactis*: evidence that lactose metabolism and proteinase activity are plasmid linked. *Applied and Environmental Microbiology* **32**, 38-44.

ELLIOTT, S. D. 1963 Teichoic acid and the group antigen of lactic streptococci (group N). *Nature* **200**, 1184.

ELLIOTT, S. D. 1966 Streptococcal infection in young pigs. 1. An immunological study of the causative agent (PM streptococcus). *Journal of Hygiene, Cambridge* **64**, 205-212.

ELLIOTT, S. D., ALEXANDER, T. J. L. & THOMAS, J. H. 1966 Streptococcal infection in young pigs. II. Epidemiology and experimental production of the disease. *Journal of Hygiene, Cambridge* **64**, 213-220.

FACKLAM, R. R. 1972 Recognition of Group D streptococcal species of human origin by biochemical and physiological tests. *Applied Microbiology* **23**, 1131-1139.

FACKLAM, R. R. 1974 Characteristics of *Streptococcus mutans* isolated from human dental plaque and blood. *International Journal of Systematic Bacteriology* **24**, 313-319.

FIELD, H. I., BUNTAIN, D. & DONE, J. T. 1954 Studies on piglet mortality. 1. Streptococcal meningitis and arthritis. *The Veterinary Record* **66**, 453-455.

FRIEDBERG, R. 1941 *Studier over ikke-haemolytiske Streptokokker*. Copenhagen: Einer Munksgaard.

FROST, W. D. & ENGELBRECHT, M. A. 1940 *The streptococci, their descriptions, classification and distribution, with special reference to those in milk*. Madison, Wisconsin: Willdof Book Co.

FROST, W. D., GUMM, M. & THOMAS, R. C. 1927 Types of hemolytic streptococci in certified milk. *Journal of Infectious Diseases* **40**, 698-705.

FUCHS, P. G., ZAJDEL, J. & DOBRZAŃSKI, W. T. 1975 Possible plasmid nature of the determinant for production of the antibiotic nisin in some strains of *Streptococcus lactis*. *Journal of General Microbiology* **88**, 189-192.

FULLER, A. T. 1938 The formamide method for the extraction of polysaccharides from haemolytic streptococci. *British Journal of Experimental Pathology* **19**, 130-138.

GARVIE, E. I. 1978 *Streptococcus raffinolactis*. (Orla-Jensen & Hansen); a group N streptococcus found in raw milk. *International Journal of Systematic Bacteriology*. In press.

GHUYSEN, J. M. 1968 Use of bacteriolytic enzymes in determination of wall structure and their role in cell metabolism. *Bacteriological Reviews* **32**, 425–464.

GRAUDAL, H. 1957 The classification of motile streptococci within the enterococcus group. *Acta Pathologica et Microbiologica Scandinavica* **41**, 403–410.

GUTHOF, O. 1955 Über eine neue serologische Gruppe alphahämolytischer Streptokokken (serologische Gruppe Q). *Zentralblatt für Bakteriologie, Parasitenkunde, Infectionskrankheiten und Hygiene* Abteilung 1 **164**, 60–69.

GUTHOF, O. 1956 Ueber pathogene "vergrunende Streptokokken". *Zentralblatt für Bakteriologie, Parasitenkunde, Infectionskrankheiten und Hygiene* Abteilung 1 **166**, 553–564.

HARDIE, J. M. & BOWDEN, G. H. 1976*a* Physiological classiciation of oral viridans streptococci. *Journal of Dental Research* **55**, Special Issue A, A166–A176.

HARDIE, J. M. & BOWDEN, G. H. 1976*b* Some serological cross reactions between *Streptococcus mutans*, *S. sanguis* and other dental plaque streptococci. *Journal of Dental Research* **55**, Special Issue C, C50–C58.

HARE, R. 1935 The classification of haemolytic streptococci from the nose and throat of normal human beings by means of precipitin and biochemical tests. *Journal of Pathology and Bacteriology* **41**, 499–512.

HARE, R. 1967 The anaerobic cocci. In *Recent Advances in Medical Microbiology*, ed. Waterson, H. P. Boston: Little, Brown & Co.

HARE, R. & FRY, R. M. 1938 Clinical observations of the beta haemolytic streptococcal infections of dogs. *The Veterinary Record* **50**, 1537–1548.

HARTMAN, P. A., REINBOLD, G. W. & SARASWAT, D. A. 1966 Indicator organisms—a review. I. Taxonomy of the fecal streptococci. *International Journal of Systematic Bacteriology* **16**, 197–221.

HEIDELBERGER, M. & ELLIOTT, S. D. 1966 Cross-reactions of streptococcal group N teichoic acid in antipneumoccocal horse sera of types VI, XIV, XVI and XXVII. *Journal of Bacteriology* **92**, 281–283.

HIRSCH, A. 1951 Growth and nisin production of a strain of *Streptococcus lactis*. *Journal of General Microbiology* **5**, 208–221.

HESS, E. L. & SLADE, H. D. 1965 An electrophoretic examination of cell free extracts from various serological types of group A hemolytic streptococci. *Biochimica et Biophysica Acta* **16**, 346–353

HITCHCOCK, C. H. 1924 Precipitation and complement fixation reactions with residue antigens in the non-hemolytic streptococcus group. *Journal of Experimental Medicine* **40**, 575–581.

HOLDEMAN, L. V. & MOORE, W. E. C. 1974 New genus, *Coprococcus*, twelve new species and emended description of four previously described species of bacteria from human faeces. *International Journal of Systematic Bacteriology* **24**, 260–277.

HOLMAN, W. L. 1916 The classification of streptococci. *Journal of Medical Research* **34**, 377–443.

HORODNICEANU, T., BOUANCHAUD, D. H., BIETH, G. & CHABBERT, Y. A. 1976 R plasmids in *Streptococcus agalactiae* (Group B). *Antimicrobial Agents and Chemotherapy* **10**, 795–801.

HUGH, R. 1959 Motile streptococci isolated from the oropharyngeal region. *Canadian Journal of Microbiology* **5**, 351–354.

JACOB, A. E. & HOBBS, S. J. 1974 Conjugal transfer of plasmid-borne multiple antibiotic resistance in *Streptococcus faecalis* var. *zymogenes*. *Journal of Bacteriology* **117**, 360–372.

JACOB, A. E., DOUGLAS, G. J. & HOBBS, S. J. 1975 Self-transferable plasmids determining the hemolysin and bacteriocin of *Streptococcus faecalis* var. *zymogenes*. *Journal of Bacteriology* **121**, 863–872.

JELÍNKOVÁ, J. & KUBÍN, V. 1974 Proposal of a new serological group ("V") of hemolytic streptococci isolated from swine lymph nodes. *International Journal of Systematic Bacteriology* **24**, 434–437.

JONES, D. 1959 Physiological and serological studies on group D streptococci. M.Sc. thesis, University of Reading.
JONES, D. & SHATTOCK, P. M. F. 1960 The location of the group antigen of Group D streptococci. *Journal of General Microbiology* 23, 335–343.
JONES, D. & SNEATH, P. H. A. 1970 Genetic transfer and bacterial taxonomy. *Bacteriological Reviews* 34, 40–81.
JONES, D., SACKIN, M. J. & SNEATH, P. H. A. 1972 A numerical taxonomic study of streptococci of serological group D. *Journal of General Microbiology* 72, 439–450.
KALINA, A. P. 1970 The taxonomy and nomenclature of enterococci. *International Journal of Systematic Bacteriology* 20, 185–189.
KANDLER, O., SCHLEIFER, K. H. & DANDL, R. 1968 Differentiation of *Streptococcus faecalis* Andrewes & Horder and *Streptococcus faecium* Orla-Jensen based on the amino acid composition of their murein. *Journal of Bacteriology* 96, 1935–1939.
KAYSER, F. H. 1974 Resistance of Gram-positive bacteria to chloramphenicol/thiamphenicol: occurrence and genetic basis. *Postgraduate Medical Journal* 50, Supplement 5, 79–83.
KEIL, P. & SKADHAUGE, K. 1973 Studies on mannitol fermenting strains of *Streptococcus bovis*. *Acta Pathologica et Microbiologica Scandinavica* 81, 10–14.
KNOX, K. W. & WICKEN, A. J. 1973 Immunological properties of teichoic acids. *Bacteriological Reviews* 37, 215–257.
KOZAK, W., RAJCHERT-TRZPIL, M. & DOBRZANSKI, W. T. 1974 The effect of proflavin, ethidium bromide and an elevated temperature on the appearance of nisin-negative clones in nisin-producing strains of *Streptococcus lactis*. *Journal of General Microbiology* 83, 295–302.
KRAUSE, R. M. 1972 The antigens of Group D streptococci. In *Streptococci and Streptococcal Diseases*, eds. Wannamaker, L. W. & Matsen, J. M. London & New York: Academic Press.
KRAUSE, R. M. & McCARTY, M. 1962 Studies on the chemical structure of the streptococcal cell wall. II. The composition of Group C cell walls and chemical basis for serological specificity of the carbohydrate moeity. *Journal of Experimental Medicine* 115, 49–62.
LANCEFIELD, R. C. 1933 A serological differentiation of human and other groups of hemolytic streptococci. *Journal of Experimental Medicine* 57, 571–595.
LANCEFIELD, R. C. 1934 A serological differentiation of specific types of bovine hemolytic streptococci (Group B). *Journal of Experimental Medicine* 59, 441–458.
LANGSTON, C. W. GUTTIERREZ, J. & BOUMA, C. 1960 Motile enterococci (*Streptococcus faecium* var. *mobilis* var. N) isolated from grass silage. *Journal of Bacteriology* 80, 714–718.
LAUGHTON, N. 1948 Canine beta haemolytic streptococci. *Journal of Pathology and Bacteriology* 60, 471–476.
LEBLANC, D. J. & HASSELL, F. P. 1976 Transformation of *Streptococcus sanguis* by plasmid deoxyribonucleic acid from *Streptococcus faecalis*. *Journal of Bacteriology* 128, 347–355.
LISTER, J. 1873 A further contribution to the natural history of bacteria and the germ theory of fermentative changes. *Quarterly Journal of Microscopical Science* 13, 380–408.
LONDON, J. 1976 The ecology and taxonomic status of the lactobacilli. *Annual Review of Microbiology* 30, 279–301.
LONDON, J. & KLINE, K. 1973 Aldolase of lactic acid bacteria: a case history in the use of an enzyme as an evolutionary marker. *Bacteriological Reviews* 37, 453–478.
LONDON, J., CHACE, N. M. & KLINE, K. 1975 Aldolase of lactic acid bacteria: immunological relationships among aldolases of streptococci and Gram-positive nonsporeing anaerobes. *International Journal of Systematic Bacteriology* 25, 114–123.
LONG, P. H. & BLISS, E. A. 1934 Studies upon minute hemolytic streptococci. 1. The isolation and cultural characteristics of minute beta hemolytic streptococci. *Journal of Experimental Medicine* 60, 619–631.

LUND, B. M. 1965 A comparison by the use of gel electrophoresis of soluble protein components and esterase enzymes of some group D streptococci. *Journal of General Microbiology* **40**, 413–419.
LUND, B. M. 1967 A study of some motile group D streptococci. *Journal of General Microbiology* **49**, 67–80.
McCARTY, M. 1958 Further studies on the chemical basis for serological specificity in Group A streptococcal carbohydrates. *Journal of Experimental Medicine* **108**, 311–321.
McKAY, L. L. & BALDWIN, K. A. 1974 Simultaneous loss of proteinase and lactose utilizing enzyme activities in *Streptococcus lactis* and reversal of loss by transduction. *Applied Microbiology* **28**, 342–346.
McKAY, L. L. & BALDWIN, K. A. 1975 Plasmid distribution and evidence for a proteinase plasmid in *Streptococcus lactis* C2-1. *Applied Microbiology* **29**, 546–548.
McKAY, L. L., CORDS, B. R. & BALDWIN, K. A. 1973 Transduction of lactose metabolism in *Streptococcus lactis* C2'. *Journal of Bacteriology* **115**, 810–815.
McKAY, L. L., BALDWIN, K. A. & EFSTATHIOU, J. D. 1976 Transductional evidence for plasmid linkage of lactose metabolism in *Streptococcus lactis* C2. *Applied and Environmental Microbiology* **32**, 45–52.
McKELL, J. 1977 A taxonomic study of the *Proteus-Providence* group with especial reference to the role of plasmids. Ph.D. thesis, University of Leicester.
MALKE, H. 1975 Transfer of a plasmid-mediating antibiotic resistance between strains of *Streptococcus pyogenes* in mixed cultures. *Zeitschrift fur Allgemeine Mikrobiologie* **15**, 645–649.
MALKE, H., STARKE, R., KOHLER, W., KOLESNICHENKO, T. G. & TOTOLIAN, A. A. 1975 Bacteriophage P13234mo-mediated intra- and intergroup transduction of antibiotic resistance among streptococci. *Zentralblatt für Bakteriologie, Parasitenkunde, Infektionskrankheiten und Hygiene* Abt. 1 **233**, 24–34.
MANN, S. O., MASSON, F. M. M. & OXFORD, A. E. 1954 Facultative anaerobic bacteria from the sheep's rumen. *Journal of General Microbiology* **10**, 142–149.
MARTIN, J. D. & MUNDT, J. O. 1972 Enterococci in insects. *Applied Microbiology* **24**, 575–580.
MATUSZEWSKI, T., PIJANOWSKI, E. & SUPINSKA, J. 1936 *Streptococcus diacetilactis*. n. sp. and its application to butter making. *Roczników Nauk Rolniczych I Lésnych* (Polish Agricultural and Forestry Annual) **36**, 1–28.
MEDREK, T. F. & BARNES, E. M. 1962 The physiological and serological properties of *Streptococcus bovis* and related organisms isolated from cattle and sheep. *Journal of Applied Bacteriology* **25**, 169–179.
MICHEL, M. F. & GOODER, H. 1962 Amino acids, amino sugars and sugars present in the cell walls of some strains of *Streptococcus pyogenes*. *Journal of General Microbiology* **29**, 199–205.
MICHEL, M. F. & WILLERS, J. M. N. 1964 Immunochemistry of Group F streptococci; isolation of group specific oligosaccharides. *Journal of General Microbiology* **37**, 381–389.
MIRICK, G. S., THOMAS, L., CURNEN, E. C. & HORSFALL, F. L. 1944 Studies on a non-hemolytic streptococcus isolated from the respiratory tract of human beings. 1. Biological characteristics of *Streptococcus* MG. *Journal of Experimental Medicine* **80**, 391–406.
MIYAMURA, S., OCHIAI, H., NITAHARA, Y., NAKAGAWA, Y. & TERAO, M. 1977 Resistance mechanism of chloramphenicol in *Streptococcus haemolyticus*, *Streptococcus pneuomoniae* and *Streptococcus faecalis*. *Microbiology and Immunology* **21**, 69–76.
MOBERG, K. & THAL, E. 1954 Beta-hämolytische Streptokokken einer neuen Lancefield Gruppe. *Nordisk Veterinärmedicin* **6**, 69–72.
MØLLER-MADSEN, A. A. & JENSEN, H. 1962 Transformation of *Streptococcus lactis*. In *Contributions to the XVIth Int. Dairy Congress, Copenhagen.*

MOOR, C. E. de 1963 Septicaemic infections in pigs caused by haemolytic streptococci of new Lancefield groups designated R, S and T. *Antonie van Leeuwenhoek* **29**, 272-280.
MOOR, C. E. de & THAL, E. 1968 Beta haemolytic streptococci of Lancefield groups E, P and U: *Streptococcus infrequens*. *Antonie van Leeuwenhoek* **34**, 377-387.
MONTAGUE, E. A. & KNOX, K. W. 1968 Antigenic components of the cell wall of *Streptococcus salivarius*. *Journal of General Microbiology* **54**, 237-246.
MOREIRA-JACOB, M. 1956 The streptococci of Lancefield's Group E: Biochemical and serological identification of the haemolytic strains. *Journal of General Microbiology* **14**, 268-280.
MOSSER, J. L. & TOMASZ, A. 1970 Choline-containing teichoic acid as a structural component of pneumococcal cell wall and its role in sensitivity to lysis by an autolytic enzyme. *Journal of Biological Chemistry* **245**, 287-298.
MUNDT, J. O. 1963a Occurrence of enterococci in animals from a wild environment. *Applied Microbiology* **11**, 136-140.
MUNDT, J. O. 1963b Occurrence of enterococci on plants in a wild environment. *Applied Microbiology* **11**, 141-144.
MUNDT, J. O. 1975 Unidentified streptococci from plants. *International Journal of Systematic Bacteriology* **25**, 281-285.
MUNDT, J. O. & GRAHAM, W. F. 1968 *Streptococcus faecium* var. *casseliflavus*. nov. var. *Journal of Bacteriology* **95**, 2005-2009.
NAKAE, N., INOVE, M. & MITSUHASHI, S. 1975 Artifical elimination of drug resistance from group A beta-haemolytic streptococci. *Antimicrobial Agents and Chemotherapy* **7**, 719-720.
NEWSOM, I. E. 1937 Strangles in hogs. *Veterinary Medicine* **37**, 137-138.
NIVEN, C. F. 1964 Microbial indices of good quality: faecal streptococci. In *Microbiological Quality of Foods*, eds Slanetz, L. W., Chichester, C. O., Gaufin, A. R. & Ordal, Z. J. New York and London: Academic Press.
NIVEN, C. F., SMILEY, K. L. & SHERMAN, J. M. 1941 The polysaccharides synthesized by *Streptococcus salivarius* and *Streptococcus bovis*. *Journal of Biological Chemistry* **140**, 104-109.
NOWLAN, S. S. & DEIBEL, R. H. 1967 Group Q Streptococci. 1. Ecology, serology, physiology and relationship to established enterococci. *Journal of Bacteriology* **94**, 291-296.
ORLA-JENSEN, S. 1919 *The Lactic Acid Bacteria*. Copenhagen: Adr. Fred Host & Son.
OSBORN, R. M., LAMBERTS, B. L., MEYER, T. S. & ROUSH, A. H. 1976 Acrylamide gel electrophoretic studies of extracellular sucrose-metabolizing enzymes of *Streptococcus mutans*. *Journal of Dental Research* **55**, 77-89.
OTTENS, H. & WINKLER, K. C. 1962 Indifferent and haemolytic streptococci possessing group-antigen F. *Journal of General Microbiology* **28**, 181-191.
PERCH, B. & KJEMS, E. 1971. Group R streptococci in man. *Acta Pathologica et Microbiologica Scandinavica* **79**, 549-550.
PERCH, B., KRISTJANSEN, P. & SKADHAUGE, K. 1968 Group R streptococci pathogenic for man. Two cases of meningitis and one of fatal sepsis. *Acta Pathologica et Microbiologica Scandinavica* **74**, 69-76.
PERCH, B., KJEMS, E. & RAVN, T. 1974 Biochemical and serological properties of *Streptococcus mutans* from various human and animal sources. *Acta Pathologica et Microbiologica Scandinavica* **82**, 357-370.
PETTE, J. W. 1955 De vorming van zwatelwaterstof in Goudse kaas, veroorzaakt door melkzuurbacterien. *Netherlands Milk & Dairy Journal* **10**, 291-302.
PRÉVOT, A. R. 1924 *Diplococcus constellatus* (n. sp). *Comptes Rendus de Scéances de la Société de Biologie* **91**, 426-428.
PRÉVOT, A. R. 1925 Les streptocoques anaérobies. *Annales de l'Institut Pasteur* **39**, 415-447.
PRÉVOT, A. R. 1933 Études de systematique bactérienne. I. Lois générales; II. Cocci anaérobius. *Annales des Sciences Naturelles* **15**, 23-260.

RAJ, H. & COLWELL, R. R. 1966 Taxonomy of enterococci by computer analysis. *Canadian Journal of Microbiology* 12, 353-362.
RAVIN, A. W. 1963 Experimental approaches to the study of bacterial phylogeny. *The American Naturalist* 97, 307-318.
RITCHEY, T. W. & SEELEY, H. W. 1976 Distribution of cytochrome-like respiration in streptococci. *Journal of General Microbiology* 93, 195-203.
ROBERTS, W. E. L. & STEWART, F. S. 1961 The sugar composition of streptococcal cell walls and its relation to haemagglutination pattern. *Journal of General Microbiology* 24, 253-260.
ROBINSON, J. A. & MEYER, F. P. 1966 Streptococcal fish pathogen. *Journal of Bacteriology* 92, 512.
ROGUINSKY, M. 1969 Reactions de *Streptococcus uberis* avec les serums G et P. *Annales de l'Institut Pasteur, Paris* 117, 529-532.
ROGUINSKY, M. 1971 Caractères biochemiques et serologiques de *Streptococcus uberis*. *Annales de l'Institut Pasteur, Paris* 120, 154-163.
ROGOSA, M. 1971 *Peptococcacae*, a new family to include the Gram-positive, anaerobic cocci of the genera *Peptococcus, Peptostreptococcus* and *Ruminococcus. International Journal of Systematic Bacteriology* 21, 234-237.
ROGOSA, M. 1974 Genus II. *Peptostreptococcus* Kleyver and van Niel 1936. In *Bergey's Manual of Determinative Bacteriology*, 8th edn, eds Buchanan, R. E. & Gibbons, N. E. Baltimore: Williams & Wilkins.
ROOP, D. R., MUNDT, J. O. & RIGGSBY, W. S. 1974 Deoxyribonucleic acid hybridization studies among some strains of Group D and Group N streptococci. *International Journal of Systematic Bacteriology* 24, 330-337.
ROSENBACH, F. J. 1884 *Mikro-organismen bei den Wund-Infections-Krankheiten des Menschen*. Weisbaden: J. F. Bergman.
SALTON, M. R. J. 1953 Studies of the bacterial cell wall. IV. The composition of the cell walls of some Gram-positive and Gram-negative bacteria. *Biochimica et Biophysica Acta* 10, 512-523.
SCHLEIFER, K. H. & KANDLER, O. 1967 Zur chemischen Zusammensetzung der Zellwand der Streptokokken. 1. Die Aminosauresequenz des Muriens von *Str. thermophilus* und *Str. faecalis. Archiv für Mikrobiologie* 57, 335-364.
SCHLEIFER, K. H. & KANDLER, O. 1972 Peptidoglycan types of bacterial cell walls and their taxonomic implications. *Bacteriological Reviews* 36, 407-477.
SCHOTTMÜLLER, H. 1903 Die Artunterscheidung der für den Menschen pathogenen Streptokokken durch Blutagar. *Münchener Medizinische Wochenschrift* 50, 849-853; 909-912.
SEELEMANN, M. 1964 *Biologie der Streptokokken*. Nürnberg: Hans Carl.
SEELEY, H. W. 1951 The physiology and nutrition of *Streptococcus uberis*. *Journal of Bacteriology* 62, 107-115.
SEELEY, H. W. & DAIN, J. A. 1960 Starch hydrolyzing streptococci. *Journal of Bacteriology* 79, 230-235.
SEYFRIED, P. L. 1968 An approach to the classification of lactobacilli using computer-aided numerical analysis. *Canadian Journal of Microbiology* 14, 313-318.
SHATTOCK, P. M. F. 1949 The streptococci of group D: the serological grouping of *Streptococcus bovis* and observations on serological refractory group D stains. *Journal of General Microbiology* 3, 80-92.
SHATTOCK, P. M. F. 1962 Enterococci. In *Chemical and Biological Hazards in Food*, eds Ayres, J. C., Kraft, A. A., Snyder, H. E. & Walker, H. W. Ames: Iowa State University Press.
SHERMAN, J. M. 1937 The streptococci. *Bacteriological Reviews* 1, 3-97.
SHERMAN, J. M. 1938 The enterococci and related streptococci. *Journal of Bacteriology* 35, 81-93.
SHERMAN, J. M. & WING, H. U. 1937 *Streptococcus durans* n.sp. *Journal of Dairy Science* 20, 165-167.

SHERMAN, J. M., NIVEN, C. F. & SMILEY, K. L. 1943 *Streptococcus salivarius* and other non-hemolytic streptococci of the human throat. *Journal of Bacteriology* **45**, 249-263.

SKADHAUGE, K. 1950 *Studies on the Enterococci with Special Reference to the Serological Properties*. Copenhagen: Einar Munksgaard.

SKADHAUGE, K. & PERCH, B. 1959 Studies on the relationship of some alpha-haemolytic streptococci of human origin to the Lancefield Group M. *Acta Pathologica et Microbiologica Scandinavica* **46**, 239-250.

SLADE, H. D. & SLAMP, W. C. 1962 Cell wall composition and the grouping antigens of streptococci. *Journal of Bacteriology* **84**, 345-351.

SLADE, H. D. & SLAMP, W. C. 1972 Peptidoglycan composition and taxonomy of group D, E and H streptococci and *Streptococcus mutans*. *Journal of Bacteriology* **109**, 691-695.

SMITH, D. G. & SHATTOCK, P. M. F. 1962 The serological grouping of *Streptococcus equinus*. *Journal of General Microbiology* **29**, 731-736.

SMITH, D. G. & SHATTOCK, P. M. F. 1964 The cellular locations of antigens in streptococci of Groups D, N and Q. *Journal of General Microbiology* **34**, 165-175.

SMITH, L. D. S. 1957 *Peptostreptococcus* Kluyver and Van Niel. In *Bergey's Manual of Determinative Bacteriology*, 7th edn, eds Breed, R. S., Murray, E. G. D. & Smith, N. R. Baltimore: Williams & Wilkins.

STABLEFORTH, A. W. 1959 Streptococcal diseases. In *Infectious Diseases of Animals*, Vol. 2, eds Stableforth, A. W. & Galloway, I. A. London: Butterworths.

STARK, P. & SHERMAN, J. M. 1935 Concerning the habitat of *Streptococcus lactis*. *Journal of Bacteriology* **30**, 639-646.

TAGG, J. R. & WANNAMAKER, L. W. 1976 Genetic basis of streptococcin A-FF22 production. Genes determining SA (formerly streptocin A), and immunity to SA, may be plasmid-borne on the same plasmid. *Antimicrobial Agents and Chemotherapy* **10**, 299-306.

TAGG, J. R., SKOJOLD, S. & WANNAMAKER, L. W. 1976 Transduction of bacteriocin determinants in group A streptococci. *Journal of Experimental Medicine* **143**, 1540-1544.

TAYLOR, R. F., IKAWA, M. & CHESBRO, W. 1971 Carotenoids in yellow pigmented enterococci. *Journal of Bacteriology* **105**, 676-678.

THAL, E. & SÖDERLIND, O. 1966 Ny serologisk grupp (Ü) av beta-hämolyserande streptokokker isolerande från svin. *Proceedings of the 10th Nordic Veterinary Congress*, Stockholm. p. 336.

THIERCELIN, E. 1899 Sur un diplocoque saprophyte de l'intestin susceptible de devenir pathogene. *Comptes Rendus des Scéances de la Société de Biologie, Paris* **51**, 269-271.

THIERCELIN, E. & JOUHAUD, L. 1903 Reproduction de l'entérocoque; taches centrales; granulations périphériques et microblastes. *Comptes Rendus des Scéances de la Société de Biologie, Paris* **55**, 686-688.

TOMURA, T., HIRANO, T., ITO, T. & YOSHIOKA, M. 1973 Transmission of bacteriocinogenecity by conjugation in group D streptococci. *Japanese Journal of Microbiology* **17**, 445-452.

TUNNICLIFFE, R. 1933 Colony formation of *Diplococcus rubeolae*. *Journal of Infectious Diseases* **52**. 39-53.

WANNAMAKER, L. W. & MATSEN, J. M. (eds) *Streptococci and Streptococcal Diseases*. New York & London: Academic Press.

WEISSMAN, S. M., REICH, P. R., SOMERSON, N. L. & COLE, R. M. 1966 Genetic differentiation by nucleic acid homology. IV. Relationships among Lancefield groups and serotypes of streptococci. *Journal of Bacteriology* **92**, 1372-1377.

WHITE, J. C. & NIVEN, C. F. 1946 *Streptococcus* S.B.E.: a streptococcus associated with subacute bacterial endocarditis. *Journal of Bacteriology* **5**, 717-722.

WHITTENBURY, R. 1963 The use of soft agar in the study of conditions affecting the utilization of fermentable substrates by lactic acid bacteria. *Journal of General Microbiology* **32**, 375-384.

WHITTENBURY, R. 1964 Hydrogen peroxide formation and catalase activity in lactic acid bacteria. *Journal of General Microbiology* **35**, 13-26.
WHITTENBURY, R. 1965a A study of some pediococci and their relationship to *Aerococcus viridans* and the enterococci. *Journal of General Microbiology* **40**, 97-106.
WHITTENBURY, R. 1965b The differentiation of *Streptococcus faecalis* and *Streptococcus faecium*. *Journal of General Microbiology* **38**, 279-287.
WINSLOW, C. E. A. & WINSLOW, A. R. 1908 *The Systematic Relationships of the Coccaceae*. New York: John Wiley.
WILKINSON, B. J. & JONES, D. 1977 A numerical taxonomic survey of *Listeria* and related bacteria. *Journal of General Microbiology* **98**, 399-421.
WILLIAMS, R. E. O. 1956 *Streptococcus salivarius* (Vel Hominis) and its relation to Lancefield's Group K. *Journal of Pathology and Bacteriology* **72**, 15-25.
WILSON, G. S. & MILES, A. A. 1975 *Topley & Wilson's Principles of Bacteriology and Immunity*, Vol. 1, 6th edn. London: Arnold.
WINDSOR, R. S. & ELLIOTT, S. D. 1975 Streptococcal infection in young pigs. IV. An outbreak of streptococcal meningitis in weaned pigs. *Journal of Hygiene* **75**, 69-78.
WITTNER, M. K. & HAYASHI, J. A. 1965 Studies on streptococcal cell walls. VII. Carbohydrate composition of Group B cell walls. *Journal of Bacteriology* **89**, 398-402.
YAGI, Y. & CLEWELL, D. B. 1976 Plasmid-determined tetracycline resistance in *Streptococcus faecalis*: tandemly repeated resistance determinants in amplified forms of pAMα1 DNA. *Journal of Molecular Biology* **102**, 583-600.
YAO, J., JACOBS, N. J., DEIBEL, R. H. & NIVEN, C. F. 1964 Group E streptococci: II. Characterisation of strains from swine cervical abscesses. *Journal of Infectious Diseases* **114**, 333-340.
YAWGER, E. S. & SHERMAN, J. M. 1937 *Streptococcus cremoris*. *Journal of Dairy Science* **20**, 205-212.

Biochemical Characteristics of *Streptococcus* Species

R. WHITTENBURY

*Department of Biological Sciences, University of Warwick,
Coventry, Warwickshire, England*

CONTENTS

1. Introduction 51
2. The aerobic nature of streptococci and other lactic acid bacteria 51
3. Soft agar culture 52
   (a) Utilization of polyols and myo-inositol in soft agar by *Streptococcus faecalis*
      and *S. faecium* 54
   (b) Utilization of hexoses by *Leuconostoc mesenteroides* 54
   (c) Utilization by *Lactobacillus brevis*, *Lact. buchneri* and *Lact. viridescens* of
      substrates separately sterilized or sterilized in the medium 54
4. The effect of adding haemin to media 56
   (a) Catalase formation 57
   (b) Cytochrome-like respiration and NADH oxidase activity in the streptococci . 58
   (c) Other properties related to the aerobic nature of streptococci 62
5. The inclusion of pediococci and aerococci within the genus *Streptococcus* . . . 64
6. References 68

1. Introduction

THERE EXISTS a wealth of published information on streptococcal biochemistry—covering aspects of both fundamental and applied importance. No attempt will be made to review this body of knowledge. Instead, it is proposed to examine two areas of current interest about which there is a certain degree of controversy: the aerobic nature of the streptococci and other lactic acid bacteria and the inclusion of pediococci and aerococci within the genus *Streptococcus*. The latter topic is a legitimate question to discuss in the context of streptococcal biochemistry as the classification of streptococci is largely rooted in biochemical characteristics.

2. The Aerobic Nature of Streptococci and Other Lactic Acid Bacteria

Streptococci and other lactic acid bacteria are conventionally considered to be facultative anaerobes with a preference for anaerobic conditions. They are unique amongst bacteria able to grow **aerobically** in that they are not able to synthesize **porphyrins**—and therefore cytochromes and catalase—and are not, apparently, capable of forming ATP via the electron transport chain. Aerobically they are presumed to carry out substrate level ATP synthesis via the fermentation mechanisms they use when growing anaerobically.

The recent descriptions of strictly anaerobic streptococci reinforce the popular notion that streptococci (and other lactic acid bacteria) are inherently anaerobic. The logic of including such organisms within the genus *Streptococcus* appears to have gone unchallenged, despite the fact that this action has created a unique bacterial genus in that it now contains both strict anaerobes and facultative anaerobes; endosporeformers, for instance, are primarily classified into genera on the basis of their aerobic/anaerobic properties. Apparently, morphology and fermentation characteristics outweigh aerobic properties in streptococcal taxonomic considerations, despite the evidence accumulating over many years (e.g. Gunsalus & Umbreit 1945) that streptococci can involve oxygen in their metabolism and, in particular, that *Streptococcus faecium* can only use glycerol aerobically. The contrary opinion to the value of oxygen as a substrate seems to lie in the conviction that H_2O_2 will always be an end product and, by definition, always a toxic end product; such a view is tacitly supported in the most recent edition of *Bergey's Manual of Determinative Bacteriology* (Buchanan & Gibbons 1974) where aerobic properties of lactic acid bacteria are treated as though they were aberrations of behaviour of normally fermentative organisms. Yet, over the past few years, it has become clear that some streptococci (and other lactic acid bacteria), grown under suitable conditions, can behave as conventional aerobes in that they are capable of oxidative phosphorylation via a cytochrome-mediated electron transport chain. Demonstration of these properties depends upon the use of suitable media and media supplements; use of these and the consequent information derived from such studies are described below.

3. Soft Agar Culture

Agar stabs and shakes, agar plates and liquid media in test tubes rarely reveal unequivocal information of the aerobic/anaerobic requirements of lactic acid bacteria. Soft agar tube cultures (Whittenbury 1963), on the other hand, have proved extremely useful in this context. A basal medium containing meat extract (Lab-Lemco), 0.5 g; peptone (Evans), 0.5 g; yeast extract (Difco), 0.5 g; Tween 80, 0.05 ml; agar (Davis), 0.15 g; tap water to 100 ml with the pH adjusted to 6.8-7.0, provided (after boiling and allowing to set) aerobic conditions in the top 0.5-1.0 cm of the medium and anaerobic conditions below that level, as judged by the reduction of methylene blue. Reducing conditions persisted in the lower half of the uninoculated controls for at least 14 days at 30 to 40°C. In media used for culture, methylene blue was omitted and bromocresol purple or bromocresol green added as a pH indicator. Method of inoculation was also important. Medium melted and cooled to 37-40°C (separately

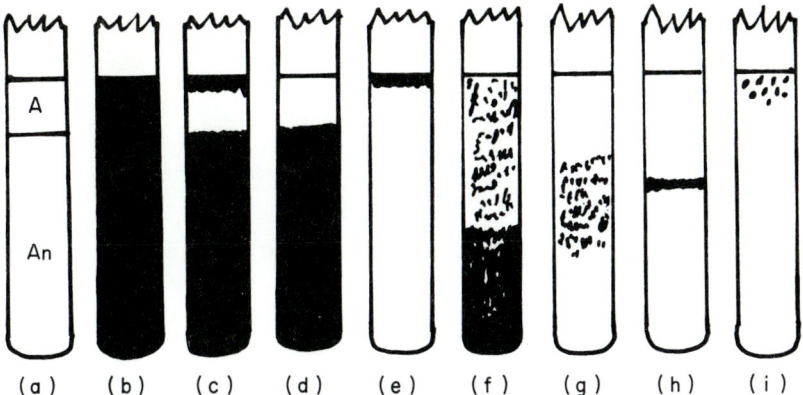

Fig. 1. Aerobic/anaerobic growth of lactic acid bacteria in soft agar. (a) Aerobic (A) and anaerobic (An) zones in basal soft agar medium (see text), (b) *S. faecalis* growing aerobically and anaerobically on glucose, (c) a strain of *Leuconostoc mesenteroides* growing on glucose —the zone of no-growth between aerobic and anaerobic growth caused by H_2O_2 diffusing downwards from aerobic growth and inhibiting initiation of later-developing anaerobic growth, (d) a strain of *Lact. plantarum* utilizing cellobiose anaerobically only, (e) a strain of *S. faecium* utilizing glycerol aerobically only, (f) a strain of *Lact. cellobiosus* utilizing glucose—growth initially anaerobic with aerobic growth occurring only as acid diffuses upward, (g) a strain of pediococcus growing on glucose in the presence of NaCl (10% w/v)—growth in absence of salt as in (b), (h) a strain of *Lact. fructovorans* utilizing glucose at 37°C—a temperature close to its growth maximum—normally growth as in (b), (i) a strain of *S. faecium* utilizing glycerol—mutant colonies developing aerobically only.

sterilized energy sources, when needed, were added aseptically at 48°C) was inoculated with a drop of liquid inoculum from an overnight culture, gently mixed and allowed to set. This process permitted the inoculum to be evenly distributed throughout the medium without making the culture highly aerobic.

In effect, this soft agar medium had both the characteristics of a liquid medium, in that bacteria could grow and spread without encountering the resistance of a normal agar which leads to localized colony development, and the characteristics of a solid medium in that mixing by convection did not occur and aerobic to anaerobic stratification persisted. Such a technique (Fig. 1) clearly revealed cultural requirements of the organisms for the use of particular substrates under different physiological conditions (added salt, elevated temperatures, etc.).

Some lactic acid bacteria exhibited marked differences in their requirements for aerobic/anaerobic conditions, dependent upon whether or not the carbohydrate was separately sterilized (by filtration or autoclaving) and added to cooled medium, or was autoclaved within the medium. Examples are cited below of this phenomenon, but at this juncture it is worth emphasizing that in trying to simulate what happens in the natural environment regard to such apparently insignificant features as sterilizing substrates within or without the medium has proved to be of paramount importance. Specific examples of the value of soft agar culture are described below.

(a) *Utilization of polyols and myo-inositol in soft agar by* Streptococcus faecalis *and* S. faecium

Most strains of *S. faecalis* able to utilize glycerol do so aerobically and anaerobically, but show a marked preference for oxygen as an external electron acceptor (Whittenbury 1963, 1965b). Unknown substances acted as an electron acceptor in the anaerobic portion of the medium and the addition of fumarate as an electron acceptor enhanced growth considerably in these instances (Deibel *et al.* 1963 encountered one strain of *S. faecalis* unable to use exogenous electron acceptors other than oxygen). *Streptococcus faecalis* strains able to use the cyclohexane derivative myo-inositol only did so aerobically unless fumarate was added to the medium. The hexitols, sorbitol and mannitol were utilized both aerobically and anaerobically by *S. faecalis* strains. On the other hand, all *S. faecium* strains able to use glycerol and the hexitols did so aerobically only and fumarate did not serve as an alternative electron acceptor to oxygen.

(b) *Utilization of hexoses by* Leuconostoc mesenteroides

Several non-dextran-forming strains of this species (Whittenbury 1966) first utilized glucose aerobically; anaerobic growth—slight by comparison with aerobic growth—developed two to three days later. An interesting phenomenon (Fig. 1) was that a zone of no-growth separated aerobic from anaerobic growth; this was found to be an effect of H_2O_2 formed aerobically, diffusing downwards and inhibiting the initiation of anaerobic growth. Inclusion of catalase or other H_2O_2-destroying agents (e.g. pyruvate) before inoculation eliminated this effect.

One other strain of this type (Whittenbury 1966) was obligately aerobic and found subsequently to be unable to reduce acetyl phosphate to ethanol. Products of glucose utilization were acetic acid (replacing ethanol), carbon dioxide and lactic acid; oxygen was the only utilizable exogeneous electron acceptor.

(c) *Utilization by* Lactobacillus brevis, Lact. buchneri *and* Lact. veridescens *of substrates separately sterilized or sterilized in the medium*

In soft agar cultures in which gluconate, glucose or other aldohexoses, or di- or trisaccharides had been autoclaved, growth of these three species was equally good both aerobically and anaerobically. However, when these substrates were separately autoclaved or filter-sterilized prior to being added to the medium,

Table 1
Effect of sterilizing substrates within or without the medium on growth

Organism	Substrate					
	Gluconate, aldohexoses, di- and trisaccharides		Fructose		Arabinose xylose mannitol	
	Sterilized separately	Sterilized in medium	Sterilized separately	Sterilized in medium	Sterilized separately or in medium	
Lact. brevis	aer or pa	fac	a/san	fac	fac	aer
Lact. buchneri	aer or pa	fac	a/san	fac	fac	aer or mic
Lact. viridescens	aer	fac	a/san	fac	Not utilized	aer

aer, aerobic only; pa, preferentially aerobic; fac, facultatively anaerobic; a/san, aerobic growth and slight anaerobic growth; mic, microaerophilic growth.
After Whittenbury (1966).

then many of the *Lact. brevis* and *Lact. buchneri* strains tested (Whittenbury 1966) were found to be obligatorily aerobic—as were the *Lact. viridescens* strains. Some *Lact. brevis* and *Lact. buchneri* strains showed slight and delayed anaerobic growth on continued fermentation (Table 1).

The importance of this observation, of course, relates to the activity of these organisms in their natural environment—a particular instance being the action of these organisms in silage fermentation. Following on these soft agar results, it was discovered (Whittenbury *et al.* 1967) that in silage the *Lact. brevis* and *Lact. buchneri* isolates utilized glucose at the expense of fructose as an external electron acceptor—forming mannitol, acetic acid, lactic acid and carbon dioxide, but not the expected ethanol in the place of acetic acid. Mannitol was formed from fructose and pure culture fermentation studies revealed that two fructose molecules were converted to mannitol per glucose molecule fermented—giving the following stoichiometry:

$$2 \text{ fructose} + 1 \text{ glucose} \rightarrow 2 \text{ mannitol} + 1 \text{ carbon dioxide} + 1 \text{ acetic acid} + 1 \text{ lactic acid}.$$

The importance of this result lies in the economy of carbohydrate conversion to conserving acids in silage; in this case the heterofermentative lactic acid bacteria were eventually shown to be extremely wasteful, a finding which was initiated by soft agar culture studies and which highlights the importance of using cultural techniques which bear some relevance to the natural environment of the organism.

These same species also showed differences in the way they fermented fructose—externally and internally sterilized. In soft agar containing separately sterilized fructose, anaerobic growth was slight by comparison with aerobic growth. The ability of strains to utilize glucose aerobically only, but fructose anaerobically as well as aerobically, reflected the property of the organisms to use fructose as an external electron acceptor in the fermentation of further fructose, i.e. two molecules of fructose were converted to mannitol in the fermentation of a further molecule of fructose to acetic acid, carbon dioxide and lactic acid.

An investigation into why *Lact. brevis* and *Lact. buchneri* strains were unable to ferment glucose anaerobically indicated (Bryan-Jones & Whittenbury, unpublished) that the mechanism for reducing acetyl phosphate to ethanol, for recycling reduced pyridine nucleotides, was either absent or not expressed.

4. The Effect of Adding Haemin to Media

Decades ago it was recognized that streptococci and other lactic acid bacteria had many nutritional deficiencies in comparison with other bacteria; a range of preformed monomers, as well as small peptides, proved necessary additives to media used to grow lactic acid bacteria. These deficiencies proved useful in the discovery and identification of many growth factors; however, the general assumption that once 'good' growth was achieved all necessary factors were available to express the full potential of the organism, is proving to be false.

The addition of haemin to media for the growth of lactic acid bacteria has revealed the presence of properties in streptococci and other lactic acid bacteria totally unexpected prior to that event (Whittenbury 1962, 1964; Bryan-Jones & Whittenbury 1969).

(a) *Catalase formation*

Streptococci and other lactic acid bacteria are unable to synthesize porphyrins and hence catalase and the haemoproteins concerned with the electron transport pathway. Consequently, the absence of catalase and cytochromes is a frequently used test in the presumptive identification of lactic acid bacteria. However, if haemin is included in the medium (Whittenbury 1964) and the cultures are incubated aerobically, many lactic acid bacteria develop a catalase-like activity. Electrophoretic and other studies (Whittenbury 1964) indicated that this activity was similar in all respects to normal bacterial catalase activity and the assumption was made that the organisms were able to synthesize the apoprotein of catalase and link it to preformed haemin. Prior addition of protoporphyrin IX inhibited the development of this catalase-like activity on later addition of haemin, indicating that the organisms were not even capable of inserting the iron in protoporphyrin IX even though they were able to link that compound with the apoprotein.

Of the streptococci tested (relatively few) only *S. faecalis* and its varieties formed catalase. Suspensions of anaerobically grown strains did not, suggesting that enzymes concerned with catalase synthesis are only induced aerobically.

Table 2
Catalase-like activities in lactic acid bacteria

Pseudocatalase	*S. faecalis*; *Streptococcus* spp; *Leuconostoc mesenteroides*; *Lact. plantarum*; *Pediococcus pentosaceous*, *P. homari* (*Aerococcus viridans*)
Catalase	*S. faecalis*; *Lact. casei*, *Lact. plantarum*; *Lact. brevis*, *Lact. fructovorans*, *Lact. viridescens*; *Leuconostoc mesenteroides*; *Pediococcus acidilactici*

No strain was found to possess both forms of activity.
Most active pseudocatalase in *Pediococcus acidilactici*—equivalent to or better than many catalase+ aerobes.

Many other lactic acid bacteria (see Whittenbury 1962, 1966) also form catalase if provided with haemin (Table 2). Some clearly benefit from this ability, e.g. some hydrogen peroxide-forming strains of *Leuconostoc mesenteroides* only survive the H_2O_2 they form aerobically if grown in the presence of haemin.

Haemin, acting as a peroxidase, seems necessary to the aerobic existence of some lactic acid bacteria. For instance, Ritchey & Seeley (1974) noted that 13 strains of *S. sanguis* required the protection of haemin to survive aerobically, and later (Ritchey & Seeley 1976) noted a similar phenomenon with several other unnamed streptococcal species considered initially to be strict anaerobes.

So, both in the case of these streptococci and of the leuconostocs mentioned earlier, addition of haemin to the medium was necessary for aerobic growth and survival. In the absence of haemin such organisms would have been deemed to be anaerobes in other circumstances—another example of how necessary it is to be cautious about the supposed adequacy of media merely because organisms will grow in them.

It could be argued that haemin effects are of little relevance to lactic acid bacteria in their natural environment. This could be a false assumption; *Lact. viridescens*, a catalase former, grows on meat surfaces with an excellent supply of haemin and, presumably, protects itself in this way from the copious amounts of H_2O_2 it forms; leuconostocs and lactobacilli isolated from silage and bruised grass are able to synthesize catalase from the free haemin available in grass juices (Whittenbury 1964).

This raises the question of how haemin may affect the metabolism of streptococci associated with pathological conditions in man and animals, e.g. *S. mutans* on teeth surfaces and *S. faecalis* infections of the heart and arterial system.

Besides being involved in catalase formation, haemin also influences the respiratory metabolism of some lactic acid bacteria; particularly, it seems, of *S. faecalis*.

(b) *Cytochrome-like respiration and NADH oxidase activity in the streptococci*

As mentioned earlier, streptococci are unable to synthesize porphoryrins and therefore cytochromes—the carriers involved in electron transport. Consequently, oxidative phosphorylation is not a property associated with streptococcal energy-yielding metabolism. Yet Gallin & Vandemark (1964) observed low levels of oxidative phosphorylation in *S. faecalis* (strain 10Cl) during the oxidation of NADH by extracts of aerobically grown organisms. Gallin & Vandemark postulated that phosphorylation may have occurred at the 'NADH-flavin' level in or during the oxidation of naphthoquinone. Later, Smalley *et al.* (1968) used molar growth yields as evidence of oxidative phosphorylation with the same organism, observing a P to O ratio of *ca* 0.6. Faust & Vandemark (1970) used fumarate as an electron acceptor for NADH oxidation by membrane fractions of this organism (*S. faecalis* 10Cl) and obtained a P to O ratio of 0.19.

This example of oxidative phosphorylation in the absence of cytochromes is unique amongst bacteria and poses questions of fundamental importance concerning oxidative phosphorylation generally and, of course, with the streptococci in particular, assuming them to be basically anaerobic in nature.

The fundamental question referred to may well be in the process of being resolved by the studies of Griffiths and his colleagues at Warwick University (e.g. Partis *et al.* 1977), who seem to be disentangling the role of the respiratory chain in ATP synthesis and respiration-driven proton translocation, ion transport and energy-linked reactions. It seems from their work that oxidative phosphorylation is analogous to substrate level phosphorylation and that lipoic acid (or some product of lipoate, D. E. Griffiths, pers. comm.) serves as a cofactor in coupling or energy transfer function between the electron transport chain and the ATP synthase complex. Such a system would eliminate the assumed necessity for electron transport via cytochromes in ATP syntheses at the oxidative level, and would also provide an explanation for the observations of Vandemark and his colleagues. An outcome could be that *S. faecalis* (strain 10Cl) would be an excellent model for studies on oxidative phosphorylation, uncomplicated as it is by the presence of cytochromes when grown in haem-free media.

Coincident with the first reports of Gallin & Vandemark (1964) on this unique instance of oxidative phosphorylation was the first report (Whittenbury 1964) of the ability of some *S. faecalis* strains to synthesize cytochromes when provided with haemin. Later (Bryan-Jones & Whittenbury 1969), a haemin-dependent oxidative phosphorylation system was demonstrated to function in some *S. faecalis* strains. Cytochromes were detected and ATP synthesis from ADP and inorganic phosphate by membrane fractions at the expense of NADH oxidation was demonstrated to occur only in strains grown aerobically in the presence of haemin. This activity was subject to inhibition by the inhibitory substances used in conventional oxidative phosphorylation studies (e.g. cyanide, azide and antimycin A). Also a yield of six moles of ATP per mole of hexose utilized was recorded for haemin-grown strains, double the amount observed in the absence of haemin. The best P to O ratio observed was 0.44 which is in line with that recorded by Ritchey & Seeley (1974) who confirmed and extended these findings with different strains of *S. faecalis*.

Perhaps the most dramatic demonstration of the powers conferred on certain strains of *S. faecalis* by haemin is the growth of the organism on lactate; the observation of ATP synthesis (Table 3) by extracts of haemin-grown *S. faecalis* at the expense of lactate and additional ATP from glucose and the marked difference in the proportion of products of aerobic metabolism in the presence and absence of haemin (Table 4).

Table 3

Molar growth yields obtained with S. faecalis and ATP yields in the presence and absence of haemin

Additions to basal medium	Anaerobic				Aerobic					
	None		50 μg haemin ml^{-1}		None		50 μg haemin ml^{-1}		50 μg protoporphyin IX ml^{-1}	
	G/M	Y$_{ATP}$	G/M	Y$_{ATP}$	G/M	Y$_{ATP}$	G/M	Y$_{ATP}$	G/M	Y$_{ATP}$
Substrate										
Glucose	16.0	1.8	16.1	1.8	26.8	3.0	55.2	6.1	31.9	3.5
Fructose	18.0	2.0	n.t.		26.2	2.9	55.2	6.1	35.9	4.0
Acetate	0.0	0.0	0.0	0.0	0.0	0.0	2.0	0.2	n.t.	
Pyruvate	0.0	0.0	0.0	0.0	5.0	0.6	5.0	0.6	n.t.	
Lactate	0.0	0.0	0.0	0.0	2.0	0.2	15.6	1.8	n.t.	

G/M, dry weight cocci in g mol^{-1} of substrate; Y$_{ATP}$, mol ATP mol^{-1} of substrate, calculated from the G/M figure assuming an average of 9.0 g dry weight cocci mol^{-1} ATP; n.t., not tested.
After Bryan-Jones & Whittenbury (1969).

Table 4
Products of glucose and lactate oxidation by washed aerobically grown S. faecalis cells

	Incubation conditions in experiment*			
Growth substrate	Anaerobic with glucose	Aerobic with glucose	Aerobic with glucose + haemin	Aerobic with lactate + haemin
Oxidation substrates				
Glucose	100	100	100	–
Lactate	–	–	–	100
Oxidation products				
Lactate	187	77	29	–
Acetate	0	64	61	38
Formate	0	0	8	0
Pyruvate	3	1	3	9
CO_2	6	130	171	91
Acetoin	–	26	43	23
C recovery (%)	96	99	95	95

* 100 mol of each substrate were added and the concentrations of the products are expressed as mol 100 mol^{-1} of substrate.
After Bryan-Jones & Whittenbury (1969).

Prior to this it was known (e.g. Ritchey & Seeley 1974) that oxygen had a beneficial influence on *S. faecalis* growth yields in that oxygen, acting as an alternative and preferred electron acceptor to pyruvate in the metabolism of glucose, spared pyruvate so permitting its dismutation to yield approximately two additional moles of ATP per mole of glucose fermented. In other words, *S. faecalis* was able, aerobically, to form, theoretically, four moles of ATP as opposed to two moles normally resulting from the homofermentative conversion of glucose to lactic acid.

Although neither cytochromes (Bryan-Jones & Whittenbury 1969) nor cytochrome-like NADH oxidase activity (Ritchey & Seeley 1974) have been detected in *S. faecium*, they have been observed (either one or both properties) in other streptococci (*S. lactis* by Sijpesteijn 1970 and Ritchey & Seeley 1976 and *S. lactis* var. *diacetylactis* by Ritchey & Seeley 1976). Other lactic acid bacteria have also been found to contain cytochromes; these include strains of *Leuconostoc mesenteroides* (Whittenbury 1964, Sijpesteijn 1970), *Lact. plantarum*, *Lact. brevis* and *Pediococcus pentosaceus* (Whittenbury 1964).

A detailed survey of NADH oxidase activity among streptococci has been recently carried out by Ritchey & Seeley (1976). Using cell-free extracts, they claimed that the distinction between cytochrome (haemin induced) and flavin

Table 5
Type of NADH oxidation in streptococci

None	Flavin-like	Cytochrome-like
S. pyogenes	S. pyogenes	S. faecalis†
S. zooepidemicus	S. equisimilis	S. faecalis var. liquefaciens ††
S. equi	S. zooepidemicus	S. faecalis var. zymogenes†
S. dysgalactiae	S. agalactiae	S. lactis*
S. sanguis	S. equisimilis	S. lactis var. diacetylactis†
S. salivarius	S. uberis	
S. bovis	S. sp. group G	
S. thermophilus	S. sp. group E	
S. mutans	S. faecium (durans)	
S. lactis*	S. faecalis†	
S. cremoris	S. faecalis var. zymogenes†	
	S. lactis*	
	S. lactis var. diacetylactis†	
	S. cremoris	

* All three categories.
† Both types of positive activity.
†† Only one activity (cytochrome-like).

systems was easily made although their criteria were somewhat arbitrary—combining levels of NADH oxidase activity and sensitivity to cyanide, azide and hydroxyquinoline-*n*-oxide. Their results (Table 5) placed streptococci in three categories—those with no oxidase activity, those with flavin systems and those with cytochrome-like systems.

Strains of *S. lactis* fell into all three categories and strains of *S. faecalis* and *S. lactis* var. *diacetylactis* fell into either the flavin or cytochrome category. No other species of streptococcus was found to possess cytochrome-like oxidase activity; and there was strain variation within species as regards flavin-like oxidase activity or no oxidase activity.

Clearly, the whole phenomenon of haemin-induced respiration in streptococci and other lactic acid bacteria still requires investigation—many questions concerning the induction of enzymes and electron carriers complementing the haemin-induced system remain unanswered, as do questions about the significance of these properties to streptococci in their natural environment.

(c) *Other properties related to the aerobic nature of streptococci*

Hydrogen peroxide production by streptococci (and other lactic acid bacteria grown aerobically) was first observed years ago (e.g. Gunsalus & Umbreit 1945, for *S. faecium*). Such an activity was widely regarded as being always deleterious and an indication of the unsuitability of aerobic conditions for the culture of lactic acid bacteria. Despite the evidence to the contrary already outlined, this view still persists (e.g. Lee *et al.* 1976, talk about "the well-known inhibitory effects of H_2O_2 on lactobacilli" without apparently realizing that many lactobacilli can cope perfectly well with H_2O_2 if provided with the right cultural conditions—in their studies on *Lact. plantarum* addition of haemin to the medium would have resulted in catalase induction and elimination of H_2O_2 accumulation).

Non-haemin-dependent mechanisms for disposing of H_2O_2 have been known to exist in lactic acid bacteria for some years. For instance, Dolin (1953) described a flavoprotein peroxidase in *S. faecalis* 10Cl and similar peroxidases have been found in various other lactic acid bacteria (Grierson & Gunsalus 1943; Douglas 1947; Seeley & Vandemark 1951). Such an enzyme does not occcur in *S. faecium*—hence the taxonomic difference between *S. faecalis* and *S. faecium* based on detectable H_2O_2 being released by *S. faecium* but not by *S. faecalis* (e.g. Whittenbury 1966).

A second H_2O_2 destroying system has also been detected in some strains of *S. faecalis* and in a variety of other lactic acid bacteria (Table 2)—this is a 'weak' form of catalase-like activity termed 'pseudocatalase' by Whittenbury (1964) to distinguish it from the haemoprotein catalase also found in lactic acid bacteria. It was first described by Felton *et al.* (1953) for some pediococci and subsequently by Dacre & Sharpe (1956) for additional strains of pediococci and strains of *Lact. plantarum* and for *S. faecalis* by Jones *et al.* (1964).

To date there have been no reports of both types of catalase activity being present in the one organism. Distinguishing features of the two catalases are shown in Table 6; as yet pseudocatalase remains undefined enzymologically and structurally.

Finally, superoxide dismutase, an enzyme disposing of the superoxide free radical (O_2^-), and regarded as being essential in bacteria possessing aerobic respiratory pathways, has been found in *S. faecalis* (Gregory & Fridovich 1973) and *S. mutans* (Vance *et al.* 1972). Superoxide is dismutated to $H_2O_2 + O_2$ and the resultant H_2O_2 can be dealt with by peroxidases and/or catalase and/or pseudocatalase.

In summing up to this point, it is now clear that streptococci and other lactic acid bacteria, far from being basically anaerobic in nature, have many properties which equip them for growth and survival in aerobic conditions. *Streptococcus faecalis* in particular appears to have a battery of enzymes to cope with H_2O_2, given the required growth supplements such as haemin, and is capable (some strains) of oxidative phosphorylation with or without the assistance of normal cytochrome electron carriers. Growth yield is greater under aerobic than anaerobic conditions—double the ATP yield per mole of glucose—and is further enhanced in the presence of haemin when a haemoprotein-mediated electron transport system is induced. In this latter instance, even lactate serves as a good energy source.

Table 6
Simple tests for distinguishing between pseudocatalase and catalase in lactic acid bacteria

	Pseudocatalase	Catalase
H_2O_2 split on 0.5% (w/v) glucose agar	+	−
H_2O_2 split on 1.0% (w/v) glucose agar	−	+
H_2O_2 splitting inhibited by 0.01 M-azide	−	+
H_2O_2 splitting activity inhibited by pretreatment with protoporphyrin IX before adding haemin	−	+

The discovery of all these properties, which imply that *S. faecalis*, for instance, has a wider array of enzymological equipment for aerobic existence than many 'normal' aerobes, has resulted mainly from the modification of techniques and media used for studying and growing lactic acid bacteria in the first instance. Such a result underlines the caution necessary in interpreting the nature of lactic acid bacteria on the basis of cultural techniques most frequently used at the present time. Merely because these organisms grow profusely under certain defined conditions in complex media should not be accepted as being in any way a true reflection of their behaviour in the natural environment or that the media themselves reflect the total environmental range of conditions in which the organisms normally grow.

5. The Inclusion of Pediococci and Aerococci within the Genus *Streptococcus*

The biochemistry of streptococci is obviously reflected in the choice and exploitation of tests used in the taxonomic description of streptococci. Biochemical tests alone are not used; their application frequently depends upon a preliminary sorting procedure involving morphology and staining tests. Consequently, such a process can lead to a distorted view of the relationship of one organism to another in that this or that series of tests is applied only to those organisms satisfying a particular set of other arbitrary criteria in the first instance. Such a situation applies to the streptococci. They can be defined at present as Gram positive, coccal-shaped organisms dividing in only one plane (therefore occurring as pairs or chains), which are catalase and cytochrome negative in media lacking haemin and which possess a glycolysis pathway resulting anaerobically in the homofermentative conversion of glucose to lactate. Such a description includes all those species formally classified within the genus

Table 7

Comparison of Streptococcus faecalis H 69D5, S. faecium 5020 and a strain of Pediococcus pentosaceus

Properties common to *P. pentosaceus*, *S. faecium* and *S. faecalis*		Properties separating *P. pentosaceus* and *S. faecium* from *S. faecalis*	*P. pentosaceus* *S. faecium*	*S. faecalis*	Properties separating *S. faecium* and *S. faecalis* from *P. pentosaceus*	*S. faecium* *S. faecalis*	*P. pentosaceus*
Fermentation of		Fermentation of			Growth at pH 9.6	+	−
Glucose	+	Melezitose	−	+	Final pH glucose broth	4.1	3.6
Fructose	+	Rhamnose	−	+	Growth on acetic acid + acetate agar	−	+
Sucrose	−	Sorbitol	−	+	Pseudocatalase	−	+
Arabinose	−	Mannitol	−	+	Tetrads formed	−	+
Maltose	+	Glycerol	aerobic	+			
Cellobiose	+	Hydrolysis of Hippurate	−	+			
Galactose	+	Vigorous reduction of					
Raffinose	−	Tetrazolium	−	+			
Melibiose	−	Litmus milk	−	+			
Mannose	+	Tolerance of 0.1% MB in Milk	−	+			
Lactose	+	Tolerance of Tellurite	−	+			
Trehalose	+	Growth with					
Inositol	−	Malate	−	+			
Salicin	+	Citrate (pH 8.0)	−	+			
Hydrolysis of		Dissimilation of malate in presence of 1.0% glucose	+	−			
Aesculin	+	Catalase (in presence of heated blood)	−	+			
Arginine	+	H_2O_2 formed on					
Growth at		Basal HBD* agar	+	−			
10°C	+	Glycerol HBD agar	+	−			
45°C	+	α-Haemolysis (greening)	+	−			
Tolerance of							
6.5% NaCl	+						
63° 30 min⁻¹	+						

* Heated blood + *o*-dianisidine agar.
After Whittenbury (1965a).

Streptococcus including strictly anaerobic species and the former *Pneumococcus* species. Excluded are pediococci and aerococci—solely on the grounds of division plane when all properties now associated with known streptococci are considered. Some species of pediococci (Whittenbury, 1965a) possess the group D antigen—namely *Pediococcus acidilactici* and *P. pentosaceus*—which is *prima facie* evidence of a relationship with the enterococci; whether or not L(+) or a mixture of D(−) and L(+) lactic acid is formed has proved to be no sure guide as to the genus of coccus to which the organism belongs.

Some strains of *S. faecalis* that normally form L(+) lactic acid have been recorded as forming inactive lactic acid (Langston & Bouma 1960) and pediococci, usually considered to be inactive lactic acid formers, have been recorded as forming solely L(+) lactic acid (Nakagawa & Kitahara 1959; Deibel & Niven 1960).

The G+C content of DNA is similar for aerococci, streptococci and pediococci. Consequently, in the absence of DNA homology studies, there appears no major feature or group of characteristics which can be used to exclude aerococci and pediococci from the genus *Streptococcus*—other than the arbitrary one of plane of cell division. That aerococci and pediococci can divide in two planes, rather than one, seems of no greater significance in taxonomy than streptococcal interspecies difference such as the ability or not to grow anaerobically.

However, in practical taxonomic exercises, the ability to form tetrads or chains has great significance in that the subsequent series of tests applied differ according to such properties. Such tests totally disguise the possible biochemical relationship of the pediococci and the streptococci—as was shown (Whittenbury 1965a) in a direct comparison of a strain of *S.faecalis* and a strain of *S. faecium* with a strain of *Pediococcus pentosaceus* (Table 7). All three organisms have properties in common and both enterococci can be distinguished from the pediococcus by an ability to grow at pH 9.6, by the final pH produced in glucose broth and ability to grow on acetic acid + acetate agar. However, *S. faecium* and the pediococcus possessed many properties in common which separated them from *S. faecalis*. Assuming that tetrad formation was overlooked in the initial screening of the organism, *P. pentosaceus* would be convincingly classified as *S. faecium* using the criteria currently used in distinguishing between *S. faecium* and *S. faecalis*.

This viewpoint on streptococcal classification is not a wholly isolated one; recently, Ritchey & Seeley (1976) concluded that the relationships of the homofermentative cocci should be reviewed, whilst Wilkinson & Jones (1977) would widen the review further to include *Erysipelothrix* and *Gemella*. Perhaps even the genus *Staphylococcus* should be drawn in; Whittenbury (1965a), for instance, commented upon the apparent close relationship of haemin-deficient mutants of *Staphylococcus aureus* to *Aerococcus viridans*.

Finally, the biochemical tests frequently used in classifying streptococci (and

Table 8
Additional tests of value in separating S. faecalis *from* S. faecium

Properties	S. faecalis	S. faecium
Fermentation of glycerol in the presence of fumarate	+	A
H_2O_2 from		
Basal medium	–	+
Glycerol	–	+*
Sorbitol	–	+*
Mannitol	–	+*
Malate as an energy source	±†	–
Gas from malate + glucose	–	+
Citrate as an energy source at pH 8.0	+	–
Catalase (in presence of heated blood)	±††	–

+, all strains positive; –, all strains negative; ±, some strains positive and some negative; A, negative or aerobic only.
* If substrate used.
† $\frac{2}{18}$ strains negative.
†† $\frac{15}{18}$ strains positive.
After Whittenbury (1965b).

lactic acid bacteria in general) tend to be used somewhat uncritically and result in either an over-simplification of interpretation or are so used as to mask real differences. Acid production from sugars—simple pH change only being recorded—is a good example of lost opportunities. Earlier in this discussion illustrations were provided of how soft agar culture techniques could provide a wealth of information about differences and similarities between organisms not revealed by the commonly used test tube liquid culture tests.

An example (Table 8) concerns the differentiation of *S. faecalis* from *S. faecium* (Whittenbury 1965b), e.g. the combination of polyol utilization and hydrogen peroxide production; the utilization or dissimilation of malate in the presence of glucose (this test, based on the diauxie effect expressed by *S. faecalis* but not by *S. faecium* which dissimilates malate in the presence or absence of glucose, was devised to separate clearly the two species) and the ability to use citrate; both species utilized citrate but only *S. faecalis* could do so at pH 8.0.

Further investigation of the biochemical properties of streptococci could lead to more precise information about the biochemical properties of individual species, and to the modification of existing tests and the devising of new tests of value to the study of the taxonomy of streptococci.

6. References

BUCHANAN, R. E. & GIBBONS, N. E. 1974 *Bergey's Manual of Determinative Bacteriology*, 8th ed. Baltimore: The Williams & Wilkins Co.

BRYAN-JONES, D. G. & WHITTENBURY, R. 1969 Haematin-dependent oxidative phosphorylation in *Streptococcus faecalis*. *Journal of General Microbiology* **58**, 247–260.

DACRE, J. C. & SHARPE, M. E. 1956 Catalase production by lactobacilli. *Nature, London* **178**, 700.

DEIBEL, R. H. & NIVEN, C. F. Jr. 1960 Comparative study of *Gaffkya homari*; *Aerococcus viridans*; tetrad forming cocci from meat curing brines, and the genus *Pediococcus*. *Journal of Bacteriology* **79**, 175–180.

DEIBEL, R. H., LAKE, D. E. & NIVEN, C. F. Jr. 1963 Physiology of the enterococci as related to their taxonomy. *Journal of Bacteriology* **86**, 1275–1282.

DOLIN, M. I. 1953 The oxidation and peroxidation of $DPNH_2$ in extracts of *Streptococcus faecalis*, 10Cl. *Archives of Biochemistry and Biophysics* **55**, 415–421.

DOUGLAS, H. C. 1947 Hydrogen peroxide in the metabolism of *Lactobacillus brevis*. *Journal of Bacteriology* **54**, 272–279.

FAUST, P. J. & VANDEMARK, P. J. 1970 Phosphorylation coupled to NADH oxidation with fumarate in *Streptococcus faecalis* 10Cl. *Archives of Biochemistry and Biophysics* **137**, 392–398.

FELTON, E. A., EVANS, J. B. & NIVEN, C. F. Jr. 1953 Production of catalase by pediococci. *Journal of Bacteriology* **65**, 481–482.

GALLIN, J. I., VANDEMARK, P. J. 1964 Evidence for oxidative phosphorylation in *Streptococcus faecalis*. *Biochemical and Biophysical Research Communications* **17**, 630–635.

GREGORY, E. M. & FRIDOVICH, J. 1973 Induction of superoxide dismutase by molecular oxygen. *Journal of Bacteriology* **114**, 543–548.

GRIERSON, E. C. & GUNSALUS, I. C. 1943 Hydrogen peroxide destruction by streptococci. *Journal of Bacteriology* **45**, 16–19.

GUNSALUS, I. C. & UMBREIT, W. W. 1945 The oxidation of glycerol by *Streptococcus faecalis*. *Journal of Bacteriology* **49**, 347–357.

JONES, D., DEIBEL, R. H. & NIVEN, C. F. Jr. 1964 Catalase activity of two *Streptococcus faecalis* strains and its enhancement by aerobiosis and added cations. *Journal of Bacteriology* **88**, 602–610.

LANGSTON, C. W. & BOUMA, C. 1960 A study of the micro-organisms from grass silage. 1. The cocci. *Applied Microbiology* **8**, 212–222.

LEE, I. H., FREDRICKSON, A. G. & TSUCHIYA, H. M. 1976 Damped oscillations in continuous culture of *Lactobacillus plantarum*. *Journal of General Microbiology* **93**, 204–208.

NAKAGAWA, A. & KITAHARA, K. 1959 Taxonomic studies on the genus *Pediococcus*. *Journal of General and Applied Microbiology* **5**, 95–126.

PARTIS, M. D., HYAMS, R. L. & GRIFFITHS, D. E. 1977 Studies of energy-linked reactions: A lipoic acid requirement for oxidative phosphorylation in *Escherichia coli*. *Febs Letters* **75**, 47–51.

RITCHEY, T. W. & SEELEY, H. W. Jr. 1974 Cytochromes in *Streptococcus faecalis* var. *zymogenes* grown in a haematin-containing medium. *Journal of General Microbiology* **85**, 220–228.

RITCHEY, T. W. & SEELEY, H. W. Jr. 1976 Distribution of cytochrome-like respiration in streptococci. *Journal of General Microbiology* **93**, 195–203.

SEELEY, H. W. & VANDEMARK, P. J. 1951 An adaptive peroxidation by *Streptococcus faecalis*. *Journal of Bacteriology* **61**, 27–35.

SIJPESTEIJN, A. K. 1970 Induction of cytochrome formation and stimulation of oxidative dissimilation by haemin in *Streptococcus faecalis* and *Leuconostoc mesenteroides*. *Antonie van Leeuwenhoek* **36**, 335–348.

SMALLEY, A. J., JAHRLING, P. & VANDEMARK, P. J. 1968 Molar growth yields as evidence for oxidative phosphorylation in *Streptococcus faecalis* strain 10Cl. *Journal of Bacteriology* **96**, 1595–1600.

VANCE, P. G., KEELE, B. B. Jr. & RAJAGOPALAN, K. V. 1972 Superoxide dismutase from *Streptococcus mutans*. *Journal of Biological Chemistry* **247**, 4782–4786.
WHITTENBURY, R. 1962 Two types of catalase-like activity in lactic acid bacteria. *Nature, London* **187**, 433–434.
WHITTENBURY, R. 1963 The use of soft-agar in the study of conditions affecting the utilization of fermentable substrates by lactic acid bacteria. *Journal of General Microbiology* **32**, 375–384.
WHITTENBURY, R. 1964 Hydrogen peroxide formation and catalase activity in the lactic acid bacteria. *Journal of General Microbiology* **35**, 13–26.
WHITTENBURY, R. 1965a A study of some pediococci and their relationship to *Aerococcus viridans* and the enterococci. *Journal of General Microbiology* **40**, 97–106.
WHITTENBURY, R. 1965b The differentiation of *Streptococcus faecalis* from *S. faecium*. *Journal of General Microbiology* **38**, 279–287.
WHITTENBURY, R. 1966 A study of the genus *Leuconostoc*. *Archiv für Mikrobiologie* **53**, 317–327.
WHITTENBURY, R., McDONALD, P. & BRYAN-JONES, D. G. 1967 A short review of some biochemical and microbiological aspects of ensilage. *Journal of Science, Food and Agriculture* **18**, 441–444.
WILKINSON, B. J. & JONES, D. J. 1977 A numerical taxonomic survey of *Listeria* and related bacteria. *Journal of General Microbiology* **98**, 399–421.

The Pattern of Streptococcal Disease in Man

M. T. PARKER

*Central Public Health Laboratory,
London, England*

CONTENTS

1. The streptococci of man 71
   (a) Classification and nomenclature 72
   (b) The streptococcal flora of man 76
2. Streptococcal disease in man 77
   (a) Clinical varieties of streptococcal sepsis 77
   (b) Non-septic streptococcal disease 82
3. Virulence of streptococci for man 85
   (a) Invasiveness 85
   (b) Toxicity 86
   (c) Localization of inflammatory lesions 87
4. Epidemiological features of streptococcal diseases in man 87
   (a) Communicability 87
   (b) Food-borne streptococcal disease 91
   (c) Infection from other animals 92
   (d) Self infection 92
   (e) Host factors in streptococcal disease 94
   (f) Antibiotic resistance 100
5. The general picture 101
6. References 102

1. The Streptococci of Man

MANY different streptococci are present on the human body surface, and nearly all of them can upon occasion cause disease. The streptococcal diseases of man are extremely various, and range from highly communicable epidemic infections in normal people to sporadic non-communicable diseases caused by streptococci normally present on the body surface that may invade the tissues of predisposed persons. The common pathogenic streptococci of man bear a general resemblance to those of other animals, but when examined closely can usually be distinguished from them. Whether this is also true of the rest of the human streptococcal flora is less certain; the necessary comparisons have not been made, and indeed would be difficult to make in the present state of our knowledge of streptococcal taxonomy.

The inflammatory lesions caused by streptococci in man and other animals are rather similar, though their anatomical distribution may differ. In man,

Table 1
The streptococci of man

Division	Common name	Specific epithet	Haemolysis*	Lancefield group antigen†
Pyogenic streptococci	Group A	*pyogenes*	β	A
	Group C	*equisimilis*	β	C
	Group G	...	β	G
	Group B	*agalactiae*	β	B
Pneumococci	Pneumococcus	*pneumoniae*	α	–
Enterococci	Faecal	*faecalis*	– (β)	D
	...	*faecium*	v	D
	Others	...	–	D
Viridans streptococci	...	*sanguis*	α (β, –)	– (H)
	...	*mitior* (*mitis*)	α (β, –)	– (O, K, M)
	Dextran-positive *mitior*-like	('*sanguis* II')	α (β, –)	–
	Unclassifiable	...	α (β, –)	–
Non-haemolytic streptococci	...	*salivarius*	–	– (K)
	...	*mutans*	–	– (E)
	...	*bovis* (typical or biotype-I)	–	D
	...	*bovis* (atypical or biotype-II)	–	D
	...	*milleri*	– (β, α)	– (A, C, F, G)
	'Slow-growing biochemically inactive'	...		

* In parenthesis, less common appearances; v, various; –, non-haemolytic.
† Characteristic of the taxon; in parenthesis, present in a minority of strains; –, not present; ..., none given.

however, one particular streptococcus—the group A streptococcus—is also responsible for specific toxic and immunological disturbances for which there appear to be no parallels among animals.

(a) *Classification and nomenclature*

No general classification of the streptococci is universally accepted. I shall therefore list and define briefly the sorts of streptococci that are commonly recognized in material from human sources, and suggest appropriate names for them (Table 1).

First, we can recognize several highly characteristic 'pyogenic' streptococci (Sherman 1937), that is to say, β-haemolytic streptococci that possess a Lancefield polysaccharide antigen. The correct specific names for them are, however, in doubt. By common usage, British workers call the pyogenic group A streptococci *Streptococcus pyogenes* (e.g. Wilson & Miles 1975, p. 712), but there is no formal authority for this. The 'large-colony-forming', usually β-haemolytic members of groups C and G are usually referred to as group C and group G streptococci. The former usually form acid from trehalose but not from sorbitol, and thus correspond to the *equisimilis* species or biotype of the veterinarians. No specific name has yet been proposed for the human or the animal group G strains. As we shall see, the A, C and G antigens are also found in the quite dissimilar species *S. milleri*, in which the colonies are very much smaller. Until the pyogenic groupable species have been re-defined, we may use the familiar names, e.g. group A streptococci, with the implication that *S. milleri* strains with the corresponding group antigen are excluded. The name *S. agalactiae* is applied to the 'human' members of group B, though there is reason to believe that they form a population distinct from the bovine group B streptococci (Ross, this volume).

The pneumococci form a distinct species easily separated from all other streptococci and present no difficulty in recognition or naming. Of the enterococci, only *S. faecalis* is an important pathogen of man; *S. faecium* is much less often associated with disease, and other enterococci that are difficult to classify are occasionally seen.

We may apply the familiar name viridans streptococci to a less well-defined group of strains that are usually but by no means always α-haemolytic. Among these, there are two reasonably well-recognized species: *S. sanguis* and *S. mitior* (or *mitis*). The former (White & Niven 1946; Colman & Williams 1972) is characterized by a number of positive characters: hydrolysis of arginine and aesculin, formation of dextran from sucrose and acidification of trehalose and inulin, but not all of these characters are present in each strain; somewhat less than one-half of *S. sanguis* strains form the Lancefield antigen H. At the other extreme are the less biochemically active strains that do not attack arginine and aesculin or form dextran, and usually do not acidify trehalose or inulin; these are referred to either as *S. mitior* (Schottmüller 1903) or *S. mitis* (Andrewes & Horder 1906), but the former name is probably preferable (see Colman 1970). However, a considerable proportion of viridans streptococci have biochemical reactions intermediate between those of *S. sanguis* and *S. mitior*. In our experience (Parker & Ball 1976, and unpublished results; see Table 2), if *S. sanguis* is defined as having at least two of the three positive characters: aesculin hydrolysis, arginine hydrolysis and formation of dextran from sucrose, it is a fairly homogeneous species containing nearly all of the group H strains; and *S. mitior*, with negative reactions in all three tests, contains no members of group H. A third group of viridans streptococci, in which dextran is formed but neither arginine nor aesculin is fermented, should probably be recognized because its pathogenic potential resembles that of *S. sanguis* rather than that of

Table 2
Cultural and biochemical characters of viridans streptococci of human origin

	Percentage frequency of the character in			
	S. sanguis (215)*	Unclassified viridans streptococci (181)*	Dextran-positive mitior-like strains (84)*	S. mitior (283)*
β-Haemolysis	14	8	5	6
α-Haemolysis	68	66	86	84
Hydrolysis of aesculin	83	20	(0)†	(0)†
Hydrolysis of arginine	89	80	(0)†	(0)†
Dextran from sucrose	70	(0)†	(100)†	(0)†
Antigen H	33	5	1	0
Acid from trehalose	94	44	46	33
Acid from raffinose	38	58	82	45
Acid from inulin	81	26	13	6
Acid from melibiose	19	42	60	32

* Number of isolates examined.
† By definition.

S. mitior. We refer to these strains as 'dextran-positive *mitior*-like' strains; they probably correspond to strains called *S. sanguis* serotype II by earlier workers (see Porterfield 1950). The remainder of the viridans streptococci—dextran-negative strains that hydrolyse either arginine or aesculin but not both—cannot be classified further by the tests we used.

Four species of predominantly non-haemolytic streptococci can be recognized in material from human sources: *S. salivarius, S. mutans, S. milleri* and *S. bovis*, and all except the first are important pathogens. *Streptococcus mutans*, though well described by Clarke in 1924, has only recently attracted the attention of dental (Carlsson 1967; Edwardsson 1968) and medical bacteriologists (Perch *et al.* 1974; Facklam 1974; Harder *et al.* 1974). Scattered accounts of the isolation of *S. bovis* from human sources have appeared for several decades, but its importance as a cause of disease in man has been recognized only in the last five years (Facklam 1972; Kiel & Skadhauge 1973; Ravreby *et al.* 1973). Two fairly distinct biotypes of *S. bovis* are found (Facklam 1972; Parker & Ball 1976); 'typical' or biotype I strains form dextran, hydrolyse starch and usually ferment mannitol and inulin; 'variant' or biotype II strains are dextran and starch negative and much less often ferment mannitol and inulin. *Streptococcus bovis* strains are often referred to in the medical literature as 'non-enterococcal group D streptococci'.

Table 3
Usual characters of four species of predominantly non-haemolytic streptococci

Character	S. salivarius	S. mutans	S. bovis biotype I	S. bovis biotype II	S. milleri
Growth enhanced by CO_2	–	(+)	–	–	(–)
Growth on 40% bile agar	(–)	–	+	+	(–)
Growth at 45°C	(–)	(–)	(+)	(+)	(–)
Hydrolysis of aesculin	+	+	+	+	(+)
Hydrolysis of arginine	–	–	–	–	(+)
Voges-Proskauer reaction	(–)	(+)	+	+	(+)
Polysaccharide from sucrose†	L+	D+	D+	–	–
Acid from					
Trehalose	(–)	+	+	(+)	(+)
Sorbitol	–	+	–	–	–
Mannitol	–	+	+	(–)	–
Raffinose	(+)	+	+	+	–
Inulin	(+)	+	+	(–)	–
Melibiose	–	(+)	+	(+)	–
Hydrolysis of starch	–	–	+	–	–

* +, > 90% positive; (+), 50–90% positive; (–), 10–50% positive, –, < 10% positive.
† D, dextran; L, laevan.
Data (except for starch hydrolysis) from Parker & Ball (1976); starch hydrolysis: unpublished results (L. C. Ball & S. A. Waitkins).

Streptococcus milleri (Guthof 1956) is an aggregated species created by Colman & Williams (1965, 1972; see also Colman 1968) to include biochemically similar strains, with a characteristic cell wall composition, that usually possess one of a series of polysaccharide type antigens (the Ottens antigens) seldom found in other streptococci. Its general acceptance has been resisted because two criteria habitually used by medical bacteriologists for the primary classification of streptococci—haemolysis and the presence of a Lancefield group antigen—are variable within the species. The majority of *S. milleri* strains are non-haemolytic and devoid of group antigen. A considerable minority, however, are β-haemolytic and a few are α-haemolytic; between one-quarter and one-third form the group antigen A, C, G or F, and indeed all streptococci with the F antigen have the biochemical characters of *S. milleri*. Haemolysis and group antigens occur independently, but there is a tendency for β-haemolytic strains to be groupable more often than α- or non-haemolytic strains. The species thus includes strains earlier classified as *S. anginosus*—β-haemolytic 'minute-colony-forming' members of groups F and G (Buchanan &

Gibbons 1974), and as '*Streptococcus* MG'—non-haemolytic and ungroupable strains isolated from the respiratory tract (Mirick *et al.* 1944).

In our experience, nearly 90% of streptococci from serious infections in man (Parker & Ball 1976) and from superficial lesions and carrier sites (unpublished) can be allocated to one of the taxa described above. The remainder include viridans streptococci that cannot be placed in one of the three named taxa, unclassifiable enterococci and poorly-growing strains with few positive biochemical activities in the tests employed. This last group, which comprises some 1% of all isolates, included no micro-aerophilic or anaerobic strains. (For fuller information about the classification of non-haemolytic streptococci see Carlsson 1968; Colman 1970; Colman & Williams 1972; Facklam 1977.)

(b) *The streptococcal flora of man*

We must try to distinguish between streptococci that are usually present as members of the normal body flora and those that are 'carried' by a minority of the population at certain times, but this is not always easy. In general, the presence of streptococci of groups A, C and G; and of pneumococci, in the upper respiratory tract is usually looked upon as carriage, because rates vary greatly in different communities, and—in the case of group A streptococci—are often high in association with outbreaks of streptococcal disease; furthermore, a person can often be shown to become a carrier after being in contact with another infected person. The carriage of group B streptococci at some sites may, however, prove to be so frequent as to constitute a normal state. The presence of viridans streptococci, *S. mutans*, *S. salivarius*, *S. milleri*, *S. bovis* and enterococci at some surface sites is so frequent that these organisms are best looked upon as members of the normal flora, though little information is available about the persistence of strains in the flora of individual persons.

The streptococci of the mouth and gut will be considered elsewhere (see Hardie and Marsh and Mead, this volume). Table 4 gives a summary of the distribution of individual streptococci at the main surface sites. The streptococci that are considered to be carried when isolated from healthy persons are usually found in the throat (less often in the nose), and in some cases there is serological evidence of subclinical infection. Occasionally these streptococci may be present in the vagina in the absence of respiratory carriage; and independent carriage of group A streptococci on the skin has been described (Section 4.a.i). Group B streptococci are found in the respiratory tract, but more often in the gut and the vagina; the relation of strains carried at multiple sites has been little studied.

The viridans streptococci and *S. mutans* are found regularly in the upper respiratory tract and the mouth, and in much larger numbers on the surfaces of the teeth; but *S. salivarius*, though present in the throat, on the tongue and in the gut, does not appear to be a normal member of the flora of the teeth.

Table 4
Presence of streptococci at various sites on the human body surface

Streptococcus	Throat	Teeth and gums	Other oral	Gut	Vagina
Groups A, C and G	(+)	−	−	−	(±)
Pneumococcus	(+)	−	−	−	(±)
Group B	(+)	−	−	(+)	(+)
Viridans	+	+	+	−	−
S. mutans	+	+	+	−	−
S. salivarius	+	−	+	+	−
S. milleri	+	+	+	+	+
Enterococci	−	−	−	+	−
S. bovis	−	−	−	+	−

+, Normal flora; (+), frequently carried; (±), occasionally carried; −, not normally present (but occasionally isolated when present in large numbers elsewhere on the body).

Studies of the distribution of *S. milleri* have been fragmentary and confined to single sites, but suggest that this organism should be looked upon as a normal inhabitant of the mouth, particularly around the teeth (Mejàre & Edwardsson 1975; Phillips *et al.* 1976) and of the gut contents in the region of the appendix (Poole & Wilson 1977); they are less often present in the vagina. The enterococci are mainly found in the gut; evidence for their presence in the mouth is contradictory; we found them on and around the teeth only after antibiotic treatment (Phillips *et al.* 1976; but see Ross 1972). *Streptococcus faecalis* is by no means the only enterococcus in human faeces; van der Weil-Korstanje & Winkler (1975) found *S. faecium* and *S. avium* almost as often as *S. faecalis* in the faeces of normal adults. They also found *S. bovis* in about 20% of the samples.

2. Streptococcal Disease in Man

(a) *Clinical varieties of streptococcal sepsis*

In most infections, the local septic lesion is the sole manifestation of disease, but in many there is a tendency for the streptococcus to spread by direct extension through neighbouring connective-tissue planes or along the lymphatics. Sometimes the streptococcus invades the bloodstream from the local lesion or after local spread, but this may also happen in the absence of detectable local inflammation. When invasion of the bloodstream is accompanied by fever and other toxic effects, the patient is said to have septicaemia; when streptococci are present in the blood in the absence of these symptoms, the conditions is described as bacteraemia. Abscesses and other inflammatory lesions may develop in an internal organ at some distance from the initial point of entry of the streptococcus. It is convenient to refer to the diseases in which streptococci are widely disseminated in the body, and cause septicaemia or inflammatory lesions in internal organs (or both) as systemic streptococcal diseases.

(i) *Local inflammation*

Streptococci vary in their ability to cause local inflammatory lesions. Acute infection of the upper respiratory tract (tonsillitis) is caused almost exclusively by group A streptococci; streptococci of groups C and G are often present in the throat in sporadic cases of tonsillitis, but it is exceedingly difficult to establish a causative relationship. Clear evidence that group G streptococci are respiratory pathogens was obtained in one almost unique food-borne outbreak of tonsillitis (Hill *et al*. 1969), but similar support for the respiratory pathogenicity of the *equisimilis* variety of group C streptococcus is lacking. Other streptococci cause sepsis in tissues adjacent to the oropharynx, but this is not preceded by an easily recognized infection of the mucous membrane; thus, pneumococci frequently cause acute middle-ear infection in infants, and *S. milleri* causes sepsis around the teeth and jaws (Guthof 1956).

Acute infection of other mucous membranes is also seen; notably of the eye with pneumococci and the vagina of children with group A streptococci. *Streptococcus faecalis* and group B streptococci cause urinary-tract infections.

Skin lesions are of two sorts: impetigo—superficial crusted sores on healthy skin or at the site of minor injuries—from which streptococci seldom spread to deeper tissues; and acute inflammation of wounds, burns, etc. without crusting but usually with pus formation and a tendency to spread through the lymphatic channels. When streptococci are implanted at the time of wounding, e.g. at surgical operation or by a puncture wound, there is often rapid generalization of infection with little pus formation. Streptococcal impetigo is caused only by group A strains; wound sepsis is usually caused by group A streptococci, and less often by members of groups C and G and pneumococci. Many other streptococci—notably members of groups B, *S. milleri* and enterococci—are found in septic wounds, but they are often present in mixed culture, and it is usually difficult to be sure that they are the cause of the inflammation. There is, however, good evidence that wounds 'silently' colonized by streptococci may form the point of entry for a systemic infection.

Streptococcal pneumonia should perhaps be looked upon as a local lesion, because the evidence suggests that the organism usually reaches the lung through the air passages. Pneumococci are by far the commonest cause, except in neonates, in whom early-onset septicaemic disease due to group B streptococci may be accompanied or perhaps preceded by multiplication of the streptococcus in the lung.

Table 5

Percentage frequency of various streptococci (and aerococci, but not pneumococci) among isolates from patients with systemic diseases (1971-1974)

	Percentage frequency of the indicated micro-organism from patients with			
Streptococcus*	Any systemic disease (719)†	Septicaemia (250)	Purulent disease (152)	Endocarditis (317)
Group A	11.0	21.6	15.1	0.6
Group C or G	5.2	8.4	6.6	2.2
Group B	10.0	12.4	22.4	2.2
S. milleri	10.3	6.0	28.3	5.4
S. faecalis	6.3	9.6	2.0	5.7
S. sanguis	10.3	5.6	5.3	16.4
S. mitior	9.9	9.6	2.6	13.2
Dextran-positive mitior-like	4.5	3.6	0	7.3
Unclassified viridans	6.7	7.6	2.6	7.9
S. mutans	6.8	1.6	0	14.2
S. bovis biotype-I	8.3	3.2	2.6	15.1
S. bovis biotype-II	3.1	4.4	2.6	2.2
Other	7.7	6.4	9.8	7.6

* All taxa constituting > 2% of the total. The rest comprised: enterococci other than *S. faecalis*, 1.7%; *S. salivarius*, 1.8%; other unidentified streptococci, 3.2%; aerococci, 1.0%.
† In parentheses: number of isolates examined.
Data from Parker & Ball (1976).

(ii) *Systemic disease*

Septicaemia and the associated toxaemia may sometimes be the sole manifestation of a systemic streptococcal disease. When an inflammatory lesion develops in an internal organ it may be either an abscess in a solid viscus, an inflammation of a body cavity (e.g. the pleura or the meninges) or a fibrinous 'vegetation' on a heart valve (endocarditis). One of three sequences of events may have occurred: the streptococcus may have entered the blood from a local lesion or after initial lymphatic spread, and subsequently have been deposited in a distant organ; it may have spread directly or via the lymphatics to the internal organ, from which the blood may subsequently be invaded; or it may have entered the bloodstream transiently from a surface site, been deposited in an organ at a distance, multiplied there, and subsequently given rise to septicaemia.

Streptococci differ greatly in the predominant pattern of the systemic diseases to which they give rise. This may depend to some extent on the point of entry of the streptococcus and its route of spread in the body, but it is clear that some streptococci have the ability to establish themselves in particular organs. Thus, the pneumococcus often causes meningitis, whether it reaches the central nervous system by direct extension from the upper respiratory tract or via the bloodstream from the lung, but it rarely causes brain abscess. Group A streptococci less often invade the central nervous system but readily cause purulent lesions in other internal organs.

Other differences in the behaviour of individual streptococci are illustrated in Tables 5 and 6. Table 5 shows the percentage frequency of various streptococci (excluding pneumococci) among isolates from patients who had (1) any systemic disease; (2) a purely septicaemic disease, (3) a purulent lesion in an internal organ or cavity, whether or not they had septicaemia and (4) endocarditis. Thus, streptococci of groups A, C, G and B were important causes of septicaemia and of purulent internal lesions, but only the group B streptococci commonly gave rise to meningitis (Table 6). The remarkable ability of *S. milleri* to form localized abscesses in a wide variety of organs, including some—such as the brain and the liver—in which other streptococci seldom become established, is elaborated in Table 7.

Enterococci and several other streptococci were isolated considerably more often from septicaemic patients than from patients with visceral pus (Table 5), but it must be admitted that the septicaemia figures for some of the viridans

Table 6

Frequency of various streptococci among isolates from patients with purulent lesions in internal organs (1971-1974)

Streptococcus	Number of isolates* from a lesion, or from blood with clinical evidence of a lesion, in				
	Meninges	Brain	Abdomen	Thorax	Bone or joint
Group A	6	1	3	8	5
Group C or G	4	1	0	4	1
Group B	33	0	0	0	1
S. milleri	8	14	14	7	1
S. faecalis or other enterococcus	2	0	2	1	0
Any viridans	6	1	2	6	2
S. bovis biotype-I or II	6	0	0	2	0
Other†	6	0	2	5	0

* Number of patients: 152; number of separate lesions: 154.
† *S. salivarius*, 4; unidentified streptococcus, 9.
Data from Parker & Ball (1976).

Table 7
Site and nature of purulent lesions in internal organs associated with S. milleri infection

Site and lesion	Number of lesions*	Site and lesion	Number of lesions*
Central nervous system	36	Abdomen	42
Brain abscess	20	Liver abscess	18
Meningitis†	12	Other biliary	5
Subdural abscess	4	Pelvic peritonitis	8
Thorax	29	Other peritoneal	2
Pleural empyema	21	Intra-abdominal abscess	6
Lung abscess	4	Perinephric abscess	2
Pericarditis	1	Acute pancreatitis	1
Pneumonia	3	Joint	
		Acute arthritis	2

* Number of patients, 105; number of lesions, 109. One patient had abscesses in liver, brain and lung, one had abscesses in liver and lung, and one in brain and pleura.
† Meningitis 'localized' in three patients.
Data in relation to cultures submitted to the Streptococcus Reference Laboratory, Colindale, in the years 1970–76.

streptococci may have been inflated—despite the precautions taken to avoid this—by the inadvertent inclusion of a few undiagnosed cases of endocarditis or of streptococcal bacteraemia in patients suffering from other serious diseases.

Almost any streptococcus can cause endocarditis, but some do so much more often than others. Endocarditis occurs rather infrequently in the course of septicaemic infections caused by the various pyogenic streptococci, *S. faecalis* and *S. milleri*, and this accounts for only a minority of the cases. Much more often, endocarditis appears as the first manifestation of streptococcal disease, and is then associated with streptococci that are not a prominent cause of other systemic diseases, and are rarely responsible for local inflammatory lesions: the various viridans streptococci (44.8%); *S. bovis* biotype I (15.1%) and *S. mutans* (14.2%). The association of certain streptococci with endocarditis is even more strikingly illustrated (Table 8) by the 'endocarditis index' (the ratio of the percentage frequency of the streptococcus in endocarditis to its percentage frequency in all other systemic infections). The highest indices are given by *S. mutans* (14:1), *S. bovis* biotype I (6:1), the dextran-forming *mitior*-like streptococci (3:1) and *S. sanguis* (3:1). All of these streptococci, and no others, habitually form dextran from sucrose. Among the viridans streptococci there is a close relation between dextran formation and endocarditis index: dextran-forming *sanguis* strains, 5:1; dextran-forming *mitior* strains, 3:1; all *sanguis* strains, 3:1; *S. mitior* and unclassified viridans strains (all dextran

Table 8
Association of individual streptococcal taxa with endocarditis

Percentage frequency in endocarditis, in rank order		'Endocarditis index',* in rank order (and percentage of dextran-forming strains in each taxon)		
S. sanguis	16.4	S. mutans	14:1	(98)
S. bovis biotype I	15.1	S. bovis biotype I	6:1	(94)
S. mutans	14.2	Dextran-forming		
S. mitior	13.2	mitior-like	3:1	(100)
Unclassified viridans	7.9	S. sanguis	3:1	(72)
Dextran-positive		S. mitior	2:1	(0)
mitior-like	7.3	Unclassified viridans	1:1	(0)
S. faecalis	5.7	S. faecalis	1:1	(0)
S. milleri	5.4	S. bovis biotype II	1:2	(0)
S. bovis biotype II	2.2	S. milleri	1:3	(0)
Group B	2.2	Group G	1:3	(0)
Group G	1.6	Group C	1:3	(0)
Group C	0.6	Group B	1:7	(0)
Group A	0.6	Group A	1:32	(0)

* Ratio of percentage frequency in endocarditis to percentage frequency in all other systemic disease (data from Parker & Ball 1976).

negative), 1:1. The difference in endocarditis index between the dextran-positive *S. bovis* biotype I (6:1) and the dextran-negative *S. bovis* biotype II (1:2) is even greater (see Section 4.2).

(b) *Non-septic streptococcal disease*

So far we have discussed the activities of streptococci as producers of acute sepsis. The group A streptococcus is by far the most important pathogen in this respect, but in addition is responsible for several diseases that are of a different nature: scarlet fever, erysipelas, rheumatic fever and acute glomerulonephritis. It is also possible that some of the clinical manifestations of bacterial endocarditis may not be direct consequences of the presence of living organisms in the affected tissues.

(i) *Scarlet fever*

A characteristic rash appears during some acute infections with group A streptococci but not with any other streptococcus. The traditional view (see Wilson & Miles 1975, p. 1910) is that this is due to a specific toxin, and that it develops when a strain of streptococcus that produces this erythrogenic toxin infects a person who does not possess the corresponding serum antibody. An infant usually acquires this antibody passively from its mother, subsequently loses it, and later develops active immunity as a result of streptococcal infection. These changes are reflected in age-related differences in the percentage of subjects who are 'Dick positive', i.e. who give an erythematous reaction to intradermally injected erythrogenic toxin.

The alternative view, that the rash is a delayed hypersensitivity reaction to group A streptococci (Dochez & Stevens 1927), has recently been revived in modified form. In the intervening years, the erythrogenic toxin had been purified and studied in experimental animals (Watson & Kim 1970); it did not give rise to a rash in young animals, but caused fever and death. Watson and Kim's view is that erythrogenic toxin ('pyrogenic exotoxin', see Section 3.b) does not cause the complete syndrome of scarlet fever. In their opinion, the toxin is combined in the bacterial cell with heat-stable materials, and hypersensitivity to this complex is a prerequisite for the development of the disease. Thus, insusceptibility to the intradermal injection of the Dick reagent may be of two types: a young child may be neither hypersensitive nor immune, and an adult after many streptococcal infections may be hypersensitive but have serum antibody to the toxin. This view is not generally accepted, but has the merit of accounting for certain anomalies in the skin-test results obtained in infants and their mothers (Cooke 1927, 1928a, b) and for the otherwise inexplicable fact that scarlet fever, which 100 years ago was a major cause of death in children, is now a clinical triviality though still moderately common.

Group A streptococci produce erythrogenic toxins of three serological types, but one predominates. Zabriskie (1964) showed that carriage of a temperate phage determined toxin production, but we know little about the frequency with which strains circulating in the community form toxin or about the relation of this to lysogenicity. Certain M types have a much greater ability to cause scarlet fever than others in particular countries, but this does not persist indefinitely. Thus, in 1964-65 the strongest association was with type 4 in Britain, type 22 in East Germany and type 1 in Holland and the USSR (Parker 1967). The predominance of type 4 as a cause of scarlet fever in Britain persisted for over 20 years but ceased after 1970; at about the same time type 22 was replaced by type 4 in East Germany.

(ii) *Erysipelas*

This is a characteristic spreading erythematous skin lesion that is seldom seen nowadays. It is clearly associated with group A streptococcal infection—usually of the upper respiratory tract—and occurs most often on the face. It is, however, less significant that streptococci are sometimes isolated from an incision into the lesion than that often they are not. A single attack confers no immunity, and indeed some persons have repeated attacks over many years, usually in the same skin area. These features suggest that erysipelas may be a hypersensitivity reaction to streptococci, but few opportunities now arise to investigate it.

(iii) *Rheumatic fever and acute glomerulonephritis*

These two late complications of infection with group A streptococci are discussed by Maxted (this volume). Both are palpably unrelated to the presence of living streptococci in the affected organs, both occur after a latent period of 1–3 weeks (usually rather longer in rheumatic fever than in nephritis) and in both only a proportion of those suffering a primary infection with a particular streptococcal strain develop the complication. These facts suggest that both diseases are related to the immune response to the streptococci. Neither disease can be reproduced in experimental animals, though histological changes said to resemble those seen in one or other disease have been described after the injection of living streptococci or their products (see Wilson & Miles 1975, pp. 1920, 1924). The pathogenesis of the two diseases appears to be quite distinct: they rarely occur together, even in large outbreaks of streptococcal infection; rheumatic fever usually follows tonsillitis, but nephritis may follow either respiratory-tract or skin infection, the histological changes in the two diseases are quite distinct; the ability to cause nephritis is restricted to members of certain M types, and some of these appear very seldom to be associated with rheumatic fever; and an attack of rheumatic fever, but not of nephritis, greatly increases the chance of another attack of the same complication if a subsequent streptococcal infection occurs.

One feature of all four of these consequences of group A streptococcal disease is their declining importance in developed countries during the last 70 years. Erysipelas, rheumatic fever and nephritis have decreased greatly in frequency, and scarlet fever in clinical severity. The progressive disappearance of rheumatic fever in Europe and North America has been related to 'improved living standards'. There are good reasons for believing that a key factor in this process has been a reduction in the total number of streptococcal attacks during childhood and adolescence. Both rheumatic fever and nephritis are now known to occur very frequently in countries with poor living standards—mainly but not exclusively in the tropics.

(iv) *Disseminated lesions in streptococcal endocarditis*

The formation of vegetations on heart valves is accompanied or followed by the development of lesions elsewhere in the body from which the causative streptococcus can rarely be isolated. Although large embolic lesions may be the result of impacted pieces of detached vegetation, the much more numerous minute focal lesions in the kidney, brain and subcutaneous tissues, and even the diffuse glomerulonephritis, that characteristically occur in endocarditis, may be a consequence of the antibody response to the streptococcus. The histological appearance of the focal lesions suggest that they may result from the deposition of immune complexes in small blood vessels, and direct evidence for the existence of these complexes in the circulation and in the kidneys has been obtained (Cordiero *et al.* 1965; Cream & Turk 1971; Keslin *et al.* 1973; Boulton-Jones *et al.* 1974; Mohammed *et al.* 1977).

3. Virulence of Streptococci for Man

Bacterial factors responsible for the virulence of streptococci for man may be looked upon as determinants for (1) *invasiveness*, the ability to gain entry to tissues and to spread from the initial focus of infection, (2) *toxicity*, (3) *localization*, with the production of inflammatory lesions in particular organs and (4) *immunogenicity*, insofar as the body's response to infection has consequences that are harmful to the affected person (Section 2.b). They have been extensively investigated only in the group A streptococci (see Maxted, this volume) and the pneumococci.

The group A streptococci produce very many substances that can be shown to harm laboratory animals—Ginsburg (1972) lists some 20 of them—but the organisms are not natural pathogens for the animals used in the experiments, and the effects of these substances on the animal body often bear little resemblance to those of natural infection. To build up a picture of the pathogenic activities of group A streptococci, we have to rely heavily on observational studies on man (usually retrospective and to some extent unplanned) or experiments on isolated human tissues or fluids, and on analogies with studies of 'animal' streptococci in their natural hosts.

(a) *Invasiveness*

We can be fairly confident that the anti-phagocytic action of the M protein is mainly responsible for the invasiveness of group A streptococci, because the results of laboratory experiments in mice (Dochez *et al.* 1919; Lancefield 1928) were supported by studies in fresh human blood (Hare 1928, 1932) and by observations on the prevention of the natural disease in man by type-specific antibody against it (Wannamaker *et al.* 1953). The additional role of M protein in promoting adhesion to mucous membranes was also established by experiments on human epithelial cells (Ellen & Gibbons 1972). There has long been a suspicion that the hyaluronic-acid capsule may also have anti-phagocytic activity, but its significance is difficult to evaluate because the balance between production and enzymic destruction of capsular material in infected tissue is not understood. Experiments on guinea-pigs with capsulated and unencapsulated strains of group C streptococci that cause natural infections in this animal species (Seastone 1939) gave suggestive results, but these may not be directly applicable to group A streptococcal infection. Several other extracellular products are leucocidal (streptolysin O, streptolysin S and nicotinamide adenine dinucleotide glycohydrolase), and streptokinase has been held to favour the spread of streptococci by dissolving fibrin clot, but their significance in natural infection is uncertain. Other factors that damage tissue cells may contribute to the formation of the local lesion; these include cellular products such as mucopeptide and group polysaccharide, as well as extracellular products listed below as general toxic agents.

It is generally agreed that the anti-phagocytic activity of the capsular polysaccharide of the pneumococcus is its most important invasive factor, and that antibody against it prevents pneumococcal infection in man (MacLeod et al. 1945; Austrian 1975), but there are unexplained differences between the ability of pneumococci of different capsular types to invade.

(b) *Toxicity*

Death from acute septicaemic infection with group A streptococci and pneumococci is due to toxaemia. In severe pneumococcal pneumonia with septicaemia, a 'point of no return' is soon reached, after which no amount of antibiotic therapy will prevent death (Austrian & Gold 1964); the same is true of the 'pure' pneumococcal septicaemia produced by intravenous inoculation in mice (Dick & Gemmell 1971). Clinical observations on septicaemia due to group A streptococci lead to similar conclusions.

The rapidity with which toxaemic symptoms appear suggest that extracellular products are responsible. Group A streptococci produce so many 'candidate' substances that it is difficult to believe that they are all of major importance, and most of our knowledge of them comes from experiments on animals. Among them are: streptolysin O and streptolysin S, which damage cell membranes and lysosomes; a proteinase and a deoxyribonuclease; and less well-defined substances that act on the kidney, liver and heart (Ginsburg 1972). An interesting possibility is that the so-called 'erythrogenic toxin' (see Section 2.b), may be responsible for toxic effects in the absence of a rash; under its modern name of 'pyrogenic exotoxin' it has been held responsible for fever, suppression of reticuloendothelial function, and triggering the Shwartzman reaction (Watson & Kim 1970).

Little is known about the factors responsible for the toxic effects of the pneumococcus. The capsular polysaccharide—like the streptococcal M protein—is non-toxic. Pneumococci produce an intracellular haemolysin serologically like streptolysin O that has lethal and dermonecrotic effects on laboratory animals (Kreger & Bernheimer 1969). Toxaemia, the cause of which is unknown, appears to be the main cause of death in systemic infections with group B streptococci.

In systemic infections with some other streptococci, toxaemia is not so marked, and death may occur for other reasons. In endocarditis the main causes of death are mechanical damage to a heart valve and non-bacterial lesions of the kidney and brain (Section 2.b).

(c) *Localization of inflammatory lesions*

Little is known about the bacterial factors responsible for this, but there are several 'marker' characteristics for streptococci that attack particular organs or tissues. These are sometimes antigens or groups of antigens. Among the group A streptococci the ability to cause impetigo and tonsillitis is related, respectively, to the possession of M antigens of two fairly distinct series (see Maxted, p. 117), and one M type (type 50) is associated with the ability to cause natural disease in laboratory mice, but apparently not in man (Hook *et al.* 1960). Among group B streptococci the ability to cause meningitis is positively correlated with the type III polysaccharide and negatively correlated with the type II polysaccharide (see Parker 1977, and Ross, p. 136). The type polysaccharide distribution among pneumococci from pneumonia of adults differs markedly from that from pneumonia in children, and from meningitis. The characteristic pathogenesis of *S. milleri* infections may not be a true organ specificity but may be related to the mode of dissemination from the primary inflammatory focus; whether this is an effect of the heat-resistant and acid-resistant protein antigen recently observed in this species (R. Lütticken, V. Wendorf, D. Lütticken, E. A. Johnson and L. W. Wannamaker, pers. comm. 1977) has yet to be determined.

The remarkable association of dextran formation *in vitro* with the ability to cause endocarditis is not yet explained. Elliott (1973) suggested, in relation to *S. sanguis*, that the dextran coating might lead to adherence to the heart valve; if this is so, it must be by virtue of dextran formed in the mouth, because continued synthesis is unlikely after the streptococcus has entered the bloodstream.

4. Epidemiological Features of Streptococcal Diseases in Man

(a) *Communicability*

The ability of a streptococcus to establish itself in a new host (its 'communicability'; see Table 9) is, with highly virulent species, recognized by the appearance of disease in the recipient. If the streptococcus is communicable and also causes disease in a considerable proportion of recipients, epidemics of streptococcal disease occur. A communicable streptococcus may cause disease only occasionally, but other recipients of it become carriers and serve as a source of infection for other persons. Many sporadic infections are, however, not caused by a recently acquired streptococcus, but arise from invasion of the body by streptococci normally present in the surface flora; these we consider to be non-communicable 'self infections'.

In this section we are concerned mainly with the spread of streptococci from

Table 9
Communicability of streptococci in man

	Spread of infection				
	Between persons causing				
	Epidemic disease	Sporadic disease	Symptomless acquisition	By food	From other animals
'Human'					
Group A	+++	+++	+++	+	+*
Group C: *equisimilis*	–	?	+	–	–
Group G	–	?	+	+	–
Group B: to neonates	?	+++	+++	–	–
Group B: to others	–	?	?	–	–
Pneumococci	–	+++	+++	–	–
S. faecalis	–	?	?	–	–
'Animal'					
Group C: *zooepidemicus*	–	–	–	+	+
Group R: ("*suis*")	–	–	–	–	+

+++, Common; +, uncommon; –, not recorded.
* Milk-borne outbreaks only; probably indirectly from a human source.

one human subject to another, because spread from other animals to man is rare (Section 4.c). Streptococci persist for a considerable time outside the body, and can be isolated from dust and surfaces in rooms for days or weeks after the removal of their human source, but do not multiply in these situations. These environmental reservoirs should not be thought of as independent 'sources' of infection. Contaminated foodstuffs (Section 4.b), on the other hand, must be considered separately because the transmission of infection appears to depend upon multiplication of the streptococcus in the food.

(i) *The spread of group A streptococci between persons*

The group A streptococcus is by far the commonest cause of streptococcal disease in man. It is highly communicable and causes disease in healthy persons who do not possess type-specific immunity to the infecting strain; thus it regularly causes epidemics, and in this respect is unique among 'human' streptococci. The sources and routes of infection by group A streptococci have been the subject of intensive investigation, the results of which will be summarized briefly (see Wilson & Miles 1975, pp. 1929-36 for detailed evidence). The epidemiological situation is rather different in three common diseases: tonsillitis; impetigo and infection of wounds, burns and the genital tract, etc., and each must be considered separately.

Streptococcal tonsillitis in the general population rests on a substratum of widespread carriage in the respiratory tract. Some 5-10% of the child population are throat carriers at any one moment, but carriage rates in individual communities (e.g. schools) and ratios of carriers to cases, vary enormously. There is little doubt that in tonsillitis infection is mainly from the upper respiratory tract of another person and that transmission is by the aerial route. Clinical cases and recent convalescents are much more infectious than long-term throat carriers. In the acute stage, streptococci are widely distributed in the upper respiratory tract, and are dispersed profusely into the environment by salivary droplets and nasal discharge; later in the disease they are confined to the tonsillar region and are dispersed little. Spread is favoured by close proximity, the infectivity of the acute case falling off progressively at distances greater than 2.5 m, yet considerable contamination of air and dust can be detected over much wider areas. This suggests that large salivary particles expelled directly from the mouth are more important than smaller dry particles (e.g. skin scales and dried nasal discharge or wound exudate) that remain suspended in the air for a long time and spread for greater distances on convectional air currents. This may be related to differences in the dose of streptococci acquired by the two routes, but there is also evidence that streptococci in dust, though viable, have undergone a considerable loss of ability to infect the respiratory tract.

Streptococcal impetigo is caused by different M types of group A streptococci from those usually responsible for tonsillitis, and the two diseases seldom occur together. Impetigo occurs in epidemics, or may be present endemically, in groups of children living under generally poor hygienic circumstances, but also in adults in military units, prisons and mental hospitals. Epidemics may develop under good hygienic conditions if physical contact is very close, as in sports teams. The causative streptococci may colonize the throat, but detailed studies of epidemics indicate that the main route of spread is from skin to skin. Although impetigo is a 'primary' streptococcal disease, in that lesions often develop on apparently unbroken skin, the presence of minor skin trauma increases the chance that lesions will develop and often determines their site. Although group A streptococci are seldom carried on the healthy skin, they may be isolated in small numbers—in the apparent absence of carriage elsewhere—from the skin for a week or so before the development of impetigo lesions.

A number of outbreaks of a disease closely resembling impetigo occurred in Britain during 1974 and 1975 among workers in abattoirs and other meat-handling establishment (Fraser *et al.* 1977). Many but by no means all of the lesions developed at the site of cuts and abrasions. Attack rates exceeding one-half of all workers in an establishment were observed. A single serotype of group A streptococci predominated in most of the outbreaks. In general, the serotypes responsible for the disease resembled those found in cases of impetigo in this country, but one (provisional type '2681') caused a number of outbreaks in widely separated areas and was seldom found except in meat handlers. Trauma to the hands and numerous opportunities for contact spread of streptococci are clearly important factors in the epidemiology of this disease; whether the meat itself plays a more specific role is uncertain (see Section 4.b).

Infection of wounds, burns and the female genital tract occur sporadically in the general population, but may be epidemic in hospitals, when the sources of infection are various. These include the respiratory tract, particularly of members of the staff. The long-term throat carrier cannot be ignored as a source of surgical wound infection, because the infective dose for direct implantation into an operation wound may be very small, but the staff member with an acute infection of the respiratory tract is the greatest danger. A rare source, but one that has been responsible for several outbreaks, is the perianal carrier, who may not harbour the streptococcus in the respiratory tract. Other patients—particularly those with acute tonsillitis or a discharging wound—are often responsible, and 'mixed' hospital outbreaks of respiratory and surgical wound infection are quite common. An additional hazard for women in maternity hospitals is the newborn baby; several outbreaks of puerperal sepsis have been superimposed on undetected endemics of umbilical colonization in the nursery.

There is little doubt that group A streptococci reach wounds by both the contact and the aerial routes. In surgical wounds, the former is probably the more important, as shown by the effect of good wound-dressing technique on sepsis rates. Nevertheless, clear instances of infection by carriers who did not touch the wound are on record. In burns, which are larger in area and more frequently exposed, the aerial route assumes greater prominence, and a filtered air supply to dressing rooms reduces sepsis rates. The evidence suggests that the aerial dissemination of dry particles of the size of skin scales is of considerably greater importance in wound than in respiratory-tract infection.

(ii) *Communicability of other streptococci*

Group B streptococci spread easily from mother to baby just before or during labour, or at parturition (Parker 1977 and Ross, p. 130). About one-half of babies born to mothers who are vaginal carriers become colonized. A small proportion of all babies—of the order of 1 per 1000—develop serious disease in the first week of life as a consequence of this, but this probably represents no more than 1 illness in 50 colonized babies and perhaps considerably less. After the first week of life, the older the neonate the less strong the evidence for infection from the mother; thus the source of infection in late-onset neonatal meningitis is unknown. There is suggestive evidence that infection may sometimes be transferred from one baby to another (Winterbauer *et al.* 1966; Steere *et al.* 1975), but little is known about the frequency of this or the route of spread.

The absence of a suitable typing system makes it difficult to assess the frequency of spread of group C and group G streptococci, particularly to the respiratory tract, and most clinical infections appear to be sporadic, but 'carrier epidemics' of colonization of burns and other wounds, and the skin of neonates occur occasionally (Drusin et al. 1973). The symptomless acquisition of pneumococci in the respiratory tract is frequently associated with interfamilial spread, particularly from donors who are suffering from colds (Gwaltney et al. 1975), but an epidemiological relationship between cases of clinical disease (e.g. pneumonia) can seldom be established. Despite the frequency of sepsis associated with *S. faecalis* in some categories of hospital patient, the spread of this organism among patients has been little investigated.

(b) *Food-borne streptococcal disease*

Disease resulting from the consumption of food containing streptococci does not usually take the form of gastroenteritis but of acute tonsillitis; this is often severe and is nearly always caused by group A streptococci. Nowadays it is uncommon, and is associated with a variety of foods, which usually appear to have been contaminated from the respiratory tract of a food handler. In the past, numerous outbreaks followed the consumption of raw milk, but in Britain reports of this had virtually ceased by 1945. The epidemiological picture in many of these outbreaks was rather different from that in outbreaks associated with other foods. The group A streptococcus was isolated from the milk of a particular cow, which usually showed evidence of mastitis. In a number of instances, however, a human carrier from whom the cow might have been infected was also identified at the farm (for references, see Wilson & Miles 1975, p. 1932). Group A streptococci appear to multiply little in normal milk at room temperature, and this probably explains the infrequency of milk-borne outbreaks caused by direct contamination from a human source. At least one outbreak due to the contamination of pasteurized milk by a worker suffering from streptococcal sore throat is on record.

I am aware of only two accounts of food-borne outbreaks of tonsillitis caused by other streptococci: an outbreak due to group G streptococci in the USA attributed to the consumption of egg salad (Hill *et al*. 1969), and a unique milk-borne outbreak in Romania caused by a group C strain of the *zooepidemicus* species or biotype, and said to have been followed by the development of acute glomerulonephritis in one-third of the patients (Duca *et al*. 1969). The suggestion that enterococci and other unidentified non-haemolytic streptococci might have been responsible for outbreaks of food-borne gastroenteritis was made by several workers between 1930 and 1950 (see Wilson & Miles 1975, p. 2094), and appeared sometimes to be supported by the results of feeding experiments on volunteers. Little evidence to support this suggestion has been presented since.

The possibility that food might be a vehicle for the transmission of septic infection is raised by observations made in meat-handling establishments (Section 4.2.i), particularly by the occurrence of associated infections in persons in closer contact with the meat than with meat handlers (e.g. meat inspectors, lorry drivers, cashiers in butcher shops and local housewives). It received some support from the evidence of A. A. Wieneke and R. J. Gilbert that streptococci isolated in connection with the outbreaks multiplied considerably in raw meat at ambient temperature (see Fraser *et al*. 1977).

(c) *Infection from other animals*

'Animal' streptococci rarely cause disease in man. Milk-borne group A streptococcal outbreaks are probably derived indirectly from a human source, but the streptococci have a limited ability to cause local infection in the cow; the evidence for considering that the human and bovine populations of group B streptococci are distinct is presented by Ross (this volume); no evidence for the association of human and animal pneumococcal infections has been presented.

There are, however, a few well-authenticated examples of human infection that are obviously derived from animals. Meningitis, or less often septicaemic disease caused by Lancefield group R streptococci—referred to by Windsor & Elliott (1975) as *S. suis* serotype 2—is a good example. These streptococci cause meningitis in weaned pigs, and can be distinguished from streptococci of group S (or *S. suis* serotype 1) which cause meningitis and arthritis in even younger piglets but have not so far been found in human disease. A number of group R infections have been reported among farmers and others who have had contact with pigs or pig products in Denmark (Perch *et al*. 1968) and Holland (Zanen & Engel 1975), and there have been at least five similar British cases.

Occasional human infections with group C streptococci of the *zooepidemicus* species or biotype have been reported. The Romanian milk-borne outbreak of tonsillitis and cervical adenitis (Section 4.b) is the most striking of these. We have seen a *zooepidemicus* strain from a case of septic arthritis in a veterinarian.

(d) *Self infection*

Many sporadic streptococcal infections are of endogenous origin; the patient may have been a carrier of the infecting strain, or it may have been a normal member of the patient's surface flora and invasion of the body may have been 'triggered' by some non-bacteriological circumstance (Section 4.e). It is, however, often difficult to decide that a particular infection is endogenous if the infecting strain is one with known powers of communicability, such as the group A streptococcus, because the carrier site is unlikely to have been examined just before the development of the lesion. In a large prospective study of sepsis in industrial injuries to the hand (Williams & Miles 1949), only 20% of subjects who later developed sepsis due to group A streptococci yielded these organisms from swabs of the throat, the wound or elsewhere on the skin soon after wounding, a proportion only slightly greater than that observed in patients who did not develop streptococcal sepsis (14;7%), and it was concluded that self-infection occurred infrequently. In streptococcal impetigo (Section 4.a), colonization of the skin may precede the development of the lesion, but the ultimate source is nevertheless exogenous. Group B streptococcal disease of the newborn is clearly exogenous in origin (Section 4.a); in adults, however, the source is usually assumed to be endogenous, and there is little or no evidence to the contrary.

When the infecting streptococcus is one that forms part of the normal body flora, and particularly when there is a history of pathological disturbance or surgical interference at or near the site at which it is harboured, self-infection is much the more likely possibility. This is so for nearly all infections with streptococci not mentioned in Table 9. Even with these, however, there are rare instances in which an extraneous source can be established. For instance, we are aware of two associated cases of meningitis due to *S. bovis* in highly susceptible patients who had been given intrathecal injections by one operator.

In Table 4 we listed the main surface sites at which various streptococci are found, and it is among these that the sources of endogenous infection must be sought. The viridans streptococci that cause endocarditis probably come from the teeth or the throat; this is supported by the fact that, immediately after dental operations and tonsillectomy, transient streptococcal bacteraemia—mainly but not exclusively of viridans strains—occurs frequently (Okell & Elliott 1935; Fischer & Gottdenker 1936; Elliott 1939), and that endocarditis develops in a proportion of susceptible subjects. A similar sequence of events may precede *S. mutans* endocarditis. Transient bacteraemia may follow operative manipulations of the genito-urinary tract in both sexes (Schottmüller 1911; Barrington & Wright 1930; Richards 1932). Enterococci that subsequently cause septicaemia or endocarditis usually first enter the body via the urinary or the genital tract (Mandell *et al.* 1970). They are assumed to have come from the gut, but the prospective studies necessary to establish this have not been made. Enterococci also enter the bloodstream after manipulation of the lower bowel, for example in sigmoidoscopy (Le Frock *et al.* 1973).

The numerous surface sites at which *S. milleri* is commonly found make it difficult to establish the point of entry of the organism in purely septicaemic infections. The distribution of purulent lesions in the internal organs (Section 2.a and Table 7), and of preceding disease and local infection in organs close to the body surface (Section 4.2) indicate that invasion occurs from a variety of sites: from the teeth to tissues of the jaw and neck (Guthof 1956); from the paranasal sinuses to the frontal region of the brain (de Louvois *et al.* 1977); from the gut to intra-abdominal organs (Parker & Ball 1976; Poole & Wilson

1977). However, the fact that several purulent lesions may develop in widely separated internal organs indicates that this argument should not be pushed too far. Finally, not all *S. milleri* lesions of internal organs may arise from direct extension through tissue. In a personal series of systemic *S. milleri* infections, 105 patients had purulent lesions in an internal organ, 32 apparently had a purely septicaemic disease, and 36 had endocarditis, but none had endocarditis and a purulent lesion. Phillips *et al.* (1976) observed that *S. milleri* was often present in the blood immediately after dental extraction, so 'primary' bloodstream dissemination of *S. milleri* to internal organs may also occur.

(e) *Host factors in streptococcal disease*

Factors in the host that affect the frequency with which streptococci invade the human body, and with which this leads to serious disease, include, on the one hand, the protective effect of specific antibody and, on the other, the predisposing effect of other factors which may be either general or local. *General predisposing factors*, such as immunodeficiency states or other severe underlying disease, are of considerable importance in some types of infection with pneumococci and group B streptococci. In infections with most other streptococci, they appear to be much less important than *local factors* which may facilitate entry to the body or determine localization in a particular organ. Thus, in our series of systemic infections other than endocarditis (which excluded pneumococci) there was evidence of severe underlying disease in 14% and of local predisposing causes in 40% of patients (Parker & Ball 1976). This was probably an overestimate of the true importance of severe underlying disease, because in a number of patients local tissue damage associated with it or with the treatment given for it appeared to be the true predisposing factor (e.g. in malignant disease and diabetes).

Several examples will be given of the effect of age on the frequency or severity of streptococcal disease; in some of these the effect appears to be direct, but in others it is a consequence of age-related differences in the state of immunity or in the frequency of local predisposing factors.

The group A streptococcus is unique among streptococci in that, under optimal conditions of exposure, a majority of normal persons may develop clinical illness, e.g. tonsillitis or impetigo. Thus, predisposing factors in the host are of small importance, and resistance to infection is attributed mainly to the presence of type-specific antibody (Maxted, this volume). The progressive accumulation of a 'library' of M antibodies against prevalent types is held to explain the infrequency of these diseases in adults except when considerable mixing of populations takes place (Rantz *et al.* 1946), and the presence of maternally acquired antibody to account for the rarity of severe disease in neonates (Zimmerman & Hill 1969). Sepsis at other sites appears more often to be determined by local factors, and the role of specific immunity in preventing it is uncertain. Septicaemic invasion by streptococci of groups A, C and G occurs at all ages, and much more often follows infections of wounds and other exposed tissues than respiratory-tract infections (Parker & Ball 1976). In hospital, surgical operations and similar procedures, such as cannulation, are prominent predisposing factors.

Pneumococci seldom cause disease in neonates, but under appropriate conditions give rise to serious disease at all other ages. The natural acquisition of protective antibodies appears to have little influence on the frequency of disease at various ages, but predisposing factors in the host are of great importance. Clinical evidence suggests that pneumonia is usually precipitated by inflammation in the respiratory tract caused by another agent, and this can sometimes be shown to be viral. In young children pneumococcal pneumonia often follows infectious fevers such as measles or whooping cough. Previous chronic damage to the lower respiratory tract or chemical irritation of it are also important predisposing factors. High fatality rates in pneumococcal pneumonia, even if adequate treatment is given, are associated with the presence of bacteraemia, age over 50 years, and infection with pneumococci of certain serotypes, notably type 3 (Austrian 1968).

In pneumococcal meningitis, the significant host factors depend upon the route of invasion: if this is directly from the respiratory tract, mechanical defects of the skull—congenital or traumatic—and pneumococcal middle-ear disease predominate; if from the bloodstream, severe underlying disease is often present. In all systemic pneumococcal infections other than those associated with pneumonia, general predisposing factors are present in many of the patients. Shapera & Matsen (1972) found evidence of significant underlying disease—notably leukaemia, reticuloendothelial or other malignancy, splenic dysfunction or autoimmune disease—in 109 of 140 such infections (78%). Invasive pneumococcal infection often occurs in the sickle-cell state, and in the absence of this is particularly severe in Negro populations, in which it often affects apparently healthy young adults.

In group B streptococcal infection, clinical attack rates appear to be low, and host factors exert strong and sometimes interrelated effects (Ross, p. 136). Meningitis occurs most often in neonates, thereafter with decreasing frequency in infancy and childhood, and least often in adults (Fig. 1). Septicaemic disease is seen in the first week of life and in adults, but rarely in the intervening years; in neonates, but not in adults, it may be associated with pulmonary infection. Thus, there are three situations: (1) early-onset neonatal disease—either meningitic or septicaemic—a maternally acquired infection that affects mainly sick infants predisposed by prematurity or difficult labour or both, (2) late-onset meningitis and meningitis in older persons, for which the source of infection is unknown and no predisposing factors have been identified and

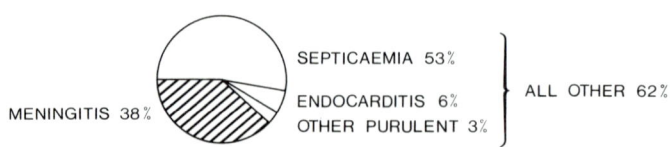

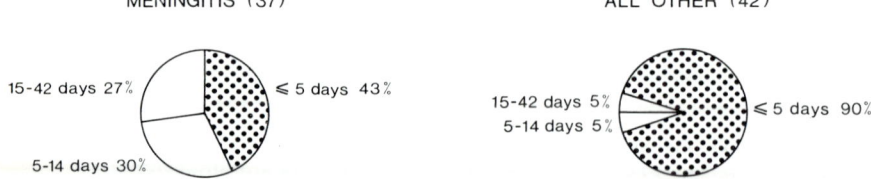

Fig. 1. Pattern of systemic infections with group B streptococci: data in respect of patients from whom cultures were sent to the Streptococcus Reference Laboratory in the years 1970 to mid-1977, expressed as percentages of the numbers of patients shown in parenthesis. All other infections: isolation from the blood or an internal organ other than the meninges. Neonate: aged six weeks or less; child: over six weeks but less than 15 years old.

(3) adult septicaemic disease which very often appears to be 'conditional' upon the presence of predisposing factors. In a personal series of 68 septicaemic infections in adults (including 14 cases of endocarditis), 44 predisposing factors were identified in 41 (60%) of the patients: childbirth or gynaecological operation in 15, lymphoma, leukaemia or Hodgkin's disease in eight, diabetes in seven; previous damage to a heart valve in four; post-operative wound infection with a group B streptococcus in three, skin sepsis in two, cholecystitis or urinary-tract infection each in one, others in four. Thus, the only group B streptococcal disease in which predisposing factors other than age are

uncommon is meningitis. This has led to the suggestion, for which there is some experimental evidence, that late-onset neonatal meningitis may occur in babies who have not received type-specific antibody passively from the mother (Baker & Kasper 1976).

Infections with faecal streptococci and *S. milleri* occur at all ages, and appear very often to be conditional upon local inflammation, trauma or surgical interference at or near a site that normally harbours the organism. The evidence suggests that *S. faecalis* rarely causes sepsis in the absence of such factors, which operate mainly in the urinary and female genital tracts and the gut itself; in hospitals, infection at the sites of minor surgical procedures must be added.

Streptococcus milleri, on the other hand, may perhaps cause local sepsis at superficial sites at which it is harboured in the absence of detectable predisposing cause. In addition to dental sepsis and sinusitis there is some evidence that it may play a part in appendicitis. Poole and Wilson (1977) isolated it from the inflamed appendix twice as often as from the normal appendix; there is little doubt that the severe bacterial complications of appendicitis are most often caused by anaerobic gram negative rods, but *S. milleri* may be concerned in the local inflammation. Whether acute appendicitis is to be looked upon as a mechanical disturbance that predisposes to bacterial infection, or as a 'primary' bacterial disease is uncertain.

In considering the causes that lead to the spread of *S. milleri* into internal organs, it is often difficult to distinguish the effects of surgical operations from those of bacterial disease, but trauma alone may lead to a systemic infection; cases have followed accidental damage to the oesophagus and the intestine, and several have been observed after road-traffic accidents without major external injury.

(i) Predisposition to streptococcal endocarditis

Damage to heart valves is the main predisposing factor for the development of bacterial endocarditis. Parker & Ball (1976) identified general predisposing factors in only 1.9%, and local factors other than valvular damage in only 4.1% of cases of streptococcal endocarditis. Until a few years ago, the natural history of endocarditis accorded well with the view that it usually developed on heart valves previously damaged by rheumatic fever or the effects of congenital malformation, and that the causative micro-organism reached the heart in bacteraemic showers from the oropharyngeal region. The patients were usually adolescents or young adults with a history of previous damage to the heart, a number of whom gave a history of recent dental treatment or tonsillectomy, and viridans streptococci were the predominant cause. Cates & Christie (1951) reported that 62% of 448 cases of endocarditis were in patients aged 15-35 years, and that there was a slight excess of females among cases in this age group; in the smaller number of older patients there was an excess of males;

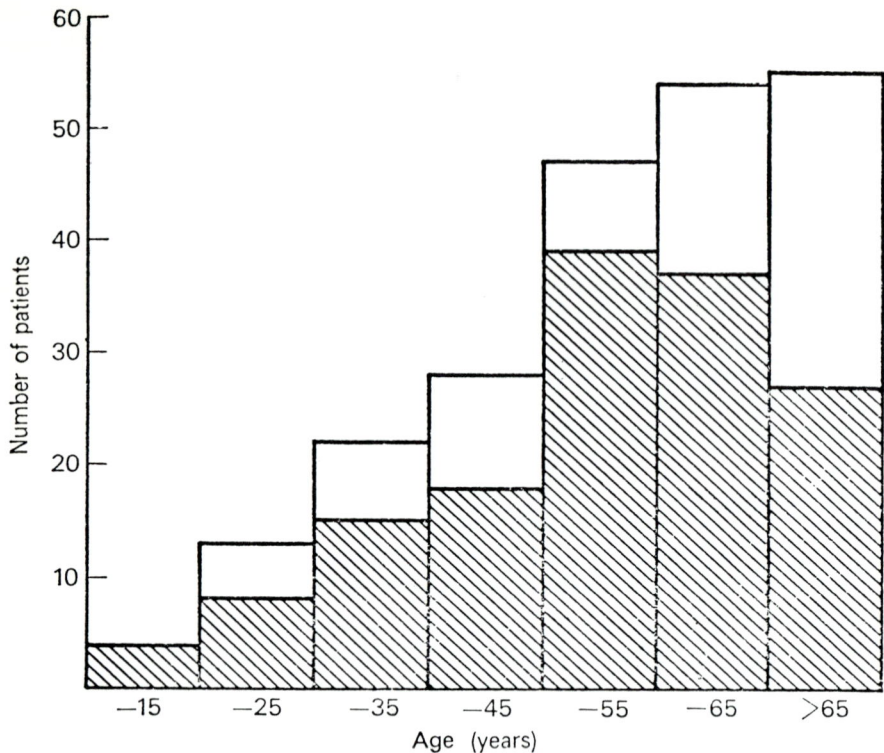

Fig. 2. Age and sex of patients with streptococcal endocarditis. (Reproduced from the *Journal of Medical Microbiology*, vol. 9, no. 3, p. 287.) Shaded areas, male patients; open areas, female patients.

94% of the isolates from the blood were streptococci and 94% of the streptococci were α-haemolytic.

Since 1960, several reports have indicated that endocarditis has become predominantly a disease of middle aged or elderly persons, and that it is caused by a greater variety of micro-organisms, including staphylococci and non-haemolytic streptococci (Uwaydah & Weinberg 1965; Hughes & Gauld 1966). In our series of streptococcal isolates from endocarditis (Parker & Ball 1976) there was a steady increase in numbers with age up to 65 years, a strong predominance of male patients (Fig. 2), and considerable age-related differences in the frequency of various streptococci among the isolates (Fig. 3). The percentage of viridans streptococci fell from 69 at ages 10–34 years to 41 at ages ⩾55, but the absolute number of viridans isolates rose slightly with increasing age. After the age of 35 years there was a considerably steeper rise in the number of isolates of

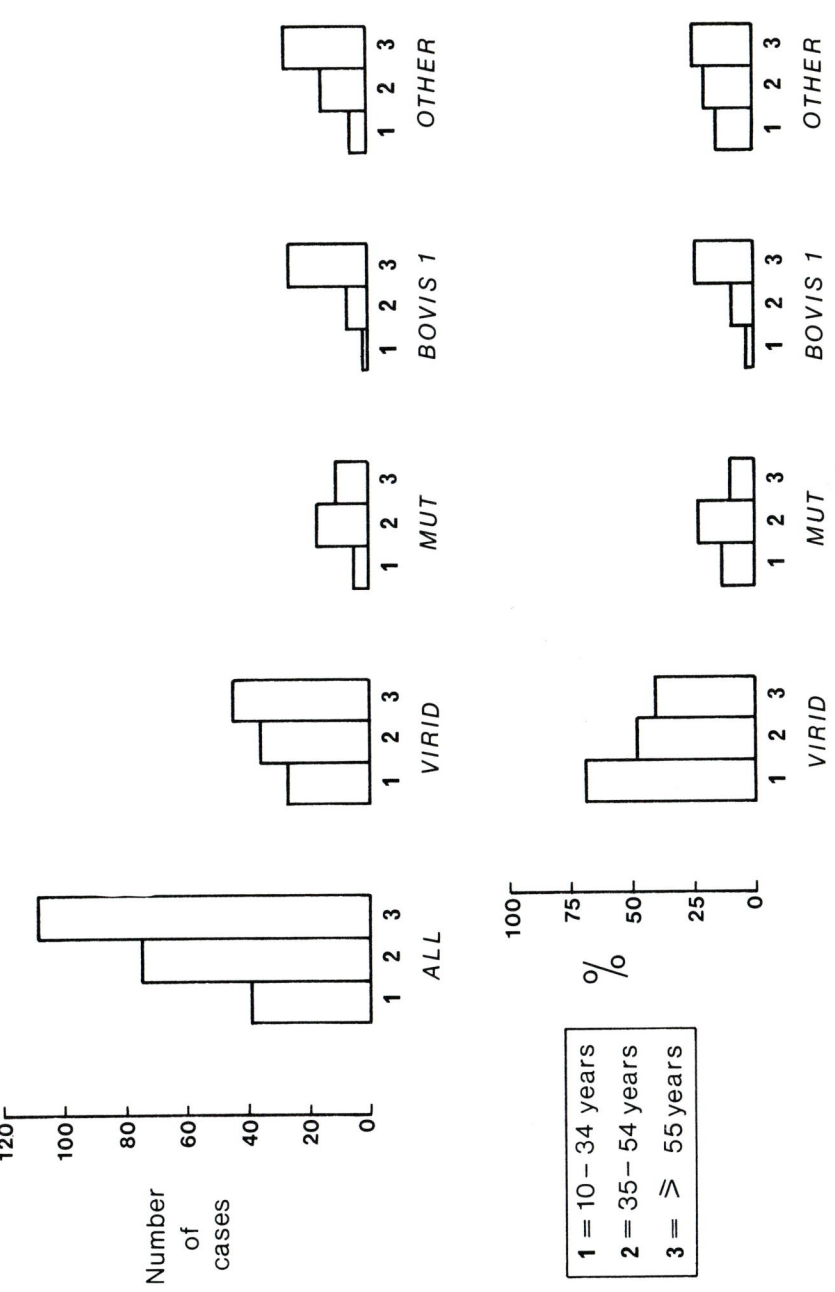

Fig. 3. Number and percentage of patients with streptococcal endocarditis in the stated age groups from whom (a) viridans streptococci (VIRID), (b) *S. mutans* (MUT), (c) *S. bovis* biotype I (BOVIS I) and other streptococci (OTHER) were isolated (data from Parker & Ball 1976).

S. bovis biotype I and of 'other' streptococci, while *S. mutans* was most frequent in the intermediate age-group 35–54 years. Thus the streptococci from patients in the youngest age group resembled those responsible for endocarditis 30 years ago: in four of five studies reported in the years 1945–49, streptococci described as *S. sanguis* or *S. mitior* were responsible for 77, 83, 86 and 92% of the isolates, and in the fifth—a small study in which *S. faecalis* predominated—the percentage was 32 (see Parker & Ball 1976). The outstanding feature of the streptococci from endocarditis in patients aged ⩾55 years is that they include a considerable minority that did not or may not have come from the oropharynx. Streptococci usually associated with the mouth (viridans strains and *S. mutans*) accounted for 56% of the isolates; the remainder included 27% of *S. bovis* biotype I and 5% each of *S. faecalis* and other enterococci that probably came from the gut, and 6% of *S. milleri* that might have come from either source; but a direct association of endocarditis with clinically overt intra-abdominal sepsis was seldom recorded.

A history of damage to heart valves before the onset of endocarditis was given twice as often in cases of endocarditis associated with viridans streptococci as in cases associated with other streptococci, and this difference was not age related. Uwaydah & Weinberg (1965) suggest that, in the elderly, streptococci become established on atheromatous valve lesions, which are often not clinically apparent. Thus elderly persons at risk for endocarditis—particularly from 'silent' intra-abdominal disease or manipulation of the colon—cannot easily be identified.

(f) *Antibiotic resistance*

Resistance to antibiotics has had little effect on the pattern of streptococcal disease. Nearly all streptococci are still sensitive to a wide range of antibiotics, but streptococcal diseases continue to cause many deaths and much disablement; this must be attributed to failure to deliver the right antibiotics to patients at the right time (*Anon.* 1968).

Streptococcus faecalis is the only important pathogenic streptococcus that is regularly resistant to a wide range of antibiotics. This may present therapeutic difficulties but does not appear to have had significant epidemiological consequences. It has been stated that enterococcal endocarditis has recently increased in frequency (Uwaydah & Weinberg 1965; Finland & Barnes 1970), but it was common in some early series, e.g. Horder (1906), 27%; Foley (1947), 56%. In any event, there is no evidence that the widespread use of antibiotics has led to the endemic prevalence of *S. faecalis* strains in hospitals. In the individual patient, however, antibiotic treatment may lead to the colonization of the mouth by *S. faecalis* (Phillips *et al.* 1976).

Other streptococci may be resistant to single antibiotics: to penicillins, tetracycline, chloramphenicol, macrolides and lincomycin-clindamycin, but 'multiple' antibiotic resistance is uncommon, and has so far not led to the endemic persistence of resistant strains in hospitals. (Since this paper was read, reports have appeared from South Africa—see *Morbidity and Mortality Weekly Reports*, Center for Disease Control, Atlanta, Ga. 2 Sept. 1977, vol. 26, pp. 285-286—of epidemics of infection caused by pneumococci that were resistant to several antibiotics, including penicillins.) Local prevalences of group A streptococci resistant to lincomycin-clindamycin occur occasionally, under conditions of excessive drug usage, both in hospital departments and extra-hospital communities. A considerable proportion of streptococci of the pyogenic groups and of pneumococci are tetracycline resistant, but the clinical significance of this is limited by the fact this antibiotic should not, in any event, be used for the treatment of streptococcal diseases. Most streptococci (other than enterococci) and all group A streptococci, are still sensitive to concentrations of benzyl-penicillin that are easily attainable in the body.

Antibiotic treatment in the individual patient is beyond the scope of this review. Briefly, the important considerations are: to detect the few streptococcal strains that are resistant to a therapeutically useful agent; to avoid selecting resistant variants in the patient's flora by ill-timed chemoprophylaxis in persons at risk for endocarditis and to determine the most effective combination of drugs for the total eradication of streptococci from an infected heart valve.

5. The General Picture

The aim of this review is to illustrate the wide diversity of streptococcal disease in man, and to attempt to relate this to the nature of the infecting streptococcus and the susceptibility of the host. The emphasis was on the role of the streptococcus as a cause of septic diseases and on the more serious consequences of these. The resulting picture tended to reflect the activities of the streptococcus as seen today in hospitals in affluent communities.

The view of the family practitioner in the same communities would be rather different. The only streptococcal diseases that would form a significant part of the daily work-load would be caused by group A streptococci (tonsillitis and its local complications, and minor sepsis of skin and wounds) and pneumococci (pneumonia, mainly in the elderly, and middle-ear disease in infants). Rheumatic fever and post-streptococcal glomerulonephritis are rarely seen. Most practitioners treat respiratory-tract infections symptomatically (with or without antibiotics according to their clinical judgement) and few aim to seek a bacteriological diagnosis or give a full course of antibiotic treatment to all patients suffering from group A streptococcal infections. The view is widely held that streptococci have now 'lost their virulence', but there is little evidence to support this. The present position is better explained by a decrease in the total incidence of streptococcal infection, particularly in children, attributable to reduced crowding and family size, and to improved personal hygiene. An additional factor of probable importance in reducing the frequency of severe pneumococcal disease of children and young adults is the effect of early 'blind' antibiotic treatment for respiratory-tract infections.

In developing countries, the important streptococcus is again the group A streptococcus and, to a lesser extent, the pneumococcus. Pneumococcal pneumonia and meningitis in young adults are prominent in certain parts of Africa. Group A streptococcal infections are of worldwide importance, but attract attention mainly as causes of rheumatic heart disease and acute glomerulonephritis. Rheumatic fever has been identified in all 'poor' countries in which it has been sought, and is associated with high titres of antibody against group A streptococci, but its relation to antecedent throat infection—and indeed the respiratory transmission of streptococci in warm climates—have been little investigated. Post-streptococcal glomerulonephritis in poor countries is associated almost exclusively with streptococcal impetigo. It is unevenly distributed geographically and seasonally, and its incidence is influenced both by the frequency of impetigo (and in turn by factors that predispose to this, such as poor personal hygiene, insect bites, and scabies) and by the prevalence of 'nephritogenic types' of group A streptococci (and this probably by the frequency of type-specific antibody against these in children in the relevant age-group).

6. References

ANDREWES, F. W. & HORDER, T. J. 1906 A study of the streptococci pathogenic for man. *Lancet* **2**, 708–713, 852–855.
ANON. 1968 Streptococcal and staphylococcal infections. *World Health Organization Technical Report. Series No. 394*. Geneva.
AUSTRIAN, R. 1968 Current status of bacterial pneumonia with especial reference to pneumococcal infection. *Journal of Clinical Pathology* **21**, suppl. 2, 93–97.
AUSTRIAN, R. 1975 Random gleanings from a life with the pneumococcus. *Journal of Infectious Diseases* **131**, 474–484.
AUSTRIAN, R. & GOLD, J. 1964 Pneumococcal bacteremia with especial reference to bacteremic pneumococcal pneumonia. *Annals of Internal Medicine* **60**, 759–776.
BAKER, C. J. & KASPER, D. L. 1976 Correlation of maternal antibody deficiency with susceptibility to neonatal group B streptococcal infection. *New England Journal of Medicine* **294**, 753–756.
BARRINGTON, F. J. F. & WRIGHT, H. D. 1930 Bacteraemia following operations on urethra. *Journal of Pathology and Bacteriology* **33**, 871–888.
BOULTON-JONES, J. M., SISSONS, J. G. P., EVANS, D. J. & PETERS, D. K. 1974 Renal lesions of subacute infective endocarditis. *British Medical Journal* **2**, 11–14.
BUCHANAN, R. E. & GIBBONS, N. E. 1974 *Bergey's Manual of Determinative Bacteriology*, 8th edn. Baltimore: Williams & Wilkins.
CARLSSON, J. 1967 Presence of various types of non-haemolytic streptococci in dental plaque and in other sites of the oral cavity in man. *Odontologisk Revy* **18**, 55–74.
CARLSSON, J. 1968 A numerical taxonomic study of human oral streptococci. *Odontologisk Revy* **19**, 137–160.

CATES, J. E. & CHRISTIE, R. V. 1951 Subacute bacterial endocarditis: review of 442 patients treated in 14 centres appointed by the Penicillin Trials Committee of the Medical Research Council. *Quarterly Journal of Medicine* 20 n.s., 93-130.

CLARKE, J. K. 1924 The bacterial factor in the aetiology of dental caries. *British Journal of Experimental Pathology* 5, 141-147.

COLMAN, G. 1968 The application of computers to the classification of streptococci. *Journal of General Microbiology* 50, 149-158.

COLMAN, G. 1970 The classification of streptococcal strains. Ph.D. Thesis, University of London.

COLMAN, G. & WILLIAMS, R. E. O. 1965 The cell walls of streptococci. *Journal of General Microbiology* 41, 375-387.

COLMAN, G. & WILLIAMS, R. E. O. 1972 Taxonomy of some viridans streptococci. In *Streptococci and Streptococcal Diseases, Recognition, Understanding and Management*, eds Wannamaker, L. W. & Matsen, J. M. New York & London: Academic Press.

COOKE, J. V. 1927 Scarlet fever; relation between antitoxin in blood and skin sensitivity to toxin in newborn infants and their mothers. *American Journal of Diseases of Children* 34, 969-978.

COOKE, J. V. 1928a Scarlet fever; development of toxin sensitivity of skin and its relation to presence of antitoxin in blood. *American Journal of Diseases of Children* 35, 762-771.

COOKE, J. V. 1928b Scarlet fever; modification of skin sensitivity as result of nonscarlatinal infections. *American Journal of Diseases of Children* 35, 772-780.

CORDIERO, A., COSTA, H. & LAGINHA, F. 1965 Immunologic phase in subacute bacterial endocarditis: a new concept and general considerations. *American Journal of Cardiology* 16, 477-481.

CREAM, J. J. & TURK, J. L. 1971 A review of the evidence for immune-complex deposition as a cause of skin disease in man. *Clinical Allergy* 1, 235-247.

DE LOUVOIS, J., GORTVAI, P. & HURLEY, R. 1977 Bacteriology of abscesses of the central nervous system: a multicentre prospective study. *British Medical Journal* 2, 981-984.

DICK, T. B. & GEMMELL, C. G. 1971 The pathogenesis of pneumococcal infection in mice. *Journal of Medical Microbiology* 4, 153-163.

DOCHEZ, A. R. & STEVENS, F. A. 1927 Studies on biology of streptococcus; allergic reactions with strains from erysipelas. *Journal of Experimental Medicine* 46, 487-495.

DOCHEZ, A. R., AVERY, O. T. & LANCEFIELD, R. C. 1919 Studies on the biology of streptococcus. I. Antigenic relationships between strains of *Streptococcus haemolyticus*. *Journal of Experimental Medicine* 30, 179-213.

DRUSIN, L. M., RIBBLE, J. C. & TOPF, B. 1973 Group C streptococcal colonisation in the newborn nursery. *American Journal of Diseases of Children* 125, 820-821.

DUCA, F., TEODOROVICI, G., RADU, C., VÎTÂ, A., TALAŞMAN-NICULESCU, P., BERNESCU, E., FELDI, C. & ROŞCA, V. 1969 A new nephritogenic streptococcus. *Journal of Hygiene, Cambridge* 67, 691-698.

EDWARDSSON, S. 1968 The characteristics of caries-inducing human streptococci resembling *Streptococcus mutans*. *Archives of Oral Biology* 13, 637-646.

ELLEN, R. P. & GIBBONS, R. J. 1972 M-protein associated adherence of *Streptococcus pyogenes* to epithelial surfaces: pre-requisite for virulence. *Infection and Immunity* 5, 826-830.

ELLIOTT, S. D. 1939 Bacteriaemia following tonsillectomy. *Lancet* 2, 589-592.

ELLIOTT, S. D. 1973 The incidence of group-H streptococci in blood cultures of patients with subacute bacterial endocardities (SBE). *Journal of Medical Microbiology* 6, xiv.

FACKLAM, R. R. 1972 Recognition of group-D streptococcal species of human origin by biochemical and physiological tests. *Applied Microbiology* 23, 1131-1139.

FACKLAM, R. R. 1974 Characteristics of *Streptococcus mutans* isolated from human dental plaque and blood. *International Journal of Systematic Bacteriology* 24, 313-319.

FACKLAM, R. R. 1977 Physiological differentiation of viridans streptococci. *Journal of Clinical Microbiology* 5, 184-201.

FINLAND, M. & BARNES, M. W. 1970 Changing aetiology of bacterial endocarditis in the antibacterial era: experience at Boston City Hospital 1933-65. *Annals of Internal Medicine* **72**, 341-348.

FISCHER, J. & GOTTDENKER, F. 1936 Über transitorische Bakterieneinschwemmung in die Blutbahn nach Tonsillektomie. *Wiener Klinische Wochenschrift* **49**, 177.

FOLEY, G. E. 1947 Further observations on the occurrence of streptococci of groups other than A in human infection. *New England Journal of Medicine* **237**, 809-811.

FRASER, C. A. M., BALL, L. C., MORRIS, C. A. & NOAH, N. D. 1977 Serological characterization of group-A streptococci associated with skin sepsis in meat handlers. *Journal of Hygiene, Cambridge* **78**, 283-296.

GINSBURG, I. 1972 Mechanisms of cell and tissue injury induced by group A streptococci: relation to post streptococcal sequelae. *Journal of Infections Diseases* **126**, 294-340, 419-456.

GUTHOF, O. 1956 Ueber pathogene "vergrunende Streptokokken". *Zentralblatt für Bakteriologie, Parasitenkunde, Infektionskrankheiten und Hygiene* Abt. I Originale **166**, 553-564.

GWALTNEY, J. M., SANDE, M. A., AUSTRIAN, R. & HENDLEY, J. O. 1975 Spread of *Streptococcus pneumoniae* in families. II. Relation of *S. pneumoniae* to incidence of colds and serum antibody. *Journal of Infectious Diseases* **132**, 62-68.

HARDER, E. J., WILKOWSKIE, C. J., WASHINGTON, J. A. III & GERACI, J. E. 1974 *Streptococcus mutans* endocarditis. *Annals of Internal Medicine* **80**, 364-368.

HARE, R. 1928 On the phagocytosis of haemolytic streptococci of high and low virulence by the blood of patients infected with that organism. *British Journal of Experimental Pathology* **9**, 337-352.

HARE, R. 1932 The production of bacteriotropins for the haemolytic streptococci by patients infected with that organism. *Journal of Pathology and Bacteriology* **35**, 701-715.

HILL, H. R., CALDWELL, G. G., WILSON, E., HAGER, D. & ZIMMERMAN, R. A. 1969 Epidemic of pharyngitis due to streptococci of Lancefield group G. *Lancet* **2**, 371-374.

HOOK, E. W., WAGNER, R. R. & LANCEFIELD, R. C. 1960 An epizootic in Swiss mice caused by group A streptococcus, newly designated type 50. *American Journal of Hygiene* **72**, 111-119.

HORDER, T. J. 1906 Infective endocarditis with an analysis of 510 cases and with special reference to the chronic form of the disease. *Quarterly Journal of Medicine* **2**, 289-324.

HUGHES, P. & GAULD, W. R. 1966 Bacterial endocarditis: a changing disease. *Quarterly Journal of Medicine* **35** n.s., 511-520.

KESLIN, M. H., MESSNER, R. P. & WILLIAMS, R. C. JR 1973 Glomerulonephritis with subacute bacterial endocarditis. *Archives of Internal Medicine* **132**, 578-581.

KIEL, P. & SKADHAUGE, K. 1973 Studies on mannitol-fermenting strains of *Streptococcus bovis*. *Acta pathologica et microbiologica scandinavica* **B81**, 10-14.

KREGER, A. S. & BERNHEIMER, A. W. 1969 Physical behaviour of pneumolysin. *Journal of Bacteriology* **98**, 306-307.

LANCEFIELD, R. C. 1928 Antigenic complex of *Streptococcus haemolyticus*: chemical and immunological properties of protein fractions. *Journal of Experimental Medicine* **47**, 469-480.

LE FROCK, J. L., ELLIS, C. A., TURCHIK, J. B. & WEINSTEIN, L. 1973 Transient bacteremia associated with sigmoidoscopy. *New England Journal of Medicine* **289**, 467-469.

MacLEOD, C. M., HODGES, R. G., HEIDELBERGER, M. & BERNHARD, W. G. 1945 Prevention of pneumococcal pneumonia by immunization with specific capsular polysaccharides. *Journal of Experimental Medicine* **82**, 445-465.

MANDELL, G. L., KAYE, D., LEVISON, M. E. & HOOK, E. W. 1970 Enterococcal endocarditis: an analysis of 38 patients observed at the New York Hospital-Cornell Medical Center. *Archives of Internal Medicine* **125**, 258-264.

MEJÀRE, B. & EDWARDSSON, S. 1975 *Streptococcus milleri* (Guthof); an indigenous organism of the human oral cavity. *Archives of Oral Biology* **20**, 757-762.
MIRICK, G. S., THOMAS, L., CURNEN, E. C. & HORSFALL, F. L. 1944. Studies on a non-hemolytic streptococcus isolated from the respiratory tract of human beings. *Journal of Experimental Medicine* **80**, 391-406, 407-430, 431-440.
MOHAMMED, I., ANSELL, B. M., HOLBOROW, E. J. & BOYCESON, A. D. M. 1977 Circulating immune complexes in subacute bacterial endocarditis and post-streptococcal glomerulonephritis. *Journal of Clinical Pathology* **30**, 308-311.
OKELL, C. C. & ELLIOTT, S. D. 1935 Bacteriaemia and oral sepsis, with special reference to the aetiology of subacute endocarditis. *Lancet* **2**, 869-872.
PARKER, M. T. 1967 International survey of the distribution of serotypes of *Streptococcus pyogenes* (group A streptococci). *Bulletin of the World Health Organization* **37**, 513-537.
PARKER, M. T. 1977 Neonatal streptococcal infections. *Post Graduate Medical Journal* **53**, 598-608.
PARKER, M. T. & BALL, L. C. 1976 Streptococci and aerococci associated with systemic infection in man. *Journal of Medical Microbiology* **9**, 275-302.
PERCH, B., KRISTJANSEN, P. & SKADHAUGE, K. 1968 Group R streptococci pathogenic for man: two cases of meningitis and one fatal case of sepsis. *Acta pathologica et microbiologica scandinavica* **74**, 69-76.
PERCH, B., KJEMS, E. & RAVN, T. 1974 The biochemical and serological properties of *Streptococcus mutans* from various human and animal sources. *Acta pathologica et microbiologica scandinavica* **B82**, 357-370.
PHILLIPS, I., WARREN, C., HARRISON, J. M., SHARPLES, P., BALL, L. C. & PARKER, M. T. 1976 Antibiotic susceptibilities of streptococci from the mouth and the blood of patients treated with penicillin or lincomycin and clindamycin. *Journal of Medical Microbiology* **9**, 393-404.
POOLE, P. M. & WILSON, G. 1977 *Streptococcus milleri* in the appendix. *Journal of Clinical Pathology* **30**, 937-942.
PORTERFIELD, J. S. 1950 Classification of the streptococci of subacute bacterial endocarditis. *Journal of General Microbiology* **4**, 92-101.
RANTZ, L. A., RANTZ, H. H., BOISVERT, P. J. & SPINK, W. W. 1946 Streptococcic and nonstreptococcic disease of the respiratory tract: epidemiologic observations. *Archives of Internal Medicine* **77**, 121-131.
RAVREBY, W. D., BOTTONE, E. J. & KEUSCH, G. T. 1973 Group D streptococcal bacteremia, with emphasis on the incidence and presentation of infections due to *Streptococcus bovis*. *New England Journal of Medicine* **289**, 1400-1403.
RICHARDS, J. H. 1932 Bacteremia following irritation of foci of infection. *Journal of the American Medical Association* **99**, 1496-1497.
ROSS, P. W. 1972 The occurrence of potentially pathogenic bacteria in the mouth. In *Host Resistance to Commensal Bacteria*, ed. MacPhee, T. Edinburgh & London: Churchill Livingstone.
SCHOTTMÜLLER, H. 1903 Die Artenunterscheidung für den Menschen pathogenen Streptokokken durch Blutagar. *Münchener Medizinische Wochenschrift* **50**, 849-853, 909-912.
SCHOTTMÜLLER, H. 1911 Zur Aetiologie des Febris puerperalis und Febris in Puerperio. *Münchener Medizinische Wochenschrift* **58**, 557-558.
SEASTONE, C. V. 1939 The virulence of group C hemolytic streptococci of animal origin. *Journal of Experimental Medicine* **70**, 361-378.
SHAPERA, R. M. & MATSEN, J. M. 1972 Host factors and capsular typing of body fluid isolates in fulminant pneumococcal infections. *Infection and Immunity* **5**, 132-136.
SHERMAN, J. M. 1937 The streptococci. *Bacteriological Reviews* **1**, 3-97.
STEERE, A. C., ABER, R. C., WARFORD, L. R., MURPHY, K. E., FEELEY, J. C., HAYES, P. S., WILKINSON, H. W. & FACKLAM, R. R. 1975 Possible nosocomial transmission of group B streptococci in a newborn nursery. *Journal of Pediatrics* **87**, 784-787.

UWAYDAH, M. M. & WEINBERG, A. N. 1965 Bacterial endocarditis—a changing pattern. *New England Journal of Medicine* **273**, 1231-1235.

WANNAMAKER, L. W., DENNY, F. W., PERRY, W. D., SIEGEL A. C. & RAMMELKAMP, C. H. 1953 Studies on immunity to streptococcal infections in man. *American Journal of Diseases of Children* **86**, 347.

WATSON, D. W. & KIM, Y. B. 1970 Erythrogenic toxins. In *Microbial Toxins*, vol. 3, eds Montie, T. C., Kadis, S. & Ajl, S. J. New York & London: Academic Press.

WEIL-KORSTANJE, J. A. A. VAN DER & WINKLER, K. C. 1975 The faecal flora of ulcerative colitis. *Journal of Medical Microbiology* **8**, 491-501.

WHITE, J. C. & NIVEN. C. F. J. JR 1946 *Streptococcus* s.b.e.: a streptococcus associated with subacute bacterial endocarditis. *Journal of Bacteriology* **51**, 717-722.

WILLIAMS, R. E. O. & MILES, A. A. 1949 Infection and sepsis in industrial wounds of the hand; a bacteriological study of aetiology and prophylaxis. *Special Report Series No. 266, Medical Research Council*. London: HMSO.

WILSON, G. S. & MILES, A. A. 1975 *Topley and Wilson's Principles of Bacteriology, Virology and Immunity*, 6th edn. London: Edward Arnold.

WINDSOR, R. S. & ELLIOTT, S. D. 1975 Streptococcal infection in young pigs. IV. An outbreak of streptococcal meningitis in weaned pigs. *Journal of Hygiene, Cambridge* **75**, 69-78.

WINTERBAUER, R. H., FORTUINE, R. & EICKHOFF, T. C. 1966 Unusual occurrence of neonatal meningitis due to group B beta-hemolytic streptococci. *Pediatrics* **38**, 661-662.

ZABRISKIE, J. B. 1964 The role of lysogeny in the production of erythrogenic toxin. In *Rheumatic Fever and Glomerulonephritis*, ed. Uhr, J. W. Baltimore: Williams & Wilkins.

ZANEN, H. C. & ENGEL, H. W. B. 1975 Porcine streptococci causing meningitis and septicaemia in man. *Lancet* **1**, 1286-1288.

ZIMMERMAN, R. A. & HILL, H. R. 1969 Placental transfer of group A type-specific streptococcal antibody. *Pediatrics* **43**, 809-814.

Group A Streptococci: Pathogenesis and Immunity

W. R. MAXTED

*Central Public Health Laboratory,
London, England*

CONTENTS

1. Introduction 107
   (a) Communicability 107
   (b) Host range 108
   (c) Diseases caused 108
2. The M proteins 109
   (a) Adhesion of group A streptococci to epithelial cells 110
   (b) Group A streptococci in the phagocytic system 111
   (c) The M antigen–antibody system in precipitation and bactericidal tests . . . 112
3. Heterogeneity among antigens of the M complex 113
   (a) M-associated protein or non-type-specific protein 113
   (b) The serum-opacity factor 114
   (c) Characteristics of the M antigen complex 117
4. M types associated with particular streptococcal diseases 118
   (a) 'Skin' and 'throat' streptococci 118
   (b) Acute glomerulonephritis (AGN) 120
   (c) Rheumatic fever 120
5. The pathogenesis of post-streptococcal glomerulonephritis and rheumatic fever. . 120
   (a) Nephritis 120
   (b) Rheumatic fever 121
6. References 123

1. Introduction

THE OUTSTANDING FEATURES of the group A streptococcus (*Streptococcus pyogenes*) as a pathogen are its ready communicability and its ability to cause epidemic disease; its almost complete adaptation to a single host—man—and the great variety of the diseases it causes.

(a) *Communicability*

As we have seen (Parker, this volume), the group A streptococcus is the only streptococcus that regularly causes epidemic disease in man. Other streptococci, e.g. pneumococci, are to some extent communicable, though they do not spread as rapidly and extensively as group A streptococci, and most infected persons become carriers. Group A streptococci are capable of causing clinical infection of the throat or skin in a large proportion of those who are infected.

(b) *Host range*

Restriction to the human host is almost total. A few instances have been recorded of sporadic infection of other animals in close contact with man, such as the occurrence of mastitis in single cows milked by a human carrier (see p. 146) and the appearance of respiratory infection in laboratory primates (Boisvert 1940). There are a few instances of naturally occurring communicable group A streptococcal disease in animals, but when carefully examined the streptococci responsible were found to differ in some way from strains normally responsible for disease in man. The most interesting examples of this are two outbreaks of cervical adenitis and pneumonia in rodents in the USA, one in the stock of a mouse farm and the other among wild mice and voles in Oregon, where there had been a recent great increase in the rodent population (Nelson 1954; Bell *et al.* 1958). The group A streptococci isolated in both outbreaks were in the first instance untypable, but the strains from the mouse farm were later shown to belong to a new M type (type 50), which has not subsequently been reported from any animal or human source.

Gourlay (1960) isolated group A streptococci from vervet monkeys suffering from a septicaemic disease in East Africa, and suggested that these organisms were endemic in local wild monkeys. These, and other monkey isolates that we have since examined, were untypable with antisera against all known M antigens (including M50).

Very occasionally, the Lancefield antigen A is present in β-haemolytic streptococci isolated from pigs and meat products, but the few strains that we have examined differed from *S. pyogenes* (and *S. milleri*) in a number of biochemical characters.

Several other β-haemolytic streptococci are restricted to a single animal host, e.g. *S. equi* to horses and *S. lentus* and *S. suis* to the pig (but see Parker, p. 92). As with the group A streptococcus, little is known of the bacterial factors responsible for host specificity. The biotypes of group C streptococci with different host ranges show small differences in biochemical characters. Many years ago it was observed that there were differences in the range of fibrinolytic activity of some of these biotypes for plasmas of different animal species (Lancefield 1941) and that 'human' strains of the *equisimilis* biotype differed from 'animal' strains of the same biotype in producing a streptokinase active on human plasma (Evans 1944). The possibility that streptokinase may be of significance in natural infection has received little recent attention (but see Krasner & Jannach 1963).

(c) *Diseases caused*

The group A streptococci are the most versatile of all the streptococci in their pathogenic activities. They cause various forms of local septic infection, and are the only or almost the only streptococcal cause of two of these, tonsillitis and impetigo. In addition they are the unique causes of scarlet fever, erysipelas, rheumatic fever and post-streptococcal glomerulonephritis, but these diseases occur in only some of the patients who suffer the initial streptococcal infection. The question arises whether this clinical diversity depends entirely on host factors, such as the route of infection, the state of immunity, the genetic constitution or the presence of a specific predisposition; or whether strains of group A streptococci differ qualitatively in their pathogenicity.

The group A streptococci form a very large number of products that exhibit toxic or aggressive action in experimental animals or give rise in them to histological changes that resemble those seen in the sequelae of group A streptococcal disease in man. The action of most of these substances has formed the basis of one or other of the innumerable hypotheses that have been devised to explain the pathogenesis of the various group A streptococcal diseases. Most of these agents are produced by most group A strains, and there is little evidence that the production of any one of them is related to the production of a particular form of disease, except that the production of erythrogenic toxin is responsible, directly or indirectly, for the rash of scarlet fever and that the presence of an M antigen is essential for invasion of the body by group A streptococci.

Nevertheless, there is a great deal of epidemiological evidence, which will be reviewed later, that particular strains do have specific pathogenic properties. Rather surprisingly, these appear in a number of instances to be related to which M antigen the strain possesses. It is not my purpose to give a comprehensive review of the pathogenesis of group A streptococcal disease, but to attempt to relate some aspects of it to recently discovered complexities of the surface-protein antigens, notably the M antigens.

2. The M Proteins

The M proteins (Lancefield 1928) are heat-stable, trypsin-sensitive protein antigens that form part of the cell surface of group A streptococci. The elegant electron micrographic studies of Swanson *et al.* (1969) have shown that they are present in a fuzz of hair-like projections around the periphery of the cell, and that these structures are not present in M negative variants. The M antigens possess type specificity in that, with very rare exceptions, a streptococcus has a single M antigen that is serologically distinct from all other M antigens. They are to be distinguished from the T antigens and the R antigens, which are also surface proteins but appear not to be of any significance in relation to pathogenicity. The distribution of M, T and R antigens among streptococci follow certain predictable patterns, so that all of them are of use in typing group A streptococci, but the T and R antigens are not restricted to this streptococcal group, and only the M antigens divide the group into a series of distinct 'packets' of strains. Over 70 different M antigens have been recognized, but this is very far from the total that exist. Even with a comprehensive set of M-typing sera, a considerable proportion of strains remain untypable, and this proportion varies widely according to the geographical source of the strains and the type of disease from which they were isolated.

Group A streptococci may lose their ability to form M antigen *in vitro*, and this loss is sometimes accompanied by colonial changes (from 'matt' to 'glossy'). Until recently it was widely believed that untypable isolates from patients were M negative variants, and that strains tended to lose their M antigens during the later stages of convalescence from streptococcal disease (e.g. Rothbard & Watson 1948). This view is now being abandoned, on the grounds that nearly all isolates from patients—at whatever stage of disease—have the ability to multiply in fresh human blood (Section 2.b) and that when determined attempts are made, nearly all isolates can be allocated to an M type (Maxted *et al.* 1973) even though it may be necessary to make additional type antisera by immunization of animals with some of the current isolates. Thus, M untypability is usually a reflection of the incompleteness of existing sets of type antisera.

The presence of the M antigen may be important in pathogenesis in a variety of ways, but there are two situations in which it is obviously a major advantage to the organism: for firm attachment to epithelial cells at the point of entry to the body when this is a mucous membrane and as a survival mechanism after entering the tissues, because it enables the streptococcus to resist phagocytosis.

(a) *Adhesion of group A streptococci to epithelial cells*

Ellen & Gibbons (1972) showed that M positive strains became attached to human buccal epithelial cells, but that M negative variants of them did not. In the presence of homologous M antiserum, the M positive streptococci did not adhere to the cells, presumably because their site of attachment was blocked by antibody. If the streptococci were first treated with trypsin, thus destroying the M protein and removing the hair-like layer from the streptococcal surface, adhesion did not occur.

Beachey & Ofek (1976) have recently produced evidence that streptococcal lipoteichoic acid (LTA) acts in a similar way. This antigen is present in the cell wall of all group A streptococci, and—like the M antigen—can be detected in the hair-like outer layer (Beachey 1975). Gentle treatment of the streptococci with pepsin removes the M protein but leaves intact the hair-like layer and the LTA (Beachey & Ofek 1976). Pepsin-treated cells adhere to the epithelial cells, and this adhesion is inhibited by the previous addition of anti-LTA serum to the streptococci. Beachey (1975) suggested that epithelial cells also have an attachment site for LTA; if epithelial cells are first treated with purified LTA, unpepsinized M positive streptococci did not adhere to the cells, but no other extracted streptococcal antigen acted in this way.

Adhesion to epithelial cells may help to explain the ready communicability of group A streptococci by the respiratory route, but it is difficult at present to form a coherent picture of its mechanism *in vivo*. Its dependence on the M type of the infecting streptococcus has not been established, and thus does not explain the association of certain M types with epidemics of tonsillitis. Inhibition of adhesion by type-specific antibody, if present as secretary immunoglobulin, might contribute to immunity to natural infection by the respiratory tract, but this possibility has not yet been investigated. It is not likely to play any part in eliminating the streptococci from the throat during convalescence, because they often persist long after the normal time for the appearance of type-specific antibody. Whether antibody to LTA leads to non-type-specific resistance to streptococcal colonization is not known.

(b) *Group A streptococci in the phagocytic system*

All M positive group A streptococci resist phagocytosis in fresh heparinized human blood that does not contain M antibody of homologous type, but M negative variants of them are phagocytosed and subsequently destroyed. Blood collected during convalescence from acute group A streptococcal infections may kill the infecting strain (Hare 1928, 1932); this is due to the presence of M antibody, which usually appears at around the fourth week after infection and may persist for many years (Lancefield 1959). Antibody with a similar action can be produced by the injection of killed streptococci into laboratory animals; it also appears slowly, and with many M types rather irregularly. The *in vitro* bactericidal test has been used extensively in two forms: the direct test in which the whole blood of the patient is used (Todd 1927), and the indirect test, in which the blood of a donor, previously shown to be free of homologous antibody, is used with the addition of a small volume of human or animal antiserum (Rothbard & Watson 1948). The test streptococcus is added—usually in quite small numbers—and the mixture is rotated at 37°C for about 4 h. Colony counts are then made by the pour-plate technique. These events may also be followed microscopically by performing opsonic counts on blood smears, but the bactericidal test is a far more satisfactory method, because it shows whether phagocytosis results finally in the destruction of the streptococci.

Suspensions of mouse peritoneal exudate to which human serum was added have been used in an opsonic system by Bergner-Rabinowitz *et al.* (1971) to demonstrate the presence of type-specific antibody in the serum of patients with acute glomerulonephritis. The results were confirmed by the bactericidal test.

The bactericidal test has considerable applications to the study of other streptococci that may be pathogenic for man, and has revealed M-like proteins among members of Lancefield's groups C and G. These organisms occasionally cause severe infections, but are much less often isolated than group A streptococci in the temperate zones. In some African countries, however, they are very common in the throat and in skin lesions, and Ogunbi (1971) believes that they may be responsible for skin sores. Strains isolated from these lesions multiply in normal human blood, and antibody against M-like protein antigens inhibits this (Lawal 1976). Maxted & Potter (1967) had observed that several group G streptococci isolated in Trinidad behaved similarly and, rather more surprisingly, that these strains possessed an M antigen indistinguishable from that of M type 12 of the group A streptococcus.

Only human blood and the blood of some monkeys have proved useful in the *in vitro* bactericidal test (Maxted 1956). This is interesting because the monkey is perhaps the only other animal in which natural infection occurs other than by direct implantation into tissues. Group A streptococci do not grow in guineapig blood; they multiply freely in rabbit blood, but the presence of M antibody does not enhance the bactericidal action of the blood.

The bactericidal test has been used in a variety of ways to study the M antigenic constitution of streptococci and the antibody response in man. Survival in normal human blood provides reliable evidence for the presence of an M antigen even when an antiserum against it has not been successfully prepared. Inhibition of the growth by antiserum is still the most reliable evidence that the serum contains M antibody, and also provides confirmation that a precipitation reaction between a streptococcus and an M antiserum of known type is due to the presence of the corresponding antigen. Repeated passage of strains through normal blood results in the selection of strongly M positive variants (Becker 1964). Inhibition of multiplication in blood is still the most sensitive and specific method of detecting M antibody in the blood of convalescents, and is useful in making a retrospective diagnosis of the infecting strain when this is in doubt.

(c) *The M antigen-antibody system in precipitation and bactericidal tests*

In routine M typing, a precipitation test is done with hot-acid-extracted M antigen and highly absorbed rabbit antiserum (Swift *et al.* 1943). Unabsorbed sera may contain precipitating antibody against other protein antigens (Wilson & Wiley 1963), but the use of the traditional test in capillary tubes was justified by the fact that all precipitation reactions not corresponding to a demonstrated bactericidal effect had been eliminated by absorption. Nowadays, less highly absorbed sera can be used in gel-diffusion tests if a control antigen is included (Rotta *et al.* 1971).

For many years it was thought that the same antibody was reponsible for precipitation and for the bactericidal effect, but we now know that this is not so, even though their specificities are identical. Brighton (1969) showed that precipitating antibody does not always cause enhancement of bactericidal action or give passive protection to mice, and that the reverse may sometimes be true. More recently, Beachey & Cunningham (1973), Russell & Facklam (1975) and Fischetti et al. (1976) each using a different method of extraction and separation of M protein have shown without doubt that the precipitating and anti-phagocytic antigens were quite separate. These important observations may have considerable influence in the future on the methods used for the preparation of M antisera, and on the choice of purified antigens to be included in vaccines for the immunization of man.

3. Heterogeneity Among Antigens of the M complex

We no longer believe that all M antigens are of the same nature except for their antigenic determinants. They can be separated into two broad categories (Section 3.c), that are associated with streptococci that tend to have rather different spectra of disease production (Section 4). The differences between the two categories are related to differences in the nature or distribution of certain other components that form part of the M protein complex (Table 1), that is to say, antigens that are present only in association with M protein, that are absent from the extracts of M negative variants, and that tend to be co-purified with the M antigen.

(a) *M-associated protein or non-type-specific protein*

Several workers described an antigenic protein constituent devoid of type specificity in preparations of M antigen. This was described as non-type-specific protein by Beachey & Stollerman (1971) and Vosti et al. (1971) and as M-associated protein (MAP) by Widdowson et al. (1971). It is present in certain extracts of all M positive strains of group A streptococci, and is absent from

Table 1
The M protein complex

Antigen	Specificity	Antibody

M antigen
  Precipitating
  Anti-phagocytic
	Complete } Parallel
	Complete
		Not protective
		Protective

Opacity factor (in certain M types)	Complete, parallel to M	Not protective
M-associated protein (in all M types)	Non-specific, but 2 main antigenic types	Not known to be protective

those of all M negative variants; it is also found in streptococci of groups C and G that have M-like antigens, but in no other streptococci.

A convenient method of detecting MAP is by performing a complement fixation test with an acid extract or purified M protein of the streptococcus and the serum of a person who has antibody against MAP (Widdowson et al. 1971) or an antiserum prepared in guineapigs (Widdowson et al. 1976). Antibody against MAP can be detected similarly by a test in which the antigen is partially purified acid-extracted M protein, preferably of a rare M type (e.g. M30) in order to avoid confusion with type specific reactions (Widdowson et al. 1971).

As we shall see, the detection of high titres of anti-MAP is of some value in the diagnosis of rheumatic fever. Earlier studies of MAP indicated that it was not type specific in that acid extracts and purified M proteins of all M types tested fixed complement in the presence of selected human sera. Later studies (J. P. Widdowson et al. 1976 and pers. comm.) with human and guineapig anti-MAP sera revealed that there are at least two main antigenic varieties of MAP, and that the antigenic form of MAP possessed by a particular strain was dependent on its M type and the presence of serum opacity factor. For the purpose of this publication these two forms of M-associated protein have been arbitrarily designated I and II (see Table 3).

(b) *The serum-opacity factor*

It has been known for some 40 years that some group A streptococci produce an opacity when grown in serum broth (Ward & Rudd 1938), and that this ability was in some way associated with the typing pattern of the strain (Gooder 1961) though the M-typing system was until recently not sufficiently developed for this relationship to be understood. It is now clear that the opacity factor (OF) is produced by all members of some M types and by none of other M

Table 2

The distribution of opacity factor among the M types of group A streptococci

Opacity factor	M type
Positive	2, 4, 9, 11, 13, 22, 25, 28, 48, 49, 58, 59, 60, 61, 62, 63, 66, 68 Provisional types: PS346, 3354, 3875, 943, 2273, 3085, 2015, 1658, 2841, 180
Negative	1, 3, 5, 6, 12, 14, 15, 17, 18, 19, 23, 24, 26, 29, 30, 31, 33, 36, 37, 39, 41, 43, 46, 47, 51, 52, 53, 54, 55, 56, 57, 64, 65 Provisional types: 3961, 1720, 2681

The provisional types are awaiting allocation of a type number.

Table 3
The M proteins of group A streptococci

Category	Molecular weight distribution of protein in crude acid extracts*	Purified M protein	Opacity factor	Antigenic category† of MAP	Antibody response to M and MAP antigens	Representative M types
(1)	Wide, no peak at void volume	Stable	Absent	I	Good	1, 5, 6, 12, 19, 30
(2)	Wide, pronounced peak at void volume	Unstable	Present	II	Poor	4, 22, 25, 49, 60, 63

* Deduced from behaviour in Sephadex G-100 columns.
† See text (Section 3.c).

types; about one-half of all known M types form it (Table 2; see Maxted *et al.* 1973). OF is a trypsin-sensitive protein that can be found both in acid extracts of streptococcal cells and in supernatant fractions from broth cultures when these are derived from M positive cultures, but is not formed extracellularly by M negative variants (Widdowson *et al.* 1970). The opacity reaction is believed to result from enzymic attack on serum α-lipoprotein, which makes lecithin more accessible to the serum enzymes responsible for the transfer of fatty acids from lecithin to cholesterol.

The most remarkable feature of OF is that it is antigenic and that its antigenic specificity—as revealed by the inhibition of opacity formation—runs exactly parallel with that of the M antigens of the strains that produce it (Widdowson *et al.* 1970). This forms the basis of an excellent supplementary typing system, because many OF positive strains give rise more easily to the production of OF antibody than M antibody both in laboratory animals and in man (Maxted *et al.* 1973).

Although the production of OF is a constant feature of all M positive strains in a particular serotype, OF positive variants have on several occasions been selected from pure cultures of an OF negative streptococcus, and have been found to have changed their M type from one that is normally OF negative to one that is normally OF positive (Maxted & Valkenburg 1969). Thus M type 62 variants (OF+) were selected from M type 12 (OF−) cultures isolated in a village in Holland in which infections with the latter type gave place to infections with the former over a period of months. This suggested that the OF+ variant had '"emerged' under selective pressure from anti-M12 antibody in the local population, but there was no opportunity to investigate this possibility. In another district, in which infections with M type 12 (OF−) and M type 22 (OF+) co-existed, OF+ variants selected from the M type 12 strain proved to belong to M type 22. In neither instance was there any change in the T-typing pattern of the streptococcus. 'Mixed' outbreaks of infection with M type 12 strains and M type 22 strains with the T antigen 12 have been seen frequently in Britain in the last few years.

M protein and OF are co-purified by the protein purification procedure usually applied to M antigen but recently Hallas and Widdowson (unpublished) in this Laboratory have succeeded in partially separating OF from M protein. OF is found in large amounts in the supernatant fractions of cultures, and might have been looked upon as an extracellular antigen had it not been possible to extract it by various methods from the cell wall and cell membrane. A consistent character of OF positive serotypes is the poor M antibody response to them both in laboratory animals given killed vaccines (Gooder 1961) and in natural infections in man (Widdowson *et al.* 1974*b*).

Because of the close association of OF and M antigen it seemed likely that the M protein might also be present in cell supernates; this has proved to be so and culture supernatant fractions of OF+ strains have proved to be a good source of 'natural' M antigen (Pinney & Widdowson 1977).

(c) *Characteristics of the M antigen complex*

It is now possible to recognize two main classes of M antigen. The characteristics that allow this are consistent features of all members of a particular M type, and are as follows: (1) the molecular size distribution as determined by column chromatography on Sephadex G-100, (2) the stability of the purified acid extracted antigen, (3) the production of opacity factor, (4) the type-specific antibody response observed in laboratory animals and (5) the antigenic properties of the MAP antigen (Table 3).

(i) *Category* (1)

The acid-extracted M proteins of this category are eluted from Sephadex G-100 columns over a wide range of fractions, suggesting a high degree of molecular-weight heterogeneity (mol. wt range 5000–$\geqslant$150,000). M proteins of this nature are usually derived from opacity factor negative strains of throat origin (see Section 4) and are stable on purification and storage. The immunological response to these M proteins is good. M proteins of this category are consistently associated with an antigenic type of MAP not found in Category (2) strains.

(ii) *Category* (2)

Acid extracts from all OF positive strains exhibit a pronounced peak of protein material at the void volume of Sephadex G-100 columns (mol. wt >150,000), with which all of the OF and the strongest M precipitin reactions are associated. Later fractions, containing small molecular weight material (5000–150,000) give positive M reactions but are devoid of opacity factor. Purified M proteins of OF positive strains and some OF negative strains commonly associated with skin infection (see Section 4) are rather unstable and difficult to purify by the methods applied to Category (1) proteins (Widdowson, unpublished). The antigenicity of Catagory (2) M protein is poor in animals and man. The MAP antigens associated with this class of M proteins are in general antigenically different from those in Category (1).

Not unexpectedly, this subdivision of M types is not entirely clear-cut, and a few types (e.g. types 41, 49 and 55) appear to have intermediate characters, in that they are largely OF negative but otherwise have the characters of Category (2) strains.

Thus, the subdivision of strains according to whether or not they produce OF gives some indication of their ability to produce antibody to surface-protein antigens as a result of natural infection in man. This conclusion was borne out by the M antibody responses seen in a number of outbreaks of streptococcal infection due to streptococci of various types (Widdowson et al. 1974b), and is true not only for M antibody but for anti-MAP. In the initial studies, on anti-MAP antibody in patients who had suffered from various streptococcal diseases, it was observed that the titre of anti-MAP antibody was almost invariably high in rheumatic fever, but not often much raised in glomerulonephritis (Widdowson et al. 1971, 1974a; Beachey et al. 1973). However, the formation of this antibody after uncomplicated streptococcal infections was variable. It was possible to study the antibody response in a number of outbreaks of infection associated with known M types, in some of which a few of the patients subsequently developed rheumatic fever (Widdowson et al. 1974b). It then became apparent that the antibody response to both M protein and to MAP tended to be higher also in patients who suffered uncomplicated infections in association with rheumatic fever cases than in patients in outbreaks in which rheumatic fever did not occur. This suggested that only some M types might be 'rheumatogenic'.

4. M types Associated with Particular Streptococcal Diseases

Epidemiological evidence suggests that several streptococcal diseases tend to be associated with certain M types or sets of M types of group A streptococci. On pp. 82–83 the temporary and geographically limited association of scarlet fever with certain types was mentioned. This will not be discussed further, because no evidence about the type distribution of the ability to produce erythrogenic toxin is available. We must, however, discuss the evidence that (1) strains that produce impetiginous lesions differ in M type from those that cause tonsillitis, (2) the ability to cause nephritis is limited to members of a few M types and (3) that there may be rheumatogenic M types.

(a) 'Skin' and 'throat' streptococci

When serious attempts were first made, in the 1950s, to study the epidemiology of streptococcal impetigo, it was found that very few of the group A streptococci isolated from the skin lesions could be allocated to an M type with existing antisera. When outbreaks in small communities were studied intensively, however, and antisera were prepared against current isolates, a large number of new M types were discovered. The previous inability to recognize these strains can be attributed partly to the fact that the M-typing system had been developed primarily for the study of respiratory-tract infections, and partly to the very poor antibody response given by many of the skin strains in rabbits and the consequent difficulty in obtaining typing sera of adequate potency.

It was soon quite clear, however, that the M types commonly associated with outbreaks of tonsillitis in temperate climates were almost never the cause of impetigo in these countries, or, indeed, in tropical and subtropical countries where impetigo was more common, though they might be responsible for wound infection and other forms of non-crusting septic lesions. These M types were commonly 'low-number' types that had been recognized for many years, and most but not all of them were OF negative.

On the other hand, the M types commonly associated with impetigo might upon occasion colonize the throat, but seldom were responsible for clinically obvious tonsillitis and when they were they seldom gave rise to epidemics of infection. Most but by no means all of these strains were OF positive. However, despite the fact that 15-20 M types associated with skin infection have been recognized in recent years, few of these are really common, and some appear to have a limited geographical distribution. Thus, a considerable proportion of skin isolates still cannot be M typed, particularly if they are isolated in countries in which intensive studies have not been made.

It must not be thought, however, that the streptococci can be allocated to two distinct classes with exclusive abilities to cause throat and skin disease. There appears rather to be a gradation of character from a preponderant association with respiratory-tract disease (e.g. in M types 1, 3, 5, 6 and 12) to a preponderant association with skin disease (e.g. in M types 31, 49, 55, 56, 57, 60 and 63) with certain strains occupying an intermediate position. Nor can the skin/throat antithesis be equated with the distinction between OF positive and OF negative serotypes, though there is a strong tendency in this direction. In

Table 4
Disease associations of different M types of group A streptococci

Disease	Site of infection	M type	Category of M protein
Pharyngitis	Throat	'throat' types	mainly (1)
Impetigo	Skin	'skin' types	mainly (2)
Acute glomerulonephritis	Throat or Skin	mainly 12 49, 55, 57, 69, 63*	(1) (2)
Rheumatic fever	Throat	1, 3, 5, 14, 17, 18† 19, 30	(1)

Category (1), OF negative, strongly antigenic M protein usually with strongly antigenic M-associated protein (MAP); *Category* (2), mainly OF positive, poorly antigenic M protein and MAP.
* Repeatedly identified from outbreaks associated with AGN.
† Good evidence of association with at least one rheumatic fever 'incident'.

the comparatively few epidemic incidents in which it has been possible to compare the antibody response to cellular-protein antigens, a poor response has tended to be related to the M protein category of the infecting strain rather than to whether the infection was of the skin rather than of the throat.

(b) *Acute glomerulonephritis (AGN)*

Acute glomerulonephritis (AGN) may follow streptococcal tonsillitis or impetigo. In economically developed countries with temperate climates it is associated almost exclusively with throat infection with streptococci of M type 12, (an OF negative type with a Category 1 protein antigen), which very rarely causes skin lesions. In tropical countries, and occasionally in non-tropical countries with poor living standards, it is mainly associated with one or other of a small range of other types, notably M types 49, 55, 57, 60 and 63, all 'skin' types with Category (2) protein antigens; but a common nephritogenic factor, present in these types and absent from others, has not been identified.

(c) *Rheumatic fever*

It was relatively easy to identify the nephritogenic M types because acute glomerulonephritis often occurs in epidemics even in the general population. Although the causative streptococcus can be isolated only from a small proportion of cases when they are admitted to hospital with nephritis, it is usually easy to secure sufficient cultures to establish for certain the M type of the causative strain. In the general population, rheumatic fever usually occurs sporadically, and members of various M types are associated with individual cases. This gave rise to the opinion that 'any streptococcus' could cause rheumatic fever. There are, however, a few well-authenticated reports of rheumatic fever in large 'closed' communities, mainly of armed-service personnel, such as the Warren Air Force Base (Rammelkamp *et al.* 1952), the the Great Lakes Naval Training Station (Stollerman 1975), the Lowry Base (James & McFarland 1971) and the Royal Air Force Base, Halton (Widdowson *et al.* 1974*a*). The M types isolated from these incidents belong to M types 1, 3, 5, 6, 12, 14, 17, 18 and 19, all of which have Category (1) proteins and are OF negative 'throat' types.

5. The Pathogenesis of Post-streptococcal Glomerulonephritis and Rheumatic Fever

(a) *Nephritis*

The fact that second attacks of post-streptococcal glomerulonephritis are rare, even in countries in which several different nephritogenic M types are circulating, made it reasonable to suppose that a nephritogenic toxin is responsible for the disease, and that antibody against it prevents subsequent attacks. Reed & Mathison (1959) have described such a toxic factor, a polypeptide present in supernates of broth cultures of nephritogenic strains of streptococci, that caused kidney lesions in rabbits but this observation has not been confirmed.

There is considerable evidence that the disease is initiated by the deposition of immune complexes in the kidneys. These have been demonstrated electron microscopically at sites corresponding to those of the early nephritic lesion (Andres et al. 1966); according to Treser et al. (1970) the streptococcal component in these complexes is a glycoprotein of the plasma membrane. These complexes also contain the C3 component of complement, and another feature of post-streptococcal glomerulonephritis is that the blood level of this substance is markedly reduced in the acute stage of the disease. Some of the disagreement about the sequence of events in nephritis may be attributed to the timing of observations. Lange et al. (1976) have demonstrated an antigen in the supernate after rupturing the streptococci, that reacts with the glomerular basement membrane of kidneys of nephritic patients, but only when the kidney biopsy material is taken in the first few days after admission to hospital. This antigen is present in all group A streptococci.

The one undoubted fact is that only members of certain M types are associated with the disease, but there is no evidence to implicate the M antigen directly in its causation. Nor is there any suggestion that the serum-antibody response to nephritogenic strains is in any way abnormal, or that the qualitative difference between M antigens that I have described bears any relation to potential nephritogenicity. In some outbreaks of nephritis, attack rates have been of the order of 10–20% of persons infected with the nephritogenic strain (see, for example, Anthony et al. 1969). In Britain, some 10% of all streptococcal infections are at present caused by members of M type 12, but nephritis is very uncommon, and outbreaks occur with exceeding rarity. It may therefore be wrong to assume that all members of nephritogenic types are nephritogenic and to expect to find common features in their M antigens.

(b) *Rheumatic fever*

Among the facts that have to be explained are (1) that most attacks of rheumatic fever follow streptococcal throat infection and few follow skin infection, (2) the interval of time between the initial streptococcal attack and the onset of rheumatic fever is somewhat longer than the corresponding period for nephritis, (3) that only early and total elimination of living streptococci from the body prevents rheumatic fever, (4) that a streptococcal infection in a person who has had a previous attack of rheumatic fever is much more likely to cause disease than is streptococcal infection in a person who has not previously suffered a rheumatic attack and (5) successive attacks in one patient, and attacks in members of the same family, tend to resemble each other clinically. It cannot be said that any of the current views of the pathogenesis of rheumatic fever account satisfactorily for all the facts. Broadly, they are that it is a cellular reaction to streptococcal cell-wall fragments that have persisted in the tissues and been translocated to distant parts of the body in inflammatory cells; a hypersensitivity phenomenon or an autoimmune reaction; but there is probably also an element of genetic predisposition in the person who develops rheumatic fever.

The view that rheumatic fever is attributable to an autoimmune reaction between streptococcal antibody and a component of host tissue is now widely held. The reaction that has attracted most attention is that between streptococcal antibody and heart muscle, and to a lesser extent with skeletal muscle (see Kaplan 1963). This has the merit of apparently linking streptococcal infection with damage to the myocardium, but this, though of great significance to the patient, is by no means the sole lesion of rheumatic fever; a constant feature is the presence of characteristic nodules in connective tissue. There are, moreover, at least five other examples of cross-reaction between streptococcal antigens and other tissue elements, and there is no reason to believe that 'molecular mimicry' is more common in streptococci than in other microorganisms.

The streptococcal product apparently responsible for the production of 'heart-reactive' antibody is a component of the streptococcal membrane. Zabriskie et al. (1970) found that it was present in large amounts in the serum of rheumatic fever patients, but that it could be detected in lesser amount in patients with nephritis and even with uncomplicated streptococcal infections. Thus the distribution of the antibody was rather similar to that of antibody to MAP (Section 3.a).

It appears that the serum-antibody response in rheumatic fever differs from that observed in many uncomplicated streptococcal infections, and in glomerulonephritis, in that there is almost invariably an active production of antibody against various cellular antigens, notably M protein, MAP and 'heart-reactive' factor. However, a similar response may occur in uncomplicated streptococcal infections caused by strains that give rise to rheumatic fever in other patients. The ability to give rise to such a response is a characteristic of strains that form Category (1) M proteins (Section 3.c).

Recent observations on the heterogeneity of antigens of the M complex raised hopes that the M antigen might be something more than a mere marker for unknown factors that determine the ability to cause a particular streptococcal disease. These hopes have not yet been fulfilled in relation to the types responsible for glomerulonephritis, impetigo, tonsillitis or scarlet fever. On the other hand, it is reasonable to believe that the M-antigen-related nature of the antibody response to the infecting strain may be of a more direct importance in the pathogenesis of rheumatic fever, though other factors, such as genetic constitution and previous experience of streptococcal disease (Parker, this volume), may determine which of the patients infected with a rheumatogenic strain develop the disease.

6. References

ANDRES, G. A., ACCINI, L., HSU, K. C., ZABRISKIE, J. B. & SEEGAL, B. S. 1966 Electron microscopic studies of human glomerulonephritis with ferritin conjugated antibody. Localisation of antigen-antibody complexes in glomerular structures of patients with acute glomerulonephritis. *Journal of Experimental Medicine* **123**, 399–412.

ANTHONY, B. F., KAPLAN, E. L., WANNAMAKER, L. W., BRIESE, F. W. & CHAPMAN, S. S. 1969 Attack rates of acute nephritis after type 49 streptococcal infection of the skin and of the respiratory tract. *Journal of Clinical Investigation* **48**, 1697–1704.

BEACHEY, E. H. 1975 Binding of group A streptococci to human oral mucosal cells by lipoteichoic acid. *Transactions of the Association of American Physicians, Philadelphia* **88**, 285–292

BEACHEY, E. H. & CUNNINGHAM, M. 1973 Type specific inhibition of pre-opsonisation versus immuno-precipitation by streptococcal M protein. *Infection and Immunity* **8**, 19–24.

BEACHEY, E. H. & OFEK, I. 1976 Epithelial cell binding of group A streptococci of lipoteichoic acid on fimbriae denuded of M protein. *Journal of Experimental Medicine* **143**, 759–771.

BEACHEY, E. H. & STOLLERMAN, G. H. 1971 Toxic effects of streptococcal M protein on platelets and polymorphonuclear leukocytes in human blood. *Journal of Experimental Medicine* **134**, 351–365.

BEACHEY, E. H., OFEK, I. & BISNO, A. L. 1973 Studies of antibodies to non-type-specific antigens associated with streptococcal M protein in the sera of patients with rheumatic fever. *Journal of Immunology* **111**, 1361–1366.

BECKER, C. G. 1964 Selection of group A streptococci rich in M protein from populations poor in M-protein. *American Journal of Pathology* **44**, 51

BELL, J. F., OWEN, C. R. & JELLISON, W. L. 1958 Group A streptococcal infections in wild rodents. *Journal of Infections Diseases* **103**, 196–203.

BERGNER-RABINOWITZ, S., OFEK, I., DAVIES, M. A. & RABINOWITZ, K. 1971 Type-specific streptococcal antibodies in pyodermal nephritis. *Journal of Infectious Diseases* **124**, 488–493.

BOISVERT, P. L. 1940 Human scarletinal streptococci in monkeys. *Journal of Bacteriology* **39**, 727–738.

BRIGHTON, W. D. 1968 Methods of determination of immunity against group A streptococci. In *Current Research on Group A Streptococci*, ed. Caravano, R. Excerpta Medica, Amsterdam: Elsevier.

ELLEN, R. P. & GIBBONS, R. J. 1972 M protein-associated adherence of streptococcus pyogenes to epithelial surfaces: pre-requisite for virulence. *Infection and Immunity* **5**, 826–830.

EVANS, A. C. 1944 Studies on haemolytic streptococci. VII. Distinguishing characters of lactose negative species of Lancefield's groups A & C. *Journal of Bacteriology* **48**, 263–267.

FISCHETTI, V. A., GOTSCHLICH, E. G., SIRIGLIA, G. & ZABRISKIE, J. B. 1976 Streptococcal M-protein extracted by nonionic detergent. *Journal of Experimental Medicine* **144**, 32–51.
GOODER, H. 1961 Association of a serum opacity reaction with serological type in *Streptococcus pyogenes*. *Journal of General Microbiology* **25**, 347–352.
GOURLAY, R. N. 1960 Septicaemia in vervet monkeys caused by *Streptococcus pyogenes*. *Journal of Comparative Pathology and Therapeutics* **70**, 339–345.
HARE, R. 1928 On the phagocytosis of haemolytic streptococci of high and low virulence by the blood of patients infected with that organism. *British Journal of Experimental Pathology* **9**, 337–352.
HARE, R. 1932 The production of bacteriotropins for haemolytic streptococci by patients infected with that organism. *Journal of Pathology and Bacteriology* **35**, 701–726.
JAMES, L. & MacFARLAND, R. B. 1971 An epidemic of pharyngitis due to a non-haemolytic group A streptococcus at Lowry Air Force Base. *New England Journal of Medicine* **284**, 750–752.
KAPLAN, M. H. 1963 Immunologic relation of streptococcal and tissue antigens. I. Properties of an antigen in certain strains of group A streptococci exhibiting an immunologic cross reaction with human heart tissue. *Journal of Immunology* **90**, 595–
KRASNER, R. I. & JANNACH, R. J. 1963 The streptokinase plasminogin system. II. Its effect on the development of local streptococcal infections in rabbit skins. *Journal of Infectious Diseases* **112**, p. 134–142.
LANCEFIELD, R. C. 1928 The antigenic complex of *Streptococcus haemolyticus*. I. Demonstration of a type-specific substance in extracts of *Streptococcus haemolyticus*. *Journal of Experimental Medicine* **47**, 91–103.
LANCEFIELD, R. C. 1941 *Harvey Lecture Series* **36**, 251–
LANCEFIELD, R. C. 1959 Persistence of type-specific antibodies in man following infection with group A streptococci. *Journal of Experimental Medicine* **110**, 271–292.
LANGE, K., AHMED, U., KLEINEBERGER, A. & TRESER, G. 1976 A hitherto unknown streptococcal antigen and its probable relation to acute post-streptococcal glomerulonephritis. *Clinical Nephrology* **5**, 207–215.
LAWAL, S. F. 1976 A study of biological characteristics of streptococci of Lancefield C and G in comparison with those associated with virulent streptococcus group A (Nigerian and British strains). Thesis presented for a Fellowship of the Institute of Science Technology).
MAXTED, W. R. 1956 The indirect bactericidal test as a means of identifying antibody to the M antigen of *Streptococcus pyogenes*. *British Journal of Experimental Pathology* **37**, 415–422.
MAXTED, W. R. & POTTER, E. V. 1967 The presence of type M12 M protein in group G streptococci. *Journal of General Microbiology* **49**, 119–125.
MAXTED, W. R. & VALKENBURG, H. A. 1969 Variation in the M-antigen of group A streptococci. *Journal of Medical Microbiology* **2**, 199–210.
MAXTED, W. R., WIDDOWSON, J. P., FRASER, C. A. M., BALL, L. C. & BASSETT, D. C. J. 1973 The use of the serum opacity reaction in the typing of group A streptococci. *Journal of Medical Microbiology* **6**, 83–90.
NELSON, J. B. 1954 Association of group A streptococci with an outbreak of cervical lymphadenitis in mice. *Proceedings of the Society for Experimental Biology* **86**, 542–545.
OGUNBI, O. 1971 A study of beta-haemolytic streptococci in throats, noses, and skin lesions in a Nigerian (Lagos) population. *Journal of the Nigerian Medical Association* **3**, 159–164.
PINNEY, A. M. & WIDDOWSON, J. P. 1977 Characteristics of the extracellular M proteins of group A streptococci. *Journal of Medical Microbiology*. **10**, 415–429.
RAMMELKAMP, C. H., DENNY, F. W. & WANNAMAKER, L. W. 1952 Studies on the epidemiology of the armed services. In *Rheumatic Fever*, ed. Thomas, L. University of Minnesota Press.

REED, R. W. & MATHISON, B. H. 1959 Experimental nephritis due to type specific streptococci. III Biological chemical and physical studies on type 12 nephritogenic substance. *Journal of Infectious Diseases* 104, 213-232.

ROTHBARD, S. 1945 I. Type specific antibodies in sera of patients convalescing from group A streptococcal pharyngitis. *Journal of Experimental Medicine* 82, 93-105.

ROTHBARD, S & WATSON, R. F. 1948 Variation occurring in group A streptococci during human infection. Progressive loss of M substance correlated with increasing susceptibility of bacteriostasis. *Journal of Experimental Medicine* 87, 521-533.

ROTTA, J., KRAUSE, R. M., LANCEFIELD, R. C., EVERLY, W. & LACKLAND, H. 1971 New approaches for the laboratory recognition of M-types of group A streptococci. *Journal of Experimental Medicine* 134, 1298-1315.

RUSSELL, H. & FACKLAM, R. R. 1975 Guanidine extraction of streptococcal M protein. *Infection and Immunity* 12, 679-686.

STOLLERMAN, G. H. 1975 Epidemiology of rheumatic fever. In *Rheumatic Fever and Streptococcal infection*, ed. G. H. Stollerman

SWANSON, J., HSU, K. C. & GOTSCHLICH, E. C. 1969 Electron microscopic studies on streptococci. 1. M antigen. *Journal of Experimental Medicine* 130, 1063-1075.

SWIFT, H. F., WILSON, A. T. & LANCEFIELD, R. C. 1943 Typing group A haemolytic streptococci by M-precipitation reactions in capillary pipettes. *Journal of Experimental Medicine* 78, 127-133.

TODD, E. W. 1927 A method of measuring the increase or decrease of the population of haemolytic streptococci in blood. *British Journal of Experimental Pathology* 8, 1-5.

TRESER, G., SEMAN, M., TY, A., SAGEL, I., FRANKLIN, M. A. & LANGE, K. 1970 Partial characterisation of antigenic streptococcal plasma membrane components in acute glomerulonephritis. *Journal of Clinical Investigation* 49, 762-768.

VOSTI, K. L., JOHNSON, R. H. & DILLON, M. F. 1971 Further characterisation of purified fractions of M protein from a strain of group A type 12 streptococcus. *Journal of Immunology* 107, 104-114.

WARD, H. K. & RUDD, G. 1938 Studies on haemolytic streptococci from human sources. I. The cultural characteristics of potentially virulent strains. *Australian Journal of Experimental Biology and Medical Science* 16, 181-192.

WIDDOWSON, J. P., MAXTED, W. R. & GRANT, D. L. 1970 The production of opacity in serum by group A streptococci and its relation with the presence of M antigen. *Journal of General Microbiology* 61, 343-353.

WIDDOWSON, J. P., MAXTED, W. R. & PINNEY, A. M. 1971 An M-associated-protein antigen (MAP) of group A streptococci. *Journal of Hygiene, Cambridge* 69, 553-564.

WIDDOWSON, J. P., MAXTED, W. R., NOTLEY, C. M & PINNEY, A. M. 1974a The antibody responses in man to infection with different serotypes of group A streptococci. *Journal of Medical Microbiology* 7, 483-496.

WIDDOWSON, J. P., MAXTED, W. R., NEWRICK, C. W. & PARKIN, D. 1974b An outbreak of streptococcal sore throat and rheumatic fever in a Royal Air Force Training camp; significance of serum antibody to M-associated protein. *Journal of Hygiene, Cambridge* 72, 1-12.

WIDDOWSON, J. P., MAXTED, W. R. & PINNEY, A. M. 1976 Immunological heterogeneity among the M-associated protein antigens of group A streptococci. *Journal of Medical Microbiology* 9, 73-88.

WILSON, A. T. & WILEY, G. G. 1963 The cellular antigens of groups A streptococci. Immuno-electrophoretic studies of the C, M, T, PGP, E, F, and E antigens of serotype 17 streptococci. *Journal of Experimental Medicine* 118, 527-556.

ZABRISKIE, J. B., HSU, K. C. & SEEGAL, B. C. 1970 Heart reactive antibody associated with rheumatic fever: Characterisation and diagnostic significance. *Clinical and Experimental Immunology* 7, 147-159.

Ecology of Group B Streptococci

P. W. ROSS

*Department of Bacteriology, University of Edinburgh,
University Medical School, Edinburgh, Scotland*

CONTENTS

1. Historical 127
2. The organism 128
3. Diseases associated with group B streptococci 129
   - (a) Neonatal disease 130
   - (b) Incidence of neonatal infections 131
4. Carriage in adults 132
5. Colonization and infection in humans 134
6. Distribution of serotypes among streptococcal isolates 135
7. Factors involved in human infections by group B streptococci 136
8. The future 138
9. References 139

1. Historical

BY THE END of the nineteenth century the genus *Streptococcus*, in particular *Streptococcus agalactiae*, was recognized as an important cause of bovine mastitis; in 1933 when Lancefield published her studies on the serological differentiation of human and other groups of haemolytic streptococci, this species was classified as a group B streptococcus on the basis of its cell wall carbohydrate antigens. Until the 1930s little attention was paid to the association of *S. agalactiae* with humans, either as part of the commensal flora of the human body or as an agent of human disease. During this period, however, this association was referred to by several workers. Lancefield & Hare (1935) described the serological differentiation of pathogenic and non-pathogenic strains of haemolytic streptococci from parturient women, Colebrook & Purdie (1937) the isolation of group B streptococci from the blood of a case of puerperal endocarditis, and Fry (1938) gave an account of three fatal infections caused by these organisms.

In the 1940s and 1950s there were few reported cases of infections associated with group B streptococci, compared with the vast number of those associated with group A organisms. Most group B infections occurred in adults, mainly in connection with childbirth and endocarditis. Towards the end of this period, however, group B streptococci were isolated from humans with increasing frequency and there was a growing awareness that these bacteria were often associated with human, and in particular neonatal, disease. In the past 15 years group B organisms and their infections have been studied widely, particularly in the USA, UK, Scandinavia and Czechoslovakia. (The relationship of group B streptococci to animal disease is considered by Wilson & Salt, this volume.)

2. The Organism

Although in former years the name *S. mastitidis* was occasionally used for the streptococci that caused mastitis in cattle this name is no longer in vogue; the name that is now generally accepted for group B streptococci, whether of human or bovine origin, is *S. agalactiae*.

Although the principal sources of group B organisms are humans and cattle they have been isolated occasionally from other species such as dogs and fish. Some early workers speculated that cattle transmitted the organisms to humans in some way, and several of them studied this possibility. Simmons & Keogh (1940) stated that human and bovine strains were dissimilar in that human strains were more virulent than bovine strains for mice and, unlike bovine strains, did not ferment lactose. El Ghoroury (1950) found serological and metabolic differences between the two strains and Butter & de Moor (1967) isolated human, but not bovine, strains from the throats of Dutch dairy farm workers. Additional serological proof of the separate identity of human and bovine strains was provided in the UK by Pattison *et al.* (1955*b*). They stated that whereas all their human strains could be classified by Lancefield's precipitation method only 75% of bovine strains could be so identified.

Following her classical work in 1933 on the group differentiation of haemolytic streptococci, Lancefield proceeded in 1934 to subdivide group B streptococci into types I, II and III (Lancefield 1934). Grouping of haemolytic streptococci is based on the carbohydrate antigen of the cell wall and, similarly, typing of group B is related to polysaccharide antigens; this is different from typing group A streptococci for which protein antigens are utilized (see Maxted, this volume). Lancefield (1938) discovered that type I contained more than one antigen and divided this into Ia and Ib. These four subdivisions of group B proved inadequate as many strains remained untypable. The situation was improved when Pattison *et al.* (1955*a*) introduced a further two protein antigens, R and X; this increased the proportion of group B strains that could be typed, despite the fact that various cross reactions were observed, such as R with X, and polysaccharide with either R or X. Wilkinson & Moody (1969) working in Atlanta on cross-reactions that occurred between Ia and Ib described intermediate strains (Ii) that possessed both the Ia and Ib antigens. Further work indicated that typical Ib strains contained an incomplete Ia antigen in addition to two Ib antigens, and this was responsible for the cross-reactions obtained with the Ia antiserum. The reason that Ib antiserum did not precipitate the complete Ia antigen found in Ia and Ii strains was that Ia specificity resided in the fraction of the Ia antigen not possessed by Ib strains. On the basis of these experiments Wilkinson and Moody described three dominant type I strains; Ia, Ib, and Ii (now named Ic). Most Ib strains contain two Ib antigens and a partial Ia antigen. Ic strains contain both Ib antigens at least one of which is complete, and a complete Ia antigen. Some Ib strains contain only one Ib antigen.

3. Diseases Associated with Group B Streptococci

Group B streptococci produce infection in adults, but it is their importance as agents of serious neonatal disease, particularly septicaemia and meningitis, that increasingly has given cause for alarm over the past decade.

Infections in adults have been less frequently reported than infections in neonates, and appear to have a bimodal age distribution: a young, previously healthy female population in which group B infection occurs as a complication of pregnancy or of the postpartum period, and an older age-group in which infection is associated with the so-called 'compromised host', i.e. the person with some underlying disease or deficiency, whether nutritional, hormonal, metabolic or immunological (Bayer et al. 1976); for example, Eickhoff and colleagues (1964) isolated group B streptococci from the peripheral gangrenous lesions of eight diabetics and from an elderly man with septic arthritis. Feingold et al. (1966) noted the association of group B infection and diabetes, with bacteraemia in adults and with urinary-tract infections in adults. Butter & de Moor (1967) described the isolation of group B streptococci from the blood of adults and Anthony & Concepcion (1975), in an analysis of clinical sources and serotypes of 490 group B strains, noted that although swabs of the female genital tract and cultures from newborn infants produced the largest number of isolates, as many as 149 were obtained from urine samples. In addition, the throat, sputum, sinuses and middle ear produced a smaller number of group B isolations, as did post-operative wounds, burns, infected lacerations and lesions of vascular insufficiency such as diabetic gangrene and ulcers. Lerner (1975) described the association of group B streptococci with meningitis in adults, and in an analysis of 24 patients from whom these organisms had been isolated, Bayer et al. (1976) described a wide spectrum of clinical illnesses, the most common of which were bacteraemic pyelonephritis, pneumonias, endometritis, and to a lesser extent meningitis and septic arthritis. Many of the patients were elderly with a predisposing condition, particularly diabetes and genito-urinary disorders, but despite the age of these patients and the severity of their infections the mortality rate was only 8%. These various reports of group B streptococcal infection in adults make it clear that a broad spectrum of diseases can be produced.

(a) *Neonatal disease*

That the pattern of neonatal bacterial infections can change over the years has been known for some time. When antibiotics were introduced in the 1940s a dramatic reduction occurred in the incidence of infections due to group A β-haemolytic streptococci. Coliform organisms replaced group A streptococci as the major aetiological agents of neonatal meningitis and septicaemia in the late 1940s. For several years this pattern remained, until the late 1950s when staphylococcal infection spread through nurseries in almost epidemic proportions. These strains of *Staphylococcus aureus* were resistant to many antibacterial agents and possessed very great virulence for neonates. The supremacy of the staphylococci did not persist for many years, however, and in the early 1960s coliforms once again assumed the role of the major neonatal pathogens. Signs of yet another shift have occurred in the past few years. According to Howard & McCracken (1974), in the 12 institutions enrolled in the Neonatal Meningitis Cooperative Study, although 38% of 131 cases of neonatal meningitis seen between July 1971 and April 1973 were caused by *Escherichia coli*, as many as 31% were caused by group B streptococci.

The vast numbers of papers published on group B streptococci and infections produced by them in the past 10 years bear witness to the fact that the role of group B streptococci in neonatal illness has changed over this period. It seems unlikely that the more sophisticated and accurate methods of laboratory diagnosis and improvements in interpretations of the results, coupled with greater clinical awareness of the neonatal syndromes produced by group B organisms alone, would have been responsible for such a greatly increased incidence of neonatal infections.

Whereas undoubtedly the most common and serious aspects of group B infection in the neonate are fulminant septicaemia, shock and meningitis, focal disease can also occur. In a series of 71 infants under 3 months of age reported on by Howard & McCracken (1974), 12 had unusual or previously unrecognized group B disease including septic arthritis of the hip joint, ethmoiditis with orbital cellulitis, pneumonia with empyema, facial cellulitis and conjunctivitis.

In 1958 Nyhan & Foussek gave the first definitive account of the link between neonatal meningitis and group B streptococci, and in 1961 Hood *et al.* described three cases of neonatal streptococcal meningitis caused by these streptococci occurring in one hospital in a three-month period, and emphasized the connection between the presence of group B organisms in the vaginal flora and the production of perinatal infections. While most previous publications were concerned with case reports of neonatal meningitis and/or septicaemia (Eickhoff *et al.* 1964; Maher & Irwin 1966; Butter & de Moor 1967; Kvittingen 1968), in the early 1970s, on the basis of reviewing major series of neonatal infections, several groups of workers came to the conclusion that there were certain common patterns in neonatal diseases caused by group B streptococci,

(Franciosi et al. 1973; Baker & Barrett 1973; Baker et al. 1973). Since then an impressive bibliography has been amassed relating group B streptococci to perinatal infection.

Neonatal disease caused by group B streptococci can be separated into two clinical syndromes, the 'early-onset' and 'late-onset'. The early-onset disease though septicaemic in type may also be meningitic, whereas the late-onset form is almost always meningitic.

The early-onset disease, frequently acute and fulminating, may occur within five days of birth, but usually occurs within the first 24–36 h. There is a close connection between early-onset infection and maternal complications, particularly premature labour and a prolonged period between rupture of the membranes and delivery; infants of low-birth weight tend to be affected. The neonate presents with acute respiratory distress, shock and apnoea; septicaemia is an invariable feature and meningitis may also be present. Clinically and radiologically the disease may resemble hyaline-membrane disease (Ablow et al. 1976). About 75% of the babies die within 24–36 h because of the fulminating nature of the infection and because from time to time the nature of the baby's illness is not recognized. Group B streptococci may be isolated from many sites including the blood, nasopharynx, skin, ears, umbilicus and meconium, but the diagnosis is sometimes made only after death. Quirante et al. (1974) diagnosed only 5 out of 17 cases of the early-onset septicaemic illness before death.

The late-onset form of infection presents usually after the tenth day of life as a purulent meningitis and can affect apparently healthy babies after normal labour. The mortality rate (15–20%) is much lower than in the early-onset form. The late-onset meningitic form can occur as late as 12 weeks after birth (McCracken 1973).

(b) *Incidence of neonatal infections*

Several reports on the incidence of infections have been made from various centres in the past few years, and, remarkably, infection rates were very similar in the various centres. From the USA Overall (1970) reported an average incidence of 0.9 per 1000 live births, Franciosi et al. (1973) and Baker & Barrett (1973) 2.1 and 2.9 per 1000 respectively and Howard & McCracken (1974) 1.4 per 1000. Reid (1975) reported a rate of 2.7 per 1000 births in Aberdeen and Finch et al. (1976) 1.1 per 1000 in London. This is not to say that these figures represent the number of babies colonized at birth; a far greater number were colonized with group B streptococci than were clinically infected; for example, in the series of Baker & Barrett (1973) a colonization rate of 262 per 1000 infants was reported with an attack rate for proved infection of 2.9 per 1000. Franciosi et al. (1973) reported a colonization rate of 12 per 1000 with an attack rate of 2.1 per 1000, and Reid a neonatal colonization rate of 19 per

1000 with a morbidity rate of 2.7 per 1000. The narrow range of reported attack rates (0.9 to 2.9 per 1000 births) is all the more surprising since reports of the colonization rates of the female genital tract vary widely from centre to centre.

4. Carriage in Adults

Lancefield & Hare (1935) reported a vaginal carrier rate of 2.3% in the puerperium, and an incidence rate in the female genital tract of 2-5% was generally accepted, until Kexel & Beck (1965) reported a much higher rate (12.4%) from vulval swabs; the vaginal and cervical swabs from the same patients produced lower isolation rates. Other carrier rates from the genital tract have been reported as follows: Bergqvist et al. (1970) 14.4%; Franciosi et al. (1973) 13.0%; Baker & Barrett (1973) 29.8%; Bevanger (1974) 18.0%; Christensen et al. (1974) 19.0%; Patterson & Hafeez (1976) (unpublished) 10%; Ferrieri et al. (1977) 8.3%. The marked differences in carrier rates may be explained partly by differences in laboratory methods and techniques. For example, the use of enrichment broths and selective media has been recommended for increasing the number of isolations (Baker & Barrett 1973) although Ferrieri et al. (1977) stated that even if their isolation rates had been doubled by the use of enrichment media these would still have been modest compared with some of the rates reported in the literature. Finch et al. (1976) obtained a carrier rate of only 6.4% even using selective and enrichment media. Other variables that may influence proper assessment of the carrier rate include the number of specimens collected from each individual and the number of sites cultured. Thom (unpublished) took multiple swabs from 150 parturient women and obtained an isolation rate of 11%. Of the 14 positive women, urethral swabbing detected six, cervical swabbing three, high vaginal and low vaginal swabbing four each, and rectal swabbing as many as 11. In an Edinburgh series (Ross, unpublished) the urethral swab proved to be superior to the cervical swab, and Anthony et al. (1975) reported that the urethral swab was more often positive than the vaginal.

There is little information available on the dynamics of group B carriage, particularly in the genital tract in pregnancy, on how long the carrier state remains, or how often a woman can convert from positive to negative, or negative to positive cultures. Ferrieri and colleagues (1977) reported that 42% of women who gave positive cultures in labour had given negative cultures during the third trimester, and 19%, who were positive then were culture negative in labour. The studies of Anthony et al. (1975) indicated that persistent carriage during pregnancy was present in only 38% of patients. Bergqvist and colleagues (1976) followed up a group of 11 women who had been group B streptococcal carriers during pregnancy four years earlier, and found that as many as 50% of these were still positive; husbands were also positive. This would appear to indicate that postpartum persistence of the carrier state may be common.

There is undoubtedly a significant carriage in the gut; Franciosi and colleagues (1973) reported an anal carrier rate of 16.8% in female hospital staff, and Patterson & Hafeez (1976) (unpublished) an isolation rate of 29.1%. Badri et al. (1976) recorded a vaginal carrier rate of 10.2% and a rectal carrier rate of 17.9% in pregnant women, and of 142 women with group B streptococci isolated from an anal swab only 61 had a positive vaginal swab. Wald et al. (1977) followed up 24 babies who had been colonized with group B organisms for three months after birth and found that most still harboured the organisms, particularly in the rectum and pharynx.

Carriage in the upper respiratory tract has been reported on several occasions. Hare (1935) described a 5% carrier rate in the throat and Butter & de Moor (1967) found that 19.3% of individuals with chronic skin conditions harboured group B streptococci in the throat: Jelinkova et al. (1970) however reported an incidence of only 1%. Franciosi et al. (1973) observed that among women who had a vaginal carrier rate of 13% there were 5.2% throat carriers. As with isolations from the vagina isolations from the throat may vary from area to area and from person to person and may also depend on sampling techniques and laboratory methodology.

The carrier rate in the female genital tract has been the subject of many reports, but fewer investigations have been carried out on the association of group B streptococci with the male genital tract. In the study of Franciosi and colleagues in 1973, 40 husbands of women with positive vaginal cultures were examined. Urethral swabs were taken and 45% proved positive, although when urine specimens were cultured only 30% were detected. In addition, Franciosi and colleagues studied 90 convicts with no history of sexual relations for a minimum of six months previously and found that 5.6% had positive urethral cultures. This may indicate that males may act as a reservoir of group B streptococci regardless of exposure to positive women. Wallin & Forsgren (1975) reported that 45% of male partners of women with vaginal carriage of group B streptococci had a urethral growth of the same streptococcal serotype. This was not related to concomitant growth of *Neisseria gonorrhoeae*, although they found that 16.4% of 457 males and 20.6% of 300 females attending a venereal disease clinic in Uppsala harboured group B organisms in the genital tract.

Although it is quite clear from the reports of various workers such as Simmons & Keogh (1940), El Ghoroury (1950) and Pattison et al. (1955) that the group B streptococcus that colonizes humans is not the same organism that causes mastitis in cattle, there are certain difficulties in deciding what the sources and reservoirs are for group B streptococci, as these organisms can be isolated from the upper respiratory tract, gut and genito-urinary tract of both males and females. Clearly the female genital tract is a most important site for colonization by group B organisms, but whether this is the source or reservoir is not immediately apparent.

The ecological conditions of the genital tract in pregnancy, the glycogen-rich mucosa and the low pH produced by the lactobacilli appear to be extremely favourable for the growth of group B streptococci, which unlike group A, are extremely acid resistant. Recent reports, however, point to the importance of the gut as a site of group B colonization. Zahradnicky and others (1972) reported a family study in which an infant died of group B sepsis. The mother had positive vaginal and cervical cultures and the father a positive urethral culture; all streptococci were of the same serotype as the dead infant. Despite prolonged antibiotic therapy the organism persisted in the gut of the parents and was considered to be responsible for the intermittent re-colonization of their genital tracts. Patterson & Hafeez (unpublished) found a faecal carrier rate of 27.3% in females which was higher than the genital carrier rate.

These reports suggest that the gut may well be the source of the organisms and that vaginal carriage is an expression of contamination from the gut. Sexual intercourse could account for urethral colonization in male partners of women with genital carriage of group B organisms but such colonization could also be the result of contamination from the gut, as in the case of females. Whether there is any relationship between upper respiratory tract and gut colonization or upper respiratory tract and vaginal colonization is not yet known.

5. Colonization and Infection in Humans

The term colonization is used in referring to adults or children who have positive cultures for group B streptococci but who do not exhibit any signs or symptoms of clinical infection.

In adults group B infection has a bimodal age distribution as previously mentioned. In the young female population septic abortion, endometritis and pyelonephritis may result from obstetric complications; infection in this instance is most likely to be endogenous, as cultures from the cervix frequently give positive results (Bayer *et al.* 1976). This is also the probable route of infection in women with gynaecological complaints who contract pelvic infections with group B organisms. In the series of Bayer *et al.* which comprised 24 patients, 11 women and 13 men, with serious group B streptococcal disease, 25% of infections were postpartum. Another 30% had a variety of urological disorders such as prostatic hypertrophy, renal cell carcinoma and renal failure, and 45% had diabetes mellitus. Bacteraemic group B streptococcal pyelonephritis was the most common syndrome, followed by pneumonia and endometritis. Meningitis, septic arthritis, cellulitis, endocarditis and tracheobronchitis also occurred; and in most cases the source of infection was apparent. For example, in two patients with meningitis the primary site of group B infection was the urinary tract and this was so with two others who had septic arthritis. In another patient endocarditis and meningitis followed group B streptococcal cellulitis. From these figures it is clear that the urogenital flora of both men and women is important in the pathogenesis of endogenous group B streptococcal infection. The throat may also be a possible source of endogenous infection (Mhalu 1976). Cross-infection may also be a main factor in Group B infections of wounds, burns or grafts and may be involved in the production of infections in other areas, such as in the upper respiratory tract.

Group B colonization of neonates at or soon after the birth is generally via the mother's genital tract (Eickhoff et al. 1964; Baker & Barrett 1973; Anthony et al. 1975). Ferrieri and her colleagues (1977) reported that nearly 50% of the infants born to culture positive mothers in labour acquired group B at birth or by the time of discharge from hospital, and the close correspondence of serotype distribution in pregnant women and their infants supports the concept of transmission from mother to infant. In early-onset infection, group B streptococci from the mother can gain access to the amniotic sac by the ascending route and the chances of this increase the longer the period between rupture of the membranes and delivery. Support for this view is that many infants who harbour group B organisms at birth have the organisms in their ear canals. Alternatively infants may acquire the organisms during passage through the maternal birth canal. Possibly, organisms may also enter the amniotic sac before rupture of the membranes, since Eickhoff and associates (1964) reported that one of the infants in their series who died after birth and from whom group B organisms were isolated at autopsy had been delivered by caesarian section.

Infants who are not colonized at birth may become colonized several days later, and in the minority this may lead to late-onset infection. The source of this type of infection is far from clear but suggestions have been made that it is nosocomial. Franciosi and colleagues (1973) considered this possibility because vaginal cultures of mothers may be negative; the cases may occur in clusters; the onset of illness may be weeks after delivery and cultures of nursery staff can reveal a large reservoir of group B streptococci. Steere et al. (1975) produced evidence that a strain that caused three cases of meningitis may have spread among healthy infants in a hospital nursery as 36% of the babies were carriers of that particular group B organism; when control measures were instituted, however, the carrier rate in the babies dropped to 9% within a month.

6. Distribution of Serotypes Among Streptococcal Isolates

There have been several accounts of the distribution of strains of group B streptococci from various types of infections. In the series of 24 adult patients studied by Bayer et al. (1976) type III was the serotype most commonly isolated. Franciosi and colleagues (1973) in Colorado reported that all 11 isolates from neonatal cases of late-onset meningitis belonged to type III. It is widely accepted that transmission of group B streptococci in early-onset infection occurs from the maternal genital tract to the neonate, illustrated by Baker & Barrett (1974) in their studies in Houston. They reported that isolates from cases with early-onset septicaemic infection (Ia, b, c, 17.7%; II, 44%; III, 33%) were similar to the isolates from the vaginally colonized mothers during a previous hospital survey (Ia, b, c, 28%; II, 35%; III, 37%). Earlier studies in Boston (Eickhoff *et al.* 1964) showed that type Ia was the single most frequent strain isolated from asymptomatic and symptomatic patients, and from neonates with meningitis, septicaemia or both, in the Netherlands (Butter & de Moor 1967). Franciosi and colleagues (1973) in Denver also stated that type Ia was the predominant type in neonates with early-onset infection. The serotype most commonly implicated in late-onset disease was type III (Baker & Barrett 1974), and suggestions have been made that type III strains have invasive properties that relate to meningeal infection. In a large series from the USA (Wilkinson *et al.* 1973) 70% of isolates from the cerebrospinal fluid and blood of patients with meningitis or sepsis, or both, were type III. These isolates were almost exclusively from children under two years of age.

British results (M. T. Parker, pers. comm.) show trends similar to the USA, though less extreme. At all ages type III strains formed 50% of isolates from meningitis and 26% from other septicaemic diseases; corresponding figures for type II were 5% and 20%. There was no evidence that age affected type distribution. Neonatal septicaemia reflected the serotypes of the maternal genital tract, as would be expected, but if late-onset infection is not related to maternal carriage, it should therefore be possible to demonstrate a predominance of type III organisms elsewhere.

7. Factors Involved in Human Infections by Group B Streptococci

In studying the causes of bacterial disease in humans at least two factors have to be examined, the organism and the infected host. As far as is known group B streptococci do not possess the number and diversity of products that contribute to the classical virulence of group A strains, and they are thought of rather as opportunist pathogens. This is not to say that they are totally devoid of factors that may contribute to virulence; for example, it has been established that the carbohydrate type antigens are related to virulence and specific antibodies to these give passive protection in mice (Lancefield 1938; Lancefield & Freimer 1966). The R and X protein antigens in contrast to the M proteins of *S. pyogenes* do not appear related to virulence, and specific antibodies do not provide protection when challenged with homologous streptococci (Wilkinson 1975). A number of strains produce hyaluronidase (Gochnauer & Wilson 1951) and a soluble haemolysin unlike streptolysin O and S (Christie *et al.* 1944). Tagg *et al.* (1975) also reported the isolation and partial purification of a bacteriocin from group B streptococci. It has also been established that types Ia and III are associated with neonatal disease and that type Ia localizes in the respiratory tract and type III in the meninges. Because of these associations, particularly the very common link of type III with the meninges, it has been suggested that type III may have special properties which produce an affinity with the central nervous system, and Baker & Kasper (1976) have speculated that capsular sialic acid may contribute to this. No work has yet shown a definitive connection between group B pathogenicity and structural components or diffusible products of the cell.

Group B streptococcal disease in adults is linked both with the compromised host, usually older persons, and with females with obstetric or gynaecological complications. The association with the latter is well known, and in the former infection has been reported in diabetes, tuberculosis, carcinoma, genito-urinary disorders and other debilitating diseases (Rantz & Keefer 1941; Hood *et al.* 1961; Eickhoff *et al.* 1964; Duma *et al.* 1969; Bayer *et al.* 1976).

In neonates low birth weight carries the risk of septicaemic infection, and obstetric complications such as prolonged labour may increase the length of time that the baby is in contact with the organisms. Because the infection rate in neonates is lower than the colonization rate by about 20-100 times, it is probable that antibody acquired from the mother may be protective.

Studies of the blood of pregnant women have indicated certain factors that may be of importance in protecting the neonate against infection. Klesius *et al.* (1973) and Mathews *et al.* (1974) studied the cellular and humoral response to group B streptococci, and also the opsonin system of these organisms, using paired sera from mother and baby. Antibodies to types Ib and III were detected in maternal sera at birth but the corresponding cord sera contained antibodies only to Ib, indicating that type III cannot cross the placenta; a serum factor against Ia was detected in the blood of neonates. Klesius and colleagues also demonstrated polymorphonuclear leucocyte activity to all serotypes except Ia in 85-100% of maternal and neonatal blood samples. Only 5.9% of mothers and 8.3% of neonates showed phagocytic activity against type Ia. This lack of non-specific serum factors, including opsonins, in the majority of sera examined by Klesius and associates may partly explain the pathogenic potential of type Ia strains. On the other hand, Baker & Kasper (1976) investigated the role of maternal antibody in neonatal group B infection, and reported that the sera from seven women who had given birth to babies who had invasive group B infection with type III strains were all deficient in antibody against this type. In contrast, sera from 22 out of 29 women who were vaginal carriers of type III strains, and whose infants were healthy, contained antibody to this type. In addition, three healthy neonates born to women with antibody also had antibody in their cord sera, confirming transplacental passage. The results of these studies are at variance with those of Klesius *et al.* (1973) and indicate that transplacental transfer of maternal antibody protects the neonate from invasive type III group B infection.

Susceptibility to infection in neonates may therefore relate to lack of passive immunity, immaturity of the immunological system or a defect in leucocytes. It may also be related to pathogenic factors in the group B streptococcus which have not been hitherto recognized, or to a combination of host and bacterial factors.

8. The Future

The incidence of serious group B streptococcal disease is generally accepted as 1-2 per 1000 live births. If this estimate is correct there should be 300-600 deaths from this cause per year in England and Wales. In fact only 100-150 deaths in children less than two months old are ascribed to septicaemia and meningitis in these countries (Mayon-White, unpublished). In the light of these figures much more research is required on the epidemiology of group B infection, particularly neonatal infection, of which the true incidence has yet to be established. This could be studied by prospective trials in areas or communities where the sample populations are large; retrospective studies may not be helpful. Laboratory isolations, hospital records and general practitioner mortality reports could all be used in any prospective surveys.

Further studies on carriage and acquisition of group B in adults and children are necessary to obtain more information on reservoirs and sources of the organisms and to attempt to elucidate the natural history of the carrier state and its relation to disease production.

Dynamics of vaginal carriage and the influence on this of the contraceptive pill and sexual intercourse is a major area of confusion, and is in need of clarification. Improvements in methodology so that group B organisms would not be missed in the laboratory would be beneficial, as well as investigations into the areas that should be swabbed to give optimal results. The two disease states— early and late infection—merit further investigation, particularly the pathogenesis of the late meningeal type and its relation to type III strains. Is there anything special about these in terms of antigenic composition that makes them predominant in meningeal infection and what is their source, if it is not the baby's mother? To what extent is hospital cross-infection of importance? What is responsible for the shift in the pathogenic role of the various group B serotypes over the last two decades?

Serological studies would help elucidate the protective roles of antibodies produced by group B streptococci, and help define the role of non-specific components of host resistance; some antibodies may play a more prominent part in resistance than others. Again, what is the role of cellular immunity as opposed to humoral immunity, particularly in regard to the prevention of group B streptococcal septicaemia? Does neonatal carriage produce antibodies that may protect? Is the detection of antibody useful in the diagnosis of group B infection?

Prophylactic measures must also be investigated in preventing spread of the organisms from mother to baby. Penicillin is not effective in eradicating the carrier state and it may be that application of a bland antiseptic substance to the vagina during or before labour may be beneficial. Can a vaccine be produced, particularly against the type III strain? Should all colonized neonates be given antibiotics?

These and many other questions have still to be answered. Research currently pursued in the USA, Europe and in the UK where, in addition to investigations carried out in individual centres, a national study has been instituted by the Public Health Laboratory Service under M. T. Parker, may produce some of these answers.

9. References

ABLOW, R. C., DRISCOLL, S. G., EFFMANN, E. L., GROSS, L., JOLLES, C. J., UAUY, R. & WARSHAW, J. B. 1976 A comparison of early-onset group B streptococcal neonatal infection and the respiratory-distress syndrome of the newborn. *New England Journal of Medicine* **294**, 65-70.

ANTHONY, B. F. & CONCEPCION, N. F. 1975 Group B streptococcus in a general hospital. *Journal of Infectious Diseases* **132**, 561-567.

ANTHONY, B. F., OKADA, D. & HOBEL, C. J. 1975 Group B streptococci in perinatal infections: natural history of maternal and neonatal colonisation. *Pediatric Research* **9**, 296 (abstract).

BADRI, M. S., ZAWANEH, S., CRUZ, A., BAER, H., SPELLACY, W. N. & AYOUB, E. M. 1976 Rectal colonisation with group B streptococci. *Pediatric Research* **10**, 394.

BAKER, C. J. & BARRETT, F. F. 1973 Transmission of Group B streptococci among parturient women and their neonates. *Journal of Pediatrics* **83**, 919-925.

BAKER, C. J. & BARRETT, F. F. 1974 Group B streptococcal infection in infants; the importance of the various serotypes. *Journal of the American Medical Association* **230**, 1158-1160.

BAKER, C. J. & KASPER, D. L. 1976 Identification of sialic acid in polysaccharide antigens of group B stretococcus. *Infection and Immunity* **13**, 284-288.

BAKER, C. J., BARRETT, F. F., GORDON, R. C. & YOW, M. D. 1973 Suppurative meningitis due to streptococci of Lancefield group B: a study of 33 infants. *Journal of Pediatrics* **82**, 724-729.

BAYER, S. A., CHOW, A. W., ANTHONY, B. F. & GUZE, L. B. 1976 Serious infections in adults due to group B streptococci. *American Journal of Medicine* **61**, 498-503.

BERGQVIST, G., HURVELL, B., THAL, E. & VACLAVINSKOVA, V. 1970 Infection of newborn infants with *Streptococcus agalactiae* (Lancefield Group B) in relation to its occurrence in the vaginal flora of term pregnant women. *Acta paediatrica scandinavica* **206** (Suppl.), 107-109.

BERGQVIST, G., HURVELL, B., THAL, E. & VACLAVINGKOVA, V. 1976 The persistence of Group B streptococci in families. *Scandinavian Journal of Infectious Diseases* **8**, 79-81.

BEVANGER, L. 1974 Carrier rate of group B streptococci with relevance to neonatal infections. *Infection* **2**, 123-126.

BUTTER, M. N. W. & DE MOOR, C. E. 1967 *Streptococcus agalactiae* as a cause of meningitis in the newborn, and of bacteremia in adults. Differentiation of human and animal varieties. *Antonie van Leeuwenhoek* **33**, 439-450.

CHRISTENSEN, K. K., CHRISTENSEN, P., FLAMHOLE, L. & RIPA, T. 1974 Frequencies of streptococci of groups A, B, C, D and G in urethra and cervix swab specimens from patients with suspected gonococcal infection. *Acta pathologica et microbiologica scandinavica* Section B **82**, 470-474.

CHRISTIE, R., ATKINS, N. E. & MUNCH-PETERSEN, E. 1944 A note on a lytic phenomenon shown by group B streptococci. *Australian Journal of Experimental Biology and Medical Science* **22**, 197-200.

COLEBROOK, L. & PURDIE, A. W. 1937 Treatment of 106 cases of puerperal fever by sulphanilamide (Streptocide). *Lancet* **2**, 1237-1242.

DUMA, R. J., WEINBERG, A. N., MEDREK, T. F. & KUNZ, L. J. 1969 Streptococcal infections. A bacteriologic and clinical study of streptococcal bacteremia. *Medicine* **48**, 87-127.

EICKHOFF, T. C., KLEIN, J. O., DALY, A. K., INGALL, D. & FINLAND, M. 1964 Neonatal sepsis and other infections due to Group B beta-haemolytic streptococci. *New England Journal of Medicine* **271**, 1221-1228.

EL GHOROURY, A. A. 1950 Comparative studies of group B streptococci of human and bovine origin: serological characters. *American Journal of Public Health* **40**, 1273-1277.

FEINGOLD, D. S., STAGG, N. L. & KUNZ, L. J. 1966 Extrarespiratory streptococcal infections—importance of the various serological groups. *New England Journal of Medicine* **275**, 356-361.

FERRIERI, P., CLEARY, P. P. & SEEDS, A. E. 1977 Epidemiology of group B streptococcal carriage in pregnant women and newborn infants. *Journal of Medical Microbiology* **10**, 103-114.

FINCH, R. G., FRENCH, G. L. & PHILLIPS, I. 1976 Group B streptococci in the female genital tract. *British Medical Journal* **2**, 1245-1247.

FRANCIOSI, R. A., KNOSTMAN, J. D. & ZIMMERMAN, R. A. 1973 Group B streptococcal neonatal and infant infections. *Journal of Pediatrics* **82**, 707-718.

FRY, R. M. 1938 Fatal infections by haemolytic streptococcus group B. *Lancet* **1**, 199-201.

GOCHNAUER, T. A. & WILSON, J. B. 1951 Production of hyaluronidase by Lancefield's group B streptococci. *Journal of Bacteriology* **62**, 405-414.

HARE, R. 1935 Classification of haemolytic streptococci from nose and throat of normal human beings by means of precipitin and biochemical tests. *Journal of Pathology and Bacteriology* **41**, 499-512.

HOOD, M., JANNET, A. & DAMERON, G. 1961 Beta hemolytic streptococcus group B associated with problems of the perinatal period. *American Journal of Obstetrics and Gynaecology* **82**, 809-818.

HOWARD, J. B. & McCRACKEN, G. H. 1974 The spectrum of group B streptococcal infections in infancy. *American Journal of Diseases of Children* **128**, 815-818.

JELINKOVA, J., NEUBAUER, M. & DUBEN, J. 1970 Group B streptococci in human pathology. *Zentralblatt für Bakteriologie, Parasitenkunde, Infektionskrankheiten und Hygiene* Abt. 1 **214**, 450-457.

KEXEL, G. & BECK, N. J. 1965 Untersuchungen über die Haufigkeit der-B-Streptokokken im Wochenbett. *Geburtshilfe und Frauenheilkunde* **25**, 1078-1085.

KLESIUS, P. H., ZIMMERMAN, R. A., MATHEWS, J. H. & KRUSHAK, D. H. 1973 Cellular and humoral immune response to group B streptococci. *Journal of Pediatrics* **83**, 926-932.

KVITTINGEN, J. 1968 Beta-haemolytic streptococcus group B causing neonatal meningitis. *Acta pathologica et microbiologica scandinavica* **74**, 143-144.

LANCEFIELD, R. C. 1933 A serological differentiation of human and other groups of hemolytic streptococci. *Journal of Experimental Medicine* **57**, 571-595.

LANCEFIELD, R. C. 1934 Serological differentiation of specific types of bovine hemolytic streptococci (group B). *Journal of Experimental Medicine* **59**, 441-458.

LANCEFIELD, R. C. 1938 Two serological types of group B hemolytic streptococci with related but not identical, type-specific substances. *Journal of Experimental Medicine* **67**, 25-40.

LANCEFIELD, R. C. & FREIMER, E. H. 1966 Type-specific polysaccharide antigens of group B streptococci. *Journal of Hygiene* **64**, 191-203.

LANCEFIELD, R. C. & HARE, R. 1935 Serological differentiation of pathogenic and non-pathogenic strains of haemolytic streptococci from parturient women. *Journal of Experimental Medicine* **61**, 335-349.

LERNER, P. I. 1975 Meningitis caused by streptococcus in adults. *Journal of Infectious Diseases* **131** (Suppl.), S9-S16.

McCRACKEN, G. H. 1973 Group B infection: the new challenge in neonatal infections. *Journal of Pediatrics* **82**, 703-706.

MAHER, E. & IRWIN, R. C. 1966 Group B streptococcal infection in infancy: a case report and review. *Pediatrics* **38**, 659-661.

MATHEWS, J. H., KLESIUS, P. H. & ZIMMERMAN, R. A. 1974 Opsonin system of the group B streptococcus. *Infection and Immunity* **10**, 1315-1320.

MHALU, F. S. 1976 Infection with *Streptococcus agalactiae* in a London hospital. *Journal of Clinical Pathology* **29**, 309-312.

NOCARD, M. & MOLLEREAU. 1887 Sur une mammite contagieuse des vaches laitières. *Annales de l'institut Pasteur, Paris* **1**, 109-126.

NYHAN, W. L. & FOUSSEK, M. 1958 Septicaemia in the newborn. *Pediatrics* **22**, 268-278.

OVERALL, J. C. 1970 Neonatal bacterial meningitis. *Journal of Pediatrics* **76**, 499-511.

PATTERSON, M. J. & HAFEEZ, A. E. B. 1976 Group B streptococci in human disease. *Bacteriological Reviews* **40**, 774-792.

PATTISON, I. H., MATTHEWS, P. R. J. & HOWELL, D. G. 1955a The type classification of group-B streptococci, with special reference to bovine strains apparently lacking in type polysaccharide. *Journal of Pathology and Bacteriology* **69**, 51-60.

PATTISON, I. H., MATTHEWS, P. R. J. & MAXTED, W. R. 1955b Type classification by Lancefield's precipitin method of human and bovine group-B streptococci isolated in Britain. *Journal of Pathology and Bacteriology* **69**, 43-50.

QUIRANTE, J., CEBALLOS, R. & CASSADY, G. 1974 Group B β-hemolytic streptococcus infection in the newborn. I. Early onset infection. *American Journal of Diseases of Children* **128**, 659-665.

RANTZ, L. A. & KEEFER, C. S. 1941 Distribution of hemolytic streptococci groups A, B, and C in human infections. *Journal of Infectious Diseases* **68**, 128-132.

REID, T. M. S. 1975 Emergence of Group B streptococci in obstetric and perinatal infections. *British Medical Journal* **2**, 533-535.

SIMMONS, R. T. & KEOGH, E. V. 1940 Physiological characters and serological types of haemolytic streptococci of groups B, C and G from human sources. *Australian Journal of Experimental Biology and Medical Science* **18**, 151-161.

STEERE, A. C., ABER, R. C., WARFORD, L. R., MURPHY, K. E., FEELEY, J. C. HAYES, P. S., WILKINSON, H. W. & FACKLAM, R. R. 1975 Possible nosocomial transmission of group B streptococci in a newborn nursery. *Journal of Pediatrics* **87**, 784-787.

TAGG, J. R., DAJANI, A. S. & WANNAMAKER, L. W. 1975 Bacteriocin of a group B streptococcus: partial purification and characterisation. *Antimicrobial Agents and Chemotherapy* **7**, 764-772.

WALD, E. R., SNYDER, M. J. & GUTBERLET, R. L. 1977 Group B beta-haemolytic streptococcal colonisation. *American Journal of Diseases of Children* **131**, 178-180.

WALLIN, J. & FORSGREN, A. 1975 Group B streptococci in venereal disease clinic patients. *British Journal of Venereal Diseases* **51**, 401-404.

WILKINSON, H. W. 1975 Immunochemistry of purified polysaccharide type antigens of group B streptococcal types Ia, Ib and Ic. *Infection and Immunity* **11**, 845-852.

WILKINSON, H. W. & MOODY, M. D. 1969 Serological relationships of type I antigens of group B streptococci. *Journal of Bacteriology* **97,** 629–634.

WILKINSON, H. W., FACKLAM, R. R. & WORTHAM, E. C. 1973 Distribution by serological type of group B streptococci isolated from a variety of clinical material over a five-year period (with special reference to neonatal sepsis and meningitis). *Infection and Immunity* **8,** 228–235.

ZAHRADNICKY, J., LUKESOVA, M., KRENUS, R., JELINKOVA, J., HEJDA, V., VALCHOVA, M. & TURKOVA, S. 1972 An unusual case of infection caused by a group B streptococcus. *Journal of Hygiene* **16,** 21–27.

Streptococci in Animal Disease

C. D. WILSON AND G. F. H. SALT

*Central Veterinary Laboratory,
Ministry of Agriculture, Fisheries and Food,
Weybridge, Surrey, England*

CONTENTS

1. Introduction 143
2. Diseases of cattle 144
   (a) Mastitis due to *Streptococcus agalactiae* 144
   (b) Mastitis due to other streptococci 146
   (c) Other streptococcal diseases of cattle 148
3. Streptococcal diseases of pigs 149
   (a) Meningo-encephalitis and arthritis 149
   (b) Cervical lymphadenitis 150
   (c) Endocarditis 151
4. Streptococcal diseases of horses 151
5. Streptococcal diseases of sheep 152
6. Streptococcal diseases of poultry 153
7. References 153

1. Introduction

THE STREPTOCOCCI are involved in a wide variety of diseases in animals and can produce acute, subacute, chronic and even subclinical conditions. Some of the species such as *Streptococcus zooepidemicus* and group G and L streptococci infect a number of hosts while others, e.g. *S. equi* are host specific. They can be isolated from the mouth, nose, throat, genital tract, skin and faeces of healthy animals and can also be found in the animal's environment.

Within the limits of this paper it is not possible to describe in detail all the diseases caused by streptococci in the various farm animals so only those diseases that are most important either from the economic or zoonotic aspects are considered, and their epidemiology discussed.

Classification of animal streptococci is based on cultural, microscopical, biochemical and serological tests. Cultures are normally made on the surface of blood agar plates. Sheep blood is commonly used in many laboratories. Certain streptococci which are β-haemolytic on horse blood agar may produce only greening or α-haemolysis when sheep blood is used. Microscopical examinations are usually made from liquid cultures to determine the morphology of the organism. Serum is added to the sugar media as many streptococci grow poorly in simple media not containing blood or serum.

Serological grouping requires potent group specific sera some of which (e.g. D and N) are very difficult to produce.

In Table 1 are given the routine tests and reactions used to identify the various streptococci recovered from diseased animals.

2. Diseases of Cattle

(a) *Mastitis due to* Streptococcus agalactiae

Bovine mastitis caused by *S. agalactiae* is still of considerable economic importance and is also of some, if diminishing, importance as a zoonosis. Prior to the war it was considered that 90% of all mastitis infections were caused by streptococci, and especially by *S. agalactiae*. However, a survey carried out in 1949-51 (Wilson 1957) in which 1500 herds were examined in Surrey and Hertfordshire indicated that only 61% of herds, and only 15% of cows, were infected with *S. agalactiae*.

Unlike most streptococci, *S. agalactiae* is largely confined to the bovine udder and does not normally multiply in other parts of the body or on the intact skin of the teats. Thus, with the advent of penicillin for mastitis treatment in 1946, eradication of this infection from infected herds could be achieved (Stableforth *et al.* 1949). The routine was to treat the infected quarters or, in badly affected herds, all quarters at the same time while general disinfection of the cows' environment indoors was carried out on the final day of treatment. This resulted in the elimination of *S. agalactiae* from most herds immediately, re-treatment of a few cows being necessary occasionally. All infected herds in Surrey were treated and the infection rate was reduced to 1% of cows but it could not be eliminated entirely in some herds. In Denmark (Olsen 1975) a more extensive campaign was launched and the prevalence of *S. agalactiae* infection reduced to less than 1% of cows in the country, but again complete eradication in a large area was not achieved. A major effort was made to do this on the island of Bornholm since this was considered to contain a closed cattle community but complete eradication was not achieved. It is alleged that the source of re-infection was the pigs since group B streptococci could be isolated from swabs of the vaginae of sows. In other areas S. J. Olsen (pers. comm.) attributed the re-introduction of *S. agalactiae* into herds from which the infection had been eradicated to a human source of infection since many of the strains freshly isolated from re-infected herds proved to be of human type.

Tolle (1975) made the point that group B infection was much more prevalent in humans drinking unpasteurized milk than in those drinking heat-treated milk although considerably less unpasteurized milk was consumed. He speculated that in the passage through the body the strains became modified and changed from bovine to human type but he offered no evidence to show that this was so.

Table 1

Differentiation of streptococci isolated from animal sources

Species of Streptococcus	Haemolysis on blood agar	CAMP test	Serogroup	Fermentation tests							Litmus milk	Methylene blue milk 1 in 1000	NH$_3$ production	Sodium hippurate hydrolysis	Aesculin hydrolysis
				Tre	Sor	Man	Raff	Sal	In	Lac					
agalactiae	β (usually narrow zone) or non-haem, blue colours	+	B	+	–	–	–	±	–	+	AC	–	+	+	–
zooepidemicus	β (wide zone)	.	C	–	+	–	–	+	–	+	A sl R	–	+	–	.
equisimilis	β (wide zone)	.	C	+	–	–	–	–	–	±	± A	–	+	–	.
equi	β (wide zone)	.	C	–	–	–	–	+	–	–	–	–	.	–	.
dysgalactiae	α	–	C	+	±	–	–	–	–	+	A ± RC	–	+	–	–
uberis	Non-haemolytic	±	–*	+	+	+	–	+	+	+	AC part R	–	+	+	+
faecalis	Greening or non-haemolytic	Growth 45°C +	D	+	+	+	–	+	–	+	ACR	RC	+	–	+
infrequens	β (wide zone)	.	E	+	±	±	–	±	–	±	± A	–	+	–	.
Group G	β (wide zone)	.	G	±	–	–	–	+	–	+	AC	–	+	–	+
Group L	β (wide zone)	.	L	+	–	–	–	±	–	+	AC	–	+	±	–
lactis	Greening or non-haemolytic	–	N	+	–	±	–	±	–	+	ACR	RC	±	±	+
Group P	β (wide zone)	–	P	+	+	+	–	+	–	±	CR	± R	+	–	+
suis type 1	Greening (ovine blood) (β equine blood)	–	D(S)	+	–	–	–	+	+	+	AR	–	+	–	+
suis type 2	Greening	–	D(R)	+	–	–	+	+	+	+	ARC	–	+	–	+
bovis	Greening	+	± D	±	–	±	+	+	+	+	A	–	–	–	+
Viridans streps. in general	Greening	+	–	+	–	–	±	±	±	±	± A	–	–	±	+
pneumoniae	Greening	'Optochin' sensitivity +													

A, acid production; C, clotting of milk; R, reduction of colour.
Tre, trehalose; Sor, sorbitol; Man, mannitol; Raff, raffinose; Sal, salicin; In, inulin; Lac, lactose.
* Some strains of *S. uberis* will react with group E serum.

C. L. Wright (pers. comm.) reported that he had examined 20 strains of *S. agalactiae* isolated from human patients in a large hospital and that all fermented lactose, but they have not been typed serologically yet. Frost *et al.* (1977) indicated that bovine strains of *S. agalactiae* possess an antigen which helps the cocci to adhere to the mucous membrane of the mammary gland and that human strains do not possess this antigen. There are, however, well authenticated cases of human types becoming established in bovine mammary glands and producing a disease indistinguishable from that produced by the bovine strains.

At the present moment, a national survey of bovine mastitis is being carried out in Great Britain and preliminary results indicate that *S. agalactiae* is still an important pathogen in this disease. Over 50% of the herds examined are still infected but, in contrast to the earlier survey, the number of cows infected is only 9%. It must be stressed that these are early figures and a change could occur as the survey progresses. While $> 50\%$ of the herds are infected there is a wide variation between herds. Some herds are heavily infected ($> 75\%$ of cows), but most herds have only one or two quarters of one cow infected. These latter herds are invariably using some form of mastitis control and it is clear that post-milking teat-dipping with a disinfectant is effective in limiting the spread of this infection. Following the Danish experience with such herds some isolates are being typed to determine whether there has been a human source of re-infection.

(b) *Mastitis due to other streptococci*

In Denmark the β-haemolytic streptococci of groups C, G and L were considered to be in the same epidemiological class as *S. agalactiae* but recently it has been decided that their eradication should not be part of a national campaign, as in the case of *S. agalactiae*.

The commonest group C streptococcus, other than *S. dysgalactiae*, in bovine mastitis is *S. zooepidemicus*. As clinical mastitis is common in this infection, the milk yield of the herd is often seriously affected, and because arthritis can also develop, eradication by antibiotic therapy of all lactating cows is sometimes advocated.

Groups G and L are similar to group B in that the infections are largely subclinical and they can spread via the milking machine in an affected herd. Only three outbreaks of mastitis due to group G have been met in 30 years; the source of infection in two outbreaks was shown to be from dogs which had constant access to the buildings where the cows were kept and milked. The source in the third outbreak has yet to be established.

In Jutland, group L streptococci are considered to occur more frequently now than group B streptococci in bovine mastitis and this is attributed by Olsen

(1964) to the fact that in small farms common in Denmark the cowman also looks after the pigs and the author showed that group L streptococci could be recovered quite frequently from swabs taken from the vaginae of sows. He deduced that the hands of the cowman, contaminated during farrowing, were transferring the infection to the cows. Group L streptococci are infrequent in mastitis in Great Britain and sows examined on farms where infection does exist have been negative for this organism. Dogs were considered to be the source of infection in one herd.

Streptococcus dysgalactiae infection is probably the third ranking streptococcus in mastitis after *S. agalactiae* and *S. uberis*. At one time it was considered to be similar to group B streptococci in being transferred during milking but a more likely explanation is that it multiplies on sores on the teats and gains entrance to the udder when the sores involve the teat orifices. In the udder it usually produces an acute mastitis without severe systemic symptoms and is also found in subclinical infections. *Streptococcus dysgalactiae*, as distinct from the other group C streptococci, produces greening on blood agar and ferments trehalose and lactose. According to Stableforth (1959) it does not split aesculin but McDonald *et al.* (1976) have shown that some strains do so under defined conditions.

This organism produces hyaluronidase and it is possible that it is this property which favours the spread of *Corynebacterium pyogenes* in dry udders since *C. pyogenes* and *S. dysgalactiae* are frequently found as a combined infection in 'summer mastitis'. This condition cannot be readily produced by injections into the udder of *C. pyogenes* alone but the addition of *S. dysgalactiae* to the inoculum increases considerably the success rate (Stuart *et al.* 1951). It is an organism which is widely dispersed in other organs of the cow and on the skin and control by eradication is not a practical possibility.

Streptococcus uberis is becoming of increasing importance in bovine mastitis mainly because cattle are now being exposed to this infection to a much greater extent as a result of the switch to loose-housing. Because *S. uberis* is present in the faeces of cattle and multiplies in the bedding, cattle which are in yards or in cubicles are exposed to a much heavier infection than before. Thus the condition is seasonal and *S. uberis* ceases to become a problem when cows are turned out to grass. The mastitis it produces in the lactating cow is usually quite mild but encapsulated forms of the organism appear to be much more pathogenic and severe systemic disturbance with pyrexia can occur. The disease also appears to be more acute in the dry cow just before calving. Control of this organism lies in improved hygiene; eradication by treatment is not possible.

Cullen (1966) carried out a three-year study on *S. uberis* and found that a number of strains did not conform to the usual biochemical criteria in that they fermented raffinose but not sorbitol. However, he was working in a single small herd and one of us (CDW) examined 100 strains of *S. uberis* which had been isolated from cases of clinical mastitis and all had the recognized reactions given in Table 1. A number of strains give a positive reaction with group E antiserum and a number are positive to the 'CAMP Test' (Christie *et al*. 1944). There are at least 11 serological types of *S. uberis*.

Pneumococcal infection in mastitis is now very rare and the two outbreaks we have encountered at Weybridge occurred over 20 years ago. In the first (Wilson & Lancaster 1949) a few cattle in a Channel Island herd were infected with a moderately severe mastitis. In the second (MacLachlan *et al*. 1958), 21 out of 49 cows were infected. The cattle were ill, the milk was grossly altered being yellow, oily and clotted and there was a marked reduction in the milk yield. The cows responded to penicillin intramammary therapy but the yield was not restored until the cows had gone through a dry period and calved again. Throat swabs of the farm staff were negative for the pneumococcus responsible for the outbreak but one cowman present at the start of the outbreak had left and could not be examined. It is probable that freedom from this infection in recent years is due to the cessation of hand-milking and hand-stripping when spitting on the hands was considered a valuable lubricant by some cowmen. An interesting point about this outbreak was that while diplococci could be seen in enormous numbers in milk films prepared from infected cows, they would not grow aerobically on primary culture but required CO_2 for growth. This may have been due to the peptone which was being used in our media at that time as a change to another peptone produced aerobic growth of pneumococci from the milk samples.

(c) *Other streptococcal diseases of cattle*

Apart from mastitis, streptococcal infection is not very important in cattle although endocarditis, abortion, urino-genital infections, in both male and female, and arthritis have been reported. They appear to be rare but occasionally a herd problem can result as was seen by one of us (CDW) in a herd with severe mastitis due to *S. zooepidemicus* and concurrent lameness in many cows.

Group D streptococci have been frequently isolated from endocarditis in cattle. In Sweden Winquist (1945) found seven out of 41 cases were due to group D streptococci whilst the others were nearly all caused by *C. pyogenes* infection. In a series of 40 cases of endocarditis in England and Wales, Evans (1957) found that 27 were infected with streptococci and 24 were in pure culture. The great majority of these were group D streptococci. The types however were different from those found in cases of human endocarditis.

Pneumococcal infection has been reported in calves in Denmark by Rømer (1948). Nearly 6% of 6000 calves examined were found to be infected (no details available) and all but one of the strains fell into the human-type classes.

3. Streptococcal Diseases of Pigs

(a) *Meningo-encephalitis and arthritis*

Gamma- and β-haemolytic streptococci were isolated from outbreaks of meningitis in piglets by McNutt & Packer (1943). Biochemical reactions of the strains were variable so that no single species appeared to be the cause of the condition. In England, Field *et al.* (1954) described disease in 2- to 6-week-old piglets which was primarily a bacteraemia with subsequent invasion of the central nervous system and/or the joints. The disease responded to prompt penicillin treatment. The streptococci were β-haemolytic on horse blood agar and α-haemolytic on sheep blood and gave clear-cut biochemical reactions, but they could not be identified by existing Lancefield sera. Antiserum prepared in rabbits reacted with strains from eight different outbreaks showing a common antigenic factor. The disease was reproduced by intravenous and subdural inoculation of broth cultures. In the Netherlands, de Moor (1963) isolated similar strains from young piglets and a different streptococcus from a similar disease in older pigs. These were designated serological groups S and R, the strains from young piglets were group S and the ones from older animals group R. Field's strains also reacted with group S. In 1966, Elliott reported that strains of group S were identical to his piglet meningitis (PM) streptococcus, that they belonged to Lancefield's group D and to a single serological type characterized by a capsular polysaccharide antigen. Elliott suggested that *S. suis* would be an appropriate name. Elliott *et al.* (1966), studying the disease in the field and in experimental work, considered the throat of the piglet as a likely portal of entry for *S. suis* and the upper respiratory tract of the sow as a possible source. Susceptible piglets could be protected against experimental infection by inoculation of serum from convalescent animals and vaccination was proposed as a practical means of control. Conventionally reared pigs older than 6-8 weeks were shown to be immune to *S. suis* by Agarwall *et al.* (1969). Investigations into the role of the palatine tonsils by Williams *et al.* (1973) demonstrated that oral infection led to establishment of *S. suis* in the tonsils, but development of clinical disease did not necessarily occur.

Windsor & Elliott (1975) studied an outbreak in weaned pigs 10-14 weeks old. The streptococci isolated appeared to be identical to de Moor's group R isolated from older pigs in the Netherlands. Examination of cultures of streptococci from brains of pigs dying from meningitis in herds in various parts of England and Wales showed that group R infections predominated and occurred in animals up to the age of at least 14 weeks.

The present day position regarding piglet meningitis streptococci appears to be:

group S = *S. suis* type 1 ⎫
⎬ Lancefield's serological group D.
group R = *S. suis* type 2 ⎭

The group T streptococci of de Moor (1963) and the associated disease in pigs needs further study before classifying under this scheme.

A number of cases of human infection with group R have occurred in persons with intensive contact with live or slaughtered pigs (Windsor & Elliott 1975; Zanen & Engel 1975).

Suppurative arthritis in young piglets is frequently associated with *S. equisimilis* (Lancefield's group C), (Ross 1972). Hare *et al.* (1942) isolated group C streptococci from vaginal secretions of sows and considered that the sow was the most probable source of infection for young piglets. Collier (1951) isolated 12 group C and one group L from 13 suppurative arthritic joints. Ten of the strains were identified as *S. equisimils*. Tittiger & Alexander (1971) isolated 50 strains of haemolytic streptococci from 478 arthritic joints of slaughtered pigs. Twenty-eight strains fell into group C, 21 into group L and one into group E. In Denmark, Nielson *et al.* (1972) studied the incidence and causes of polyarthritis in 805 litters of pigs. Haemolytic steptococci accounted for 67% of the isolates, 42% were group C and 21% were group L.

(b) *Cervical lymphadenitis*

In North America this economically important disease entity which affects pigs aged 10–20 weeks old is principally caused by group E streptococci. The disease may be endemic on some farms and abscesses are detected when pigs are slaughtered. Abscesses of the cervical lymph nodes in pigs were associated with haemolytic streptococci by Newsom (1937). These streptococci were shown to belong to group E. Collier (1956) examined 492 abscesses from the pharyngeal region of market pigs and isolated group E streptococci from 331 (67%). He carried out experimental work in succeeding years (Collier 1965) and proposed that *S. suis* was a more appropriate name. However the older name for group E streptococci, *S. infrequens* seems to have persisted. Armstrong & Payne (1969) examined 451 samples of abscess exudate from naturally infected pigs. Group E streptococci were recovered from 293 (65%) and *C. pyogenes* from 94 (20.8%). Several other bacteria were isolated but none in appreciable numbers. Nearly 90% of the group E isolates fell into serological type IV. Oral administration of an avirulent group E strain protected pigs of 10 weeks or more against subsequent challenge with virulent strains (Engelbrecht & Dolan 1968). The incorporation of chlortetracycline at 50g t^{-1} of complete feed was shown to be useful in preventing the formation of abscesses in laboratory experiments and in the field (Gouge *et al.* 1957).

In the UK, group E streptococci are an infrequent cause of cervical lymphadenitis in pigs. We certainly have group E streptococci but not associated with a disease problem.

(c) *Endocarditis*

Although *Erysipelothrix rhusiopathiae* is the organism most commonly isolated from valvular lesions in pigs, several species of streptococci have been associated with this condition. Hont & Banks (1944) showed that endocarditis could be produced by dosing with haemolytic group D streptococci. Elliott *et al*. (1966*b*) also produced the disease in young pigs by intravenous inoculation of group D. Jones (1968) isolated four strains of group L and two of group C from cases of endocarditis in sows. In 500 six-month-old bacon pigs, he found a single case of streptococcal endocarditis. In 1969 he reproduced endocarditis in six of 10 pigs by the intravenous inoculation of a group L streptococcus which had been isolated from a natural case. These experimental pigs also developed arthritis. Streptococci seem to account for a small proportion of cases of endocarditis in pigs. The most usual streptococci involved appear to be groups D, L and C.

A wide range of streptococci has been isolated from various tissues of pigs affected by several disease conditions. Many apprently normal pigs also yield streptococci. The pathogenicity of a number of the streptococci in certain diseases of the pig has been conclusively established by experimental work. Doubtless further studies will determine specific relationships of streptococci with other diseases of the pig.

4. Streptococcal Diseases of Horses

Streptococcal infections in horses are caused principally by *S. equi* and *S. zooepidemicus*. Both organisms give wide zones of haemolysis on blood agar and, serologically, fall into Lancefield's group C. *Streptococcus equi* is found only in horses and is the causative organism of 'strangles', an acute contagious disease of young horses up to about two years of age; in older animals the disease is usually milder. The disease is characterized by abscess formation in the submaxillary lymph nodes in the pharyngeal region which is preceded by inflammation of the upper respiratory mucous membranes and a muco-purulent nasal discharge. In Australia immunological studies by Bazeley (1940) led to the preparation of an effective vaccine by using young encapsulated organisms from rapidly growing cultures which had been killed by a minimum amount of heat. The vaccine was used extensively in army horses with considerable success. In further experimental work (Bazeley 1947) he showed that an antiserum produced by a single strain of *S. equi* protected mice against 32 strains. Serum agglutination tests had demonstrated previously that a single serological type antigen was involved (Bazeley 1942*b*).

Streptococcus zooepidemicus has been isolated from a variety of disease conditions in many animals. It may be isolated from various sites in the horse and has been associated with respiratory infections, infertility and abortion in

Table 2
Differentiation of equine streptococcal pathogens

	Haemolysis	Trehalose	Sorbitol	Lactose	Sero-group
S. zooepidemicus	Wide zone	−	+	+	C
S. equi	Wide zone	−	−	−	C
S. equisimilis	Wide zone*	+	−	±	C

* Some strains of *S. equisimilis* produce variable haemolysis on sheep blood agar, some colonies producing β-haemolysis and some greenish colouration around a zone of haemolysis. Subcultures of single colonies produce variable haemolytic reactions in many instances.

mares. In horses, wound infections are in many cases due to *S. zooepidemicus*. Thal & Moberg (1953) isolated Lancefield's group C strains from cervical mucus of mares and from pneumonia, septicaemia or local lesions. Most of the streptococci isolated were *S. zooepidemicus*, a few *S. equisimilis*, a few belonged to group D and one to group E. Miller & Francis (1974) contended that a mare harbouring haemolytic streptococci or other potentially pathogenic organisms in the reproductive tract is significantly less likely to produce a foal than an uninfected one. Hughes (1975) re-examined the possibility that infection with *S. zooepidemicus* is the cause of much equine infertility and queried their conclusions. Bryans & Moore (1972) in an antigenic study of 164 strains of *S. zooepidemicus* from horses placed 77 of the strains in 15 serotypes. The type antigen was a trypsin-labile protein.

Streptococcus zooepidemicus emerges as the most common streptococcus isolated from the horse but its importance as a primary cause of disease in this species is open to doubt. *Streptococcus equi* infection, which is confined solely to the equine species, is clearly responsible for the disease 'strangles'.

Streptococcus equisimilis has been isolated occasionally from horses (Thal & Moberg 1953; Mahaffey, quoted by Stableforth 1959).

The three streptococci are easily identified and differentiated by haemolysis, fermentation reactions and serological grouping (Table 2).

5. Streptococcal Diseases of Sheep

Diseases in sheep associated with streptococci do not appear to pose problems of major economic importance.

Cornell & Glover (1925) and Blakemore (1939) found that suppurative polyarthritis of lambs was associated with streptococcal infection. In 1941 Blakemore *et al.* reported that a bacteraemia occurred frequently and acute endocarditis occurred in 20% of cases. The streptococci resembled *S. dysgalactiae*, produced haemolysis on blood agar and serologically fell into Lancefield's group C.

Streptococcal endocarditis in lambs which had no joint lesions was described by Jamieson (1950). The infecting streptococcus was *S. faecalis* and was insensitive to penicillin. Thal & Moberg (1953) found groups D and L in addition to group C associated with a variety of disease conditions in sheep.

Vegetative endocarditis associated with *S. faecalis* and *C. pyogenes* infection was described by Paliwal *et al*. (1974). *Streptococcus zooepidemicus* was isolated from sheep with fibrinous pericarditis and from lambs with fibrinous pleuritis and pneumonia. Experimental inoculation into the trachea reproduced the lesions, deaths occurring in 6-7 days.

A streptococcus culture isolated from the milk of a ewe with mastitis was identified as *S. agalactiae* group B type III (CVL, Weybridge).

6. Streptococcal Diseases of Poultry

Only a few streptococcal infections in poultry have been reported. *Streptococcus zooepidemicus* has been associated with considerable losses in chickens in Canada (Genest & Nadeau 1944) in England (Buxton 1952) and in America (Peckham 1966).

Thal & Moberg (1953) isolated streptococci from birds. Most strains fell into groups C and D but some fell into groups E, G and L. Moberg & Thal (1954) isolated a single strain of group P from a chicken with anaemia.

In Italy, Agrimi (1956) isolated five strains of *S. faecalis* and one strain of *S. zooepidemicus* from six cases of streptococcosis in poultry.

Greening or non-haemolytic streptococci isolated from poultry and examined at CVL, Weybridge have almost invariably belonged to the enterococcus division. It seems likely that these streptococci are secondary invaders and are not a primary cause of disease. The majority of these strains have failed to react with commercially produced group D and group Q antisera.

One might have thought that with the advent of antibiotics, and particularly penicillin, streptococci would have diminished in importance as a cause of disease in animals but this does not appear to be the case. Despite the fact that apart from the enterococci, streptococci are very sensitive to penicillin and the development of resistance is not a problem, attempts to eradicate various species of streptococci have been only partially successful. There always appears to be a source of re-infection that makes it impossible or at least uneconomical to eradicate streptococcal infection completely.

7. References

AGARWALL, K. K., ELLIOTT, S. D. & LACHMANN, P. J. 1969 Streptococcal infection in young pigs, III. The immunity of adult pigs investigated by the bactericidal test. *Journal of Hygiene, Cambridge* **67**, 491-503.

AGRIMI, P. 1956 Studio sperimentale su alcuni focolai di streptococcosi nel pollo. *Zooprofilassi* 11, 491-501.
ARMSTRONG, C. H. & PAYNE, J. B. 1969 Bacteria recovered from swine affected with cervical lymphadenitis (jowl abscess). *American Journal of Veterinary Research* 30, 1607-1612.
BAZELEY, P. L. 1940 Studies with equine streptococci. 2. Experimental immunity to *Streptococcus equi*. *Australian Veterinary Journal* 16, 243-259.
BAZELEY, P. L. 1942a Studies with equine streptococci. 3. Vaccination against strangles. *Australian Veterinary Journal* 18, 141-145.
BAZELEY, P. L. 1942b Studies with equine streptococci. 4. Cross immunity to *Streptococcus equi*. *Australian Veterinary Journal* 18, 189-194.
BLAKEMORE, F. 1939 Joint-ill (polyarthritis) of lambs in East Anglia. *Veterinary Record* 51, 1207-1218.
BLAKEMORE, F., ELLIOTT, S. D. & HART-MERCER, J. 1941 Studies on suppurative polyarthritis (joint ill) in lambs. *Journal of Pathology and Bacteriology* 52, 57-83.
BRYANS, J. T. & MOORE, B. O. 1972 Group C streptococcus infections of the horse. In *Streptococci and Streptococcal Diseases*, eds Wannamaker, L. W. & Matsen, J. M. New York & London: Academic Press.
BUXTON, J. C. 1952 Disease in poultry associated with *Streptococcus zooepidemicus*. *Veterinary Record* 64, 221-223.
CHRISTIE, R., ATKINS, N. E. & MUNCH-PETERSEN, E. 1944 A note on a lytic phenomenon shown by Group B streptococci. *Australian Journal of Experimental Biology and Medical Science* 22, 197-200.
COLLIER, J. R. 1951 A survey of beta-haemolytic streptococci from swine. *Proceedings: 88th Annual Meeting of American Veterinary Medical Association,* 169-172.
COLLIER, J. R. 1956 Abscesses of the pharyngeal region of swine. Bacteriological examination of exudates. *American Journal of Veterinary Research* 17, 640-642.
COLLIER, J. R. 1965 Abscesses of swine. *Journal of the American Veterinary Medical Association* 146, 344-347.
CORNELL, R. L. & GLOVER, R. E. 1925 Joint-ill in lambs. *Veterinary Record* 5, 833-839.
CULLEN, G. A. 1967 Classification of *Streptococcus uberis* with biochemical tests. *Research in Veterinary Science* 8, 83-88.
ELLIOTT, S. D. 1966 Streptococcal infections in young pigs. I. An immunochemical study of the causative agent (PM Streptococcus). *Journal of Hygiene, Cambridge* 64, 205-212.
ELLIOTT, S. D., ALEXANDER, T. J. L. & THOMAS, J. H. 1966 Streptococcal infections in young pigs. II. Epidemiology and experimental production of the disease. *Journal of Hygiene, Cambridge* 64, 213-220.
ENGELBRECHT, H. & DOLAN, M. 1968 Vaccination of swine for jowl abscesses. Oral administration of Group E streptococcus vaccine (live culture-modified). *Veterinary Medicine and Small Animal Clinician* 63, 872-875
EVANS, E. T. R. 1957 Bacterial endocarditis of cattle. *Veterinary Record* 69, 1190-1202.
FIELD, H. I., BUNTAIN, D. & DONE, J. T. 1954 Studies on piglet mortality. I. Streptococcal meningitis and arthritis. *Veterinary Record* 66, 453-455.
FROST, A. J., WANASINGHE, D. D. & WOOLCOCK, J. B. 1977 Some factors affecting selective adherence of micro-organisms in the bovine mammary gland. *Infection and Immunity* 15, 245-253.
GENEST, P. & NADEAU, J. D. 1944 Observations chez la poule d'une épizootie due à *Streptococcus zooepidemicus*. *Canadian Journal of Comparative Medicine and Veterinary Science* 8, 342-349.
GOUGE, H. E., BROWN, R. G. & ELLIOTT, R. F. 1957 The control of laboratory-induced cervical (jowl) abscesses in swine by the continuous feeding of various levels of chlortetracycline. *Journal of American Veterinary Medical Association* 131, 324-326.
HONT, S. & BANKS, A. W. 1944 Streptococcal endocarditis in young pigs. *Australian Veterinary Journal* 20, 206-210.

HUGHES, K. L. 1975 *Streptococcus zooepidemicus* and infertility in horses. *Australian Veterinary Journal* **51**, 281-282.
JAMIESON, S. 1950 Recent investigations into certain diseases of sheep. *Veterinary Record* **62**, 772-774.
JONES, J. E. T. 1968 The cause of death in sows: a one year survey of 106 herds in Essex. *British Veterinary Journal* **124**, 45-55.
JONES, J. E. T. 1969 Experimental production of streptococcal endocarditis in the pig. *Journal of Pathology* **99**, 307-318.
MacLACHLAN, G. K., WILSON, C. D. & STUART, P. 1958 A herd outbreak of mastitis associated with *Streptococcus pneumoniae* infection. *Veterinary Record* **70**, 987-989.
McDONALD, J. S., McDONALD, J. T. & FREEMAN, B. A. 1976 Esculin hydrolysis by *Streptococcus dysgalactiae*. *American Journal of Veterinary Research* **37**, 1115-1117.
McNUTT, S. H. & PACKER, R. A. 1943 A study of some cases of *Streptococcus* infection in swine. *Veterinary Student* **6**, 68-69 and 95-97.
MILLAR, R. & FRANCIS, J. 1974 Relation of clinical and bacteriological findings to fertility in thoroughbred mares. *Australian Veterinary Journal* **50**, 351-355.
MOBERG, K. & THAL, E. 1954 Beta haemolytische Streptokokken einer neuen Lancefield-Gruppe. *Nordisk Veterinaermedicin* **6**, 69-72.
MOOR, C. E. DE, 1963 Septicaemic infections in pigs by haemolytic streptococci of new Lancefield groups designated R.S. and T. *Antonie van Leeuwenhoek* **29**, 272-280.
NEWSOM, I. E. 1937 Strangles in hogs. *Veterinary Medicine* **32**, 137-138.
NIELSEN, N. C., BILLE, N. & LARSEN, J. L. 1972 Incidence and causes of polyarthritis in suckling pigs. *Proceedings of the 2nd Congress of the International Pig Veterinary Society*, 1972, p. 119.
OLSEN, S. J. 1964 Undersøgelser over Gruppa L-Streptokokker. Forekamst og infektioner saerlig hos kvaeg og svin. *Veterinary Bulletin* **34**, 509 abstract 3185.
OLSEN, S. J. 1975 A mastitis control system based upon extensive use of control laboratories. *Proceedings of International Dairy Federation Seminar on Mastitis Control*, Reading University, 1975, pp. 410-421. International Dairy Federation, Brussels (Document 85).
PALIWAL, O. P., KRISHNA, L. & KULSHRESTHA, S. B. 1974 Vegetative endocarditis in sheep. *Indian Veterinary Journal* **51**, 42-44.
PECKHAM, M. C. 1966 An outbreak of streptococcosis (apoplectiform septicaemia) in White Rock chickens. *Avian Diseases* **10**, 413-421.
RØMER, O. 1948 Typebestemmelse of Pneumokokker paavist hos Dyr, specielt Kalve. *Medlemsblad for den Danske Dyrlaegeforening* **31**, 316-325
ROSS, R. F. 1972 Streptococcal infections in swine. In *Streptococci and Streptococcal Diseases*, eds Wannamaker, L. W. & Matsen, J. M. New York & London: Academic Press.
STABLEFORTH, A. W., HULSE, E. C., WILSON, C. D., CHODKOWSKI, A. & STUART, P. 1949 Herd eradication of *Str. agalactiae* by simultaneous treatment of all cows with 5 doses of 100,000 units of penicillin at daily intervals and disinfection. *Veterinary Record* **61**, 357-362.
STABLEFORTH, A. W. 1959 *Streptococcal diseases. I. Infectious Diseases of Animals Due to Bacteria*, Vol. 2, eds Stableforth, A. W. & Galloway, I. A. London: Butterworths Scientific Publications.
STUART, P., BUNTAIN, D. & LANGRIDGE, R. G. 1951 Bacteriological examination of secretions from cases of "Summer Mastitis" and experimental infection of non-lactating bovine udders. *Veterinary Record* **63**, 451-453.
THAL, E. & MOBERG, K. 1953 Serolígsche Gruppenbestimmung der bei Tieren vorkommenden beta-haemolytschen Streptokokken. *Nordisk Veterinaermedicin* **5**, 835-846.
TITTIGER, F. & ALEXANDER, D. C. 1971 Studies on the bacterial flora of condemned portions from arthritic hogs. *Canadian Journal of Comparative Medicine and Veterinary Science* **35**, 244-248.

TOLLE, A. 1975 Mastitis. The disease in relation to control methods. *Proceedings of International Dairy Federation Seminar on Mastitis Control*, Reading University, 1975, pp. 3-15. International Dairy Federation, Brussels. (Document 85).

WILLIAMS, D. M., LAWSON, G. H. K. & ROWLAND, A. C. 1973 Streptococcal infection in piglets. The palatine tonsils as portals of entry for *Streptococcus suis*. *Research in Veterinary Science* **15**, 352-362.

WILSON, C. D. 1957 The control of bovine mastitis. *Tijdschrift voor Diergeneeskunde* **82**, 615-630

WILSON, R. A. S. & LANCASTER, J. E. 1949 Bovine mastitis caused by *Diplococcus pneumoniae* (Pneumococcus). *Veterinary Record* **61**, 349.

WINDSOR, R. S. & ELLIOTT, S. D. 1975 Streptococcal infections in young pigs IV. An outbreak of streptococcal meningitis in weaned pigs. *Journal of Hygiene, Cambridge* **75.** 69-78

WINQUIST, G. 1945 Fibros and ulcerative endocarditis in domestic animals. *Scandinavisk Veteriner-Tidskrift* **35**, 575-585.

ZANEN, H. C. & ENGEL, H. W. B. 1975 Porcine streptococci causing meningitis and septicaemia in man. *Lancet* **1** (7919) 1286-1288.

Streptococci and the Human Oral Flora

J. M. HARDIE AND P. D. MARSH

*Oral Microbiology Department and MRC Dental Epidemiology Unit,
The London Hospital Medical College,
London, England*

CONTENTS

1. Introduction 157
2. Streptococci found in the mouth 158
   (a) Taxonomy and nomenclature 158
   (b) Isolation and identification 164
   (c) Serology of oral streptococci 167
   (d) Distribution of streptococcal species in the mouth . . 171
   (e) Adherence to surfaces 172
3. Some aspects of the biochemistry and metabolism of oral streptococci . . . 175
   (a) Acid production 175
   (b) Uptake of glucose and action of fluoride 176
   (c) Extracellular and intracellular polysaccharides . . . 177
4. Microbial interactions involving oral streptococci . . . 181
5. The relationship of oral streptococci to disease 185
   (a) Dental caries 185
   (b) Other oral infections 188
   (c) Dental bacteraemia and infective endocarditis 189
   (d) Other systemic infections 192
6. References 193

1. Introduction

THE MICROBIAL FLORA of the mouth is extremely complex and varies in composition from site to site (Socransky & Manganiello 1971; Hardie & Bowden 1974*a*). Streptococci comprise a significant proportion of the microflora, representing on average 45% of the total viable count of samples from the dorsal surface of the tongue, 46% of saliva, 28% of dental plaque and 29% of the gingival crevice flora (Socransky & Manganiello 1971). They are amongst the earliest colonizers of the oral cavity after birth, although some species such as *Streptococcus sanguis* and *S. mutans* fail to become established until after the first deciduous teeth have erupted (McCarthy *et al.* 1965; Carlsson *et al.* 1970*a,b*, 1975).

The microbial composition of dental plaque has received considerable attention because it is believed to be the main aetiological factor in the two major dental diseases, dental caries and periodontal disease (Hardie & Bowden 1975; Theilade & Theilade 1976). Streptococci have featured particularly in the search for a specific pathogen in dental caries. One species, *S. mutans*, has been shown to induce caries in experimental animals and there is also evidence for an association between this species and human caries. Largely because of the potential relationship of this streptococcus to dental caries a great deal of research has been carried out on many of the properties of *S. mutans* and the literature is very extensive. Other streptococcal species present in the mouth have received far less attention in recent years, although several of them may on occasions be involved in serious systemic infections such as infective endocarditis.

No attempt will be made in this review to cover all possible aspects of the subject in detail: to do justice to such a venture would require a book of its own. Of necessity, the subjects discussed reflect our own special interests and predilections.

2. Streptococci Found in the Mouth

(a) *Taxonomy and nomenclature*

Most of the streptococci isolated from the mouth are of the greening (viridans) or non-haemolytic (indifferent) type on blood agar and would be considered to fall within the 'viridans'-group of Sherman (1937). However, haemolysis does not seem to be a particularly useful or reliable property of these streptococci since all types of haemolysis (α, β and γ) occur amongst oral isolates (Colman 1976).

Classification of the viridans streptococci has traditionally been difficult, but the situation has greatly improved in recent years, thanks to the studies of Colman & Williams (1972), Carlsson (1968) and others. Much of the original literature has been reviewed recently and will not be repeated here in full (Colman 1976; Hardie & Bowden 1976a). Two extensive surveys of large numbers of streptococci from human clinical sources, using the more recent classification schemes, have been reported during the last year (Parker & Ball 1976; Facklam 1977).

The species of facultative streptococci most commonly found in the oral cavity are *S. mitior*, *S. sanguis*, *S. mutans*, *S. milleri* and *S. salivarius*. Since the same nomenclature is not used by all workers, a list of some of the alternative names is given in Table 1. Other species are also isolated on occasions, but usually in lower numbers. Enterococci, including *S. faecalis*, *S. faecium* and *S. durans*, can be found and their distribution in the mouth has been reported (Gold *et al.* 1975). **Streptococcus bovis** has not been isolated with certainty from the human mouth, although it can be isolated from dental plaque samples from some herbivorous animals (V. Dent, pers. comm.). The species listed in Table 1 generally account for most streptococcal isolates from the mouth, but other strains are encountered which do not appear to fit within existing taxa.

Some of the oral streptococci, especially certain strains of *S. mutans*, require

Table 1
Nomenclature of streptococci commonly found in the mouth

Colman and Williams (1972)	Carlsson (1968)	Facklam (1977)	Parker and Ball (1976)
S. mutans	S. mutans (Group II)	S. mutans	S. mutans
S. sanguis	S. sanguis (Group I:B)	S. sanguis I	S. sanguis
S. mitior	S. sanguis (Group I:A)	S. sanguis II	S. mitior Dx + mitior
	S. mitis (Group IV, V)	S. mitis	'viridans'
S. milleri	?	S. MG-intermedius	S. milleri
		S. anginosus-constellatus	
S. salivarius	S. salivarius (Group III)	S. salivarius	S. salivarius

Dx + mitior, dextran positive *S. mitior*.

carbon dioxide for growth on solid media. Relatively little information is available about the prevalence, distribution and properties of strictly anaerobic species in the mouth. Anaerobic Gram positive cocci have been reported in cultural studies on dental plaque (Socransky *et al.* 1977), and have also been isolated from deep carious dentine (Edwardsson 1974), infected pulp chambers and root canals (Berg & Nord 1973; Kantz & Henry 1974; Wittgow & Sabiston 1975; Sundquist 1976) and dental abscesses (Sabiston & Gold 1974). It would appear that elaborate anaerobic isolation techniques are necessary for recovery of anaerobic streptococci from oral material. In our own studies on supragingival plaque, using conventional anaerobic jar techniques with prereduced media, anaerobic streptococci have rarely (if ever) been isolated. There is clearly a need for further detailed studies in this area.

(i) Streptococcus mutans

Although this species is not listed in the latest edition of *Bergey's Manual of Determinative Bacteriology*, it is generally recognized as a well-defined species with a distinctive set of characteristics (Clarke 1924; Edwardsson 1968; Facklam 1974; Perch *et al.* 1974; Hardie & Bowden 1976*a*). Strains of *S. mutans* typically ferment mannitol and sorbitol, produce extracellular glucans (dextran) from sucrose, hydrolyse aesculin but not arginine (with the exception of serotype b strains) or starch, produce acetoin from glucose, do not usually produce hydrogen peroxide, tolerate 6.5% sodium chloride or grow at 45°C. Numerical

Table 2
Some characteristics of streptococci resembling S. mutans

Proposed species name	DNA base ratio* (mol% G+C)	Serotype†	Cell wall†† carbohydrates
S. *mutans*	36–38	c, e, f	Glucose, rhamnose
S. *rattus*	41–43	b	Galactose, rhamnose
S. *cricetus*	42–44	a	Glucose, galactose, rhamnose
S. *sobrinus*	44–46	d, g	Glucose, galactose, rhamnose
S. *ferus*	43–45	c	Not known

* Coykendall (1970), Dunny *et al.* (1972).
† Bratthall (1970), Perch *et al.* (1974), Coykendall *et al.* (1976).
†† Hardie & Bowden (1974).
Data from Coykendall (1977).

taxonomic studies on oral streptococci have consistently shown a discrete cluster of strains corresponding to *S. mutans* (Carlsson 1968; Colman 1968; Drucker & Melville 1971).

As discussed later, several serotypes of *S. mutans* have been described (Bratthall 1970; Perch *et al.* 1974) and these can be correlated with the carbohydrate composition of the cell walls (Hardie & Bowden 1974b). Studies on the DNA base content and DNA hybridization between strains of *S. mutans*, together with other biochemical and serological data, have led to the proposal that different named subspecies be recognized (Coykendall 1974). More recently Coykendall (1977) has suggested that the proposed subspecies be elevated to species status, namely *S. mutans*, *S. rattus*, *S. cricetus*, *S. sobrinus* and *S. ferus*. Some of the distinguishing features of these 'species' are shown in Table 2. It is not clear at present how widely acceptable and valuable these proposals will prove to be in practice. In the first instance it would perhaps be more helpful if all laboratories and reference books were to recognize strains resembling *S. mutans* (collectively) before attempting to introduce the complexity of adopting five separate names for the different genetic varieties. Some biochemical characteristics correlate quite well with the different subdivisions of *S. mutans*, such as the production of ammonia from arginine by serotype b strains (*S. rattus*) and such observations have led to the development of a biochemical scheme for the recognition of different serotypes (Shklair & Keene 1974). This scheme was originally formulated on the basis of the reactions of a small number of reference strains and it is not certain how closely these are matched by wild-type strains.

Further subtyping of *S. mutans* strains may be useful for epidemiological studies on human populations and typing by bacteriocins or 'bacteriocin-like substances' has been described by several workers (see Section 4).

(ii) Streptococcus sanguis

This species was originally isolated from blood of patients with bacterial endocarditis, but is now recognized to be one of the most common and numerous streptococci in the mouth (Carlsson 1965, 1967). The earlier literature on *S. sanguis* and the relationship to Lancefield group H streptococci has been discussed elsewhere (Colman & Williams 1972; Colman 1976; Hardie & Bowden 1976).

Streptococcus sanguis only appears in the mouth of infants at about the age of six months, corresponding with the time that the first teeth erupt (Carlsson *et al*. 1970*b*). Thereafter this species constitutes a significant proportion of the streptococci present in dental plaque and appears to be one of the earliest colonizers of the clean tooth surface (Tinanoff *et al*. 1976; Socransky *et al*. 1977). By comparison, only low levels of *S. sanguis* have been found in human faeces (van Houte *et al*. 1971), so it would appear that the mouth is the most likely source of this organism when it occurs in cases of endocarditis. Strains of *S. sanguis* have also been isolated from soil (Gledhill & Casida 1969).

Strains of *S. sanguis* typically hydrolyse arginine and aesculin, produce hydrogen peroxide and form extracellular glucan (dextran) when grown in the presence of sucrose. The latter property gives rise to the production of characteristic hard, adherent colonies on 5% sucrose agar. Non-polysaccharide-producing strains resembling *S. sanguis* are also found occasionally.

There is some confusion over which strains should be called *S. sanguis*, as indicated in Table 1. Early studies on the serology of this species indicated that there were different serotypes designated I, II and I/II (Washburn *et al*. 1946), represented by strains NCTC 7863 (ATCC 19556), NCTC 7864 (ATCC 10557) and NCTC 7865 (ATCC 10558). Strains resembling NCTC 7864 (type II) formed separate clusters from type I sanguis strains in the numerical taxonomic studies of Colman (1968) and Carlsson (1968), and this evidence, together with cell wall, serological and genetic differences, provides a strong argument for separating these into another species (*S. mitior*). In the recent report by Parker & Ball (1976) the names *S. sanguis* and *S. mitior* are used to describe these two groups of organisms, whereas Facklam (1977) at the Streptococci Reference Laboratory in America prefers to retain the terms *S. sanguis* biotypes I and II. Most workers find occasional 'intermediate' strains which do not fall clearly into either of these two species or biotypes. In the study quoted above (Parker & Ball 1976), 37% of strains designated *S. sanguis* reacted with group H antiserum.

(iii) Streptococcus mitior

The name *S. mitior* orginates from Schottmuller (1903) and as mentioned above has been proposed by Colman & Williams (1972) for a group of strains which formed a homogeneous cluster in a numerical taxonomic study (Colman

1968). These strains were found to be genetically related in transformation tests (Colman 1969). They possess a distinctive cell wall sugar pattern from which rhamnose is absent but ribitol teichoic acid present (Colman & Williams 1965). Strains allocated to this species produce hydrogen peroxide, fail to hydrolyse arginine or aesculin and give variable reactions in the Voges-Proskauer test. Some strains produce extracellular glucan (dextran) and these often give colonies on sucrose agar which are indistinguishable from those of *S. sanguis*.

If both dextran positive and negative strains are accepted within the definition of *S. mitior*, this species corresponds to the clusters of strains designated I:A (*S. sanguis*) and V:A in the numerical study of Carlsson (1968). The terms *S. mitior* and dextran positive mitior have been used by Parker & Ball (1976) but other workers prefer alternative names to describe similar strains. According to the scheme used by Facklam (1977), *S. sanguis* II is the name given to strains which ferment raffinose, whilst raffinose negative isolates are designated as *S. mitis*. Dextran formation is regarded as a variable characteristic in both species.

Streptococcus mitior is widely distributed in the mouths of man and in animals (V. Dent, pers. comm.), and is also commonly isolated from the blood in dental bacteraemias and from cases of bacterial endocarditis (see below). This species probably corresponds to many of the isolates described in the literature as *S. viridans* or *S. mitis* (when quoting from original papers where the authors have used the name *S. mitis*, this name has been retained). There is a need for further taxonomic and serological studies on strains resembling *S. mitior*, in order to clarify whether it should be regarded as one, rather heterogeneous, species or subdivided into two or more taxa.

(iv) Streptococcus salivarius

Strains of *S. salivarius* (Sherman *et al.* 1943) usually produce extracellular levan when grown in sucrose (Niven *et al.* 1941a, b). This property results in the production of characteristic large, mucoid colonies on sucrose agar. There does not appear to be any disagreement amongst different workers concerning the physiological properties and nomenclature of this species, and numerical taxonomic studies have confirmed its homogeneity (Colman 1968; Carlsson 1968).

Although the physiological characteristics of *S. salivarius* are well described (e.g. Cowan & Steele 1974), difficulty can occasionally be experienced when attempting to identify strains with a limited number of tests if the characteristic mucoid colonies are not observed on sucrose agar. The ability of some strains to hydrolyse urea (Colman 1976) may prove to be a useful additional property.

(v) Streptococcus milleri

This species, first described by Guthof (1956), is another example where differences of opinion exist as to the correct terminology. The name *S. milleri* has been used by several investigators recently and detailed descriptions of its characteristics have been published (Colman & Williams 1972; Mejare & Edwardsson 1975; Hardie & Bowden 1976; Parker & Ball 1976). The species as described by these workers includes strains which have in the past been designated *Streptococcus* MG (Mirick *et al.* 1944). Small colony-forming haemolytic streptococci possessing Lancefield group F or G antigens also fall into this classification. Facklam (1977) has reported the properties of strains resembling *S. milleri*, but prefers to divide these into *S. MG-intermedius* (lactose +) and *S. anginosus-constellatus* (lactose −).

Strains of *S. milleri* usually hydrolyse arginine and aesculin and produce acetoin from glucose. They do not produce extracellular polysaccharides from sucrose, form hydrogen peroxide or ferment mannitol or sorbitol. *Streptococcus milleri* isolates are commonly resistant to sulphonamides and can consequently grow on the selective media designed by Carlsson (1967) for isolation of *S. mutans* (Mejare & Edwardsson 1975) (see also pp. 380-383).

As pointed out by Colman (1976), *S. milleri* is a reasonably homogeneous species based on physiological properties, but is very heterogeneous serologically.

Comparatively few numerical taxonomic studies have been carried out on collections of viridans and non-haemolytic streptococci (Colman 1968; Carlsson 1968; Drucker & Melville 1971). These studies have supported the division of oral streptococci into the taxa listed in Table 1, but do not solve the problems of nomenclature of strains resembling *S. mitior* and *S. milleri*. A numerical study on a large number of oral streptococci from a variety of sources and including physiological tests, cell wall analysis and serology, is currently being undertaken in this laboratory. Hopefully this will shed some light on the classification of 'intermediate' strains which do not fit exactly into the presently recognized species.

Chemotaxonomic methods have not been used extensively for studies on the oral streptococci. Colman & Williams (1956) demonstrated that a variety of cell wall carbohydrate patterns exists amongst the streptococci and that certain features, such as absence of rhamnose in *S. mitior,* can be correlated with particular species. In strains of *S. mutans* it has been found that differences in cell wall carbohydrate composition can be related to serotype (Hardie & Bowden 1974*b*). Fatty acid fingerprints of *S. mutans* have been examined by Drucker and colleagues, and shown to be dependent upon cultural conditions such as temperature, age of culture, substrate, dilution rate (in continuous culture), aeration, pH and concentration of micronutrients (Drucker *et al.* 1976; Drucker & Veazey 1977). Characterization of whole cells by pyrolysis gas chromatography has been tried as a possible rapid method for identification of oral streptococci, but has not so far resulted in a scheme which can be generally adopted (Stack *et al.* 1973). Other techniques such as electrophoresis and isoelectrofocusing of whole cell lysates may also prove to be of value in classification of streptococci. When used with representative strains of *S. mutans*, this approach yielded patterns of stained protein bands which could be related to existing subdivisions within this species (Russell 1976).

Examination of the lipid composition of streptococci may be a useful area for future chemotaxonomic studies. This type of analysis has proved to be extremely valuable in the classification of coryneform bacteria (Goodfellow *et al.* 1976; Minnikin *et al.* 1977).

(b) *Isolation and identification*

The appropriate method to be used for collection of bacteriological samples from the mouth depends very much upon the particular site or habitat under investigation, as has been discussed previously (Hardie & Bowden 1974*a*, 1976*b*). The choice of a suitable method for dispersion of plaque samples in quantitative studies is a problem, since there is a danger of destroying certain bacteria, but facultative streptococci seem to survive most of the commonly used methods, including sonication, and are not generally sensitive to oxygen exposure. For recovery of strictly anaerobic streptococci, however, precautions need to be taken to avoid exposure of the samples to atmospheric oxygen during processing.

Oral streptococci can be isolated on non-selective media such as blood agar, but selection of streptococcal colonies may be difficult when they are overgrown by the many other organisms which are commonly present. In addition it is difficult to distinguish between different species of oral streptococci from their colonial appearance on blood agar. Several selective media have been devised for isolation of streptococci from the mouth and throat, and some of these are described in detail elsewhere in this symposium (pp. 380-383). Since some oral streptococci produce extracellular polysaccharides from sucrose, a number of the media include high concentrations of this sugar. On such media strains of *S. mutans*, *S. sanguis*, *S. mitior* and *S. salivarius* tend to produce characteristic colonial forms, and these have often been used as a basis for presumptive identification in epidemiological investigations. However, colonial morphology alone can be misleading and it is desirable that such identification should be confirmed by other criteria.

The streptococci commonly present in the mouth can usually be identified to species level by conventional biochemical and physiological tests (Cowan & Steel 1974). In our laboratory a short set of seven tests is routinely used for screening fresh isolates (Hardie & Bowden 1976). The usual reactions of oral species in selected physiological tests are shown in Table 3; the data used for compiling this table were derived from our own observations together with

Table 3
Some physiological reactions of streptococci found in the mouth

Tests	S. mutans	S. sanguis	S. mitior	S. milleri	S. salivarius
Fermentation of					
mannitol	+	−	−	−	−
sorbitol	+	−	−	−	−
Hydrolysis of					
arginine	−	+	−	+	−
aesculin	+	+	−	+	+
Production of					
acetoin	+	−	V	+	V
H_2O_2	−	+	+	−	−
Polysaccharide	+	+	V	−	+
Fermentation of					
raffinose	+	V	V†	−	+
inulin	+	+	−	−	+
lactose	+	+	+	+††	+
trehalose	+	−/V*	V	+	+/V*
melibiose	+/V*	V	V	−	−
salicin	+	+	V	+	+
Growth in					
6.5% NaCl	−	−	−	−	−
10% bile	V*	V*	−*	V*	V*
40% bile	V*	V*	−	−/V*	−/V*

+, 85–100% positive; −, 0–15% positive; V, 16–84% positive.
* Variable reports from different authors.
† Facklam (1977) divides strains into *S. sanguis* II and *S. mitior* on basis of raffinose.
†† Facklam (1977) divides strains into *S. MG-intermedius* and *S. anginosus-constellatus* on basis of lactose.
For references see text.

several published reports (Carlsson 1967, 1968; Colman & Williams 1972; Mejare & Edwardsson 1975; Parker & Ball 1976; Facklam 1977).

The first seven tests listed in Table 3 are those employed as a short set for routine identification. This scheme has now been used to examine 2808 streptococcal isolates from approximal dental plaque during the course of epidemiological studies on 12- to 14-year-old school children (Bowden *et al.* 1975, 1976; Hardie *et al.* 1977). Over 85% of the strains could be identified to one of five named species by means of this system (Table 4). Dextran positive and negative strains of *S. mitior* have not been distinguished in this table. Occasional isolates of enterococci (Lancefield group D) have also been found but are not shown in Table 4.

A few variations from the typical combinations of reactions given in Table 3 have been found consistently amongst streptococci isolated on repeated

Table 4
Identification of streptococci from approximal dental plaque

Number of subjects	S. mutans	S. mitior	S. sanguis	S. milleri	S. salivarius	Unidentified	Total*
10	103	103	49	51	67	49	422
20	80	106	81	66	50	92	475
39	357	575	365	227	117	211	1911
Total number of strains	540	784	495	344	234	352†	2808

Summary of results from three epidemiological studies.
* The total number of strains includes a few enterococci which are not listed separately.
† The unidentified strains represent 12.5% of the total number examined.

Table 5
Differentiation of S. mutans *from some other species*

Tests	S. mutans	S. bovis 1	S. bovis 2	S. faecalis	S. uberis
Fermentation of					
mannitol	+	+	–/V	+	+
sorbitol	+	–	–	+	+
Hydrolysis of					
arginine	–	–	–	+	V
aesculin	+	+	+	+	+/V
starch	–	+	–	–	V
hippurate	–	–	–	V	+/V
Production of					
acetoin	+	+	+	V(?)	V
H_2O_2	–	–	(?)	–	–
Polysaccharide	+	+	–	–	–
Growth in/at					
45°C	V	+	+	+	–
60°C/30 min	–	–	–	+	–
40% bile	–	+	+	+	+
6.5% NaCl	–	–	–	+	V

+, 85–100% positive; –, 0–15% positive; V, 16–84% positive; (?), reactions uncertain.

occasions from particular subjects. These include strains resembling *S. sanguis* which ferment sorbitol, and mannitol negative *S. mutans* strains. Within *S. mutans* it is known that some serotypes give slightly different biochemical reactions (Shklair & Keene 1974; Perch *et al.* 1974; Hardie & Bowden 1976). Up until now we have not isolated any examples of the arginine positive, serotype b strains of *S. mutans*.

Although *S. mutans* is quite easily distinguishable from the other streptococci commonly isolated from the human mouth, it could be confused with streptococci from some other sources, such as *S. bovis*, *S. faecalis* and *S. uberis*. Some of the characteristics which should enable *S. mutans* to be differentiated from these species are listed in Table 5.

(c) *Serology of oral streptococci*

Several early attempts at producing an all-embracing serological classification of the viridans streptococci, along the lines developed by Lancefield for the haemolytic streptococci, failed to produce a satisfactory scheme, (e.g. Lancefield 1925*a*, *b*; Selbie *et al.* 1949; Soloway 1942; Williamson 1964). It is obvious from these studies that the viridans and non-haemolytic streptococci are antigenically heterogeneous. The early studies were probably frustrated by the lack of clear taxonomic bases from which to start serological investigations. Since the recognition of several reasonably well-defined species, it has been proved possible to develop more useful serological subdivisions within some of these species.

(i) Streptococcus mutans

The existence of serological differences between strains of *S. mutans* was first observed by Zinner & Jablon (1968) and Crousaz & Guggenheim (1966). Subsequently, Bratthall (1970) reported the present of five distinct serological types (originally called 'groups') within the species. These serotypes, designated a–e, were demonstrated by immunoelectrophoresis and immunodiffusion of acid extracted antigens. The strains designated serotype e were found to give precipitin reactions with Lancefield group E sera, as do some strains of *S. uberis* (Cullen 1967). It has been shown that *S. mutans* type e antigen is not identical with the Lancefield group E antigen (Perch *et al.* 1974; Linzer 1976).

Bratthall's serotypes have now been extended to 7, by the recognition of types f and g (Perch *et al.* 1974). Cell wall agglutination tests and cell wall carbohydrate patterns have been found to correlate well with the serotypes based on precipitin tests (Hardie & Bowden 1974). Several groups of workers have developed fluorescent antibody techniques for the identification of the serotypes of *S. mutans* (Jablon & Zinner 1966; Bratthall 1972; Grenier *et al.* 1973; Jablon *et al.* 1976; McKinney & Thacker 1976; Thomson *et al.* 1976; Hamada *et al.* 1976).

In our laboratory, a detailed study of the antigenic composition and serological interrelationships among the oral streptococci is currently in progress, with the ultimate aim of developing a standardized precipitin typing scheme. Studies on *S. mutans* have been based on (1) confirming the serotype-specific nature of the cell wall carbohydrate antigens, not just among a few type strains but also in a range of fresh isolates; (2) devising a standardized method for producing antisera which will react with these carbohydrate antigens; (3) comparing various extraction techniques to see which yield the carbohydrate antigens and (4) the identification of the serological cross-reactions resulting from the chosen extraction technique (Marsh *et al.*, in preparation). Preliminary studies have shown that autoclave extracts of whole cells yield both the specific and cross-reacting carbohydrate antigens of *S. mutans*, together with negatively charged antigens, which are possibly teichoic acid, that cross-react widely with streptococci of other serotypes and species (Marsh *et al.* 1975, 1977). The occurrence of serological cross-reactions between strains of *S. mutans* and other streptococci due to cell wall teichoic acid or membrane lipoteichoic acid has been noted by several investigators (Markham *et al.* 1975, Chorpenning *et al.* 1975; Knox *et al.* 1976, Knox & Wicken 1976; Hardie & Bowden 1976).

Strong cross-reactions are found between *S. mutans* serotypes a, d and g,

Table 6
Type specific carbohydrate antigens of S. mutans *serotypes a, d and g*

Components (mg 100 ml^{-1})	a (HS6)	d (B13)	g (6715)
Rhamnose	0.0	0.9	0.0
Galactose	54.0	62.0	60.6
Glucose	10.4	33.0	10.3
Galactosamine	5.4	0.0	0.0
Phosphorus	0.3	0.3	0.4
Protein	5.0	1.6	9.5
Major determinant	D-glucose	D-galactose	D-galactose
Minor determinant	D-galactose (a–d determinant)	?	(D-glucose ?) ?

Data from Mukasa & Slade (1973), Linzer & Slade (1974), Iacono *et al.* (1976), Linzer (1976).

and between types c and e (Bratthall 1972, Perch *et al.* 1974). Antisera can be made type specific by reciprocal absorptions, although this may lead to a considerable drop in homologous antibody titre. At present commercially prepared antisera against *S. mutans* are not available.

In the last few years there have been a number of detailed studies on the immunochemistry of representative strains of *S. mutans* and in each case a specific carbohydrate type antigen has been defined. The results of these studies have been reviewed recently and some of the characteristics of the type antigens are summarized in Tables 6 and 7 (Linzer 1976; Linzer *et al.* 1976; Iacono *et al.* 1976).

Table 7
Type specific carbohydrate antigens of S. mutans *serotypes b, c, e and f*

Components (mg 100 ml^{-1})	b (FAI)* 1	b (FAI)* 11	c (Ingbritt)	e (MT703)	f (OMZ 175)
Rhamnose	47	0.3	69	56	49
Galactose	27	29	–	–	–
Glucose	0.5	5.1	29	37	47
Galactosamine	1.9	3.0	–	–	–
Phosphorus	2.3	1.3	0.5	0.3	Trace
Protein	5.4	39	0.5	5.0	Trace
Major determinant	Galactose	Galactose	α-D-glucose	β-D-glucose ?-D-glucose	α-D-glucose 6-D-glucose

* 1 and 11 represent different fractions from strain FAI.
Data from: Wetherell & Bleiweis (1975), Hamada *et al.* (1976), Linzer (1976), Linzer *et al.* (1976).

Epidemiological studies on various population groups show some variation in the distribution of *S. mutans* serotypes, but types c and d appear to be the most prevalent in almost every community examined (Bratthall 1972; Hamada *et al.* 1976; Qureshi *et al.* 1977).

(ii) Streptococcus sanguis

In the original serological subdivision of *S. sanguis* by Washburn *et al.* (1946) three types were described (I, II and I/II) but no common group antigen was identified. As mentioned previously, the type II strains are now regarded as belonging to a separate species, *S. mitior*. The relationship between *S. sanguis* and group H streptococci (Hare 1935) was studied by several early workers but not completely resolved (for reviews see Colman 1970; Colman & Williams 1972; Hardie 1975). Studies on cell wall sugar composition and cell wall agglutination reactions of strains of *S. sanguis* (excluding those now allocated to *S. mitior*) indicate the existence of at least two or three distinct types which cannot be distinguished by physiological tests (Hardie 1975).

Rosan (1973) analysed the antigens present in autoclave extracts (Rantz & Randall 1955) of 45 strains of *S. sanguis* and *S. mitior* by immunodiffusion and immunoelectrophoresis, using antiserum against strain M-5. Five antigens were detected (designated a–e), and strains were allocated to type I, type II or a 'heterogeneous group' according to the combination of these antigens present. It has been proposed that the a antigen, which is thought to be a glycerol teichoic acid, be considered as group H antigen (Rosan 1976). Both membrane and cell wall lipoteichoic acids have also been detected in the *S. sanguis* strain ATCC 10556 (Chiu *et al.* 1974).

Practical problems still remain when attempting to identify streptococci as group H, since the choice of immunizing strains used by commercial antiserum producers has not been standardized. Thus 'English group H' is not the same as 'American group H'.

(iii) Streptococcus mitior

Reference has already been made to the physiological, serological and cell wall differences between *S. mitior* and *S. sanguis*, but there is little information available concerning the antigenic structure of *S. mitior*. Kalonaros & Bahn (1965) described two serological groups within strains identified as *S. mitior*, but since all the strains examined contained glucose and rhamnose in their cell walls, they do not appear to be the same as *S. mitior* defined by Colman & Williams (1972). Some strains of *S. mitior* are known to cross-react in precipitin tests with sera of Lancefield groups O, M and K (Colman 1976) and recently a cross-reaction with group B serum has been observed with some animal isolates (V. Dent, pers. comm.). Precipitating antigens common to strains of *S. sanguis* and *S. mitior* have also been recorded (Washburn *et al.* 1946; Dodd 1949; Rosan 1973; Hardie & Bowden 1976).

Preliminary observations on a collection of oral isolates of *S. mitior*, including dextran positive and negative strains, indicate that they are serologically heterogeneous (Woods & Hardie, unpublished).

(iv) Streptococcus milleri

Although this species is physiologically homogeneous, there appears to be a considerable degree of serological heterogeneity and cross-reactions are known to occur with Lancefield groups A, C, F, G and K sera (Colman 1976). In the study by Parker & Ball (1976), 81 strains were identified as *S. milleri*, of which five reacted with group A serum, five with group C, three with group G and seven with group F. The comparable study by Facklam (1977) showed that of 231 strains allocated to the species *S. MG-intermedius*, nine reacted with group A, six with group C, 53 with group F, three with group G, two with group K and six with group A variant serum. Strains of *anginosus-constellatus* also yielded reactions with sera of groups A, C, F and G.

A more detailed antigenic analysis of a representative selection of strains of *S. milleri* would be extremely valuable.

(v) Streptococcus salivarius

This is another example of a species which is physiologically homogeneous but serologically heterogeneous (Colman 1976). Strains designated type I (Sherman *et al.* 1943) usually react with Lancefield group K serum and may also react with antiserum against streptococcus MG (Mirick *et al.* 1944; Williams 1956; Montague & Knox 1968). Type II strains of *S. salivarius* fail to precipitate with either of these antisera (Montague & Knox 1968). Out of 81 strains of *S. salivarius* examined by Facklam (1977), 35 reacted with group K and three with group F sera, whilst Parker & Ball (1976) reported 7/14 strains group K positive.

(d) *Distribution of streptococcal species in the mouth*

The relative numbers of different streptococcal species vary from site to site within the mouth. Thus, for example, *S. salivarius* is found in high numbers on the dorsal surface of the tongue and in saliva, whereas it only constitutes a small proportion of the streptococci present in dental plaque (Krasse 1954; Socransky & Manganiello 1971). The preferred habitat of *S. mutans* appears to be the tooth surface, but even within dental plaque this species is usually localized to particular areas (Gibbons *et al.* 1974). A summary of the proportional distribution of streptococci within the oral cavity, based on several published reports, is given in Table 8 (Gibbons & van Houte 1975*b*; Gold *et al.* 1975; Mejare & Edwardsson 1975; Carlsson 1967; Liljemark & Gibbons 1972).

Table 8
The prevalence of streptococcal species in various parts of the mouth

Species	Prevalence in				
	Saliva	Tongue	Cheek	Plaque	Gingival crevice
S. salivarius	High	High	Low or moderate	Low	Low
S. mitior (mitis)	High	High	High	High	Moderate
S. sanguis	Moderate	Low or moderate	Moderate	High	Moderate or high
S. milleri	Low	Low	Low	Moderate	High
S. mutans	Low	Low	Low	Usually low sometimes high	Moderate
Enterococci	Low	Low	Low	Low	Low

High, > 15% of total viable count; moderate, 4–15% of total viable count; low, < 4% of total viable count.

Various ecological factors may be of significance in determining the distribution and relative proportions of different bacteria within the oral cavity. One property which has attracted much attention recently is the ability of various species to adhere to different types of surface (Gibbons & van Houte 1975b).

(e) *Adherence to surfaces*

The mouth has two distinctly different surfaces to which bacteria adhere. First, teeth, which are hard, non-shedding surfaces and which, consequently, allow the accumulation of large masses of bacteria together with their extracellular products (dental plaque). Secondly, epithelial surfaces (gums, cheek, tongue) which are constantly being shed and which, therefore, do not allow such large bacterial deposits to accumulate.

As shown in Table 8, the prevalence of streptococci in the mouth varies according to the surface sampled. Oral streptococci have been found to adhere preferentially to particular surfaces *in vitro*. This *in vitro* selective adherence correlates closely with the *in vivo* distribution of these organisms (Gibbons & van Houte 1971, 1975b; Liljemark & Gibbons 1972). For example, a recent study compared the selective ability of indigenous and pathogenic bacteria to adhere to and colonize the dorsum of the tongue of humans and rats (Gibbons *et al.* 1976). *Streptococcus salivarius* and *S. sanguis* were isolated from the tongues of humans but not from rat tongues, whereas *S. faecalis* and a serum-requiring diphtheroid were found only on rat, but not human tongues. When adherence of these strains to human and rat dorsal tongue surfaces was compared, *S. salivarius* adhered in highest proportions to the human cells and *S. faecalis* and the diphtheroid to rat cells, both *in vivo* and *in vitro*. By means of scanning electron microscopy, the colonization was found not only to be very specific but also highly localized.

The mechanisms involved in bacterial adherence are poorly understood, but two recent approaches may yield a better understanding of the phenomenon.

Olsson *et al.* (1976*a*) studied the electrophoretic mobilities of 28 batch-grown strains of *S. mutans*, *S. sanguis* and *S. salivarius* under standardized conditions. From this study, the zeta potential for each strain was calculated (a theoretical value expressing the charge density at the surface of particles). Significant differences were found both between different species and different serotypes, although some strains of the same serotypes of *S. mutans* had identical or almost identical potentials. A standardized method for studying the adherence of oral streptococci to solid surfaces was devised (Olsson & Krasse 1976). Using this method it was found that strains of oral streptococci belonging to the same species but with different zeta potentials adhered with different strengths. However, a strain of *S. sanguis* was found to adhere more strongly than a strain of *S. salivarius* even though both strains had approximately identical zeta potentials. Thus, it would appear that once electrostatic forces have been overcome, species or type-specific properties of the cell surface may be of determinant importance in adherence (Olsson *et al.* 1976*b*).

Rutter & Abbott (1978) have used a rotating disc method to study the adherence of batch-grown strains of *S. mitior* and *S. salivarius* to glass and polystyrene surfaces from salt solutions. Because the initial interaction between the bacterium and the substrate depends on the physical properties of both surfaces, the well-established principles of colloid science have been used by these authors. The Derjaguin-Landau and Verway-Overbeek theory postulates that interactions between a particle and a surface involves both attractive and repulsive components. Rutter and Abbott calculated all the forces involved enabling them to derive approximate Hamaker constants for both species (values directly related to the attractive interaction property of two surfaces). When these constants were compared to that of polystyrene latex particles the bacteria were found to have a higher collection efficiency by the rotating disc. The authors concluded that additional factors had to be taken into account, such as extracellular polymers associated with the bacterial cell wall absorbing to the surface involved and forming some sort of polymer bridging. Such a system, involving an initial reversible phase depending on a balance between van der Waals attractive forces and electrostatic repulsive forces between the bacterium and the surface, together with a time-dependent irreversible phase involving polymer bridging, has been proposed for the attachment of some marine bacteria, to hard surfaces (Marshall *et al.* 1971).

Rutter and Abbott found *S. salivarius* had a greater tendency to deposit than *S. mitior* under the conditions employed. The *S. salivarius* cells possessed a 'fuzzy coat' when viewed under the electron microscope, and when this layer was removed with trypsin treatment the cells behaved similarly to polystyrene particles.

These studies suggest that the outer layers of the bacterial cell envelope are important in adhesion. As growth conditions can influence the composition of cell walls a chemostat would be useful in providing standardized cells for these valuable experiments.

If a two-stage adherence system does operate in the mouth then the initial electrostatic interaction between bacteria and surfaces could also involve teichoic acid. Sucrose-grown *S. mutans* strains have an increased negative charge (Kelstrup & Funder-Nielsen 1972; Melvaer *et al.* 1974) which is probably due to the production of extracellular teichoic acid (Markham *et al.* 1975) which can form complexes with insoluble polysaccharides in the bacterial coat. Blood group substances have been detected on the tooth surface. Both teichoic acid (Wicken & Knox 1975) and glucosyl transferases (cited by Rolla 1976) react with red blood cells and salivary blood-group substances respectively, and so by this method contribute to the initial reversible phase of adhesion. Calcium bridging between the negatively charged bacteria and the negatively charged pellicle on the tooth surface may also be involved in this initial electrostatic interaction. The evidence for this has been reviewed by Rolla (1976).

Insoluble extracellular glucans synthesized from sucrose by glucosyl transferases (see Section 3.c) appear to be involved in the irreversible attachment of bacteria to the tooth surface and much of the evidence for this has been reviewed by McCabe *et al.* (1976). These authors propose that in both cell to cell, and in cell to surface adherence, glucosyl transferases serve to synthesize glucan, while the actual binding of glucan occurs at a unique cell surface site. The nature of bacterial surface components and the role of polymers in adherence has also been reviewed by Gibbons & van Houte (1975*b*), van Houte (1976) and Rolla (1976).

Burkhardt & Guggenheim (1976) proposed that the Roseman hypothesis (Roseman 1970) may explain many of the adherent interactions involved in bacterial colonization of the mouth. In this hypothesis, cell-associated enzymes and substrates are essential for selective cell recognition and adherence. However, Schachtele *et al.* (1976), reported that between 4 and 19% of the total salivary flora from schoolchildren possessed dextranase-producing activity with most of the active strains being similar to *S. mitis*. These strains of *S. mitis* were capable of blocking sucrose-dependent adherence between both *S. mutans* and *S. sanguis*, and between these streptococci and solid surfaces, by competing for cell surface glucan binding sites in plaque, destroying glucans capable of functioning as cell binding sites, and by suppressing the synthesis of water-insoluble glucan from sucrose. Thus, these indigenous bacteria may be capable of preventing the irreversible colonization of teeth by *S. mutans* and *S. sanguis*.

Bacterial adherence to surfaces is of fundamental importance in many areas of microbiology. An understanding of the mechanisms involved may allow the development of measures to prevent adherence. Studies on negative chemotaxis has enabled surfaces to be treated with a variety of non-toxic chemicals, such as tannins, which have prevented all bacterial accumulation (reviewed by Mitchell 1976). As adherence is essential for colonization by pathogenic micro-organisms in all fields, not just oral microbiology, investigations such as those described in this section could have far-reaching implications.

3. Some Aspects of the Biochemistry and Metabolism of Oral Streptococci

A great deal of literature has been published on the metabolism of oral streptococci in an attempt to gain greater insight into the properties that may be related to dental caries. Space will only allow a few of these studies to be discussed here. The unique advantages afforded by continuous culture techniques have been exploited in many of these studies and the relevant literature on this aspect has been reviewed recently (Ellwood 1976; Ellwood & Hunter 1976).

(a) *Acid production*

Several groups of workers have studied the acid end products of glucose metabolism when *S. mutans* has been grown in a chemostat under different conditions. In glucose-limited complex media, Ellwood et al. (1974) found the conversion of glucose to lactate by *S. mutans* Ingbritt was only 8.5% at low dilution rates ($D = 0.05\,h^{-1}$) but was 98% at high dilution rates ($D = 0.5\,h^{-1}$). Carlsson & Griffith (1974), found that at a constant dilution rate ($D = 0.125\,h^{-1}$), *S. mutans* JC2 and *S. sanguis* 804 (NCTC 10904) produced predominantly lactate under conditions of glucose excess, but under glucose limitation the main fermentation products were formate, acetate and ethanol, with only small amounts of lactate. *Streptococcus salivarius* ATCC 13419 produced lactate as the major acid end product under both conditions. Similar results were found by Mikx & van der Hoeven (1975) for *S. mutans* C67-1, grown under glucose limitation. At low dilution rates, acetate, formate and ethanol were the major acid end products whereas at high dilution rates lactate was predominantly produced.

The regulation of two enzymes, lactate dehydrogenase (EC 1.1.1.27) and pyruvate formate-lyase (EC 2.7.1.40), controls the conversion of pyruvate into fermentation products in *S. mutans* JC2 (Yamada & Carlsson 1975; 1976). Under conditions of glucose excess intracellular concentrations of fructose-1,6-diphosphate (which activates lactate dehydrogenase) and D-glyceraldehyde-3-phosphate (which inactivates pyruvate formate-lyase) are high and *S. mutans* produces predominantly lactate. Under glucose limitation cellular levels of pyruvate formate-lyase will increase and the intracellular levels of fructose-1,6-diphosphate and D-glyceraldehyde-3-phosphate will be low. This stops the lactate dehydrogenase activation and releases pyruvate formate-lyase inhibition, resulting in the production of mainly formate, acetate and ethanol. Thus, when studying dental plaque it is necessary to know the identity and concentration of bacteria at a particular site; also, it is desirable to have some idea of the metabolism of these bacteria. As growth conditions can affect the acid end products of *S. mutans* the pathogenic expression of this or any other species will vary in response to the particular micro-environment.

Cooney *et al.* (1976) studied acid production by *S. mutans* PR89 in a glucose-pulsed, glucose-limited chemostat. The object was to simulate *in vivo* conditions more closely so that long periods of 'steady-state' activity were interrupted by short bursts of active growth. Although only slight cell growth and accumulation occurred during the periods after pulsing, there was an immediate increase in lactic acid production leading to a transient accumulation of lactic acid. This acid persisted for several hours at levels higher than those found under steady-state conditions, which may have important *in vivo* implications.

(b) *Uptake of glucose and action of fluoride*

Many strains of oral streptococci take up sugars via the phosphoenol pyruvate (PEP) phosphotransferase system (Schachtele & Mayo 1973; Hamilton 1977). The activity of this system in *S. mutans* Ingbritt was studied in a glucose-limited chemostat at different dilution rates (see Ellwood 1976). At low growth rates ($D = 0.05\,h^{-1}$) PEP phosphotransferase activity was high but this activity fell as the growth rate increased. Thus at slow growth rates the cell has a highly active sugar uptake system.

Variations in the fluoride sensitivity of *S. mutans* Ingbritt was found with cells grown at different growth rates in a glucose-limited chemostat (Ellwood 1976). At slow growth rates ($D = 0.05\,h^{-1}$) as little as $15\,mg\,l^{-1}$ of fluoride inhibited acid production by *S. mutans* Ingbritt whereas up to $100\,mg\,l^{-1}$ of fluoride had no effect on the same number of cells grown at higher growth rates ($D = 0.5\,h^{-1}$). The effect was attributed to fluoride inhibition of enolase, which converts 2-phosphoglycerate into PEP, so that slow growing cells with a high requirement of PEP for glucose uptake were most susceptible. However, Hamilton & Ellwood (1977) have reported further studies on sugar transport in the presence of fluoride by cells of *S. mutans* Ingbritt grown at different pH values under conditions of glucose excess ($D = 0.05\,h^{-1}$). Cells grown at pH 5.5

were more resistant to fluoride than cells grown at higher pH values. This suggested that in these cells there is an, as yet, unknown transport system which is less sensitive to fluoride than the PEP phosphotransferase system.

For a detailed discussion of the effects of fluoride on bacterial carbohydrate metabolism, and on the metabolism of the oral flora in general, the recent publications by Hamilton (1977) and Kleinberg et al. (1977), respectively, are recommended.

(c) *Extracellular and intracellular polysaccharides*

Oral streptococci can produce a variety of extracellular polysaccharides from sucrose (Table 9). The literature has recently been reviewed by Newbrun (1976). Much attention has been paid both to the polymers themselves and to the enzyme systems responsible for their synthesis, because of their role in the aggregation, adherence and pathogenicity of oral bacteria.

(i) *Composition and structure of extracellular polysaccharides*

Some years ago, Jordan & Keyes (1966), Gibbons & Banghart (1967) and Guggenheim & Newbrun (1969), showed that when cariogenic strains of *S. mutans* were grown in the presence of sucrose large amounts of extracellular glucans and fructans were produced which were considered essential for the colonization of smooth tooth surfaces. Baird et al. (1973) made a detailed study

Table 9
Extracellular polysaccharides produced from sucrose by some oral streptococci

Species	Type of polymer (with predominant linkages)
S. mutans	Water-insoluble glucan (mutan) $\alpha\text{-}(1 \rightarrow 3), \alpha\text{-}(1 \rightarrow 3) + \alpha\text{-}(1 \rightarrow 6)$
	Water-soluble glucan (dextran) $\alpha\text{-}(1 \rightarrow 6)$
	Fructan $\beta\text{-}(2 \rightarrow 1)$
S. sanguis	Water-insoluble glucan $\alpha\text{-}(1 \rightarrow 3) + \alpha\text{-}(1 \rightarrow 6)$
	Water-soluble glucan (dextran) $\alpha\text{-}(1 \rightarrow 6)$
S. salivarius	Fructan (levan) $\beta\text{-}(2 \rightarrow 6) + \beta(2 \rightarrow 1)$

For references, see text.

of the polysaccharides produced by *S. mutans* Ingbritt A from sucrose. To avoid contamination of their analyses with carbohydrate of cellular origin, the polysaccharides were prepared using glycosyl transferases present in the culture filtrate. Water-insoluble glucan comprised 25% of the total glucan produced and contained both α-(1 $\rightarrow$ 6) and α-(1 $\rightarrow$ 3) linkages. The results did not indicate a simple mixture of two homogeneous polymers but rather a mixed polymer of both linkage types, which was highly branched. The water-soluble glucan was also highly branched but consisted predominantly of α-(1 $\rightarrow$ 6) linkages. Water soluble fructans with a β-(2 $\rightarrow$ 1) linkage, as opposed to the β-(2 $\rightarrow$ 6) fructofuranoside structure found in the levans of other bacteria, were also detected. Recent work reviewed by Newbrun (1976) has shown that the insoluble polysaccharides consist of essentially pure linear α-(1 $\rightarrow$ 3) glucan as well as mixed polymers with both α-(1 $\rightarrow$ 6) and α-(1 $\rightarrow$ 3) linkages in varying proportions. The water-insoluble glucans are not homogeneous in structure or molecular weight, variations occur depending on the growth conditions. Thus it is important in any study of these polysaccharides to use well-defined, tightly controlled conditions for polymer production. The term 'mutan' has been proposed for the water-insoluble glucan produced by *S. mutans* (Guggenheim 1970). The characteristic watery exudate produced by *S. mutans* colonies on sucrose-containing agar media consists mainly of the water-soluble glucan (Donkersloot *et al.* 1976).

Arnett & Mayer (1975) showed that *S. sanguis* ATCC 10558 synthesized a branched, water insoluble polymer comprising a mixture of α-(1 $\rightarrow$ 6) and α-(1 $\rightarrow$ 3) linkages and concluded that several enzymes were involved in their synthesis. *Streptococcus sanguis* produces a water-soluble glucan but does not produce a fructan.

The polymer produced from sucrose by *S. salivarius* is a fructan (levan) which has β-(2 $\rightarrow$ 6) linkages and β-(2 $\rightarrow$ 1) branching. It has a high molecular weight (31.5×10^6), and small amounts of fatty acids and proteins have been found to be associated with the polymer (Ehrlich *et al.* 1975). These authors concluded that the high molecular weight and low intrinsic viscosity of this polysaccharide indicated a branched and spherically shaped molecule which would aid its retention in plaque where it could be readily hydrolysed by other bacteria. This point is discussed in the microbial interactions section of this review. Long *et al.* (1975) studied the effect of pH on levan production by *S. salivarius*. The levans produced at pH 5.1 had considerably lower molecular weights (*ca*. 10^6) and were less branched than levans produced at higher pH again demonstrating the need for standardized conditions to be used in the study of these polymers.

(ii) *Enzymic synthesis of extracellular polysaccharides*

Two recent articles have reviewed the relevant literature (Newbrun 1976;

Ciardi 1976). Extracellular polysaccharide production from sucrose is mediated by glucosyl transferases (GTF) (EC 2.4.1.5) and fructosyl transferases (EC 2.4.1.10). GTF has been demonstrated, isolated and purified from *S. sanguis* strain 804. The enzyme was found both cell bound and extracellularly (Sharma *et al.* 1974). Although several workers have used *S. sanguis* in this type of study, the majority of research has been performed on the GTF of *S. mutans*. Glucosyl transferases are constitutive enzymes and occur cell bound or extracellularly, though both forms appear to be similar (Guggenheim & Newbrun 1969; Kuramitsu 1974). The work of Ceska *et al.* (1972) and Fukui *et al.* (1974a) implied two distinct GTF activities, one with α-(1 → 6) specificity and one with α-(1 → 3) specificity. The isolation of branched polymers with various amounts of both linkages implies an interaction of both these GTF in polysaccharide formation, although the existence of separate branching enzymes cannot be disregarded. The results of several workers (Ciardi & Wittenberger 1974; Germaine *et al.* 1974; Ciardi *et al.* 1976; Martin & Cole 1977) imply that GTFs are glycoproteins which could be important in cell to cell and cell to surface interactions. Wide variations in the specific activities of GTFs isolated from different serotypes of *S. mutans* and comparisons of the properties of these enzymes have been tabulated by Ciardi (1976) and Newbrun (1976). True comparisons are difficult because the properties of isolated GTF depend not only on the serotype of the producer strain but also on the isolation procedures used and the type and pH of the culture media. These parameters were not always standardized in the studies referred to above. However, enzyme-associated carbohydrate, the presence of two different enzyme activities synthesizing either water-soluble or water-insoluble polymers, K_m, pH optima (range 5.0–7.0) and isoelectric points, were all similar features.

Differences in the possible chemical composition or physical structure of the GTF from different serotypes of *S. mutans* have been deduced from serological studies (Genco *et al.* 1974; Fukui *et al.* 1974b; Ciardi 1976) and biochemical investigations using polyacrylamide gel electrophoresis (Ciardi 1976; Osborne *et al.* 1976). Antisera rasied against purified GTF from *S. mutans* serotypes a and d showed similarities between these two serotypes but not with GTF from serotypes b, c and e. The antisera also inhibited GTF activity in isolates of serotypes a and d and prevented sucrose mediated cell adherence to glass surfaces. One study reported weak cross-reactions with antisera raised against purified serotype d GTF and serotype b enzyme (cited by Ciardi 1976). Polyacrylamide gel electrophoresis demonstrated certain similarities between GTF from serotypes a and d, and c and e, with serotype b having components common to both groups. However, each pattern was distinctive, emphasizing the complex nature of these enzymes.

Growth conditions have been shown to affect the production of GTF. Activity was found to be highest for both *S. sanguis* (Carlsson & Erlander 1973) and *S. mutans* (Ellwood et al. 1974) when grown glucose-limited in complex media in a chemostat. Ellwood (1976) found pH to be a critical factor in that while maximal activity was at pH 6.5, a rise or fall of 0.1 pH led to a 4- to 5-fold loss of enzyme activity. The immunogenicity of GTF and its role in adherence has aroused interest in the possibility of using such enzymes as immunogens for inhibiting colonization of the teeth by *S. mutans*.

(iii) *Importance of extracellular polysaccharides in pathogenicity*

Several workers have derived mutants of *S. mutans* lacking the ability to produce water-insoluble glucans from sucrose (de Stoppelaar et al. 1971; Tanzer et al.1974; Michalek et al. 1975). The mutants were compared with the parent strains in their ability to produce caries in gnotobiotic rats. Less plaque was formed by mutants and the caries scores were lower on smooth surfaces in the animals infected with the mutants. Michalek et al. (1975) demonstrated a definite correlation between the level of insoluble glucan synthesis and the pathogenicity and adherence of *S. mutans*. Mutants with higher or lower levels of GTF activity had a higher or lower virulence *in vivo* than the parent strain. The role of GTF and glucan synthesis in adherence has been discussed in Section 2(e).

These findings led to several studies on the effect of mouth rinses containing α-(1 $\rightarrow$ 3) hydrolases on the *in vivo* colonization of teeth by *S. mutans*. The results of Guggenheim et al. (1972) showed reductions in both the amount of plaque and in the caries score in rats undergoing the enzyme treatment, while those of Kelstrup et al. (1973) with humans, showed that although the amount of plaque remained constant, the levels of *S. mutans* were lowered. These results confirm the importance of the water-insoluble glucans in the pathogenicity of *S. mutans*, and have stimulated interest in the possible use of such enzymes for plaque control.

(iv) *Intracellular polysaccharide synthesis*

Another factor implicated in the pathogenicity of *S. mutans* is the synthesis of intracellular polysaccharides (IPS). Epidemiological studies have indicated an association between the presence of IPS producers in plaque and caries experience (van Houte et al. 1969). Seven mutants of *S. mutans* NCTC 10449S and Ingbritt-1600 (both serotype c) defective in IPS synthesis have been tested for pathogenicity in specific pathogen-free rats (Tanzer et al. 1976). These mutants were able to colonize and persist on tooth surfaces but differed from the wild types in their diminished virulence on smooth surfaces and in fissures. This difference in virulence was attributed to the inability to produce acid from endogenous IPS stores in the absence of exogenous supplies of carbohydrate. However, caution must be taken in making a general association between IPS synthesis and virulence as strains of *S. mutans* serotype d have low IPS levels and yet are highly virulent.

The synthesis and breakdown of IPS has been investigated by Hamilton (1976). *Streptococcus salivarius* ATCC 25975 produced and degraded glycogen much faster than *S. sanguis* ATCC 10556 and various strains of *S. mutans* in the presence of excess glucose. With all strains, glycogen degradation occurred in the presence of exogenous glucose and sucrose. Detailed studies of the enzymology of *S. mutans* strains showed enzyme activities capable of supporting greater rates of glycogen metabolism than those observed with intact cells suggesting the operation of negative regulatory systems within these cells. There was considerable variation in glycogen metabolism between randomly isolated strains of *S. mutans* from rats although the overall results suggested the strains of *S. mutans* and *S. sanguis* may not have the inherent capacity for significant intracellular carbohydrate storage.

This apparent discrepancy between the conclusions from these two studies probably stems from the nature of the IPS material. The study of Hamilton was concerned specifically with glycogen while that of Tanzer *et al.* (1976) dealt with polysaccharide material in general. Thus, *S. mutans* may have a non-glycogen polysaccharide as a major intracellular storage compound.

4. Microbial Interactions Involving Oral Streptococci

The complex nature of the normal oral flora has aroused interest in the possible interactions, both stimulatory and inhibitory, that may occur between microorganisms and which may, therefore, influence the composition, metabolic activity and consequently, pathogenicity of the flora at a specific site.

As discussed in Section 3, oral streptococci are able to synthesize a variety of glucans and fructans from sucrose. Levans derived from *S. salivarius* were readily metabolized by plaque bacteria (Wood 1964; Parker & Creamer 1971; Leach *et al.* 1972). Almost 50% of 28 fresh isolates (van Houte & Jansen 1968) and 37% of 57 fresh isolates of oral streptococci (De Costa & Gibbons 1968) were capable of metabolizing a *S. salivarius* levan. It has been considered that the catabolism of levans could extend the duration of acid production by plaque (Manly & Richardson 1968). Indeed, a levan from *S. salivarius* and an amylopectin from a strain of *Neisseria* were utilized *in vitro* more readily than was sucrose by oral streptococci and a strain of *Lactobacillus* (Parker & Creamer 1971). However, streptococcal glucans do not appear to be as readily utilized by oral bacteria (Wood 1969; Parker & Creamer 1971), although recently, Nakamura *et al.* (1976), found 18 freshly isolated strains designated *Bacteroides ochraceus* possessing streptococcal polyglucan-splitting activity. Mikx *et al.* (1976) studied the interactions of an exo-dextranase producing *S. mutans* OMZ 176 and an endo-dextranase producing *S. mitis* S3 on the hydrolysis of dextran in pure and mixed batch cultures. Greatest acid production was found in the mixed culture suggesting a symbiotic relationship between the two streptococci. The *in vivo* implications of greater metabolic activity and therefore greater caries activity due to this synergistic interaction was tested in gnotobiotic rats. However, no difference was found in rats mono- or diassociated with the two streptococci. A possible explanation was the inaccessibility of the dextran to hydrolysis.

Recently Schachtele *et al.* (1976) reviewed the literature on dextranase production by oral bacteria. *Streptococcus mutans, S. mitis, S. salivarius, Actinomyces israelii, Bacteroides ochraceus* and *Fusobacterium fusiformis* all possessed either exo- and/or endo-hydrolytic activity. The possible role of the exohydrolytic dextranase from *S. mitis* in the prevention of colonization by *S. mutans* and *S. sanguis* is discussed in the section on adherence.

In a series of papers, Carlsson (1970a, b, 1971a, 1972) described the growth requirements of *S. mutans, S. sanguis* and *S. salivarius*. He found that under anaerobic conditions a strain of *S. mutans* required p-aminobenzoic acid for growth, and that in a mixed batch culture system using a medium free of this growth factor, a strain of *S. sanguis* was able to support the growth of the *S. mutans* strain (Carlsson 1971b).

Another end product of streptococcal metabolism is lactic acid, believed to be responsible for the demineralization of tooth enamel. Veillonellae, which commonly occur in dental plaque, are unable to metabolize glucose but can ferment lactate. An *in vitro* symbiotic relationship has been demonstrated between *S. mutans* and *Veillonella alcalescens* (Mikx & van der Hoeven 1975). In mixed continuous culture, *V. alcalescens* degraded all the lactate produced by *S. mutans* from the metabolism of glucose. This finding has important implications since this type of interaction could reduce the damaging effect of lactate. Some evidence for this was obtained using combinations of *S. mutans, S. sanguis* and *V. alcalescens* in a gnotobiotic animal system (Mikx *et al.* 1972). Less caries was detected in animals infected with mixed cultures of a streptococcus with the *Veillonella* strain than in animals mono-infected with either of the streptococci. These findings have been extended in a recent publication of Mikx *et al.* (1976). The combination of studying microbial interrelationships under defined conditions in the chemostat and then assessing the significance of these *in vitro* findings in the *in vivo* situation using gnotobiotic animals appears to be a valuable approach to the understanding of microbial interactions.

Few other attempts have been made to study the interactions of oral microorganisms from dental plaque in a chemostat. One reason is probably the complex and consequently poorly understood theory of mixed continuous culture growth. However, Ellwood *et al.* (1972) inoculated a glucose-limited chemostat with dental plaque. Samples were taken for chemical analysis and microscopic examination at different dilution rates ($D = 0.04\,h^{-1}$ to $0.6\,h^{-1}$). As the dilution rate increased the levels of acid in the culture supernatant rose with lactic only present in appreciable amounts above $D = 0.2\,h^{-1}$. However the data collected were predominantly biochemical (see Ellwood & Hunter 1976) and a combined bacteriological and biochemical study of plaque grown under controlled conditions would be of great value. Such studies might allow the ecologically important factors controlling the balance of the oral flora to be recognized.

Parker (1970) studied combinations of seven oral bacteria in paired batch cultures. While no one species inhibited or stimulated all others, the ultimate beneficiary was usually one of the streptococcal strains. However, the nature of these inhibitory or stimulatory interactions was not determined.

Many *in vitro* examples of antagonism among members of the oral flora have been described. Indeed, early workers considered using antagonistic members of the oral flora to prevent wounds in the oral cavity from being infected (Hugenschmidt 1896) and in reducing the incidence of dental caries (Scrivener *et al*. 1950; Scrivener 1955; Rutter *et al*. 1961).

Recently, examples of the nature of the antagonistic agents produced *in vitro* by various streptococci have been described. *S. mutans*, *S. sanguis*, *S. mitior* and *S. salivarius* were found to produce concentrations of lactic and acetic acids on agar supplemented with glucose which were inhibitory to a wide range of bacteria (Donoghue & Tyler 1975). Hydrogen peroxide produced by *S. sanguis* (Holmberg & Hallander 1973; Donoghue & Tyler 1975) and *S. mitior* (Donoghue 1975; Lebien & Bromel 1975) has a broad spectrum of activity against other oral Gram positive cocci and rods including lactobacilli and *Actinomyces* sp. (Holmberg & Hallander 1972).

A strain of *S. salivarius* isolated from a throat culture inhibited the growth of a whole range of streptococci, staphylococci, corynebacteria and unusually, peptococci and peptostreptococci (Bill & Washington 1975). However, the cause of the inhibition was not determined.

The production of bacteriocins and bacteriocin-like substances has been detected *in vitro* in strains of *S. mutans* (Kelstrup & Gibbons 1969a; Rogers 1972, 1976; Donoghue 1975; Yamamoto *et al*. 1975, Hamada & Ooshima 1975), *S. sanguis* (Schlegel & Slade 1972; Donoghue & Tyler 1975; Kolstad 1976, Dajani *et al*. 1976a) and *S. mitis* (Dajani *et al*. 1976a; Vernazza & Melville 1977). Names including 'mutacins', 'streptocins' and 'viridins' have been used to describe these agents.

Subdividing streptococcal strains for epidemiological studies by bacteriocin-typing rather than serology has been proposed by several workers (Kelstrup & Gibbons 1969a; Berkowitz & Jordan 1975; Rogers 1975; Kolstad 1976). Using such a system (Rogers 1975) found the same bacteriocin-type of *S. mutans* in all members of a particular family and that type was not found in any other unrelated subjects. Also, Rogers (1975) found that only one bacteriocin-type of *S. mutans* predominated in a mouth. Kelstrup *et al*. (1970) finger-printed strains of *S. salivarius* and were able to demonstrate their transfer from mother to babies.

With very few exceptions, bacteriocin activity has only been demonstrated on agar, and most workers have had difficulty in isolating the active agent. However, Dajani *et al.* (1976a, b) successfully isolated, characterized and purified viridins from two strains of *S. mitis* and one strain of *S. sanguis* by disrupting cells in a tissue homogenizer. Every other method, including sonication of whole cells, and the freezing and thawing of inhibition zones on agar plates, failed to yield the active agent. These viridins were found to be unique among those of Gram positive bacteria in that they inhibited many Gram negative organisms including *Neisseria* spp. Other Gram negative bacteria of oral significance, such as *Veillonella* and *Bacteroides* species were not tested for sensitivity.

All of the above bacteriocin and bacteriocin-like agents have activity against other streptococci likely to be found in plaque, and some have wide ranges of activity against other oral bacteria. For example, a bacteriocin-like agent from a strain of *S. mitis* was active against strains of *S. mutans, S. sanguis, Actinomyces naeslundii, Rothia dentocariosa* and oral staphylococci (Vernazza & Melville 1977). It has been proposed that the ability to produce such inhibitors may confer an ecological advantage on these organisms. However, because all these interactions have only been demonstrated *in vitro* their significance *in vivo* remains speculative. There is little evidence available on whether bacteriocins would be active under the conditions known to exist in plaque. Earlier reports indicated that some bacteriocins of *S. mutans* were inactivated by proteolytic enzymes in saliva (Kelstrup & Gibbons 1969b) and that bacteriocin-sensitive strains of *S. mutans* and *S. salivarius* became resistant when grown in the presence of sucrose due to the production of extracellular polysaccharides (Rogers 1974). Conversely, Delisle (1976) found that bacteriocins were still produced by *S. mutans* strains BHT and GS-5 and that indicator strains remained sensitive when tested on sucrose-containing medium, and that human saliva had no effect on bacteriocin activity.

Recently van der Hoeven *et al.* (1977) provided possibly the best evidence yet for the *in vivo* demonstration of antagonistic activity. Gnotobiotic rats were diassociated with a bacteriocin-sensitive strain of *Actinomyces viscosus*, (Ny-1), and either a bacteriocin-producing (C67-1, or T2) or a non-producing (OMZ 176) strain of *S. mutans*. The bacteriocin-producing strains prevented colonization by *A. vicosus* Ny-1, while the latter became established when infected with non-bacteriocin-producing or with mutagen-induced, non-bacteriocin-producing strains of C67-1 and T2. The authors suggested that bacteriocins may therefore be of importance on a micro-ecological scale in plaque but would not necessarily lead to the elimination of susceptible strains from the site.

Preliminary reports of a longitudinal study of the microbiology of approximal dental plaque have shown that some strains of *S. mutans* can dramatically increase in viable count during the 13 week periods between sample collection (Bowden *et al*. 1976; Hardie *et al*. 1977). The ability of a species to dominate a site in this way when growing in competition with a wide range of different bacteria is of great ecological significance. The possible *in vivo* production of bacteriocins by such strains is one explanation of this phenomenon currently being explored.

5. The Relationship of Oral Streptococci to Disease

Oral streptococci are frequently involved in disease processes within the mouth, such as dental caries and dental abscesses, and also, less commonly, at other sites in the body. Micro-organisms usually reach the more distant sites via the blood stream. These streptococci may be regarded as opportunist pathogens which cause disease when they get into situations outside their normal habitat.

(a) *Dental caries*

Since the last century it has been widely accepted that dental caries is initiated by acid-producing organisms as a result of fermentation of dietary carbohydrates. Based on the observation that oral bacteria could produce sufficient acid *in vitro* to decalcify tooth enamel, Miller (1890) was able to propose his famous 'Chemico-Parasitic Theory' for the aetiology of caries. It was also recognized at that time that bacteria accumulate on the tooth surface in the form of dental plaque and that this material is an important aetiological factor in dental caries.

Many of the bacteria commonly present in dental plaque are saccharolytic and therefore potentially damaging to the teeth. Streptococci and actinomycetes are invariably present in high numbers in supragingival plaque, and lactobacilli may also be found, although their numbers are usually low (Bowden *et al*. 1975).

The most direct proof of the cariogenic potential of specific bacteria has come from experiments with gnotobiotic animals. It was shown by Orland *et al.* (1955) that completely germ-free rats remained caries-free when fed a high sucrose diet, whereas animals inoculated with an enterococcus and a proteolytic rod developed caries. Many workers have subsequently shown that caries can be induced in experimental animals by mono-infecting them with particular streptococci and that the disease can be transmitted from one animal to another (Keyes 1960; Fitzgerald 1968).

Most of the strains of streptococci which have been shown to be highly cariogenic in animals belong to the species *S. mutans*, and caries can be induced by this organism in rats, hamsters, gerbils and monkeys. ***Streptococcus mutans*** was originally isolated from carious human teeth by Clarke (1924). The caries produced in experimental animals by *S. mutans* usually involves both fissures and smooth surfaces of the teeth, and may result in almost complete destruction of the hard tissues. No obvious differences in cariogenicity between the different serotypes of *S. mutans* have been reported, but pathogenicity does appear to be related to the ability of strains to produce insoluble extracellular polysaccharide. As mentioned previously, mutant strains which have lost the ability to produce such polymers show diminished cariogenic activity (de Stoppelaar *et al.* 1971; Tanzer *et al.* 1974; Michalek *et al.* 1975). Virulence of *S. mutans* strains may also be related to the ability to produce intracellular polysaccharide (Tanzer *et al.* 1976).

In most animal caries experiments a high sucrose diet is employed, which enhances the establishment of *S. mutans* in the mouth. However, animals maintained on diets containing other carbohydrates may also develop caries (Colman *et al.* 1977).

Although it has been established conclusively that *S. mutans* is highly cariogenic in experimental animals, other streptococcal species, and even other genera, can also induce caries. Thus a few strains of *S. salivarius*, *S. mitis*, *S. sanguis*, *S. milleri* and enterococci, as well as some *Actinomyces* and *Lactobacillus* species, have been shown to possess some caries-inducing activity in gnotobiotic animals. The amount of caries produced by these organisms is usually less than that found with *S. mutans* and is generally confined to the occlusal fissures of the teeth (Gibbons & van Houte 1975*b*).

Thus from the experimental caries studies, it appears that there is a spectrum of cariogenicity amongst the strains of streptococci that have been examined, with *S. mutans* being the most pathogenic species. Some streptococci may only be able to colonize certain animals, an example of this being *S. sanguis* which can readily be established in rats but not in hamsters (Krasse & Carlsson 1970). An interesting recent development has been the introduction of mixtures of bacteria into gnotobiotic animals, showing that the cariogenic potential of one strain can be modified by the presence of other organisms such as *Veillonella alcalescens* (Mikx *et al.* 1972, 1976).

Since the demonstration of the cariogenic potential of *S. mutans* in animals, several investigators have looked for an association between this streptococcus and human caries. Because caries usually develops in relatively inaccessible parts of the tooth, such as the pits and fissures and at approximal areas between adjoining teeth, early diagnosis is difficult. Once actual cavitation has developed, which is easy to detect, the disease has reached an advanced stage and the bacterial flora present may not be the same as that which initiated the caries process (Hardie & Bowden 1976*c*). It is largely because of technical problems in diagnosing the very earliest stages of caries that it is difficult to make a precise assessment of microbiological factors which are associated with initiation of the disease.

There are basically two types of epidemiological survey which can be undertaken to study the relationship between bacteria and dental caries, 'cross-sectional' and 'longitudinal'. In cross-sectional studies the prevalence or isolation frequency of various species can be related to the prevalence of caries in the population under investigation at a particular time. Numerous studies of this type have shown an association between the presence and numbers of *S. mutans* and the number of decayed, missing and filled (DMF) teeth (e.g. Krasse *et al.* 1968; Jordan *et al.* 1969; Hoerman *et al.* 1972; Littleton *et al.* 1970; Schamschula & Barmes 1970; Rogers 1973; Shklair *et al.* 1974; Gibbons *et al.* 1974; Loesche *et al.* 1975). In one of these studies, the correlation between *S. mutans* and caries was best seen when plaque samples from individual tooth surfaces were examined; less good associations were shown when pooled plaque or saliva samples were used (Loesche *et al.* 1975).

Individual exceptions to the association between *S. mutans* and caries have been recorded in several studies. Thus caries has sometimes been diagnosed at sites where no *S. mutans* could be detected and, conversely, not all infected surfaces show signs of caries.

Longitudinal surveys are more appropriate to the study of a relatively slowly developing disease such as dental caries. In such studies a population is examined over a period of months or years so that the development of caries can be monitored clinically and radiographically. Samples for microbiological analysis can be collected periodically from caries-susceptible sites and the results correlated with the incidence of disease at these particular sites. This type of study is labour intensive and expensive to undertake and for this reason comparatively few large scale longitudinal studies have been reported.

Some evidence for a relationship between *S. mutans* and the development of dental caries has been produced (Edwardsson *et al.* 1972; Ikeda *et al.* 1973; Keene & Shklair 1974; De Stoppelaar 1971; Woods 1971; Swenson *et al.* 1976), but not all the published information supports these findings (Mikkelsen & Poulsen 1976).

In our own studies, which are still in progress, no clear evidence for an increase in any individual microbial species has been found prior to the detection of caries at specific target sites in a group of school children after two years of observation (Bowden *et al.* 1976; Hardie *et al.* 1977). A few sites which developed caries appeared to be free of detectable *S. mutans*. The general trend so far is for the isolation frequency and relative numbers of *S. mutans* to increase at the target sites after caries has been detected rather than before. This study involves the regular examination of plaque samples from 100 sites in 50 children, but since the data are incomplete at this stage the preliminary results should be viewed with caution. However, it does seem likely that the development of caries may depend upon a combination of interrelated microbial factors rather than the presence or absence of one particular species such as *S. mutans*.

Although the evidence incriminating *S. mutans* as the sole causative agent in caries is inconclusive, there is sufficient data to indicate that it may at least be one of the important aetiological factors. There is also immunological evidence to suggest that antibody levels to *S. mutans* antigens vary in subjects with high and low caries experience (Lehner 1975). Since dental caries is an ubiquitous disease which is expensive to treat, and because existing preventive methods, although effective, are difficult to implement on a nationwide scale, several groups of workers are investigating the possibility of immunizing against dental caries with *S. mutans* vaccines. Experiments have been carried out in rodents and primates, using a variety of types of vaccine, including whole cells, cell walls and crude glucosyl transferase preparations. Mixed success has been achieved in the rodent experiments, but encouraging results have been reported from trials on monkeys (Bowen 1969; Bowen *et al*. 1975; Lehner *et al*. 1976; Caldwell *et al*. 1977).

Much of the published work on caries vaccines has been reviewed recently (Bowen *et al*. 1976). It is clear that further investigation is required before human clinical trials can be contemplated.

(b) *Other oral infections*

Viridans streptococci are very frequently isolated from infected dental root canals and are the organisms most often isolated from oral pus. In one series, viridans streptococci were recovered from 901 out of 1000 consecutive specimens of pus from various oral lesions (Sims 1974). In the majority of published reports, no attempt has been made to identify the streptococcal isolates to species level. Sims (1974) has indicated that many clinical isolates resemble *S. mitior*, with lesser numbers of *S. sanguis*.

Information on the distribution of individual species in various common oral conditions, such as dental abscesses, deep periodontal pockets, pericoronitis and dry sockets, is fragmentary at present. Since the precise identity of the streptococci involved in such conditions is unlikely to affect the treatment prescribed in the majority of cases, it is arguable that such information is only of academic interest. However, in view of the potential involvement of *S. milleri* in more serious purulent conditions elsewhere in the body (see Section 5.c,d), it would be interesting to study the distribution of this species in greater detail. Isolation of *S. milleri* from infected root canals (Mejare & Edwardsson 1975), deep dentinal caries (Edwardsson 1974) and dental abscesses (Guthof 1956) has been reported, but the frequency with which this streptococcus occurs in oral infections is not known at present.

(c) *Dental bacteraemia and infective endocarditis*

Since the early investigations of Okell & Elliott (1935), it has been recognized that a transient bacteraemia is produced whenever teeth are extracted. Detailed studies on such bacteraemias have been carried out by many workers since that time (Jokinen 1970). Although dental extractions constitute the greatest risk, oral bacteria can also be introduced into the blood stream by a variety of other operative procedures (Lineberger & De Marco 1973). Almost any oral micro-organisms may be present in such bacteraemias, but streptococci have been reported most frequently (e.g. Phillips *et al*. 1976; Symington 1975). It is likely that failure to isolate other types of micro-organisms, especially anaerobes, from dental bacteraemias in many of the earlier studies was related to the cultural methods employed (Crawford *et al*. 1974).

In many of the published reports on streptococci isolated from the blood following dental treatment, and from patients with bacterial endocarditis, the organisms have not been identified to species level using the criteria described earlier in this review. Often the streptococci have been referred to as *S. viridans* and consequently it is difficult to make detailed comparisons between different reports. In one recent study, strains of *S. mitior,* intermediate viridans streptococci and *S. milleri* were most commonly isolated from the blood of patients following tooth extractions. *Streptococcus sanguis* and *S. mutans* were found less frequently, whilst no strains of *S. salivarius* or enterococci were recorded (Phillips *et al*. 1976).

The significance of a transient bacteraemia, whatever the origin of the bacteria, is that patients with predisposing conditions may develop infective endocarditis. In this serious condition, bacteria colonize the surface of the heart valves giving rise to infected vegetations. Patients at risk include those who have rheumatic heart damage, those with congenital heart abnormalities of various types, and following cardiac surgery, especially after insertion of prosthetic heart valves. A great variety of micro-organisms have been known to cause infective endocarditis, but streptococci are the organisms most commonly involved. In the subacute form of the disease, the streptococci are usually of the viridans or non-haemolytic type.

Two major surveys of streptococci isolated from infective endocarditis and other systemic conditions have recently been reported, one from either side of the Atlantic (Parker & Ball 1976; Facklam 1977). Notwithstanding differences in nomenclature, the streptococci most frequently isolated were *S. sanguis* and *S. mitior.* These two species together constituted over 60% of the strains in both series, thus confirming the observations of previous workers (White & Niven 1946; Hehre 1948; Porterfield 1950; Farmer 1953). The next most common species was *S. mutans,* being found in almost 20% of cases. Endocarditis due to this species was first recorded by Abercrombie & Scott (1928) and several case reports have appeared more recently (Perch *et al.* 1974; Lockwood *et al.* 1974; Harder *et al.* 1974; Neefe *et al.* 1974). Strains of *S. milleri* and *S. salivarius* were only occasionally isolated from cases of endocarditis. The series of Parker & Ball (1976) included a significant number of strains of enterococci and *S. bovis*, but such strains were not included in Facklam's report. *Streptococcus bovis* has not been shown to comprise part of the normal human oral flora, and although enterococci are found in low numbers in the mouth it is likely that the main source of Lancefield group D streptococci in endocarditis cases is some site other than the mouth.

The clinical evidence which implicates oral organisms as the cause of endocarditis following a dental bacteraemia is invariably circumstantial and retrospective since the disease may not become apparent until several weeks or even months after the initiating event. However, using an animal model system, described originally by Garrison & Freedman (1970) it has been demonstrated that streptococcal endocarditis can be produced in the rabbit as a result of oral manipulations such as tooth extraction (McGowan & Hardie 1974). In this model system, non-infected lesions are created by insertion of a plastic cannula into the heart, prior to introducing test organisms into the blood stream. The pathology of streptococcal endocarditis has been studied extensively in the rabbit model by Durack and his colleagues (Durack & Beeson 1972*a,b*, Durack *et al.* 1974; Durack 1975*a*). Most of these experimental studies have been carried out using a dextran positive strain of *S. mitior* (*S. sanguis* II) isolated originally from a patient with endocarditis. The model has also been employed to test the efficacy of potential prophylactic and chemotherapeutic methods (Durack & Petersdorf 1973; Durack *et al.* 1974; Southwick & Durack 1974; Pelletier *et al.* 1975; Sande & Irvin 1974; Hooke *et al.* 1975). McGowan (1975 and pers. comm.) has shown that strains of *S. sanguis*, *S. mitior*, *S. salivarius*, *S. mutans*, *S. milleri* and *S. faecalis* are all capable of inducing endocarditis if administered in high doses intravenously to cannulated rabbits.

(i) *Antibiotic susceptibility of oral streptococci*

In view of the need to provide antibiotic cover for patients who are at risk to infective endocarditis following dental extraction and other operative procedures, it is important to gather information about the susceptibility of oral streptococci to antibiotics. Several recent surveys of strains isolated either from the mouth or from the blood of patients following dental extractions have been reported (Drucker & Jolly 1971; Sukchotiratana *et al.* 1975; Phillips *et al.* 1976). Most of the 301 strains examined by Phillips *et al.* (1976), including representatives of *S. sanguis*, *S. mitior*, *S. mutans*, *S. milleri* and *S. salivarius*, were inhibited by $0.12 \mu g \, ml^{-1}$ or less of penicillin, although some more resistant strains were found in each species. Predictably, strains of *S. faecalis* were found to be more resistant than the other streptococci. Most isolates, other than *S. faecalis*, were sensitive to clinamycin, erythromycin and cephaloridine, but only 62% were fully sensitive to $1.2\,\mu g\,ml^{-1}$ of tetracycline. Pre-operative penicillin (600 mg intramuscularly 1 h before extractions) was found to suppress detectable bacteraemia completely, whilst lincomycin reduced the incidence by 60%.

Detailed studies on the antibiotic susceptibility of *S. mutans* indicate that the most effective agents (with MICs below $0.1\,\mu g\,ml^{-1}$) are penicillin, ampicillin, erythromycin, cephalothin and methicillin. Rifampin, lincomycin, spiromycin, tetracycline, vancomycin, chloramphenicol, gentamicin and some other agents were found to be effective against this species at concentrations between $0.1\,\mu g$ and $10\,\mu g\,ml^{-1}$ (Baker & Thornsberry 1974; Ferretti & Ward 1976). Syngergy between penicillin and aminoglycosides was found against 2/9 strains of *S. mutans*, whilst antagonism occurred in eight strains with mixtures of clindamycin and gentamicin (Snyder *et al.* 1975).

As has been discussed by Phillips *et al.* (1976), antibiotic therapy regularly leads to the emergence of resistant strains in the mouth. Sukchotiratana *et al.* (1975) found that resistant streptococci were detectable within 24 h of commencing oral penicillin and were still present eight weeks after discontinuing the antibiotic. Fewer resistant strains were found following cephalexin therapy, and strains resistant to clindamycin were not observed in 10/11 subjects examined.

In one study (Bentley *et al.* 1971) 60% of 15 control subjects not exposed to recent penicillin therapy were found to have penicillin-resistant streptococci in their saliva, as did 61% of subjects on regular benzathine penicillin injections and 97% of those taking oral penicillin V. Drucker & Jolly (1971) reported that over 25% of subjects not on antibiotic treatment had penicillin-resistant streptococci in their mouth. These authors also demonstrated that the proportion of resistant organisms detected could be markedly increased if several oral sites, including the gingival crevice, were sampled in addition to saliva.

Long-term therapy with either benzathine pencillin or sulphadiazine for the prevention of recurrent attacks of rheumatic fever has not been found to have any significant effect on levels of *S. mutans* in dental plaque in a group of Canadian children (Weld & Sandham 1976). Previously, it had been reported that long-term prophylactic use of penicillin reduces the incidence of dental caries (Handelman *et al.* 1966). Topical treatment for five days with kanamycin has been shown to reduce the proportions of streptococci in dental plaque from 30 to 3% in institutionalized mentally retarded subjects (Loesche *et al.* 1977). Several other poorly absorbed antibiotics, including niddamycin, vancomycin and bacitracin have also been considered as possible agents for control of potentially pathogenic dental plaque bacteria (Newbrun *et al.* 1976).

(ii) *Prophylaxis against bacterial endocarditis*

Patients who are at risk of contracting infective endocarditis when exposed to a transient bacteraemia require prophylaxis prior to dental and other forms of surgery. The principle of such prophylaxis is that a high blood level of the chemotherapeutic agent should be attained immediately before the operative procedure and maintained for a few hours. The optimum length of time for continuing the administration of antibiotic has not been clearly established. In order to achieve the required blood level at the critical time, it is preferable to give the antimicrobial agent parenterally shortly before surgery.

Traditionally penicillin has been the first choice drug for prophylaxis before dental operations, with erythromycin as a commonly used alternative for patients who are allergic to penicillin (Jolly & Drucker 1971). Bacteristatic agents such as tetracycline should not be used for this purpose (Southwick & Durack 1974). The recent experimental studies mentioned above have indicated that better protection against streptococcal endocarditis is achieved by means of synergistic combinations of penicillin with an aminoglycoside, as commonly used for treatment of enterococcal infection (Sande & Irvin 1974; Durack *et al.* 1974; Pelletier *et al.* 1975). Prior to dental manipulations, Durack (1975*b*) recommends the following prophylactic regimes:

First choice. Benzylpenicillin (2 million units) plus procaine penicillin (600,000 units) plus streptomycin 1.0 g, intramuscularly 30 min before treatment.

Alternative. Vancomycin 1.0 g intravenously, 5 min before treatment. Cephazolin (1 g) with streptomycin (1 g) may also be used (D. A. McGowan, pers. comm.). In practice it is difficult to evaluate the different prophylactic regimes that have been proposed. Ideally a prospective clinical trial would be carried out, including a placebo-treated control group, but such a study is difficult to justify on ethical grounds (Pogrel & Welsby 1975).

(d) *Other systemic infections*

The two recent surveys of streptococci isolated from a variety of clinical sources indicate that streptococci of possible oral origin may be involved in several systemic conditions in addition to bacterial endocarditis, including purulent lesions in the brain, abdomen and thorax. In particular, *S. milleri* (or *S. MG-intermedius* and *S. anginosus-constellatus*) was commonly found in brain abscesses, accounting for 13/16 cases in the British series (Parker & Ball 1976) and 23/28 of the American series (Facklam 1977). Altogether *S. milleri* accounted for 29.3% of the streptococci isolated from 116 cases of purulent conditions of various organs (Parker & Ball 1976).

Streptococcus milleri was originally the name given to streptococci isolated from dental abscesses and other lesions around the mouth (Guthof 1956). It is not known at present if this species has any natural habitat other than the oral cavity and upper respiratory tract.

6. References

ABERCROMBIE, G. F. & SCOTT, W. M. 1928 A case of infective endocarditis due to *Streptococcus mutans*. *Lancet* **2**, 697-699.
ARNETT, A. T. & MAYER, R. M. 1975 Structural characteristics of native and enzymically formed dextran of *S. sanguis* ATCC 10558. *Carbohydrate Research* **42**, 339-345.
BAIRD, J. K., LONGYEAR, V. M. C. & ELLWOOD, D. C. 1973 Water insoluble and soluble glucans produced by extracellular glycosyltransferases from *Streptococcus mutans*. *Microbios* **8**, 143-150.
BAKER, C. N. & THORNSBERRY, C. 1974 Antimicrobial susceptibility of *Streptococcus mutans* isolated from patients with endocarditis. *Antimicrobial Agents and Chemotherapy* **5**, 268-271.
BENTLEY, D. W., FRANK, T. & HAHN, J. J. 1971 Penicillin-resistant oral streptococci. *Antimicrobial Agents and Chemotherapy* 1970. 277-279.
BERG, J. D. & NORD, C. E. 1973 A method of isolation of anaerobic bacteria from endodontic specimens. *Scandinavian Journal of Dental Research* **81**, 163-166.
BERKOWITZ, R. J. & JORDAN, H. V. 1975 Similarity of bacteriocins of *Streptococcus mutans* from mother and infant. *Archives of Oral Biology* **20**, 725-730.
BILL, N. J. & WASHINGTON, J. A. 1975 Bacterial interference by *Streptococcus salivarius*. *American Journal of Clinical Pathology* **64**, 116-120.
BOWDEN, G. H., HARDIE, J. M. & SLACK, G. L. 1975 Microbial variations in approximal dental plaque. *Caries Research* **9**, 253-277
BOWDEN, G. H., HARDIE, J. M., McKEE, A. S., MARSH, P. D., FILLERY, E. D. & SLACK, G. L. 1976 The microflora associated with developing carious lesions of the distal surfaces on the upper first premolars in 13-14 year old children. Proceedings *Microbial Aspects of Dental Caries*, eds Stiles, H. M., Loesche, W. J. & O'Brien, T. C. Special Supplement to *Microbiology Abstracts* **1**, 223-241.
BOWEN, W. H. 1969 A vaccine against dental caries. *British Dental Journal* **126**, 159--160.
BOWEN, W. H., COHEN, B., COLE, M. F. & COLMAN, G. 1975 Immunization against dental caries. *British Dental Journal* **139**, 45-58
BOWEN, W. H., GENCO, R. J. & O'BRIEN, T. C. 1976 (eds) *Immunologic Aspects of Dental Caries*. Special Supplement to *Immunology Abstracts*. Washington, DC & London: Information Retrieval Inc.
BRATTHALL, D. 1970 Demonstration of five serological groups of streptococcal strains resembling *Streptococcus mutans*. *Odontologisk Revy* **21**, 143-152.
BRATTHALL, D. 1972 Immunofluorescent identification of *Streptococcus mutans*. *Odontologisk Revy* **23**,181-196.
BUCHANAN, R. E. & GIBBONS, N. E. 1974 *Bergey's Manual of Determinative Bacteriology*, 8th ed. Baltimore: Williams and Wilkins Co.
BURCKHARDT, J. J. & GUGGENHEIM, B. 1976 Interactions of antisera, sera, and oral fluid with glucosyltransferases. *Infection and Immunity* **13**, 1009-1022.
CALDWELL, J., CHALLACOMBE, S. J. & LEHNER, T. 1977 A sequential bacteriological and serological investigation of *Rhesus* monkeys immunised against dental caries with *Streptococcus mutans*. *Journal of Medical Microbiology* **10**, 213-224.
CARLSSON, J. 1965 Zooglea-forming streptococci, resembling *Streptococcus sanguis*, isolated from dental plaque in man. *Odontologisk Revy* **16**, 348-358.
CARLSSON, J. 1967 Presence of various types of non-haemolytic streptococci in dental plaque and in other sites of the oral cavity in man. *Odonotologisk Revy* **18**, 55-74.
CARLSSON, J. 1968 A numerical taxonomic study of human oral streptococci. *Odontologisk Revy* **19**, 137-160.

CARLSSON, J. 1970a Chemically defined medium for growth of *Streptococcus sanguis*. *Caries Research* **4**, 297-304.
CARLSSON, J. 1970b Nutritional requirements for *Streptococcus mutans*. *Caries Research* **4**, 305-320.
CARLSSON, J. 1971a Nutritional requirements of *Streptococcus salivarius*. *Journal of General Microbiology* **67**, 69-76.
CARLSSON, J. 1971b Growth of *Streptococcus mutans* and *Streptococcus sanguis* in mixed culture. *Archives of Oral Biology* **16**, 963-965.
CARLSSON, J. 1972 Nutritional requirements of *Streptococcus sanguis*. *Archives of Oral Biology* **17**, 1327-1332.
CARLSSON, J. & ERLANDER, B. 1973 Regulations of dextransucrase formation by *Streptococcus sanguis*. *Caries Research* **7**, 89-101.
CARLSSON, J., GRAHNEN, H., JONSSON, G. & WIKNER, S. 1970a Early establishment of *Streptococcus salivarius* in the mouths of infants. *Journal of Dental Research* **49**, 415-418.
CARLSSON, J., GRAHNEN, H., JONSSON, G. & WIKNER, S. 1970b Establishment of *Streptococcus sanguis* in the mouths of infants. *Archives of Oral Biology* **15**, 1143-1148.
CARLSSON, J., GRAHNEN, H. & JONSSON, G. 1975 Lactobacilli and streptococci in the mouth of children. *Caries Research* **9**, 333-339.
CARLSSON, J. & GRIFFITH, C. J. 1974 Fermentation products and bacterial yields in glucose-limited and nitrogen-limited cultures of streptococci. *Archives of Oral Biology* **19**, 1105-1109.
CESKA, M., GRANATH, K., NORMAN, B. & GUGGENHEIM, B. 1972 Structural and enzymatic studies on glucans synthesised with glucosyltransferases of some strains of oral streptococci. *Acta chemica scandinavica* **26**, 2223-2230.
CHIU, T. H., EMDUR, L. I. & PLATT, D. 1974 Lipoteichoic acids from *Streptococcus sanguis*. *Infection and Immunity* **118**, 471-479.
CHORPENNING, F. W., COOPER, H. R. & ROSEN, S. 1975 Cross-reactions of *Streptococcus mutans* due to cell wall teichoic acid. *Infection and Immunity* **12**, 586-591.
CIARDI, J. E. & WITTENBERGER, C. L. 1974 Glucosyltransferase and polysaccharide production by *Streptococcus mutans*. *Federation Proceedings* **33**, 1435.
CIARDI, J. E., HAGEAGE, G. J. & WITTENBERGER, C. L. 1976 The multi-component nature of the glucosyltransferase system of *S. mutans*. *Journal of Dental Research* **55** (Spec. Iss. C), 87-96.
CIARDI, J. E. 1976 Glucosyltransferases of *Streptococcus mutans*. Proceedings *Immunologic Aspects of Dental Caries*, eds Bowden, W. H., Genco, R. J. & O'Brien, T. C. Special Supplement to *Immunology Abstracts*, 101-110.
CLARKE, J. K. 1924 On the bacterial factor in the aetiology of dental caries. *British Journal of Experimental Pathology* **5**, 141-147.
COLMAN, G. 1968 The application of computers to the classification of streptococci. *Journal of General Microbiology* **50**, 149-158.
COLMAN, G. 1969 Transformation of viridans-like streptococci. *Journal of General Microbiology* **57**, 247-255.
COLMAN, G. 1970 The classification of streptococcal strains. Ph.D. Thesis, University of London.
COLMAN, G. 1976 The viridans streptococci. In *Selected Topics in Clinical Bacteriology*, ed. De Louvois, J. London: Baillière Tindal.
COLMAN, G. & WILLIAMS, R. E. O. 1965 The cell walls of streptococci. *Journal of General Microbiology* **41**, 375-387.
COLMAN, G. & WILLIAMS, R. E. O. 1972 Taxonomy of some human viridans streptococci. In *Streptococci and Streptococcal Disease*, eds Wannamaker, L. & Matsen, J. M. London & New York: Academic Press.

COLMAN, G., BOWEN, W. H. & COLE, M. F. 1977 The effects of sucrose, fructose, and a mixture of glucose and fructose on the incidence of dental caries in monkeys (*M. fascicularis*). *British Dental Journal* **142**, 217–221.

COONEY, C. L., LEUNG, J. & SINSKEY, A. J. 1976 Growth and physiology of *Streptococcus mutans* during transients in continuous culture. Proceedings *Microbial Aspects of Dental Caries,* eds Stiles, H. M., Loesche, W. J. & O'Brien, T. C. Special Supplement to *Microbiology Abstracts* **3**, 799–807.

COWAN, S. T. 1974 Cowan and Steel's Mannual for the identification of medical bacteria. Cambridge: Cambridge University Press.

COYKENDALL, A. L. 1974 Four types of *Streptococcus mutans* based on their genetic, antigenic and biochemical characteristics. *Journal of General Microbiology* **83**, 327–338.

COYKENDALL, A. L. 1977 Proposed to elevate the subspecies of *Streptococcus mutans* to species status, based on their molecular composition. *International Journal of Systematic Bacteriology* **27**, 26–30.

CRAWFORD, J. J., SCONYERS, J. R., MORIARTY, J. D., KING, R. C. & WEST, J. F. 1974 Bacteraemia after tooth extractions studied with the aid of prereduced anaerobically sterilized culture media. *Applied Microbiology* **27**, 927–932.

CROUSAZ, Ph.De & GUGGENHEIM, B. 1966 Immunochemical studies on cariogenic streptococci. *Helvetica Acta Odontologica* **10**, 38–40

CULLEN, G. A. 1967 Classification of *Streptococcus uberis* with biochemical tests. *Research in Veterinary Science* **8**, 83–88.

DAJANI, A. S., TOM, M. C. & LAW, D. J. 1976a Viridins, bacteriocins of alpha-haemolytic streptococci: Isolation, characterisation and partial purification. *Antimicrobial Agents and Chemotherapy* **9**, 81–88.

DAJANI, A. S., LAW, D. J., BOLLINGER, R. O. & ECKLUND, P. S. 1976b Ultrastructural and biochemical alterations effected by viridin B, a bacteriocin of alpha-haemolytic streptococci. *Infection and Immunity* **14**, 776–782.

DE COSTA, T. & GIBBONS, R. J. 1968 Hydrolysis of levan by human plaque streptococci. *Archives of Oral Biology* **13**, 609–617.

DELISLE, A. L. 1976 Activity of two *Streptococcus mutans* bacteriocins in the presence of saliva, levan and dextran. *Infection and Immunity* **13**, 619–626.

DE STOPPELAAR, J. D. 1971. The occurrence of *Streptococcus mutans* in dental plaque: an epidemiological survey with special reference to caries activity. In *Tooth Enamel* Vol. 2, eds Fearnhead, R. & Stack, M. V. Bristol: Wright.

DE STOPPELAAR, J. D., KONIG, K. G., PLASSCHAERT, A. J. M. & VAN DER HOEVEN, J. S. 1971 Loss of cariogenicity of a mutant of *Streptococcus mutans*. *Archives of Oral Biology* **16**, 971–975.

DODD, R. 1949 Serologic relationship between streptococcus group H and *Streptococcus sanguis*. *Proceedings of the Society of Experimental Biology and Medicine* **70**, 598–599.

DONKERSLOOT, J. A., DE LEON, H. A., CHASSY, B. M. & KRICHEVSKY, M. I. 1976 Analysis of the exudate produced by *Streptococcus mutans* SL–1 colonies on sucrose-containing agar media. *Applied and Environmental Microbiology* **32**, 448–450.

DONOGHUE, H. D. & TYLER, J. E. 1975 Antagonisms amongst streptococci isolated from the human oral cavity. *Archives of Oral Biology* **20**, 381–387.

DRUCKER, D. B. & MELVILLE, T. H. 1971 The classification of some oral streptococci of human or rat origin. *Archives of Oral Biology* **16**, 845–853.

DRUCKER, D. B. & JOLLY, M. 1971 Sensitivity of oral micro-organisms to antibiotics. *British Dental Journal* **131**. 442–444.

DRUCKER, D. B. & VEAZEY, F. J. 1977 Fatty acid fingerprints of *Streptococcus mutans* NCTC 10832 grown at various temperatures. *Applied and Environmental Microbiology* **33**, 221–226.

DRUCKER, D. B., GRIFFITH, C. J. & MELVILLE, T. H. 1976 Fatty acid fingerprints of some chemostat-grown streptococci with computerized data analysis. *Microbios Letters* **1**, 31–34.

DURACK, D. T. 1975a Experimental bacterial endocarditis. IV Structure and evolution of very early lesions. *Journal of Pathology* **115**, 81–89.
DURACK, D. T. 1975b Current practice in prevention of bacterial endocarditis. *British Heart Journal* **37**, 478–481.
DURACK, D. T. & BEESON, P. B. 1972a Experimental bacterial endocarditis. I. Colonization of a sterile vegetation. *British Journal of Experimental Pathology* **53**, 44–49.
DURACK, D. T. & BEESON, P. B. 1972b Experimental bacterial endocarditis. II. Survival of bacteria in endocardial vegetations. *British Journal of Experimental Pathology* **53**, 50–53.
DURACK, D. T., BEESON, P. B. & PETERSDORF, R. G. 1973 Experimental bacterial endocarditis. III. Production and progress of the disease in rabbits. *British Journal of Experimental Pathology* **54**, 142–151.
DURACK, D. T. & PETERSDORF, R. G. 1973 Chemotherapy of experimental streptococcal endocarditis. I. Comparison of commonly recommended prophylactic regimens. *The Journal of Clinical Investigation* **52**, 592–598.
DURACK, D. T., PELLETIER, L. L. Jr. & PETERSDORF, R. G. 1974 Chemotherapy of experimental streptococcal endocarditis. II. Synergism between penicillin and streptomycin against penicillin-sensitive streptococci. *The Journal of Clinical Investigation* **53**, 829–833.
EDWARDSSON, S. 1968 Characteristics of caries-inducing human streptococci resembling *Streptococcus mutans*. *Archives of Oral Biology* **13**, 637–646.
EDWARDSSON, S. 1974 Bacteriological studies on deep areas of carious dentine. *Odontologisk Revy* **25**, Suppl. 32.
EDWARDSSON, S., KOCH, G. & OBRINK, M. 1972 *Streptococcus sanguis*, *Streptococcus mutans* and *Streptococcus salivarius* in saliva. Prevalence and relation to caries increment and prophylactic measures. *Odontologisk Revy* **23**, 279–296.
EHRLICH, J., STIVALA, S. S., BAHARY, W. S., GARG, S. K., LONG, L. W. & NEWBRUN, E. 1975. Levans: 1. Fractionation, solution viscosity, and chemical analysis of levan produced by *Streptococcus salivarius*. *Journal of Dental Research* **54**, 290–297.
ELLWOOD, D. C. 1976 Chemostat studies of oral bacteria. Proceedings *Microbial Aspects of Dental Caries*, eds Stiles, H. M., Loesche, W. J. & O'Brien, T. C. Special Supplement to *Microbiology Abstracts*, **3**, 785–798.
ELLWOOD, D. C. & HUNTER, J. C. 1976 The Mouth as a Chemostat. In *Continuous culture 6. Applications and New Fields*, eds Dean, A. C. R., Ellwood, D. C., Evans, C. G. T. & Melling, J. Chichester: Ellis Horwood Limited.
ELLWOOD, D. C., LONGYEAR, V. M. C. & HUNTER, J. R. 1972 Growth of mixed cultures of organisms derived from human dental plaque in a chemostat. *Journal of General Microbiology* **71**, (Abstract).
ELLWOOD, D. C., HUNTER, J. R. & LONGYEAR, V. M. C. 1974 Growth of *Streptococcus mutans* in a chemostat. *Archives of Oral Biology* **19**, 659–664.
FACKLAM, R. R. 1974 Characteristics of *Streptococcus mutans* isolated from human dental plaque and blood. *International Journal of Systematic Bacteriology* **24**, 313–319.
FACKLAM, R. R. 1977 Physiological differentiation of viridans streptococci. *Journal of Clinical Microbiology* **5**, 184–201.
FARMER, E. D. 1953 Streptococci of the mouth and their relationship to subacute bacterial endocarditis. *Proceedings of the Royal Society of Medicine* **46**, 201–208.
FERRETTI, J. J. & WARD, M. 1976 Susceptibility of *Streptococcus mutans* to antimicrobial agents. *Antimicrobial Agents and Chemotherapy* **10**, 274–276.
FITZGERALD, R. J. 1968 Dental caries research in gnotobiotic animals. *Caries Research* **2**, 139–146.
FUKUI, K., FUKUI, Y. & MORIYAMA, T. 1974a Purification and properties of dextran sucrase and invertase from *Streptococcus mutans*. *Journal of Bacteriology* **118**, 796–804.

FUKUI, K., FUKUI, Y. & MORIYAMA, T. 1974b Some immunochemical properties of dextransucrase and invertase from *Streptococcus mutans*. *Infection and Immunity* **10**, 985-990.
GARRISON, P. K. & FREEDMAN, L. R. 1970 Experimental endocarditis I. Staphylococcal endocarditis in rabbits resulting from placement of a polyethylene canula in the right side of the heart. *Yale Journal of Biology and Medicine* **42**, 394-410.
GENCO, R. J., EVANS, R. T. & TAUBMAN, M. A. 1974 Specificity of antibodies to *Streptococcus mutans*; significance in inhibition of adherence. In *The Immunoglobulin A System*, eds Mestecky, J. & Lawton, A. R. New York: Plenum Publishing Corporation.
GERMAINE, G. R., CHLUDZINSKI, A. M. & SCHACHTELE, C. F. 1974 *Streptococcus mutans* dextransucrase. Requirement for primer dextran. *Journal of Bacteriology* **120**, 287-294.
GIBBONS, R. J. & BANGHART, S. B. 1967 Synthesis of extracellular dextran by cariogenic bacteria and its presence in human dental plaque. *Archives of Oral Biology* **12**, 11-24.
GIBBONS, R. J. & VAN HOUTE, J. 1971 Selective bacterial adherence to oral epithelial surfaces and its role as an ecological determinant. *Infection and Immunity* **3**, 567-573.
GIBBONS, R. J. & VAN HOUTE, J. 1975a Dental Caries. *Annual Review of Medicine* **26**, 121-136.
GIBBONS, R. J. & VAN HOUTE, J. 1975b Bacterial adherence in oral microbial ecology. *Annual Reviews of Microbiology* **29**, 19-44.
GIBBONS, R. J., DE PAOLA, R. P., SPINELL, D. M. & SKOBE, Z. 1974 Interdental localization of *Streptococcus mutans* as related to dental caries experience. *Infection and Immunity* **9**, 481-488.
GIBBONS, R. J., SPINELL, D. M. & SKOBE, Z. 1976 Selective adherence as a determinant of the host tropisms of certain indigenous and pathogenic bacteria. *Infection and Immunity* **13**, 238-246.
GLEDHILL, W. E. & CASIDA, L. E. Jr. 1969 Predominant catalase negative soil bacteria. 1. Streptococcal population indigenous to soil. *Applied Microbiology* **17**, 208-213.
GOLD, O. G., JORDAN, H. V. & VAN HOUTE, J. 1975 The prevalence of enterococci in the human mouth and their pathogenicity in animal models. *Archives of Oral Biology* **20**, 473-477.
GOODFELLOW, M., COLLINS, M. D. & MINNIKIN, D. E. 1976 Thin-layer chromatographic analysis of mycolic acid and other long-chain components in whole-organism methanolysates of coryneform and related taxa. *Journal of General Microbiology* **96**, 351-358.
GRENIER, E. M., EVELAND, W. C. & LOESCHE, W. J. 1973 Identification of *Streptococcus mutans* serotypes in dental plaque by fluorescent antibody techniques. *Archives of Oral Biology* **18**, 707-715.
GUGGENHEIM, B. 1970 Enzymatic hydrolysis and structure of water-insoluble glucan produced by glucosyltransferases from a strain of *Streptococcus mutans*. *Helvetica Odontologika Acta* **14**, Suppl. V, 89-108.
GUGGENHEIM, B. & NEWBRUN, E. 1969 Extracellular glucosyltransferase activity of an HS strain of *Streptococcus mutans*. *Helvetica Odontologika Acta* **13**, 84-97.
GUGGENHEIM, B., REGOLATI, B. & MUHLEMAN, H. R. 1972 Caries and plaque inhibition by mutanase in rats. *Caries Research* **6**, 289-297.
GUTHOF, O. 1956 Ueber pathogene 'Vergrunende Streptokokken'. Streptokokken— Befunde bei dentogenen Abszessen und Infiltraten im Bereich der Mundhohle. *Zentralblatt für Bakteriologie, Parasitenkunde, Infektionskrankheiten und Hygiene* Abt. I **166**, 553-564.
HAMADA, S. & OOSHIMA, T. 1975 Inhibitory spectrum of a bacteriocin-like substance (Mutacin) produced by some strains of *Streptococcus mutans*. *Journal of Dental Research* **54**, 140-145.

HAMADA, S., MASUDA, N., OOSHIMA, T., SOBUE, S. & KOTANI, S. 1976 Epidemiological survey of *Streptococcus mutans* among Japanese children. *Japanese Journal of Microbiology* **20**, 33–44.

HAMILTON, I. R. 1976 Intracellular polysaccharide synthesis by cariogenic microorganisms. Proceedings *Microbial Aspects of Dental Caries*, eds Stiles, H. M., Loesche, W. J. & O'Brien, T. C. Special Supplement to *Microbiology Abstracts* **3**, 683–701.

HAMILTON, I. R. 1977 Effects of fluoride on enzymatic regulation of bacterial carbohydrate metabolism. *Caries Research* **11** (Suppl. 1), 262–291.

HAMILTON, I. R. & ELLWOOD, D. C. 1977 Growth of *Streptococcus mutans* in the chemostat. Effect of pH on the inhibition of carbohydrate metabolism by fluoride. International Association for Dental Research. British Division. Pre-printed abstract 115.

HANDELMAN, S. L., MILLS, J. R. & HAWES, R. R. 1966 Caries incidence in subjects receiving long-term antibiotic therapy. *Journal of Oral Therapeutics and Pharmacology* **2**, 338–345.

HARDER, E. J., WILKOWSKE, C. J., WASHINGTON, J. A. II & GERACI, J. E. 1974 *Streptococcus mutans* endocarditis. *Annals of Internal Medicine* **80**, 364–368.

HARDIE, J. M. 1975 Studies on *Streptococcus sanguis* and *Streptococcus mutans*. Ph.D. Thesis, University of London.

HARDIE, J. M. & BOWDEN, G. H. 1974a The normal microbial flora of the mouth. In *The Normal Microbial Flora of Man*, eds Skinner, F. A. & Carr, J. G. London & New York: Academic Press.

HARDIE, J. M. & BOWDEN, G. H. 1974b Cell wall and serological studies on *Streptococcus mutans*. *Caries Research* **8**, 301–316.

HARDIE, J. M. & BOWDEN, G. H. 1975 Bacterial flora of dental plaque. *British Medical Bulletin* **31**, 131–136.

HARDIE, J. M. & BOWDEN, G. H. 1976a Physiological classification of oral viridans streptococci. *Journal of Dental Research* **55**, A166–A176.

HARDIE, J. M. & BOWDEN, G. H. 1976b Some serological cross-reactions between *Streptococcus mutans*, *S. sanguis* and other dental plaque streptococci. *Journal of Dental Research* **55**, C50–C58.

HARDIE, J. M. & BOWDEN, G. H. 1976c The microbial flora of dental plaque: bacterial succession and isolation considerations. Proceedings *Microbial Aspects of Dental Caries*, eds Stiles, H. M., Loesche, W. J. & O'Brien, T. C. Special Supplement to *Microbiology Abstracts* **1**, 63–87.

HARDIE, J. M., THOMSON, P. L., SOUTH, R. J., MARSH, P. D., BOWDEN, G. H., McKEE, A. S., FILLERY, E. D. & SLACK, G. L. 1977 A longitudinal epidemiological study on dental plaque and the development of dental caries—interim results after two years. *Journal of Dental Research* **56**, C90–C99

HARE, R. 1935 The classification of hemolytic streptococci from the nose and throat of normal human beings by means of precipitin and biochemical tests. *Journal of Pathology and Bacteriology* **41**, 499–512.

HEHRE, E. J. 1948 Dextran-forming streptococci from the blood in subacute endocarditis and from the throats of healthy persons. *Bulletin of the New York Academy of Medicine* **24**, 543–544.

HOERMAN, K. C., KEENE, H. J., SHKLAIR, I. L. & BURMEISTER, J. A. 1972 The association of *Streptococcus mutans* with early carious lesions in human teeth. *Journal of the American Dental Association* **85**, 1349–1352.

HOLMBERG, K. & HALLANDER, H. O. 1972 Interference between Gram-positive microorganisms in dental plaque. *Journal of Dental Research* **51**, 588–595.

HOLMBERG, K. & HALLANDER, H. O. 1973 Production of bactericidal concentrations of hydrogen peroxide by *Streptococcus sanguis*. *Archives of Oral Biology* **18**, 423–434.

HOOK, E. W. III, ROBERTS, R. B. & SANDE, M. 1975 Antimicrobial therapy of experimental enterococcal endocarditis. *Antimicrobial Agents and Chemotherapy* **8**, 564–570.

HUGENSCHMIDT, A. C. 1896 Experimental study of the different modes of protection of the oral cavity against pathogenic bacteria (saliva). *Dental Cosmos* **38**, 797.
IACONO, V. J., TAUBMAN, M. A., SMITH, D. J., GARANT, P. R. & POLLOCK, J. J. 1976 Structure and function of the type-specific polysaccharide of *Streptococcus mutans* 6715. In *Immunologic Aspects of Dental Caries*, eds Bowen, W. H., Genco, R. J. & O'Brien, T. C. Washington DC & London: Information Retrieval Inc.
IKEDA, T., SANDHAM, H. J. & BRADLEY, E. L. Jr. 1973 Changes in *Streptococcus mutans* and lactobacilli in plaque in relation to the initiation of dental caries in Negro children. *Archives of Oral Biology* **18**, 555-566.
JABLON, J. M., FERRER, T. & ZINNER, D. D. 1976 Identification and quantitation of *Streptococcus mutans* by the fluorescent antibody technique. *Journal of Dental Research* **55**, A76-A79.
JOKINEN, M. A. 1970 Bacteraemia following dental extraction and prophylaxis. *Suomen Hammaslaakariseuran Toimituksia* **66**, 69-100.
JOLLY, M. & DRUCKER, D. B. 1971 Prevention of bacterial endocarditis. Some dental aspects. *British Dental Journal* **131**, 539-542.
JORDAN, H. V. & KEYES, P. H. 1966 *In vitro* methods for the study of plaque formation and carious lesions. *Archives of Oral Biology* **11**, 793-801.
JORDAN, H. V., ENGLANDER, H. R. & LIM, S. L. 1969 Potentially cariogenic streptococci in selected population groups in the Western hemisphere. *Journal of the American Dental Association* **78**, 1331-1335.
KALONAROS, I. V. & BAHN, A. N. 1965 Antigenic composition of the cell wall of *Streptococcus mitis*. *Archives of Oral Biology* **10**, 625-633.
KANTZ, W. E. & HENRY, C. A. 1974 Isolation and classification of anaerobic bacteria from intact pulp chambers of non-vital teeth in man. *Archives of Oral Biology* **19**, 91-96.
KEENE, H. J. & SHKLAIR, I. L. 1974 Relationship of *Streptococcus mutans* carrier status to the development of carious lesions in initially caries free recruits. *Journal of Dental Research* **53**, 1295-1296.
KELSTRUP, J. & FUNDER-NIELSEN, T. D. 1972 Molecular interactions between the extracellular polysaccharides of *Streptococcus mutans*. *Archives of Oral Biology* **17**, 1659-1670.
KELSTRUP, J., FUNDER-NIELSEN, T. D. & MØLLER, E. N. 1973 Enzymatic reduction of the colonisation of *Streptococcus mutans* in human dental plaque. *Acta Odontologica Scandinavica* **31**, 249-253.
KELSTRUP, J. & GIBBONS, R. J. 1969a Bacteriocins from human and rodent streptococci. *Archives of Oral Biology* **14**, 251-258.
KELSTRUP, J. & GIBBONS, R. J. 1969b Inactivation of bacteriocins in the intestinal canal and oral cavity. *Journal of Bacteriology* **99**, 888-890.
KELSTRUP, J., RICHMOND, S., WEST, C. & GIBBONS, R. J. 1970 Fingerprinting human oral streptococci by bacteriocin production and susceptibility. *Archives of Oral Biology* **15**, 1109-1116.
KEYES, P. H. 1960 The infectious and transmissable nature of experimental dental caries. *Archives of Oral Biology* **1**, 304-320.
KLEINBERG, I., CHATTERJEE, R., REDDY, J. & CRAW, D. 1977 Effects of fluoride on the metabolism of the mixed oral flora. *Caries Research* **11** (Suppl. 1), 292-320.
KNOX, K. W. & WICKEN, A. J. 1976 Grouping and cross-reacting antigens of oral lactic acid bacteria. *Journal of Dental Research* **55**, A116-A122.
KNOX, K. W., MARKHAM, J. L. & WICKEN, A. J. 1976 Formation of cross-reacting antibodies against cellular and extracellular lipoteichoic acid of *Streptococcus mutans*. BHT. *Infection and Immunity* **13**, 647-652.
KOLSTAD, R. A. 1976 Strain typing of oral streptococci by the use of bacterial antagonism. *Journal of Dental Research* **55**, Special Issue A, A154-A165.
KRASSE, B. 1954 The proportional distribution of *Streptococcus salivarius* and other streptococci in various parts of the mouth. *Odontologisk Revy* **5**, 203-211.

KRASSE, B. & CARLSSON, J. 1970 Various types of streptococci and experimental caries in hamsters. *Archives of Oral Biology* **15**, 25-32.
KRASSE, B., JORDAN, H. V. & EDWARDSSON, S. 1968 The occurrence of certain "Caries-inducing" streptococci in human dental plaque material with special reference to frequency and activity of caries. *Archives of Oral Biology* **13**, 911-918.
KURAMITSU, H. K. 1974 Characterisation of cell-associated dextransucrase activity from glucose-grown cells of *Streptococcus mutans*. *Infection and Immunity* **10**, 227-235.
LANCEFIELD, R. 1925*a* The immunological relationships of *Streptococcus viridans* and certain of its chemical fractions. I. Serological reactions obtained with antibacterial sera. *Journal of Experimental Medicine* **42**, 377-395.
LANCEFIELD, R. 1925*b* The immunological relationships of *Streptococcus viridans* and certain of its chemical fractions. II. Serological reactions obtained with antinucleoprotein sera. *Journal of Experimental Medicine* **42**, 397-412.
LEACH, S. A., APPLETON, J., DADA, O. & HAYES, M. L. 1972 Some factors affecting the metabolism of fructan by human oral flora. *Archives of Oral Biology* **17**, 137-145.
LEBIEN, T. W. & BROMEL, M. C. 1975 Antibacterial properties of a peroxidogenic strain of *Streptococcus mitior* (mitis). *Canadian Journal of Microbiology* **21**, 101-103.
LEHNER, T. 1975 Immunological aspects of dental caries and periodontal disease. *British Medical Bulletin* **31**, 125-130.
LEHNER, T., CHALLACOMBE, S. J. & CALDWELL, J. 1976 Immunologic basis for vaccination against dental caries in Rhesus monkeys. *Journal of Dental Research* **55**, C166-C180.
LILJEMARK, W. F. & GIBBONS, R. J. 1972 Proportional distribution and relative adherence of *Streptococcus mitior* (mitis) on various surfaces in the human oral cavity. *Infection and Immunity* **6**, 852-859.
LINEBERGER, L. T. & DE MARCO, T. J. 1973 Evaluation of transient bacteraemia following routine periodontal procedures. *Journal of Periodontology* **44**, 757-761.
LINZER, R. 1976 Serotype polysaccharide antigens of *Streptococcus mutans*: composition and serological cross-reactions. In *Immunological Aspects of Dental Caries*, eds Bowen, W. H., Genco, R. J. & O'Brien, T. C. Washington DC & London: Information Retrieval Inc.
LINZER, R., GILL, K. & SLADE, H. D. 1976 Chemical composition of *Streptococcus mutans* type *c* antigen: comparison to type *a*, *b* and *d* antigens. *Journal of Dental Research* **55**, A109-A115.
LITTLETON, N. W., KAKEHASHI, S. & FITZGERALD, R. J. 1970 Recovery of specific caries-inducing streptococci from carious lesions in the teeth of children. *Archives of Oral Biology* **15**, 461-463.
LOCKWOOD, W. R., LAWSON, L. A., SMITH, D. L., McNEILL, K. M. & MORRISON, F. S. 1974 *Streptococcus mutans* endocarditis. Report of a case. *Annals of Internal Medicine* **80**, 369-370.
LOESCHE, W. J., ROWAN, J., STRAFFON, L. H. & LOOS, P. J. 1975 Association of *Streptococcus mutans* with human dental decay. *Infection and Immunity* **11**, 1252-1260.
LOESCHE, W. J., HOCKETT, R. N. & SYED, S. A. 1977 Reductions in proportions of dental plaque streptococci following a 5 day topical kanamycin treatment. *Journal of Periodontal Research* **12**, 1-10.
LONG, L. W., STIVALA, S. S. & EHRLICH, J. 1975 Effect of pH on the biosynthesis of levan and on the growth of *Streptococcus salivarius*. *Archives of Oral Biology* **20**, 503-507.
McCABE, M. M., HAYNES, A. U. & HAMELIK, R. M. 1976 Cell adherence of *Streptococcus mutans*. Proceedings *Microbial Aspects of Dental Caries*, eds Stiles, H. M., Loesche, W. J. & O'Brien, T. C. Special Supplement to *Microbioloby Abstracts* **2**, 413-424.
McCARTHY, C., SNYDER, M. L. & PARKER, R. B. 1965 The indigenous oral flora of man. 1. The newborn to the 1 year old infant. *Archives of Oral Biology* **10**, 61-70.

McGOWAN, D. A. 1975 Experimental endocarditis in prepared rabbits infected with oral micro-organisms. *Journal of Dental Research* **54**, Special Issue A, L101 (abstract no. L406).

McGOWAN, D. A. & HARDIE, J. M. 1974 Production of bacterial endocarditis in prepared rabbits by oral manipulation. *British Dental Journal* **137**, 129–131.

McKINNEY, R. M. & THACKER, L. 1976 Improvement in specificity of immunofluorescent reagents for identifying *Streptococcus mutans* by DEAE–cellulose-bacterial cell column immunosorption methods. *Journal of Dental Research* **55**, A50–A57.

MARKHAM, J. L., KNOX, K. W., WICKEN, A. J. & HEWETT, J. 1975 Formation of extracellular lipoteichoic acid by oral streptococci and lactobacilli. *Infection and Immunity* **12**, 378–387.

MANLY, R. S. & RICHARDSON, D. T. 1968 Metabolism of levan by oral samples. *Journal of Dental Research* **47**, 1080–1086.

MARSH, P. D., HARDIE, J. M. & BOWDEN, G. H. 1975 Extraction of antigens from *Streptococcus mutans*. *Journal of Dental Research* **54**, Special Issue A, L132 (abstract L530).

MARSH, P. D., HARDIE, J. M., McKEE, A. S. & BOWDEN, G. H. 1977 Specific and shared antigens within *Streptococcus mutans*. *Caries Research* **11**, 121–122 (abstract).

MARSHALL, K. C., STOUT, R. & MITCHELL, R. 1971 Mechanism of the initial events in the sorption of marine bacteria to surfaces. *Journal of General Microbiology* **68**, 337–348.

MARTIN, M. V. & COLE, J. A. 1977. Are the glucosyltransferases from *Streptococcus mutans* glycoproteins? International Association for Dental Research. British Division. Pre-printed abstract 100.

MEJARE, B. & EDWARDSSON, S. 1975 *Streptococcus milleri* (Guthof); an indigenous organism of the human oral cavity. *Archives of Oral Biology* **20**, 757–762.

MELVAER, K. L., HELGELAND, D. & ROLLA, G. 1974 A charged component in purified polysaccharide preparations from *Streptococcus mutans* and *Streptococcus sanguis*. *Archives of Oral Biology* **19**, 539–546.

MICHALEK, S. M., SHIOTA, T., IKEDA, T., NAVIA, J. M. & McGHEE, J. R. 1975 Virulence of *Streptococcus mutans*: biochemical and pathogenic characteristics of mutant isolates. *Proceedings of the Society for Experimental Biology and Medicine* **150**, 498–502

MIKKELSEN, L. & POULSEN, S. 1976 Microbiological studies on plaque in relation to development of dental caries in man. *Caries Research* **10**, 178–188.

MIKX, F. H. M., VAN DER HOEVEN, J. S., KONIG, K. G., PLASSCHAERT, A. J. M. & GUGGENHEIM, B. 1972 Establishment of defined microbial ecosystems in germ free rats. 1. The effect of the interaction of *Streptococcus mutans* or *Streptococcus sanguis* with *Veillonella alcalescens* on plaque formation and caries activity. *Caries Research* **6**, 211–223.

MIKX, F. H. M. & VAN DER HOEVEN, J. S. 1975 Symbiosis of *Streptococcus mutans* and *Veillonella alcalescens* in mixed continuous culture. *Archives of Oral Biology* **20**, 407–410.

MIKX, F. H. M., VAN DER HOEVEN, J. S. & WALKER, G. J. 1976 Microbial symbiosis in dental plaque studied in gnotobiotic rats and in the chemostat. Proceedings *Microbial Aspects of Dental Caries*, eds Stiles, H. M., Loesche, W. J. & O'Brien, T. C. Special Supplement to *Microbiology Abstracts* **3**, 763–771.

MILLER, W. D. 1890 *The Micro-organisms of the Human Mouth*. Reprinted by Basel: S. Karger, 1973.

MINNIKIN, D. E., GOODFELLOW, M. & COLLINS, M. D. 1977 Lipid composition in the classification and identification of coryneform and related taxa. In *Biology of the Coryneform Bacteria*, eds Bousefield, I. J. & Cally, A. G. London & New York: Academic Press.

MIRICK, G., THOMAS, L., CURNEN, E. & HORSFALL, F. Jr. 1944 Studies on a non-haemolytic streptococcus isolated from the respiratory tract of human beings III. Immunological relationship of *Streptococcus MG* to *Streptococcus salivarius* Type 1. *Journal of Experimental Medicine* **80**, 431-440.

MITCHELL, R. 1976 Mechanism of attachment of micro-organisms to surfaces. Proceedings *Microbial Aspects of Dental Caries*, eds Stiles, H. M., Loesche, W. J. & O'Brien, T. C. Special Supplement to *Microbiology Abstracts* **1**, 47-53.

MONTAGUE, E. A. & KNOX, K. W. 1968 Antigenic components of the cell walls of *Streptococcus salivarius*. *Journal of General Microbiology* **54**, 237-246.

NAKAMURA, T., SUGINAKA, Y. & OBATA, T. 1976 Enzymatic action of oral *Bacteroides* against the dental plaque forming substance from streptococci. *Bulletin of the Tokyo Dental College* **17**, 107-122.

NEEFE, L. I., CHRETIEN, J. H., DELAHA, E. C. & GARAGUSI, V. I. 1974 *Streptococcus mutans* endocarditis. Confusion with enterococcal endocarditis by routine laboratory testing. *Journal of the American Medical Association* **230**, 1298-1299.

NEWBRUN, E. 1976 Polysaccharide synthesis in plaque. Proceedings *Microbial Aspects of Dental Caries*, eds Stiles, H. M., Loesche, W. J. & O'Brien, T. C. Special Supplement to *Microbiology Abstracts* **3**, 649-664.

NEWBRUN, E., FELTON, R. A. & BULKACZ, J. 1976 Susceptibility of some plaque micro-organisms to chemotherapeutic agents. *Journal of Dental Research* **55**, 574-579.

NIVEN, C. F. Jr., SMILEY, K. L. & SHERMAN, J. M. 1941a The production of large amounts of a polysaccharide by *Streptococcus salivarius*. *Journal of Bacteriology* **41**, 479-484.

NIVEN, C. F. Jr., SMILEY, K. L. & SHERMAN, J. M. 1941b The polysaccharides synthesized by *Streptococcus salivarius* and *Streptococcus bovis*. *Journal of Biological Chemistry* **140**, 105-109.

OKELL, C. C. & ELLIOTT, S. D. 1935 Bacteraemia and oral sepsis with special reference to the aetiology of subacute endocarditis. *Lancet* **229**, 869-872.

OLSSON, J. & KRASSE, B. 1976 A method for studying adherence of oral streptococci to solid surfaces. *Scandinavian Journal of Dental Research* **84**, 20-28.

OLSSON, J., GLANTZ, P-O. & KRASSE, B. 1976a Electrophoretic mobility of oral streptococci. *Archives of Oral Biology* **21**, 605-609.

OLSSON, J., GLANTZ, P-O. & KRASSE, B. 1976b Surface potential and adherence of oral streptococci to solid surfaces. *Scandinavian Journal of Dental Research* **84**, 240-242.

ORLAND, F. J., BLAYNEY, J. R., HARRISON, R. W., REYNIERS, J. A., TREXLER, P. C., ERVIN, R. F., GORDON, H. A. & WAGNER, M. 1955 Experimental caries in germ-free rats inoculated with enterococci. *Journal of the American Dental Association* **50**, 259-272.

OSBORNE, R. M., LAMBERTS, B. L., MEYER, T. S. & ROUSH, A. H. 1976 Acrylamide gel electrophoretic studies of extracellular sucrose-metabolising enzymes of *Streptococcus mutans*. *Journal of Dental Research* **55**, 77-84.

PARKER, M. T. & BALL, L. C. 1976 Streptococci and aerococci associated with systemic infection in man. *Journal of Medical Microbiology* **9**, 275-302.

PARKER, R. B. 1970 Paired culture interaction of the oral microbiota. *Journal of Dental Research* **49**, 804-809.

PARKER, R. B. & CREAMER, H. R. 1971 Contribution of plaque polysaccharides to growth of cariogenic micro-organisms. *Archives of Oral Biology* **16**, 855-862.

PELLETIER, L. L. Jr., DURACK, D. T. & PETERSDORF, R. G. 1975 Chemotherapy of experimental streptococcal endocarditis. IV. Further observations on prophylaxis. *The Journal of Clinical Investigation* **56**, 319-330.

PERCH, B., KJEMS, E. & RAVN, T. 1974 Biochemical and serological properties of *Streptococcus mutans* from various human and animal sources. *Acta pathologica et microbiologica scandinavica* (B), **82**, 357-370.

PHILLIPS, I., WARREN, C., HARRISON, J. M., SHARPLES, P., BALL, L. C. & PARKER, M. T. 1976 Antibiotic susceptibilities of streptococci from the mouth and blood of patients treated with penicillin or lincomycin and clindamycin. *Journal of Medical Microbiology* **9**, 393-404.

POGREL, M. A. & WELSBY, P. D. 1975 The dentist and prevention of infective endocarditis. *British Dental Journal* **139**, 12-16.

QURESHI, J. V., GOLDNER, M., RICHE, W. H. le. & HARGREAVES, J. A. 1977 *Streptococcus mutans* serotypes in young schoolchildren. *Caries Research* **11**, 141-152.

RANTZ, L. A. & RANDALL, E. 1955 Use of autoclaved extracts of hemolytic streptococci for serological grouping. *Stanford Medical Bulletin* **13**, 290-291.

ROGERS, A. H. 1972 Effect of the medium on bacteriocin production among strains of *Streptococcus mutans*. *Applied Microbiology* **24**, 294-295.

ROGERS, A. H. 1973 The ecology of *Streptococcus mutans* in carious lesions and on caries-free surfaces of the same tooth. *Australian Dental Journal* **18**, 226-228.

ROGERS, A. H. 1974 Bacteriocin production and susceptibility among strains of *Streptococcus mutans* grown in the presence of sucrose. *Antimicrobial Agents and Chemotherapy* **6**, 547-550.

ROGERS, A. H. 1975 Bacteriocin types of *Streptococcus mutans* in human mouths. *Archives of Oral Biology* **20**, 853-858.

ROGERS, A. H. 1976 Bacteriocinogeny and the properties of some bacteriocins of *Streptococcus mutans*. *Archives of Oral Biology* **21**, 99-104.

RÖLLA, G. 1976 Inhibition of adsorption—general considerations. Proceedings *Microbial Aspects of Dental Caries*, eds Stiles, H. M., Loesche, W. J. & O'Brien, T. C. Special Supplement to *Microbiology Abstracts* **2**, 309-324.

ROSAN, B. 1973 Antigens of *Streptococcus sanguis*. *Infection and Immunity* **7**, 205-211.

ROSAN, B. 1976 Relationship of the cell wall composition of Group H streptococci and *Streptococcus sanguis* to their serological properties. *Infection and Immunity* **13**, 1144-1153.

RUSSELL, R. R. B. 1976 Classification of *Streptocccus mutans* strains by SDS gel electrophoresis. *Microbios Letters* **2**, 55-59.

RUTTER, P. R. & ABBOTT, A. A study of the interaction between oral streptococci and hard surfaces.

RUTTER, P. R., RUEFENACHT, W. G., CHAMBERLAIN, C. R., THOMASSEN, P. R., ROSE, M. & SCRIVENER, C. A. 1961 The principle of bacterial antagonism applied as an aid in the reduction of dental caries. *Journal of Dental Research* **40**, 1112-1115.

SABISTON, C. B. & GOLD, W. A. 1974 Anaerobic bacteria in oral infections. *Oral Surgery* **38**, 187-???.

SANDE, M. A. & IRVIN, R. G. 1974 Penicillin-aminoglycoside synergy in experimental *Streptococcus viridans* endocarditis. *The Journal of Infectious Diseases* **129**, 572-576.

SCHACHTELE, C. F. & MAYO, J. A. 1973 Phosphoenolpyruvate-dependent glucose transport in oral streptococci. *Journal of Dental Research* **52**, 1209-1215.

SCHACHTELE, C. F., HARLANDER, S. K. FULLER, D. W., ZOLLINGER, P. K. & LEUNG, W-L. S. 1976 Bacterial interference with sucrose-dependent adhesion of oral streptococci. Proceedings *Microbial Aspects of Dental Caries*, eds Stiles, H. M., Loesche, W. J. & O'Brien, T. C. Special Supplement to *Microbiology Abstracts* **2**, 401-412.

SCHAMSCHULA, R. G. & BARMES, D. E. 1970 A study of the streptococcal flora of plaque in caries free and caries active primitive peoples. *Australian Dental Journal* **15**, 377-382.

SCHLEGEL, R. & SLADE, H. D. 1972 Bacteriocin production by transformable group H streptococci. *Journal of Bacteriology* **112**, 824-829.

SCHOTTMÜLLER, H. 1903 Die Artunterscheidung der Für den Menschen pathogenen Streptokokken durch Blutagar. *Münchener Medizinische Wochenschrift* **50**, 849, 909.

SCRIVENER, C. A. 1955 A possible dental caries prevention by micro-organisms—bacterial antagonists employed intra-orally. *Journal of California State Dental Association and Nevada Dental Society* **31**, 151.

SCRIVENER, C. A. MYERS, H. L., MOORE, N. A. & WAINER, B. W. 1950 Bacterial antagonism in the prevention of dental caries. *Dental Items of Interest* **72**, 1239.
SELBIE, F. R., SIMON, R. D. & ROBINSON, R. H. M. 1949 Serological classification of viridans streptococci from subacute endocarditis, teeth and throats. *British Medical Journal* **2**, 667–672.
SHARMA, M., DHILLON, A. S. & NEWBRUN, E. 1974 Cell-bound glucosyltransferase activity of *Streptococcus sanguis* strain 804. *Archives of Oral Biology* **19**, 1063–1072.
SHERMAN, J. M. 1937 The streptococci. *Bacteriological Reviews* **1**, 3–97.
SHERMAN, J. M., NIVEN, C. F. Jr. & SMILEY, K. L. 1943 *Streptococcus salivarius* and other non-hemolytic streptococci of the human throat. *Journal of Bacteriology* **45**, 249–263.
SHKLAIR, I. L. & KEENE, H. J. 1974 A biochemical scheme for the separation of the five varieties of *Streptococcus mutans*. *Archives of Oral Biology* **19**, 1079–1081.
SHKLAIR, I. L., KEENE, H. J. & SIMONSON, L. G. 1971 Distribution and frequency of *Streptococcus mutans* in caries active individuals. *Journal of Dental Research* **51**, 882.
SHKLAIR, I. L., KEENE, H. J. & CULLEN, P. 1974 The distribution of *Streptococcus mutans* on the teeth of two groups of naval recruits. *Archives of Oral Biology* **19**, 199–202.
SIMS, W. 1974 The clinical bacteriology of purulent oral infections. *British Journal of Oral Surgery* **12**, 1–12
SNYDER, R. J., WILKOWSKE, C. J. & WASHINGTON, J. A. II. 1975 Bactericidal activity of combinations of gentamicin with penicillin or clindamycin against *Streptococcus mutans*. *Antimicrobial Agents and Chemotherapy* **7**, 333–335.
SOCRANSKY, S. S. & MANGANIELLO, A. D. 1971 The oral microbiota of man from birth to senility. *Journal of Periodontology* **42**, 485–494.
SOCRANSKY, S. S., MANGANIELLO, A. D., PROPAS, D., ORAM, V. & VAN HOUTE, J. 1977 Bacteriological studies of developing supragingival dental plaque. *Journal of Periodontal Research* **12**, 90–106.
SOLOWAY, M. 1942 A serological classification of viridans streptococci with special reference to those isolated from subacute bacterial endocarditis. *Journal of Experimental Medicine* **76**, 109–126.
SOUTHWICK, F. S. & DURACK, D. T. 1974 Chemotherapy of experimental streptococcal endocarditis. III. Failure of a bacteriostatic agent (tetracycline) in prophylaxis. *Journal of Clinical Pathology* **27**, 261–264.
STACK, M. V., DONOGHUE, H. D., TYLER, J. E. & MARSHALL, M. 1973 Identification of oral streptococci with the use of standardized data from pyrolysis gas chromatography. *Journal of Dental Research* **52**, 969 (abstract no. 145).
SUKCHOTIRATANA, M., LINTON, A. H. & FLETCHER, J. P. 1975 Antibiotics and the oral streptococci of man. *Journal of Applied Bacteriology* **38**, 277–294.
SUNDQUIST, G. 1976 Bacteriological studies of necrotic dental pulps. *Umea University Odontological Dissertations*, No. 7.
SWENSON, J. I., LILJEMARK, W. F. & SCHUMAN, L. M. 1976 A longitudinal epidemiologic evaluation of the association between the detection of plaque streptococci and development of dental caries in children. Proceedings *Microbial Aspects of Dental Caries*, eds Stiles, H. M., Loesche, W. J. & O'Brien, T. C. Special Supplement to *Microbiology Abstracts* **1**, 211–222.
SYMINGTON, J. M. 1975 Streptococci isolated from post-extraction bacteraemias. *British Journal of Oral Surgery* **13**, 91–94.
TANZER, J. M., FREEDMAN, M. L., FITZGERALD, R. J. & LARSON, R. H. 1974 Diminished virulence of glucan synthesis-defective mutants of *Streptococcus mutans*. *Infection and Immunity* **10**, 197–201.
TANZER, J. M., FREEDMAN, M. L., WOODIEL, F. N., EIFERT, R. L. & RINEHIMER, L. A. 1976 Association of *Streptococcus mutans* virulence with synthesis of intracellular polysaccharide. Proceedings *Microbial Aspects of Dental Caries*, eds Stiles, H. M., Loesche, W. J. & O'Brien, T. C. Special Supplement to *Microbiology Abstracts* **3**, 597–616.

THEILADE, E. & THEILADE, J. 1976 Role of plaque in the etiology of periodontal disease and caries. *Oral Sciences Reviews* **9**, 23-63.
THOMSON, L. A., LITTLE, W. & HAGEAGE, G. J. 1976 Application of Fluorescent antibody methods in the analysis of plaque samples. *Journal of Dental Research* **55**, A80-A86.
TINANOFF, N., GROSS, A. & BRADY, J. M. 1976 Development of plaque on enamel. Parallel investigations. *Journal of Periodontal Research* **11**, 197-209.
VAN DER HOEVEN, J. S., ROGERS, A. H. & MIKX, F. H. M. 1977 Inhibition of *A. viscosus* by bacteriocin-producing gnotobiotic rats. *Journal of Dental Research* **56**, Special Issue A: A132. Pre-printed abstract 357.
VAN HOUTE, J. 1976 Oral bacterial colonisation: Mechanisms and implications. Proceedings *Microbial Aspects of Dental Caries*, eds Stiles, H. M., Loesche, W. J. & O'Brien, T. C. Special Supplement to *Microbiology Abstracts* **1**, 3-32.
VAN HOUTE, J. & JANSEN, H. M. 1968 Levan degradation by streptococci isolated from human dental plaque. *Archives of Oral Biology* **13**, 827-830.
VAN HOUTE, J., BACKER-DIRKS, O., DE STOPPELAAR, J. D. & JANSEN, H. M. 1969 Iodophilic polysaccharide-producing bacteria and dental caries in children consuming fluoridated and non-fluoridated drinking water. *Caries Research* **3**, 178-189.
VAN HOUTE, J., JORDAN, H. V. & BELLACK, S. 1971 Proportions of *Streptococcus sanguis*, an organism associated with subacute bacterial endocarditis, in human faeces and dental plaque. *Infection and Immunity* **4**, 658-659.
VERNAZZA, T. R. & MELVILLE, T. H. 1977 Bacteriocin-like activity in strains of *Streptococcus mitis*. International Association for Dental Research. British Division. Preprinted abstract. 76.
WASHBURN, M. R., WHITE, J. C. & NIVEN, C. F. Jr. 1946 Streptococcus S.B.E.: immunological characteristics. *Journal of Bacteriology* **51**, 723-729.
WELD, H. G. & SANDHAM, H. J. 1976 Effect of long-term therapies with penicillin and sulfadiazine on *Streptococcus mutans* and lactobacilli in dental plaque. *Antimicrobial Agents and Chemotherapy* **10**, 200-204.
WETHERELL, J. R. & BLEIWEIS, A. S. 1975 Antigens of *Streptococcus mutans*: characterization of a polysaccharide antigen from walls of strain GS-5. *Infection and Immunity* **12**, 1341-1348.
WHITE, J. C. & NIVEN, C. F. Jr. 1946 Streptococcus S.B.E.: A streptococcus associated with subacute bacterial endocarditis. *Journal of Bacteriology* **51**, 717-722.
WICKEN, A. J. & KNOX, K. W. 1975 Lipoteichoic acid: a new class of bacterial antigen. *Science* **187**, 1161-1167.
WILLIAMS, R. E. O. 1956 *Streptococcus salivarius* (vel. *hominis*) and its relations to Lancefield's group K. *Journal of Pathology and Bacteriology* **72**, 15-25.
WILLIAMSON, C. K. 1964 Serological classification of viridans streptococci from the respiratory tract of man. In *Taxonomic Biochemistry and Serology*, ed. Leone, C. A. New York: Ronald Press Co.
WITTGOW, W. C. Jr. & SABISTON, C. B. Jr. 1975 Micro-organisms from pulpal chambers of intact teeth with necrotic pulps. *Journal of Endodontics* **1**, 168-171.
WOOD, J. M. 1964 Polysaccharide synthesis and utilisation by dental plaque. *Journal of Dental Research* **43**, 955 (Abstract).
WOOD, J. M. 1969 The state of hexose sugar in human dental plaque and its metabolism by the plaque bacteria. *Archives of Oral Biology* **14**, 161-168.
WOODS, R. 1971 A dental caries susceptibility test based on the occurrence of *Streptococcus mutans* in plaque material. *Australian Dental Journal* **16**, 116-121.
YAMADA, T. & CARLSSON, J. 1975 Regulation of lactate dehydrogenase and change of fermentation products in streptococci. *Journal of Bacteriology* **124**, 55-61.
YAMADA, T. & CARLSSON, J. 1976 The role of pyruvate formate-lyase in glucose metabolism of *Streptococcus mutans*. Proceeedings *Microbial Aspects of Dental Caries*, eds Stiles, H. M., Loesche, W. J. & O'Brien, T. C. Special Supplement to *Microbiology Abstracts* **3**, 809-819.

YAMAMOTO, T., IMAI, S., NISIZAWA, T. & ARAYA, S. 1975 Production of, and susceptibility to, bacteriocin-like substances in oral streptococci. *Archives of Oral Biology* **20**, 389-391.

ZINNER, D. D. & JABLON, J. M. 1968 Human Streptococcal strains in experimental caries. In *The Art and Science of Dental Caries Research*, ed. Harris, R. S. London & New York: Academic Press.

Streptococci in the Alimentary Tract of the Ruminant

M. J. LATHAM AND D. J. JAYNE-WILLIAMS

*National Institute for Research in Dairying,
Shinfield, Reading, Berkshire, England*

CONTENTS

1. Introduction 207
2. Streptococci in the alimentary tract of the healthy ruminant 208
   (a) Rumen of preruminant animals 208
   (b) Rumen of adult domestic animals 213
   (c) Rumen of wild species 215
   (d) Gastrointestinal tract of young animals 216
   (e) Gastrointestinal tract of adult domestic and wild ruminants 217
3. Involvement of *Streptococcus bovis* in alimentary disorders 219
   (a) Bloat 219
   (b) Acidosis 221
4. Aspects of the physiological and biochemical characteristics of *Streptococcus bovis* and other group D streptococci of significance to rumen function . . . 224
   (a) Nitrogen metabolism 224
   (b) Carbohydrate metabolism 225
   (c) Chemical composition 228
5. Interrelationships between *Streptococcus bovis* and other rumen micro-organisms . 230
6. Conclusions 232
7. References 233

1. Introduction

THE ECONOMIC IMPORTANCE of domestic cattle and sheep has stimulated extensive research into the physiology and, particularly during the last 25 years, the microbiology of the rumen. Hungate (1966), Hobson (1971) and others have described in detail the unique mode of digestion in the ruminant, and only a brief outline is presented here.

All herbivorous mammals possess an enlarged segment of their alimentary tract in which the opportunity exists for micro-organisms to digest and ferment components of the diet that the host is unable to degrade. In some herbivores, such as the horse, the caecum becomes enlarged whereas in others, such as cattle and sheep, part of the oesophagus enlarges to form the rumen (Fig. 1). Thus the characteristic feature of digestion in the ruminant is that microbial fermentation occurs in a site anterior to the true stomach. The caecum of the ruminant is also enlarged and provides a secondary site for microbial fermentation, but, compared with the rumen, this is of lesser nutritional importance.

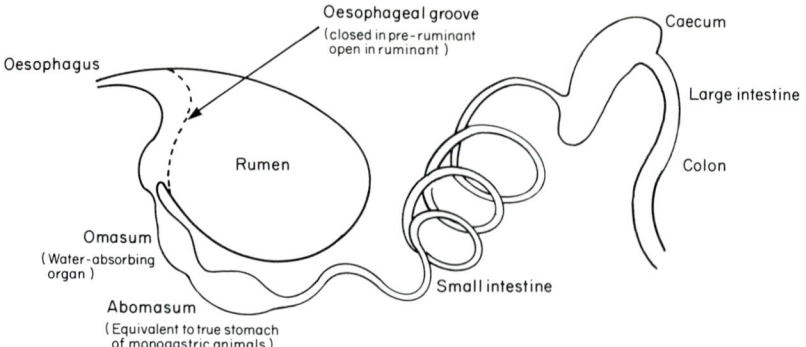

Fig. 1. The ruminant alimentary tract.

Volatile fatty acids (VFA) are the major products of microbial fermentation in these organs. These acids are absorbed through the surrounding epithelial tissue (Fig. 2) and provide the most important single source of carbon available to the host for biosynthetic and energy-yielding purposes. Fermentation gases which collect above the rumen digesta are vented by eructation. After leaving the rumen, undigested food residues and microbial cells pass through a water-absorbing organ (the omasum) and are then subjected to digestion in the true stomach (the abomasum) and more distal parts of the gut. Much of the material which remains undigested is eventually diverted into the caecum where it undergoes a further stage of microbial digestion. Contents of the caecum are periodically discharged and are voided in the faeces after passage through the colon.

A modification of this process occurs in the preruminant animal given liquid (e.g. milk) or semi-solid diets. Ingestion of diet stimulates the reflex closure of a structure known as the oesophageal groove (Fig. 1). When closed this groove maintains the oesophagus as a continuous tube which permits most of the food to flow directly to the abomasum, thus bypassing the undeveloped rumen. Ingress of diet into the rumen due to malfunctioning of the groove results in the pH of the contents dropping from the normal values of *ca*. 7.5 to *ca*. 6.5 or lower (Smith 1961). Operation of the groove mechanism and factors which affect it have been studied by numerous workers (e.g. Ørskov *et al*. 1970: Guilhermet *et al*. 1973; Newhook & Titchen 1976). The ingestion of roughage or solid food by the more mature animal causes the groove to open thus allowing the material to enter the rumen. When this happens many changes take place. A large mixed population of strictly anaerobic bacteria and protozoa develops, the rumen enlarges and its epithelium becomes keratinized and densely papillated.

The high energy requirements of sheep and cattle reared under modern

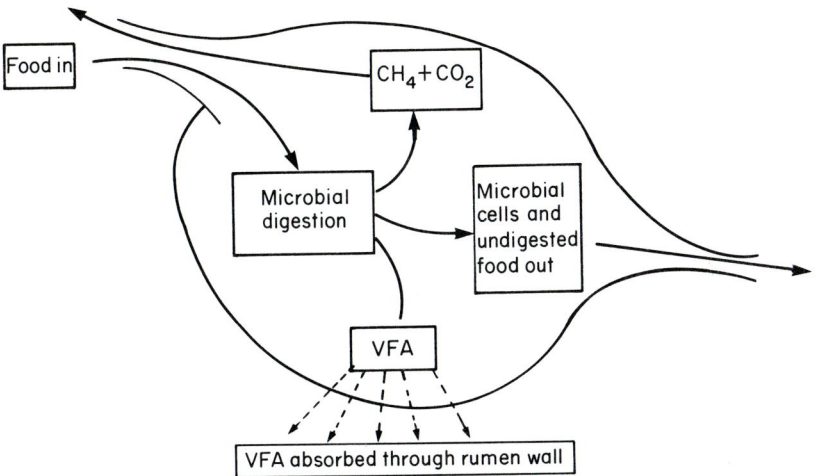

Fig. 2. The digestion of food by rumen organisms.

conditions of animal husbandry demand a correspondingly high intake of diet. Despite the large capacity of the rumen relative to the total body weight, the bulkiness of many roughage rations restricts the amount that can be consumed and hence restricts the intake of energy. To overcome this problem, grain, which is readily digested and has a high energy density, is frequently fed as the major component of a ration. The overfeeding of any readily fermentable foodstuff be it grain or lush pasture can, however, result in serious alimentary disorders stemming from changes in the metabolism of the rumen microbial population.

Although warm-blooded animals provide a major habitat for most species of the genus *Streptococcus*, only the group D and some of the viridans (Sherman 1937) streptococci can be regarded as truly commensal organisms of the alimentary tract of the ruminant. Inevitably much of the information on the qualitative and quantitative nature of the rumen microflora has been obtained from studies using domestic animals, often maintained under rigorously controlled dietary and environmental conditions. Information on the distribution of micro-organisms in other segments of the alimentary tract of domestic ruminants, or in the digestive system of wild ruminants as a whole, is surprisingly scant.

In this paper an attempt is made to survey the occurrence of the main species of streptococci in the ruminant gut, to examine the involvement of streptococci in certain alimentary disorders and to discuss some aspects of the physiology and chemistry of these organisms and their interactions with other rumen micro-organisms which may be of significance to rumen function.

Details of the media and methods used by different workers to enumerate and isolate streptococci from the alimentary tract have generally been omitted for the sake of brevity. However, the influence of cultural conditions on the growth of some of the group D streptococci (e.g. *S. bovis* and *S. equinus*) should be borne in mind since aerobic incubation on primary isolation (Hungate 1957; Dubos *et al.* 1965; Jayne-Williams, unpublished) and many of the selective reagents suitable for more tolerant species such as *S. faecalis* and *S. faecium* (Sabbaj *et al.* 1971; Switzer & Evans 1974) can be inhibitory.

Unidentifiable or aberrant strains of streptococci are frequently isolated from gut and faecal samples, but it is beyond the scope of the present article to describe their ecological or taxonomic significance.

2. Streptococci in the Alimentary Tract of the Healthy Ruminant

(a) *Rumen of preruminant animals*

Information on the numbers and species of streptococci occurring in the rumen of preruminant (i.e. liquid-fed) animals is very scant. In their study of the microflora of the rumen of young calves, Bryant *et al.* (1958) paid particular attention to certain groups of organisms characteristic of the adult animal and although facultatively anaerobic amylolytic streptococci were encountered they were not enumerated. Mann & Oxford (1955) isolated amylolytic streptococci (*S. bovis*) typical of the adult animal from the rumen and abomasum of calves up to eight weeks of age, but the strains found most frequently were deemed atypical in being mannitol positive, non-haemolytic and forming large white colonies. It was concluded that the typical strains did not become properly established until the pH of the rumen approached neutrality after several weeks. However, it should be noted that the low pH values of 5.7–6.0 they recorded for rumen contents of calves 10–11 days old are lower than the near neutral values reported by others for animals of this age (Ziolecki & Briggs 1961; data in Table 1). Mann & Oxford (1955) also found Gram positive cocci in the rumen contents of very young goats but these were not enumerated or identified.

Rumen samples taken from groups of three calves on four different rearing regimens, but all providing access to solid food, were studied by Ziolecki & Briggs (1961). A total of 309 samples from animals from birth to three months of age were examined by dilution counts in a non-selective broth medium followed by isolation and identification of the organisms present. Counts ($\log_{10}$) of streptococci were in the range 6–8, depending on diet, with higher counts being generally observed in samples from younger animals. The presence of chlortetracycline in the diet of one group of calves resulted in somewhat lower counts. The majority (*ca.* 83%) of 247 isolates were classified as *S. bovis* among which four physiological types were recognized. The incidence of these types

Table 1

The incidence of streptococci in the rumen of liquid-fed calves

Calf no.	Rearing system*	Age when sampled (days)	pH of sample	Viable count ($\log_{10}$) ml^{-1} of sample					
				Total organisms	Total streptococci	S. bovis	S. faecalis	S. faecium	S. equinus
1	A	15	6.5	7.7	4.4	4.4	3.4	‡	‡
2	A	{ 28	¶	9.3	6.8	6.8	5.2§	‡	‡
		135	7.2	8.6	6.8	6.8	5.1	4.7	‡
3	B	49	7.0	9.6	9.0	9.0	‡	‡	7.0
4	B	4	6.9	9.9	6.2	6.2	‡	4.2	‡
5	C	{ 4	6.7	10.0	7.2	7.0	5.1	6.5	6.4
		63	7.3	8.0	6.9	6.9	‡	2.8	5.0
		112†	8.0	7.7	6.0	6.0	‡	‡	‡
6	D	40	7.1	9.5	9.1	9.1	4.5	3.8	5.0
7††	D	7	4.8	9.7	9.4	9.2	6.8	9.0	7.7
8††	D	6	5.3	9.5	9.5	9.5	‡	8.4	‡

* All calves received colostrum, then: A, liquid milk direct from a bucket; B, liquid milk direct from a bucket through a rubber teat attached to a bucket; C, allowed to suckle a nurse cow; D, milk replacer diet containing alkane-grown yeast protein (ca. 39% Toprina; British Petroleum Co. Ltd) direct from a bucket. Calves were not given solid food but calf 5 had access to the nurse cow's diet from time to time. Calves 1 and 2 had rumen fistulae and samples were taken through the fistula; samples from the remainder were obtained by means of a tube passed down the oesophagus.

† Repeated attempts were made before a sample was obtained; it contained a high proportion of saliva which increased the pH and reduced the viable counts.

†† These calves were suffering from severe bloat when sampled. Calf 8 bloated late in the day and sufficient time was available only for a dilution count (duplicate tubes/dilution) to be performed in ETSB; positive tubes were then streaked on the media normally used for plate counts and different colony types enumerated and isolated for identification.

‡ Not detected.

§ S. faecalis and subsp. liquefaciens in about equal proportions.

¶ Not done.

appeared to be influenced by the age of the animal. Diet was found to have no effect on the composition of the streptococcal population, but there was a tendency for species other than *S. bovis* to be more common in very young animals.

Streptococcus bovis was included as one of the components used by Lysons *et al.* (1971, 1976) to establish a rumen microflora in gnotobiotic lambs and in one instance *S. faecalis* was also included. Successful establishment of these organisms was shown by microscopical examination but viable counts were not made.

Work is at present in progress at the Institute to provide more information on the numbers and types of bacteria occurring in the calf rumen, and to determine to what extent the microflora is influenced by methods of feeding, type of diet and failure of the oesophageal groove to function correctly. Some of the results obtained for streptococci are outlined in the following section.

In liquid-fed calves the oesophageal groove bypass mechanism ensures that most of the diet ingested passes directly to the abomasum. Nevertheless the rumen contains an appreciable volume of liquor—*ca.* 1 l in one-month-old calves increasing to *ca.* 6 l in eight-month-old calves (Smith 1959). The liquor comprises sloughed off epithelial cells, saliva and other secretions, and supports a large and complex population of bacteria, including streptococci. A mean figure for the rate of flow of fluid into and out of the calf rumen (during the period when the animal is not feeding) has been calculated by Smith (1959) to be $255\,\text{ml}\,\text{h}^{-1}\,100\,\text{kg}^{-1}$ of body weight. Thus although it is non-functional with respect to the digestion of dietary constituents, the rumen acts as a potent source of organisms which continuously drain into the abomasum and more

The numbers and identities (determined by the methods described by Jayne-Williams, 1976) of streptococci occurring in the rumen of calves on different rearing systems and diets are shown in Table 1. The counts given represent the highest obtained irrespective of the medium on which they occurred. On four occasions streptococci constituted more than 10% of the total viable population; two of these samples were from calves (Nos 7 & 8) suffering from severe bloat due to the ingress of yeast-containing diet into the rumen. The rumen contents of the bloating animals had low pH values.

Streptococcus bovis was the predominant species encountered irrespective of the age of the animals or system of rearing. Although the majority of these isolates were raffinose and mannitol positive, some were atypical in other respects, particularly with regard to the fermentation of melibiose. The significance of these biotypes is not known. The incidence of other species bore no relation to age of the animal (cf. Ziolecki & Briggs 1961) or to system of rearing except that both samples from bloating calves contained high proportions of species other than *S. bovis*. On the basis of these findings, therefore, it would seem that the major factor influencing the streptococcal populations of the rumen of young animals is not age or system of rearing but malfunction of the oesophageal groove permitting ingress of diet into the rumen.

(b) *Rumen of adult domestic animals*

Group D streptococci are the only facultative anaerobes of any numerical significance that have been regularly isolated from the rumen of cattle and sheep (Hungate 1966), water buffalo (Srivastava & Srivastava 1974) and goats (Dehority & Grubb 1977). They form a relatively small proportion of the total bacterial population and comprise mainly *S. bovis* with smaller numbers of *S. faecium* and *S. faecalis*. Mann *et al.* (1954) found that out of 300 facultative anaerobes isolated from sheep, 120 were group D streptococci, of which 82% were *S. bovis* and 6% *S. faecalis*. A similar distribution of species was observed in water buffalo by Srivastava & Srivastava (1974).

The physiological characteristics of *S. bovis* enable it to be placed in two important functional groups of rumen bacteria: those that hydrolyse starch (amylolytic) and those that produce lactic acid as their major fermentation product (Table 2). Neither *S. faecium* nor *S. faecalis* can be assigned to functional groups with certainty, though the ureolytic activity of some strains of *S. faecium* (to be discussed later) makes them important organisms in the metabolism of non-protein nitrogen.

Fermentation rates in the rumen of animals fed roughage rations are largely determined by the activities of organisms capable of degrading the insoluble cellulose, hemicellulose and pectin, contained in plant cell walls to soluble carbohydrate derivatives. *Streptococcus bovis* has only a limited capacity to degrade insoluble cell wall substances and is therefore in constant competition with other organisms for the limited amounts of soluble carbohydrate available.

Table 2
Some important rumen bacteria having similar functions

Organism	Function
Streptococcus bovis *Bacteroides amylophilus* *Bacteroides ruminicola* *Succinomonas amylolytica*	Amylolytic: hydrolysis and fermentation of starch
Streptococcus bovis *Lactobacillus* spp. *Bifidobacterium* spp.	Lactogenic: formation of lactic acid as major fermentation product
Megasphaera elsdenii *Selenomonas ruminantium* *Veillonella alcalescens*	Lactic acid-utilizing

Kistner and coworkers examined the bacterial populations in the rumen of sheep given poor or good quality hay and found that the amylolytic bacteria were unaffected by variations in the level of intake of very poor quality teff hay (Gilchrist & Kistner 1962) but increased in numbers with more digestible lucerne hay (Kistner *et al*. 1962). Irrespective of diet, however, *S. bovis* constituted only a minor part of the amylolytic flora. Thorley *et al*. (1968) also found that diets of ground grass, unlike those containing long grass, increased the proportions of lactic acid-producing and -utilizing bacteria in the rumen of cows, but did not significantly affect the proportion of *S. bovis*. In other studies direct microscopic or viable counts of streptococci in rumen contents of various domestic animals receiving high roughage diets e.g. grazing cows (Higginbottom & Wheater 1954; Wolstrup *et al*. 1974) or cows stall-fed with hay or pelleted grass (Higginbottom & Wheater 1954; Hungate 1957; Latham *et al*. 1971; Wolstrup *et al*. 1974), sheep (Macpherson 1953) and Zebu cattle and Murrah buffalo (Panjarathinam & Laxminarayana 1975) all indicate that the numbers of streptococci and, where determined, numbers of *S. bovis* deviate very little from the range $\log_{10} 6-7$ g^{-1}.

Because starch is a primary energy substrate for *S. bovis*, it is to be expected that increasing the amount of this material in the diet would favour the development of large populations of streptococci. Increased numbers of unidentified Gram positive streptococci following such a change in diet have frequently been observed (Gall & Huhtanen 1951; Hungate *et al*. 1952; Maki & Foster 1957). These changes tend to be transitory, however, and not necessarily typical of animals which have been fed such diets for some time. Latham *et al*. (1971) compared the rumen bacterial populations of cows receiving diets of all hay with those of the same animals some weeks after a transition to high grain diets of rolled barley or flaked maize and found only small increases in the numbers and proportion of *S. bovis*. The increase was more marked with rolled barley than with flaked maize, which suggested that the type of grain might be a determining factor. Observations of a similar nature were made by Slyter *et al*. (1970) who noted that *S. bovis* was more numerous in the rumen of animals fed wheat than in those fed maize. More variable increases were observed in lactating cows with a high level of food intake during gradual adaptation to high grain, milk-fat-depressing diets (Latham *et al*. 1974). Although in these studies *S. bovis* was often the major amylolytic species present, it was never the predominant lactic acid-producing organism. These results, and those of other workers (Hungate 1957; Bryant & Robinson 1961; Krogh 1963; Eadie *et al*. 1967, 1970; Slyter *et al*. 1974) suggest that in animals which have had sufficient time to adapt to grain diets the mean daily counts of *S. bovis* will not be very much greater than those observed in animals given diets high in roughage. There are a few exceptions to this since large populations of streptococci or unidentified Gram positive homofermentative cocci have been observed in animals apparently adapted to high grain rations (Perry et al. 1955; Bauman & Foster 1956; Whitelaw et al. 1972). The occurrence of streptococci as part of the majority flora together with the lactate-utilizing anaerobe, *Megasphaera elsdenii*, in sheep receiving dietary supplements of minerals (Thomson et al. 1977) suggests that the maintenance of a high buffering capacity within the rumen, either by manipulation of the diet (e.g. the addition of mineral salts) or by increased secretion of saliva by the host, may significantly influence the populations of streptococci.

Group D streptococci other than *S. bovis*, *S. faecium* and *S. faecalis* have not been isolated regularly from the rumen of the adult animal. Krogh (1963) recovered *S. inulinaceus* from the rumen of sheep suffering from acidosis but there have been no further reports of this species having been isolated. Neither *S. equinus*, which predominates in the equine caecum, nor *S. avium* have been shown to occur in the rumen of the adult animal. This is surprising in view of the physiological similarity of the former to *S. bovis* and the latter to *S. faecium*. A pyogenic species, *S. uberis*, which causes bovine mastitis has been isolated from the rumen (Cullen & Little 1969) but its low numbers and other factors connected with its distribution on the skin of the animal indicate that it is not a rumen commensal. The serological group N streptococcus, *S. lactis*, has been isolated from rumen samples in sufficient numbers (10^4 g^{-1}, M. E. Sharpe, pers. comm.) to suggest that the organism is capable of multiplication in the rumen. This species, together with *S. faecalis* subsp. *liquefaciens* (Mundt et al. 1962) and another group N species, *S. diacetylactis* (Matuszewski et al. 1936 cited by Sandine et al. 1972), can form part of the epiphytic flora of a wide variety of plants; such organisms may thus enter the rumen on foodstuffs. Mann et al. (1954) found *S. faecalis* in significant numbers on stored hay being fed to sheep but not on fresh hay, and concluded that contamination arose from the dust in the animal house.

(c) *Rumen of wild species*

Few studies have been made on the rumen microflora of animals other than cattle and sheep. *Streptococcus bovis* has been reported as the predominant streptococcal species in the rumen of the mule deer (Pearson 1969), white-tailed deer (Pearson 1965), red deer (Prins & Mulder 1969; Hobson et al. 1976), fallow deer (Prins & Mulder 1969), buffalo (Pearson 1967) and elk (McBee et al. 1969). The numbers of this organism occurring in red and fallow deer were reported by Prins & Mulder (1969) to be about 10- to 100-fold greater than in cows. Amylolytic cocci showing marked seasonal fluctuations in numbers were observed in the chamois by von Brüggermann et al. (1967) and Gram positive cocci resembling *S. bovis* were reported to form a minor part of the rumen population of the Scottish reindeer by Hobson et al. (1976). Anaerobic bacteria similar or identical to bovine rumen organisms have been found in the rumen of African Zebu cattle, gazelle, eland, suni and camel (Hungate et al. 1959). Notwithstanding the absence of data on streptococci it seems unlikely that these organisms did not form one of the components of the microflora.

(d) *Gastrointestinal tract of young animals*

The age of the animal from birth to 12 days appears to have little effect on the counts of streptococci, though Smith (1965) noted marked variations in counts between calves for all regions of the gut. The highest counts ($\log_{10} g^{-1}$) obtained increased from *ca*. 7.0 in the first segment of the small intestine to *ca*. 8.5 in the caecum. Mylrea (1969) used re-entrant cannulae sited at six different levels of the small intestine of two- to four-week-old calves fed whole milk and showed that counts ($\log_{10}$) of streptococci g^{-1} of chyme from the pylorus to the ileum ranged from 5.1 to 6.5 which suggested that the sources of many of the bacteria occurring in the small intestine, and possibly large intestine also, are the forestomachs and abomasum. Smith (1962, 1971) also examined the incidence of streptococci in the abomasum and more distal regions of the gut of both healthy calves and those suffering from neonatal diarrhoea during the first 14 days after birth and found no difference between the two groups of animals nor between colostrum-fed and -colostrum-deprived calves. The median counts ($\log_{10}$) of streptococci in 13 healthy animals were 4.9 in the abomasum, 3.7–5.7 in the anterior and posterior sections of the small intestine, 7.4 in the large intestine and 7.3 in the rectum. The commonest species occurring in the small intestine (as judged by the appearance on the TITG medium of Barnes 1956) were *S. faecium* and *S. faecalis* (including subsp. *liquefaciens*) whereas these were outnumbered in lower sections of the gut by other, unspecified, streptococci. Similar counts were obtained by Contrepois & Gouet (1973) in the gut of healthy calves up to 11 days old either suckled by the dam or given whole milk by bucket.

Using animals fitted with re-entrant fistulae Hartman et al. (1966) examined the influence of diet and age on the microflora of ileal contents and faeces of calves fed whole milk alone, or milk supplemented with either sucrose or starch. With one exception the mean counts ($\log_{10}$) of streptococci in the ileal contents were within the range 6.8–7.9 irrespective of diet. However, the numbers of streptococci in the faeces of calves fed milk alone were approximately ten times greater than in the ileum. Little or no difference between faecal and ileal counts of streptococci were observed with animals given milk plus sucrose, and the differences with starch-fed animals were intermediate. It was suggested that these differences (which were also observed with other components of the microflora) might be related either to the higher pH values of the gut contents which obtained when milk alone was fed, or to the more rapid transit time of digesta in animals given the carbohydrate supplements. Both the ileal and faecal counts of streptococci decreased by about 3 log cycles in the period 31 to 128 days in calves given milk plus sucrose.

Several workers have attempted to assess the effect of age, dietary antibiotic and other factors on the numbers and types of streptococci occurring in the calf gut by examining faeces only. Haenel (1960), using media selective for enterococci, examined the faeces from six calves and obtained counts ($\log_{10}$) in the range 5.0-7.5 g^{-1}. Kjellander (1960) reported similar counts of streptococci in the faeces of 10 calves and counts ($\log_{10}$) of sorbitol-fermenting streptococci (*S. faecalis*) of 3.9 g^{-1}.

Smith & Crabb (1961) and Smith, H. W. (1961) showed that the counts of streptococci in the faeces of calves and lambs were considerably greater in the faeces of young calves and lambs than in older animals, dropping from nearly $\log_{10}$ 10 to *ca.* $\log_{10}$ 6 g^{-1} in 31 and 48 weeks, respectively. As judged by the appearance of the colonies on TITG agar, *S. faecium* and *S. faecalis* (and subspecies) predominated in the faeces of neonatal calves but were then replaced by other, unspecified streptococci, which grew very poorly on TITG agar. Samples of faeces obtained at intervals from 71 calves on a variety of artificial diets following early weaning were examined for streptococci by Gedek (1969). A non-selective medium incubated aerobically was used and although the isolates were referred to as enterococci they would presumably include *S. bovis*. The results indicated that the counts ($\log_{10}$) of streptococci were in the range 6.0-9.0 g^{-1} of faeces, but in contrast to the studies of Smith, H. W. (1961) lower counts were generally associated with younger animals. Species were not identified.

The live weight gain of non-ruminating calves is stimulated if antibiotics are incorporated into their ration. Neither aureomycin (Rusoff *et al.* 1953) nor bacitracin (Ellsworth *et al.* 1953) were found to have any effect on the numbers of streptococci in calf faeces. However, a report by Radisson *et al.* (1956) that the incorporation of terramycin in the diet increased the weight gain of calves and at the same time reduced the count of 'enteric streptococci' in the faeces, led to the suggestion that streptococci may be implicated in the growth depression of young animals.

(e) *Gastrointestinal tract of adult domestic and wild ruminants*

Information about the microflora of the large intestine of the adult ruminant has been largely derived from examination of faecal samples. In an early comparative investigation, Winslow & Palmer (1910) found that bovine faeces contained many more raffinose-fermenting strains of streptococci and fewer mannitol-fermenting strains than did human faeces. Raffinose-fermenting strains, which also attacked starch, were given the species name *bovis* by Orla-Jensen

(1919). The observations of Winslow & Palmer (1910) were later confirmed by Ayres & Mudge (1923) who identified the predominant streptococcus in bovine faeces as *S. bovis*. Considerable variation in the incidence of this species in bovine and ovine faeces has been reported subsequently (Cooper & Ramadan 1955; Buttiaux 1958; Wilssens & Buttiaux 1958) and has been ascribed to regional differences. Final confirmation of the numerical importance of *S. bovis* in the microflora of the faeces and colon of the ruminant and of the relatively low level of phenotypic variation between bovine isolates came from the work of Medrek & Barnes (1962*a, b*) and Mieth (1962).

The ability to hydrolyse starch is an unusual characteristic among the streptococci possessed only by *S. bovis* and *S. equinus*. Starch-hydrolysing strains were isolated by Seeley & Dain (1960) from bovine and ovine faeces and also from samples from the goat, dromedary camel, red deer, American elk, Arabian gazelle and alpaca. Streptococci were not isolated from the faeces of the llama and vicuna, but this was attributed to mishandling of the specimens.

Group D streptococci unable to hydrolyse starch (*S. faecalis* and *S. faecium*) constitute only a small proportion of the micro-organisms in bovine faeces (Mieth 1962; Medrek & Barnes 1962*a, b*) but are more numerous in sheep faeces (Medrek & Barnes 1962*a*). These species appear to be numerically less important in herbivores such as cows, horses guinea-pigs and rabbits than in man, dogs, rats and poultry (Haenel & Müller-Beuthow 1956).

Bacteria are present throughout the alimentary tract of ruminants, the numbers being lowest in the abomasum and increasing progressively down the gut (Ulyatt *et al.* 1975). In view of the large numbers of bacteria continuously leaving the rumen, the numbers and types of bacteria in the small intestine must be largely determined by the relative resistance of the various organisms to the bactericidal action of the acid secretions of the abomasum. Gram positive organisms seem to be better able to withstand these conditions than Gram negative bacteria (Pounden *et al.* 1950). These authors observed, but did not identify, Gram positive cocci in various segments of the ruminant intestine, and data presented by Kern *et al.* (1974) indicate that the counts ($\log_{10}$) of Gram positive cocci in the abomasum, ileum, caecum and terminal colon are of the order of 5.7, 6.4, 7.7 and 6.5 g^{-1}, respectively. Higginbottom & Wheater (1954) isolated *S. bovis* from all segments of the gut, the numbers being lowest in the abomasum presumably due to the high acidity. *Streptococcus bovis* and raffinose-fermenting strains of *S. faecium* similar to those isolated by Medrek & Barnes (1962*b*) have also been isolated from the duodenum, jejunum and ileum of ruminating calves (Latham & Sharpe, unpublished). It seems probable that the raffinose-fermenting strains isolated by Winslow & Palmer (1910) may have comprised *S. faecium* as well as *S. bovis*.

Information on the normal microflora of the ruminant caecum is scant. Mann & Ørskov (1973) obtained viable counts ($\log_{10}$) of 8.0 g^{-1} caecal contents of sheep fed dried grass cubes but found that *S. bovis* and *S. faecalis* formed only a small proportion of the flora. Sharpe *et al.* (1976) reported the presence of streptococci in the bovine caecum but the species were not identified. However, subsequent work on these and other isolates from the caeca of calves has shown them to comprise *S. bovis* and several unidentified species (Latham & Sharpe, unpublished).

Increasing the amount of grain in a diet results in increasing proportions escaping digestion in the rumen and reaching more distal parts of the gut. This particuarly applies to diets containing ground maize (Lindsay 1970). The effect that the presence of undigested grain in the lower gut has on the bacterial flora was shown by Allison *et al.* (1975) who overfed cattle and sheep with cracked maize by introducing it into the rumen, and found that streptococci and lactobacilli became the predominant organisms in the caecum. Conversely, Mann & Ørskov (1973) observed little change in the streptococcal populations in the caecum of sheep fed various carbohydrates either conventionally (thus allowing fermentation in the rumen) or by bottle feeding (which stimulated closure of the oesophageal groove and hence bypass of the rumen). The work of both these groups involved the use of unnatural feeding regimes and it would be of great interest to know if, in animals allowed time to adapt naturally to normal grain diets, the fermentation of carbohydrates (starch in particular) by streptococci in the caecum proved to be of nutritional significance to the host.

As with the rumen, the occurrence in other parts of the gut of species of streptococci other than those mentioned above is of doubtful significance. Although *S. uberis* was shown to occur in large numbers on the lips of cows (Cullen & Little 1969) its incidence in the bovine rectum is low, suggesting that it is unable to survive passage through the gut (Sharma & Parker 1970). *Streptococcus salivarius* was isolated from the bovine mouth by Andrewes & Horder (1906), *S. bovis* from the back of the mouth by Ayers & Mudge (1923) and *S. dysgalactiae* from the tonsils of cows by Zeidler *et al.* (1968). No further observations on the microflora of the bovine oral cavity appear to have been reported but V. Dent (pers. comm.) has recently isolated *S. mutans* and other streptococci from this site. In view of the physiological similarity between *S. bovis* and *S. mutans*, it is possible that the latter species might have occurred among the strains isolated by Ayers & Mudge (1923).

3. Involvement of *Streptococcus bovis* in Alimentary Disorders

(a) *Bloat*

Bloat is an alimentary disorder of the rumen associated with the ingestion of readily fermentable materials. Its incidence, possible causes and treatment have been fully reviewed by Clarke & Reid (1974). Gases formed during microbial fermentation in the rumen normally accumulate freely above the digesta and are vented from time to time by eructation. If gas becomes trapped as small bubbles in the rumen digesta a stable foam develops which increases the volume occupied by the rumen digesta and results in a high intra-ruminal pressure. This pressure may prevent the cardiac and pharyngo-oesophageal sphincters from opening thus exacerbating the condition. Finally the circulation fails and the animal becomes comatose and dies. A crude way of relieving this pressure, which is frequently the only one possible, is to puncture the rumen from the outside to allow the gas and digesta to escape. Bloat occurs with animals fed on leguminous green fodder or lush young grass rich in soluble carbohydrates (pasture bloat) and in cattle intensively fed with cereals (feedlot bloat). The causes of these two types of bloat are poorly understood. Plant constituents such as proteins, saponins and pectic substances in the case of pasture bloat and animal or microbial factors in the case of feedlot bloat have all been suggested as contributing to the occurrence of the disorder.

The role of streptococci in pasture bloat is not clear. There appear to be no large differences in the protozoal (Clarke 1954, 1965) or bacterial populations (Bryant *et al.* 1960) of bloating compared with non-bloating animals. Streptococci were shown to be more numerous in the bloating rumen, however, and to be susceptible to penicillin, which proved to be an effective treatment (Bryant *et al.* 1960); but in view of their low numbers in comparison to the total count they were considered to be unimportant. Hartman *et al.* (1962) also found that the numbers of streptococci increased, but only after the symptoms of bloat had subsided.

Salivary mucin has been shown to be an effective anti-foaming agent when mixed with foaming rumen digesta (Bartley & Yadava 1961). Since secretion of saliva is reduced in animals fed succulent rations (Meyer *et al.* 1964) the amount of mucin left in the rumen after degradation by bacteria may be too low to prevent the formation of stable foams.

Streptococcus bovis (Robertson *et al.* 1940) and several other species of rumen bacteria (Mishra *et al.* 1968) secrete powerful mucinolytic enzymes. When *S. bovis* was inoculated into the rumen of cows given a bloat-inducing ration bloating was more intense and persistent than in uninoculated animals (Mishra *et al.* 1968).

Excess production of polysaccharide slime by bacteria in the rumen was proposed by Hungate *et al.* (1955) as a factor contributing to bloat. *Streptococcus bovis* and several other rumen bacteria, including *Butyrivibrio fibrisolvens* and *Bacteroides ruminicola*, produce copious amounts of slime under certain nutritional conditions. Dain *et al.* (1956) and Bailey & Oxford (1958) studied the production of dextran from sucrose by *S. bovis* and suggested that, because high concentrations of this sugar occur in certain pasture plants, these organisms may play an important role in bloat. However, this seems to be unlikely since dextran slimes do not accumulate in bloating animals (Bailey 1959b). Moreover, an organism resembling *Bifidobacterium bifidum* which actively degrades the dextran formed by the streptococci has been isolated from the rumen of a bloating animal in numbers roughly equal to those of *S. bovis* (Clarke 1959).

Antibiotics appear to have a therapeutic effect on pasture bloat (Emery *et al.* 1958) but administration of 50-150 mg of penicillin day^{-1} (equivalent to a final maximum concentration of 2 μg ml^{-1} of rumen fluid) had relatively little effect on the numbers of streptococci (Wiseman *et al.* 1960) although lower levels of the antibiotic inhibit the growth of these organisms in pure culture (El Akkad & Hobson 1966; Fulgham *et al.* 1968).

In contrast to pasture bloat, microbial slime is considered to be largely responsible for increasing the viscosity, and hence the foaming properties, of the rumen contents of animals suffering from feedlot bloat. Gutierrez *et al.* (1961) found that the appearance of carbohydrate-rich, ethanol-precipitable slime, in rumen contents coincided with symptoms of bloat. Unlike the slime associated with legume bloat, that of feedlot bloat was also rich in DNA indicating a microbial origin. Jacobson *et al.* (1957) noted a highly significant correlation between an increase in capsulated organisms and the occurrence of bloat. Counts of encapsulated *S. bovis* (Gutierrez *et al.* 1959; Mishra *et al.* 1967) and of lactic acid-utilizing *Megasphaera* (*Peptostreptococcus*) *elsdenii* (Gutierrez *et al.* 1959) have been observed to increase substantially in bloating animals. The megasphaerae produce copious amounts of gas and a complex mixture of VFA including *n*-valeric acid from the fermentation of glucose, lactic acid or amino acids (Rogosa 1971) and it is tempting to speculate that streptococci and megasphaerae grow associatively under dietary conditions which predispose an animal to bloat. *Streptococcus bovis* produces a capsule in ordinary media (Hobson & MacPherson 1954) and large quantities of dextran slime when provided with high levels of sucrose (Bailey & Oxford 1958). The capsule and dextran appear to be morphologically distinct (Cheng *et al.* 1976) but whether or not they are chemically distinct seems open to question (Kane & Karakawa, 1971).

Agglutinating antibodies against streptococcal capsular material are produced by the host and gain access to the rumen via the saliva (Horacek *et al.* 1977). These antibodies agglutinate chains of streptococci *in vitro*. Should such agglutination also occur *in vivo* the viscosity of the rumen contents could be increased. Thus, variation between animals in their susceptibility to feedlot bloat may be due to differences in production of antibody.

(b) *Acidosis*

Excessive consumption of grain or an abrupt change from a high roughage to a high grain ration generally gives rise to an alimentary disorder known as acidosis. Various aspects of this disorder which may occur in an acute or chronic form have been comprehensively reviewed by Dunlop (1972) and Slyter (1976) and some of the clinical and pathological symptoms described by Dirksen (1970). Clinical signs of acidosis are sudden loss of appetite, reduced contractions or even stasis of the rumen, muscular tremors and diarrhoea. Numerous fermentable substances will give rise to acidosis but wheat, barley and flaked maize seem particularly active. Although there are no obvious microbiological explanations for the difference, sheep appear to be able to consume relatively more cereal than cattle without ill effect.

The condition is brought about by the accumulation of lactic acid in the rumen. Since the estimated dissociation constant of lactic acid (pKa = 3.7) in the rumen (Dunlop 1972) is lower than that of the VFA, the pH of the rumen falls rapidly as the concentration of H^+ increases. Absorption of lactic acid by the rumen epithelium, which occurs more efficiently at pH 5 and below (Williams & Mackenzie 1965), is not sufficiently rapid to prevent the accumulation of lactic acid. Increased salivation by the animal partially neutralizes the acidity but once the blood alkali reserves become depleted, salivary flow diminishes. Ulcerative ruminitis is a common sequela and as the epithelium of the rumen begins to desquamate, organisms such as *Sphaerophorus necrophorus* (*Fusobacterium necrophorum*) (Jensen et al. 1954) invade the tissue and ulcerative infections of the liver—from which *Sphaero. necrophorus* has been isolated (Calkins & Scrivener 1967)—ensue.

Large diurnal variations in bacterial numbers have been observed in the rumen of animals fed concentrate rations (Bryant & Robinson 1961) and if concentrates are fed in such a way that high levels of soluble carbohydrate are maintained, *S. bovis* will multiply rapidly and attain numbers in excess of 10^9 ml^{-1} (Hungate et al. 1952). In the work of Hungate, acidosis was induced in sheep by infusing the rumen with glucose or by feeding cracked maize in excess. Later studies by Krogh (1959, 1960, 1961) in which dietary supplements of lactose, sucrose and starch were used to induce acidosis in sheep, confirmed the important role of *S. bovis* in initiating the disorder. Although *S. bovis* is capable of rapid growth—under favourable conditions it has a doubling time of only 12 min (von Brüggeman & Giesecke 1965)—its numbers, as discussed previously, are generally low. One possible explanation of this seeming paradox was put forward by Schwartz & Gilchrist (1975) who suggested that, because the organism has a very high substrate saturation constant, conditions permitting it to express its full growth potential only occur for a brief period after the animal has ingested a diet rich in soluble sugars. The fact that lactic acid accumulates in the rumen suggests that it is not normally produced in significant amounts (Hungate 1966) and its rapid formation by *S. bovis* and later by the lactobacilli exceeds the limited capacity of other rumen bacteria, such as *Selenomonas ruminantium*, *Megasphaera elsdenii* and *Veillonella alcalescens*, to metabolize it.

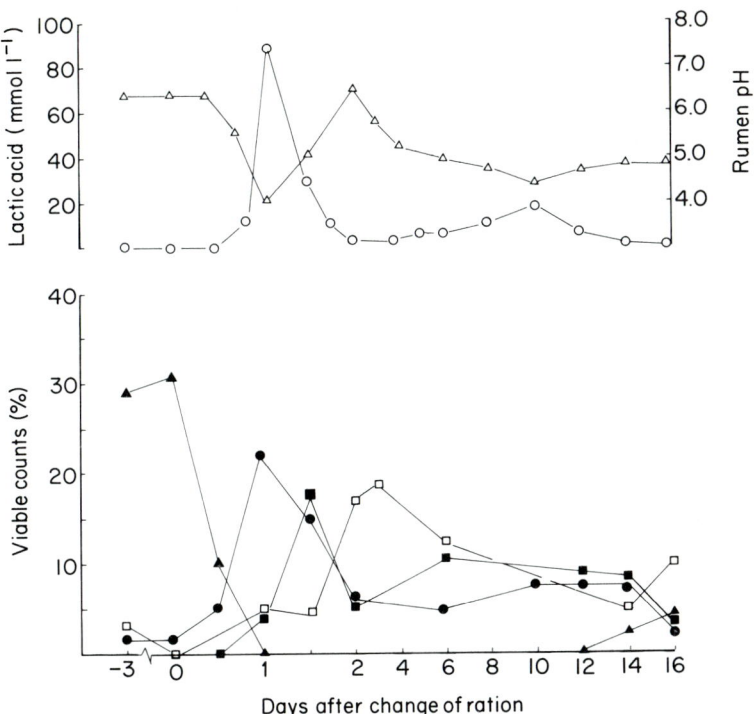

Fig. 3. Changes in the concentration of lactic acid, pH and bacterial flora in the rumen of a cow before, during and after the onset of acidosis induced by an abrupt change of rations from all hay to all flaked maize. ○, lactic acid; △, pH; □, lactic acid-utilizing bacteria (*Selenomonas ruminantium* and *Megasphaera elsdenii*); ●, *Streptococcus bovis*; ▲, *Butyrivibrio* spp; ■, lactobacilli and bifidobacteria.

Changes in the rumen pH, lactic acid concentration and some components of the rumen microflora in a cow in which acidosis was induced by an abrupt change of ration from all hay to all flaked maize are presented in Fig. 3 (Latham & Sutton, unpublished). The rise in the population of *S. bovis* and, a short while later, in the population of lactobacilli and bifidobacteria exemplifies the speed with which these lactogenic species can multiply in the rumen after readily fermentable starchy feedstuffs have been ingested. Under the prevailing acid conditions the production of microbial α-amylase continues and as growth of many of the glucose-utilizing strict anaerobes (such as butyrivibrios) is inhibited, the concentration of free glucose may reach as much as 5 μmoles ml^{-1} of rumen fluid (Ryan 1964). This glucose provides an ideal substrate for the further growth of aciduric lactobacilli and persistence of large populations of lactobacilli after the growth and acid production of *S. bovis* has become self-inhibitory has been observed by Krogh (1963), Allison et al. (1975) and Dirksen (1970). Once animals have been established on high grain rations both the proportion of lactic acid-utilizing bacteria (Ogimoto & Geisecke 1974) and the rate of lactic acid utilization (Jayasuriya & Hungate 1959) may be considerably greater than in animals fed high roughage diets. The transfer of rumen contents from sheep established on cereal diets to animals newly started on a wheat ration can reduce or even eliminate acidosis (Allison et al. 1964) emphasizing the value of allowing time for the rumen microflora to adapt to high cereal diets.

4. Aspects of the Physiological and Biochemical Characteristics of *Streptococcus bovis* and other Group D Streptococci of Significance to Rumen Function

(a) *Nitrogen metabolism*

Ruminants conserve nitrogen by recycling the urea formed in the liver to the rumen via the saliva. Urea is also a common non-protein nitrogen component of diets for domestic ruminants. The ureolytic activity of rumen contents appears to be confined to the bacterial (Jones et al. 1964) rather than the protozoal (Naga & El Shazly 1968) fractions of rumen fluid. Several species of strictly anaerobic ureolytic bacteria have been isolated from the rumen (Gibbons & Doetsch 1959; Slyter et al. 1968; Elias, 1971; John et al. 1974) but the majority of actively ureolytic organisms so far isolated have proved to be facultative anaerobes. Among these, Gram positive cocci have been isolated by Heald et al. (1953); Mann et al. (1954); Blackburn & Hobson (1962); Slyter et al. (1968); van Wyk & Steyn (1975) and Cook (1976a). Slyter et al. (1966) presumptively identified ureolytic cocci as strains of *S. bovis* but there have been no further reports of this species possessing this character. Cook (1976a) isolated a strongly ureolytic strain of *S. faecium* by enrichment. This strain is of additional interest in that its urease activity appears to be plasmid coded (Cook 1976b). Although under normal conditions Gram positive cocci constitute only a minor part of the total rumen population, their collective activity may be sufficient to account for the ureolytic activity observed in rumen fluid, there being no salivary activity. Reports concerning the effect dietary urea has on urease activity in the rumen are conflicting (Caffrey et al. 1967; Chalupa et al. 1970) but the ureases of *S. faecium* (Cook 1976a) and the strict anaerobe, *Selenomonas ruminantium* (John et al. 1974) are both inducible. They differ in that the specific activity of the *Selenomonas ruminantium* enzyme is markedly reduced by the accumulation of NH_4^+ whereas the urease of *S. faecium* does not appear to be similarly affected.

Ammonia provides the major source of nitrogen available to the rumen bacteria *in vivo* and may arise from salivary urea or dietary protein or non-protein nitrogen. Unlike the other group D streptococci, many strains of *S. bovis* are capable of utilizing ammonia as their sole source of nitrogen (Wolin et al. 1959; Prescott et al. 1959). Fixation of ammonia in the rumen is mediated by glutamic dehydrogenase which catalyses the formation of glutamate from α-ketoglutarate and NH_4^+ (Allison 1970). The activities of many enzymes of intermediary metabolism in various rumen bacteria have been surveyed by Joyner & Baldwin (1966) and a highly active NADP-linked glutamic dehydrogenase was detected in S. bovis and in another important amylolytic species, Bacteroides amylophilus. Most of the glutamic dehydrogenases of rumen bacteria are NADP-linked rather than NAD-linked, although both types of enzyme can be detected in extracts of mixed rumen bacteria.

Preformed amino acids often stimulate the growth of S. bovis, and are taken up by a few strains in preference to being synthesized de novo (Bryant & Robinson 1961). Glutamine alone (Wolin et al. 1959), arginine alone (Niven et al. 1948; Prescott & Stutts 1955), a mixture of glutamic acid and arginine or asparagine alone (Niven et al. 1948) can satisfy the nitrogen requirements of many strains. Polyoxyethylene sorbitan mono-oleate (Tween 80) promotes the uptake of aspartic acid (Paul 1961) presumably by increasing cell wall permeability. Arginine supports the best growth of any single amino acid though resting cells seem unable to utilize it (Hungate 1966). Carbon dioxide, which is required for the initiation of growth (Dain et al. 1956) is rapidly incorporated into aspartic and glutamic acids, threonine and some purines (Paul 1961). Incorporation of CO_2 into amino acids is far greater in defined than in complex media (Hayashi & Kitahara 1976).

The Gram positive cocci are not considered to be important in the breakdown of protein in the rumen (Allison 1970), although some strains of S. faecalis subsp. liquefaciens are highly proteolytic in vitro (Deibel 1964). Streptococcus bovis is not proteolytic nor is it regarded as having any catabolic action towards amino acids. However, Scheifinger et al. (1976) have recently made an observation which suggests that S. bovis may catabolize amino acids in vivo, since fresh isolates of S. bovis were shown to deaminate various amino acids when the acids were included in a complete medium at the very low concentrations typical of those found in the rumen. One strain degraded all the neutral, basic and aromatic amino acids tested and also released appreciable amounts of the S-amino acid, methionine, into the culture supernatant. These findings contrast with the earlier observation of Allison & Bucklin (cited by Allison 1970) in which a single laboratory strain of S. bovis was found to be incapable of deaminating various branched amino acids.

(b) *Carbohydrate metabolism*

The group D streptococci possess neither cytochromes nor a tricarboxylic acid cycle and their primary mechanism for glucose oxidation and energy generation is the Emden-Meyerhof pathway. It is now well established that this pathway is the main route of hexose metabolism in mixed rumen populations and that the utilization of pentoses occurs via the synthesis of hexoses (Hungate 1966). The presence of the Emden-Meyerhof pathway in *S. bovis* is indicated by the occurrence of aldolase in cell extracts (Joyner & Baldwin 1966). Unlike *S. faecalis*, which exhibits considerable metabolic versatility in fermenting substrates such as pyruvate, citrate, malate, glycerol and gluconate (Deibel 1964), the energy-yielding activities of *S. bovis* are restricted to the fermentation of hexoses and a few pentoses.

The inability to use terminal electron acceptors other than pyruvate or compounds derived from pyruvate places a severe constraint upon bacteria growing in anaerobic environments such as the rumen. For organisms like *S. bovis*, which reduce most of their pyruvate to lactic acid in the course of re-oxidizing the reduced pyridine nucleotides formed during glycolysis, the net yield of ATP is restricted to only 2 moles of ATP for every mole of glucose metabolized. This relatively inefficient generation of energy is thought to place *S. bovis* at an energetic disadvantage compared with other rumen bacteria and may explain its relatively minor contribution to the rumen bacterial population of the normal animal (Hungate 1966). Although *S. bovis* is usually regarded as being homofermentative, acetate, formate and carbon dioxide in addition to lactate have been detected in batch cultures grown on glucose or starch (Geisecke 1960). It is well known that the formation of formate, acetate and ethanol by *S. faecalis* subsp. *liquefaciens* is affected by pH (Gunsalus & Niven 1942) but preliminary experiments (Latham, unpublished) show that in pH-controlled (pH 6.8), glucose-limited continuous cultures of *S. bovis* the ratio of fermentation products is dependent upon growth rate (or dilution rate, D) of the culture. At low dilution rates ($D = 0.05$ h^{-1}) acetate, formate and ethanol are the main products of fermentation and account for over 60% of the glucose metabolized, whereas lactate is the main product at dilution rates in excess of 0.3 h^{-1}. In the absence of pH control or when nitrogen is limited, lactic acid is the main product at all dilution rates. Fructose-1, 6-diphosphate (FDP) acts as an allosteric activator of the lactic dehydrogenase (LDH) of *S. bovis*, *S. faecalis* and other streptococci (Wolin 1964) and variations in the intracellular concentrations of FDP have been implicated in growth rate-dependent changes in fermentation products by *Lactobacillus casei* (de Vries et al. 1970). Variations in the FDP pool size may also account for the effect of growth rate on lactate formation by *S. bovis*. The continuous culture data suggest that under normal conditions *in vivo* *S. bovis* may adopt a heterolactic rather than a homolactic type of metabolism. Consequently the net yield of ATP may well be greater than the 2 moles per mole of glucose mentioned earlier.

Starch and glycogen are important storage compounds of plants and animals.

Starch is composed of varying proportions of amylose (an unbranched α-1, 4 glucoside) and amylopectin (an α-1, 4 glucoside with 1, 6 branching), and can be degraded by either α or β-amylases. α-amylase (α-1, 4 glucan 4-glucanohydrolase) attacks the amylose and, to a lesser extent, the amylopectin in starch with the initial formation of dextrins and later a mixture of maltose and glucose in the ratio 6 to 1 (Sokatch 1969).

With the exception of some *Bacillus* spp, most bacteria produce α-amylases. Starch hydrolysis is a characteristic common to both *S. bovis* and *S. equinus* but their amylases are quite different. *Streptococcus equinus* is unusual in that it can only attack starch in the presence of small concentrations of fermentable monosaccharide; dextrins (but not reducing sugars) accumulate and hydrolysis is inhibited by anaerobic conditions (Dunican & Seeley 1962). It is most unlikely, therefore, that this species can act as part of the amylolytic flora in the caeca of herbivores. In contrast the amylases of *S. bovis* appear to be conventional and of the α-type (Hobson & MacPherson 1952) most actively hydrolysing amylose but having less effect on starch or amylopectin. The products of starch hydrolysis—glucose, maltose and maltotriose—all support good growth of the organism. Although potato and maize starch are rapidly degraded in the ovine rumen, the *S. bovis* α-amylase is much less active towards potato starch than maize starch granules (Walker & Hope 1964). These workers also found that the enzyme adsorbs readily to maize starch at 0°C but unlike ovine salivary amylase adsorbs less readily at the temperature of the rumen, 39°C. A more detailed study of the α-amylase of *Bacteroides ruminicola* has been reported by McWethy & Hartman (1977) and, in view of the importance of starch digestion to the nutrition of the host, further studies on the factors influencing the activity of the streptococcal amylases seem warranted.

Streptococcus bovis also secretes several enzymes which may enable it to scavenge various sugars from the cell sap and walls of plant material. Many pasture plants contain α-linked galactose in the form of oligosaccharides (such as raffinose) or the digalactosyl lipids of chloroplast lipid. Most rumen bacteria are unable to ferment raffinose or melibiose *in vitro* which suggests that they do not possess an α-galactosidase but Bailey (1962, 1963) detected α-galactosidase activity both in mixed rumen organisms and in *S. bovis*. The enzyme secreted by *S. bovis* is capable of hydrolysing a range of sugars including melibiose, raffinose, stachyose and digalactosyl glycerol. However, the presence of fructose or glycerol (as in raffinose and digalactosyl glycerol, respectively) linked to the reducing carbon of an oligosaccharide markedly reduces the rate of hydrolysis. The streptococcal enzyme is constitutive and compared with α-galactosidases from other sources has a very narrow pH range of activity of 5.6–6.3. Moreover, it is restricted in the substrates it can attack to those possessing a 1, 6-linked galactoside. Williams & Doetsch (1960) detected a β-galactosidase, which released galactose from galactomannan, in an unnamed, amylolytic, Gram positive coccus isolated from the rumen. Mannans of various types are very widespread in Nature and form a significant proportion of the total carbohydrate in the endosperm of the seeds of many common fodder plants.

Pectin is digested in the rumen with the production of VFA. Polygalacturonidase is present in rumen fluid (Smart et al. 1964), in several species of rumen bacteria (Tomerska 1971; Wojciechowicz & Tomerska 1971) and, together with pectin methyl esterase, in the rumen protozoa (Wright 1960). Ziolecki et al. (1972) found that 94% of strains of S. bovis from the sheep rumen degraded pectin to unsaturated lower oligogalacturonides (pectin lyase activity) and galacturonic acid and both cell-bound and cell-free pectin lyase and pectin methyl esterase activities have been found in continuously cultured S. bovis (Latham, unpublished). Although neither the unsaturated galacturonides nor the galacturonic acid can be utilized for growth by S. bovis (Ziolecki et al. 1972) the importance of pectin digestion may lie in the unmasking or release of fermentable sugars which might otherwise be inaccessible to the organism.

(c) Chemical composition

The microbial biomass of the rumen which subsequently becomes available for digestion is of considerable nutritional significance to the host. Microscopical evidence suggests that the cell walls of Gram positive rumen organisms are more resistant to degradation by the acid and pepsin secreted in the absomasum than those of Gram negative bacteria (Pounden et al. 1950). Much of the information on the cell walls of S. bovis derives from work done on their immunochemistry. In common with other group D streptococci, S. bovis contains the glycerol teichoic acid group antigen, which is located between the cell wall and the cell membrane (Archibald & Baddily 1966). Although the exact location of the numerous type specific antigens has not been determined with certainty, the evidence available suggests that they lie in the cell wall or capsule, but not in the dextran formed from sucrose.

Dextran, the formation of which is considerably enhanced by the presence of carbon dioxide (Dain et al. 1956), consists mainly of an α-1,6-linked glucoside (Bailey 1959a) and has a mol. wt of ca. 90,000 (Greenwood 1954). It is synthesized by a constitutive transglycosidase, dextran sucrase (Bailey 1959b) and up to 87% of the sucrose utilized can be accounted for in the dextran formed (Bailey & Oxford 1958). The immunological similarity observed between dextrans from different strains is probably due to the presence of isomaltose as a common antigenic determinant (Kane & Karakawa 1971). Hobson & MacPherson (1954) and Bailey & Oxford (1958) demonstrated the presence of glucose, galactose, rhamnose and uronic acids in capsular material extracted from S. bovis grown on starch or glucose. Some of these capsular carbohydrates may, however, have originated from the cell walls (Kane *et al.* 1972)· Lactosyl, glucosyl and related components of the cell wall glycans might be expected to stimulate immune responses and may thus give rise to serological heterogeneity of the sort observed with *S. bovis* by Medrek & Barnes (1962*b*). The repeating units of the main chain of the *S. bovis* cell wall glycan have been shown to have the following configuration: L-rhamnosyl-1, 3 D-galactosyl-1, 2 L-rhamnosyl-1, 3 6-deoxy-L-talosyl-1, 3 ... which is similar to that of many other streptococcal glycans (Pazur *et al.* 1976). The cell wall mucopeptides of *S. bovis* consist of *N*-acetyl glucosamine and *N*-acetyl muramic acid with alanine, glutamic acid, lysine and threonine present in the ratio 3 to 1 to 1 to 1. The occurrence of *N*-lysylthreonine dipeptide in *S. bovis* indicates that the interpeptide bridge contains threonine (Kane *et al.* 1969) in contrast to the lys-(ala)$_3$ and lys-asp bridges characteristic of *S. faecalis* and *S. faecium*, respectively (Gooder 1970). *Meso*-diaminopimelic acid, which is used as a marker of bacterial protein in the rumen (Weller *et al.* 1962), does not appear to be present in the cell walls of either *S. bovis* or *S. faecalis* (Synge 1953).

Most of the soluble carbohydrate present in the diet is fermented in the rumen and hence does not reach the small intestine, whereas much of the intracellular polysaccharide synthesized by the micro-organisms does escape rumen fermentation, and may provide the host with a significant source of α-linked glucose polymers. Some, but not all, strains of *S. bovis* produce intracellular iodophilic material. Hobson & Mann (1955) showed that this material accumulated in the cells of *S. bovis* grown in media containing dextrins of amylose or amylopectin, maltose, maltobiose, glycogen or amylose glycollate, but not in media containing glucose, sucrose, trehalose, glucose-1-phosphate, glucose-6-phosphate, cellobiose or amylose. It appeared that the maltose and maltobiose formed during hydrolysis of starch was utilized for the synthesis of polysaccharide. Accumulation of intracellular polysaccharide, however, was not associated with rapid growth of the organism and the material did not appear to be similar to the α-linked storage compounds from other biological sources.

Lipid of bacterial origin generally only constitutes about 4% of the total lipid present in rumen digesta. The contribution of lipid from *S. bovis* to the total lipid pool is accordingly insignificant. In comparison with other rumen organisms, *S. bovis* has only a very simple long chain fatty acid composition comprising mainly C16:0, C16:1, C18:1 and C14:0 acids with only a small percentage of the odd-numbered and branched fatty acids (Ikfovits & Ragheb 1968). The organism is not involved in lipolysis or hydrogenation reactions *in vivo* (Viviani 1970). The contribution of *S. bovis* to the protein synthesized by bacteria in the rumen is likewise very small. A. B. McAllan (pers. comm.) has recently determined the amounts of DNA-nitrogen, RNA-nitrogen and total nitrogen in pure cultures of *S. bovis*; these were 4, 14 and 113 g kg^{-1} of dry cells, respectively, values similar to those found for other pure cultures of rumen bacteria. *Streptococcus bovis* synthesizes only small amounts of vitamin B_{12} (*in vitro*) and would appear to contribute very little to the synthesis of vitamins *in vivo* (Hardie 1952).

5. Interrelationships Between *Streptococcus bovis* and other Rumen Micro-organisms

In the rumen as in any other microbial ecosystem, the breakdown of complex organic molecules to simple substances such as CO_2, CH_4 and H_2O depends on the combined action of all the micro-organisms present. Various aspects of the interrelationships between rumen micro-organisms have been reviewed by Prins & van den Vorstenbosch (1975), Wolin (1975) and Coleman (1975).

As discussed previously, lactic acid does not normally accumulate in the rumen and there are several species of bacteria in the rumen which can utilize lactic acid under normal dietary conditions (Table 2). Lactic acid resulting from the fermentation of starchy rations is generally metabolized to propionate (Satter & Esdale 1968). Propionate may be formed from lactate via acrylate (acrylate pathway) by *Megasphaera elsdenii* or via dicarboxylic acid intermediates (randomizing pathway) by *Selenomonas ruminantium* and *Veillonella alcalescens* (Fig. 4). The contribution of the acrylate pathway to propionate formation and hence, by inference, the activity of *M. elsdenii*, is greatest with diets rich in carbohydrate (Wallnöfer *et al.* 1966) and may account for 57–88% of the rumen propionate (Prins & van der Meer 1976). *Megasphaera elsdenii* is often the most numerous of the lactate-utilizing bacteria in the rumen of cattle adapted to high grain rations (Ogimoto & Geisecke 1974). The occurrence of large populations of lactic acid bacteria, either lactobacilli (Hobson *et al.* 1958; Eadie *et al.* 1959) or streptococci (Gutierrez *et al.* 1959; Thomson *et al.* 1977), together with large populations of *M. elsdenii* suggests some form of synergy between the two, possibly involving lactic acid utilization. An interaction of this type has been shown to occur between *S. mutans* and *V. alcalescens* (Mikx & van der Hoeven 1975) but observations by Prins & van der Meer (1976) suggest that *M. elsdenii* may only ferment lactate *in vivo* after other more readily fermentable substrates, such as amino acids, become limiting.

Medrek & Barnes (1962b) drew attention to the possibility that the species *S. bovis* may be divided into two biotypes on the basis of mannitol fermentation and other characteristics. Mannitol-fermenting strains are frequently associated with human systemic infections (Parker & Ball 1975; Gross *et al.* 1975), whereas the majority of rumen strains are unable to ferment mannitol and appear to be non-pathogenic. Bacteriocins have been described for many bacterial species including the enterococci (Brock *et al.* 1963) and *S. bovis* has been found to be sensitive to bacteriocins produced by other group D streptococci (Pleçeas *et al.* 1971). Bacteriocins are produced by some strains of *S. bovis*, particularly

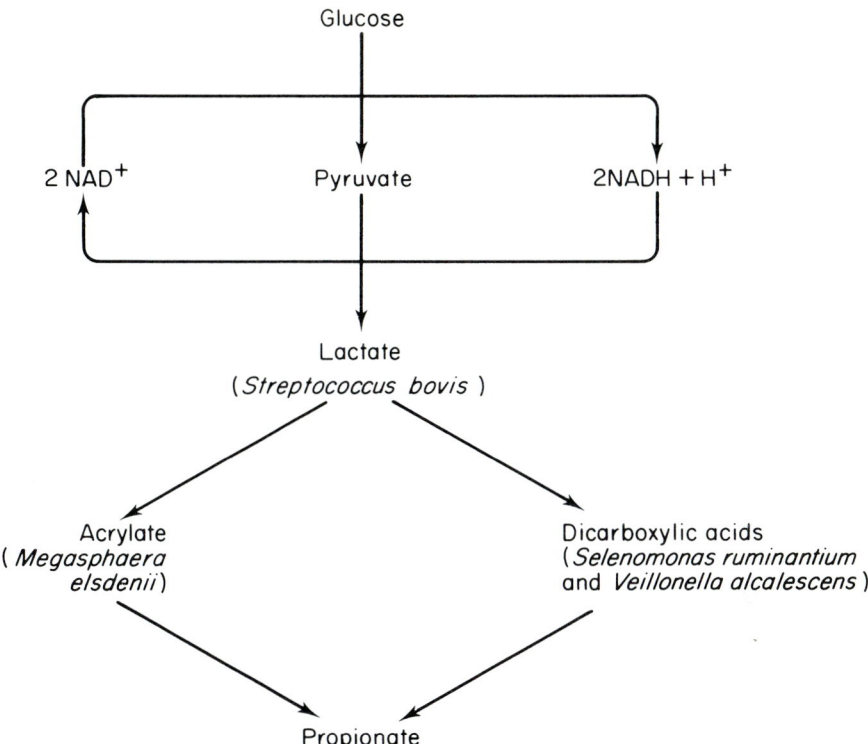

Fig. 4. Pathways of glucose fermentation to propionate involving possible interactions between lactic acid-forming *Streptococcus bovis* and lactic acid-utilizing *Megasphaera elsdenii*, *Selenomonas ruminantium* and *Veillonella alcalescens*.

those strains isolated from cases of human endocarditis, and there appears to be a correlation between the ability to ferment mannitol and to produce bacteriocins (Iverson & Millis 1976c). The low incidence of bacteriocin-producing strains among the ruminant strains tested by these authors suggests that bacteriocins to *S. bovis* are probably not common in the ruminant and are not an important ecological determinant.

Three groups of micro-organisms, ciliate protozoa, mycoplasmas and phage, are present in the rumen and are capable of marked antimicrobial activity. The protozoa readily engulf rumen bacteria often in proportion to the numbers of bacteria in the fluid surrounding the protozoa (Coleman 1975). *Streptococcus bovis* and, where tested, *S. faecalis*, are engulfed by the protozoal species, *Entodinium caudatum* (Coleman 1964), *Ento. simplex* (Coleman 1972) and *Epidinium* spp. (Coleman & Laurie 1974). The rate at which bacteria are killed once they have been ingested by the protozoa varies with the species of bacteria and protozoa but in the case of *Ento. caudatum* only 1–7% of *S. bovis* cells were viable $2\frac{1}{2}$ h after ingestion.

An obligately anaerobic mycoplasma, some strains of which are capable of lysing bacterial cells, has been isolated from the bovine rumen in numbers of 10^{5-7} ml^{-1} (Hungate 1970; Robinson & Hungate 1973; Robinson *et al.* 1975). Lytic strains of this mycoplasma readily digest cell walls of Gram negative rumen bacteria but those of Gram positive bacteria including *S. bovis* are resistant. Electron microscopy has revealed the presence of many morphological types of bacteriophage associated with a variety of different rumen bacterial types (Hoogenraad *et al.* 1967; Paynter *et al.* 1969; Hoogenraad & Hird 1970). In the studies of Paynter *et al.* (1969) samples from a cow fed alfalfa were examined over a period of one year and found to contain a minimum of 10^7 phages ml^{-1} of rumen fluid. Bacteriophage capable of lysing *S. durans* (Brailsford & Hartman 1968) and *S. bovis* (Brailsford & Hartman 1968; Iverson & Millis 1976*a*) have been isolated from rumen fluid. Phage has also been isolated from lysogenic strains of *S. bovis* (Tarakanov 1974; Iverson & Millis 1976*b*). The bacteriophage isolated by Iverson & Millis (1976*a*) comprised two types; one which resembled the *S. durans* phage and phage to other group D streptococci, and a second which had not been previously described. The ecological significance of *S. bovis* phages remains obscure, however. Tarakanov (1974) reported that 48% of the bovine strains of *S. bovis* he examined were lysogenic which contrasts sharply with the results of Iverson & Millis (1976*b*) who found only one lysogenic strain among 23 rumen strains they examined. Various technical considerations related to the detection of phage may account for these conflicting results but the data of Iverson & Millis together with the low frequency of isolation of *S. bovis* phage from rumen fluid suggests that virulent phage, and lysogeny in this species, may not be an important ecological determinant.

6. Conclusions

In this review an attempt has been made to survey some major aspects of the ecology and physicochemical characteristics of the streptococci in the gut of the ruminant. The ecology of the streptococci in the rumen of the adult is, for the most part, well understood but there is a dearth of information on all other regions of the gut both in young and adult animals alike. Nevertheless, the available evidence indicates that *S. bovis, S. faecalis* (and its subspecies) and *S. faecium* are present soon after birth and are thereafter numerically the most important of the streptococci in all regions of the alimentary tract. Other species of streptococci do not appear to be true commensals.

The role of streptococci in the metabolism of the rumen of the adult animal is, in certain respects, clear cut. For example, the ability of *S. bovis* to multiply rapidly and to ferment some dietary starches to lactic acid, establishes it as one of the most important organisms in the aetiology of lactic acidosis. However, its interrelationships with lactate-utilizing species are poorly understood and its assumed role as a major lactic acid-producing species in the normal animal may be open to question. As the streptococci comprise only a small proportion of the total rumen population, their contribution to the total metabolic activity of the rumen is probably small, but the ability of *S. bovis* to hydrolyse starch and of some strains of *S. faecium* to hydrolyse urea are activities of particular nutritional importance to the host. Although the latter activity continues to attract a great deal of interest, it is most surprising that starch hydrolysis by *S. bovis* has received so little attention. As the amylases of only a very few strains have been characterized there is a need to establish if these characteristics are indeed typical of the species as a whole. In addition, those factors which result in varying rates of hydrolysis of different cereal starches should be identified.

The undeveloped rumen of the young animal is a potent source of organisms for the rest of the alimentary tract. Although these organisms may provide a barrier against the establishment of pathogens there is evidence that the streptococci are involved in conditions of bloat in the young animal. *Streptococcus bovis* has been shown to form a numerically important part of the caecal microflora under some unusual dietary conditions but the contribution of this species to the normal microflora of this organ has yet to be determined. It is not possible to assess the nutritional significance of streptococci in the gastrointestinal tract of the preruminant and ruminant animal until we have a more complete understanding of the ecology of other species. Future work with mixed culture systems *in vitro* or with gnotobiotic preruminant and ruminant animals may provide many of the answers we seek.

7. References

ALLISON, M. J. 1970 Nitrogen metabolism of ruminal microorganisms. In *Physiology of Digestion and Metabolism in the Ruminant*, ed. Phillipson, A. T. Newcastle, England: Oriel Press.

ALLISON, M. J., BUCKLIN, J. A. & DOUGHERTY, R. W. 1964 Ruminal changes after overfeeding wheat and the effect of intraruminal inoculation on adaptation to a ration containing wheat. *Journal of Animal Science* **23**, 1164–1171.

ALLISON, M. J., ROBINSON, I. M., DOUGHERTY, R. W. & BUCKLIN, J. A. 1975 Grain overload in cattle and sheep: changes in microbial populations in the caecum and rumen. *American Journal of Veterinary Research* **36**, 181–185.

ANDREWES, F. W. & HORDER, T. J. 1906 A study of the streptococci pathogenic for man. *Lancet* **2**, 708–713.

ARCHIBALD, A. R. & BADDILY, J. 1966 The teichoic acids. *Advances in Carbohydrate Chemistry* **21**, 323–375.

AYRES, S. H. & MUDGE, C. S. 1923 Streptococci of the faeces and mouth of cows. V. Studies of the streptococci. *Journal of Infectious Diseases* **33**, 155–160.

BAILEY, R. W. 1959a Transglucosidase activity of rumen strains of *Streptococcus bovis*: structure of the dextran produced from sucrose. *Biochemical Journal* **71**, 23-26.
BAILEY, R. W. 1959b The concentration of soluble polysaccharide in the rumen contents of cows fed on red clover. *New Zealand Journal of Agricultural Research* **2**, 355-364.
BAILEY, R. W. 1962 α-galactosidase activity of rumen bacteria. *Nature, London* **195**, 79-80.
BAILEY, R. W. 1963 The intracellular α-galactosidase of a rumen strain of *Streptococcus bovis*. *Biochemical Journal* **86**, 509-514.
BAILEY, R. W. & OXFORD, A. E. 1958 A quantitative study of the production of dextran from sucrose by rumen strains of *Streptococcus bovis*. *Journal of General Microbiology* **19**, 130-145.
BARNES, E. M. 1956 Methods for the isolation of faecal streptococci (Lancefield group D) from bacon factories. *Journal of Applied Bacteriology* **19**, 193-203.
BAUMAN, H. E. & FOSTER, E. M. 1956 Characteristics of organisms isolated from the rumen of cows fed high and low roughage rations. *Journal of Bacteriology* **71**, 333-338.
BARTLEY, E. E. & YADAVA, I. S. 1961 Bloat in cattle. 4. The role of bovine saliva, plant mucilages and animal mucins. *Journal of Animal Science* **20**, 648-653.
BLACKBURN, T. H. & HOBSON, P. N. 1962 Further studies on the isolation of proteolytic bacteria from the sheep rumen. *Journal of General Microbiology* **29**, 69-81.
BRAILSFORD, M. D. & HARTMAN, P. A. 1968 Characterisation of *Streptococcus durans* bacteriophages. *Canadian Journal of Microbiology* **14**, 397-402.
BROCK, T. D., PEACHER, B. & PIERSON, D. 1963 Survey of the bacteriocins of enterococci. *Journal of Bacteriology* **86**, 702-707.
BRYANT, M. P. & ROBINSON, I. M. 1961 An improved nonselective culture medium for ruminal bacteria and its use in determining diurnal variation in numbers of bacteria in the rumen. *Journal of Dairy Science* **44**, 1446-1456.
BRYANT, M. P., SMALL, N., BOUMA, C. & ROBINSON, I. 1958 Studies on the composition of the ruminal flora and fauna of young calves. *Journal of Dairy Science* **41**, 1747-1767.
BRYANT, M. P., BARRENTINE, B. F., SYKES, J. F., ROBINSON, I. M., SHAWVER, C. V. & WILLIAMS, L. W. 1960 Predominant bacteria in the rumen of cattle on bloat-provoking ladino clover pasture. *Journal of Dairy Science* **43**, 1435-1444.
BRYANT, M. P. ROBINSON, I. M. & LINDAHL, I. L. 1961 A note on the flora and fauna in the rumen of steers fed a feedlot bloat-provoking ration and the effect of penicillin. *Applied Microbiology* **9**, 511-515.
BUTTIAUX, R. 1958 Les streptocoques fécaux des intestins humains et animaux. *Annales L'Institute Pasteur* **94**, 778-782.
CAFFREY, P. J., HATFIELD, E. E., NORTON, H. W. & GARRIGUS, U. S. 1967 Nitrogen metabolism in the ovine. 1. Adjustment to a urea-rich diet. *Journal of Animal Science* **26**, 595-600.
CALKINS, H. E. & SCRIVENER, L. H. 1967 Isolation of *Sphaerophorus necrophorus* from bovine liver abscesses. *Applied Microbiology* **15**, 1492-1493.
CHALUPA, W., CLARK, J., OPLIGER, P. & LAVKER, R. 1970 Ammonia metabolism in rumen bacteria and mucosa from sheep fed soy protein or urea. *Journal of Nutrition* **100**, 161-169.
CHENG, K-J., HIRONAKA, R., JONES, G. A., NICAS, T. & COSTERTON, J. W. 1976 Frothy bloat in cattle: production of extracellular polysaccharides and development of viscosity in cultures of *Streptococcus bovis*. *Canadian Journal of Microbiology* **22**, 450-459.
CLARKE, R. T. J. 1954 Studies on the flora and fauna of the bovine rumen. Ph.D. thesis, Massey University, New Zealand.
CLARKE, R. T. J. 1959 A dextran-fermenting organism from the rumen closely resembling *Lactobacillus bifidus*. *Journal of General Microbiology* **20**, 549-553.

CLARKE, R. T. J. 1965 Role of the rumen ciliates in bloat in cattle. *Nature, London* **205**, 95-96.
CLARKE, R. T. J. & REID, C. S. W. 1974 Foamy bloat of cattle. A review. *Journal of Dairy Science* 57, 753-785.
COLEMAN, G. S. 1964 The metabolism of *Escherichia coli* and other bacteria by *Entodinium caudatum*. *Journal of General Microbiology* 37, 209-223.
COLEMAN, G. S. 1972 The metabolism of starch, glucose, amino acids, purines, pyrimidines and bacteria by the rumen ciliate *Entodinium simplex*. *Journal of General Microbiology* **71**, 117-131.
COLEMAN, G. S. 1975 The interrelationship between rumen ciliate protozoa and bacteria. In *Digestion and Metabolism in the Ruminant*, eds McDonald, I. W. & Warner, A. C. I. Sydney, Australia: The University of New England Publishing Unit.
COLEMAN, G. S. & LAURIE, J. I. 1974 The metabolism of starch, glucose, amino acids, purines, pyrimidines and bacteria by three *Epidinium* spp isolated from the rumen. *Journal of General Microbiology* **85**, 244-256.
CONTREPOIS, M. & GOUET, P. 1973 La microflore du tube digestif du jeune veau pré-ruminant: dénombrement de quelques groupes bactériens à différents niveaux du tube digestif. *Annales de Recherches Vétérinaires* **4**, 161-170.
COOK, A. R. 1976a Urease activity in the rumen of sheep and the isolation of ureolytic bacteria. *Journal of General Microbiology* **92**, 32-48.
COOK, A. R. 1976b The elimination of urease activity in *Streptococcus faecium* as evidence for plasmid-coded urease. *Journal of General Microbiology* **92**, 49-58.
COOPER, K. E. & RAMADAN, F. M. 1955 Studies on the differentiation between human and animal populations by means of faecal streptococci. *Journal of General Microbiology* **12**, 180-190.
CULLEN, G. A. & LITTLE, T. W. A. 1969 Isolation of *Streptococcus uberis* from the rumen of cows and from soil. *Veterinary Record* **85**, 114-118.
DAIN, J. A., NEAL, A. L. & SEELEY, H. W. 1956 The effect of carbon dioxide on polysaccharide production by *Streptococcus bovis*. *Journal of Bacteriology* **72**, 209-213.
DEHORITY, B. A. & GRUBB, J. A. 1977 Characterisation of the predominant bacteria occurring in the rumen of goats (*Capra hircus*). *Applied and Environmental Microbiology* **33**, 1030-1036.
DEIBEL, R. H. 1964 The Group D streptococci. *Bacteriological Reviews* **28**, 330-366.
DE VRIES, W., KAPTEIJN, W. M. C., VAN DER BEEK, E. G. & STOUTHAMER, A. H. 1970 Molar growth yields and fermentation balances of *Lactobacillus casei* L3 in batch cultures and in continuous cultures. *Journal of General Microbiology* **63**, 333-345.
DIRKSEN, G. 1970 Acidosis. In *Physiology of Digestion and Metabolism in the Ruminant*, ed. Phillipson, A. T. Newcastle, England: Oriel Press.
DUBOS, R., SCHAEDLER, R. W., COSTELLO, R. & HOET, P. 1965 Indigenous, normal and autochthonous flora of the gastrointestinal tract. *Journal of Experimental Medicine* **122**, 67-76.
DUNICAN, L. K. & SEELEY, H. W. 1962 Starch hydrolysis by *Streptococcus equinus*. *Journal of Bacteriology* **83**, 264-269.
DUNLOP, R. H. 1972 Pathogenesis of ruminant lactic acidosis. *Advances in Veterinary Science and Comparative Medicine* **16**, 259-301.
EADIE, J. M., HOBSON, P. N. & MANN, S. O. 1959 A relationship between some bacteria, protozoa and diet in early weaned calves. *Nature, London* **183**, 624-625.
EADIE, J. M., HOBSON, P. N. & MANN, S. O. 1967 A note on some comparisons between the rumen content of barley-fed steers and that of young calves also fed on a high concentrate ration. *Animal Production* **9**, 247-250.
EADIE, J. M., HYLDGAARD-JENSEN, J., MANN, S. O., REID, R. S. & WHITELAW, F. G. 1970 Observations on the microbiology and biochemistry of the rumen of cattle given different quantities of a pelleted barley ration. *British Journal of Nutrition* **24**, 157-177.

EL AKKAD, I. & HOBSON, P. N. 1966 Effects of antibiotics on some ruminal and intestinal bacteria. *Nature, London* **209,** 1046-1047.
ELIAS, A. 1971 The rumen bacteria of animals fed on a high molasses-urea diet. Ph.D. thesis, University of Aberdeen.
ELLSWORTH, S. A., HUFFMAN, C. F., SMITH, C. K. & RALSTON, N. P. 1953 Effect of feeding antibiotics to dairy calves. 1. Aureomycin and bacitracin feed supplements. *Quarterly Bulletin of the Michigan Agricultural Experimental Station* **36,** 60-66.
EMERY, R. S., SMITH, C. K. & HUFFMAN, C. F. 1958 Feeding penicillin for control of bloat in grazing cattle and its effect on milk production. In *US Department of Agriculture Report* p. 9 Chicago: USDA.
FULGHAM, R. S., BALDWIN, B. B. & WILLIAMS, P. P. 1968 Antibiotic susceptibility of anaerobic ruminal bacteria. *Applied Microbiology* **16,** 301-307.
GALL, L. S. & HUHTANEN, C. N. 1951 Criteria for judging a true rumen organism and a description of five rumen bacteria. *Journal of Dairy Science* **34,** 353-362.
GEDEK, B. 1969 Bakteriologische Faecesanalysen bei mit milchaustauschern ernahrten Kälbern. *Zentralblatt für Bakteriologie, Parasitenkunde, Infektionskrankheiten und Hygiene* Abt. I Originale A **209,** 244-261.
GIBBONS, R. J. & DOETSCH, R. N. 1959 Physiological study of an obligatory anaerobic ureolytic bacterium. *Journal of Bacteriology* **77,** 417-428.
GIESECKE, D. 1960 Untersuchungen am *Streptococcus bovis* und einer gelb wachsenden variante auf dem Rinderpansen. *Zentralblatt für Bakteriologie, Parasitenkunde, Infektionskrankheiten und Hygiene* Abt. I Originale **179,** 448-455.
GILCHRIST, F. M. C. & KISTNER, A. 1962 Bacteria of the ovine rumen. I. Composition of the population on a diet of poor teff hay. *Journal of Agricultural Science* **59,** 77-83.
GOODER, H. 1970 Cell wall composition in the classification of streptococci. *International Journal of Systematic Bacteriology* **20,** 475-482.
GREENWOOD, C. T. 1954 A physico-chemical examination of the capsular polysaccharide from an amylolytic sheep rumen streptococcus. *Biochemical Journal* **57,** 151-153.
GROSS, K. C., HOUGHTON, M. P. & SENTERFIT, L. B. 1975 Presumptive speciation of *Streptococcus bovis* and other group D streptococci from human sources by using arginine and pyruvate tests. *Journal of Clinical Microbiology* **1,** 54-60.
GUILHERMET, R., MATHIEU, C. M. & TOULLEC, R. 1973 Observations sur le transit des aliments liquides au niveau de la gouttière oesophagienne chez le veau préruminant et ruminant. *Annales de Biologie Animale Biochimie Biophysique* **13,** 715-718.
GUNSALUS, J. C. & NIVEN, C. F. 1942 The effect of pH on the lactic acid fermentation. *Journal of Biological Chemistry* **145,** 131-136.
GUTIERREZ, J., DAVIS, R. E., LINDAHL, I. L. & WARWICK, E. J. 1959 Bacterial changes in the rumen during the onset of feed-lot bloat of cattle and characteristics of *Peptostreptococcus elsdenii* n.sp. *Applied Microbiology* **7,** 16-22.
GUTIERREZ, J., DAVIS, R. E. & LINDAHL, I. L. 1961 Some chemical and physical properties of a slime from the rumen of cattle. *Applied Microbiology* **9,** 209-212.
HAENEL, H. 1960 Aspekte der mikroökologischen Beziehungen des Mikroorganismus. *Zentralblatt für Bakteriologie, Parasitenkunde, Infektionskrankheiten und Hygiene* Abt. I Referate **176,** 305-425.
HAENEL, H. & MÜLLER-BEUTHOW, W. 1956 Vergleichende quantitative Untersuchungen über Keimzahlen in den Faeces des Menschen und einiger Wirbeltiere. *Zentralblatt für Bakteriologie, Parasitenkunde, Infektionskrankheiten und Hygiene* Abt. I, Originale **167,** 123-133.
HARDIE, W. B. 1952 Vitamin B_{12} production by *Streptococcus bovis* from the bovine rumen. M.S. thesis, Washington State University, Pullman, Washington.
HARTMAN, P. A., JOHNSON, R. N., BROWN, L. R., JACOBSON, N. L., ALLEN, R. S., SHELLENBERGER, P. R. & VAN HORN, H. H. 1962 Relationship of rumen facultative anaerobes to feedlot and pasture bloat. *Iowa State Journal of Science* **36,** 217-221.

HARTMAN, P. A., MORRILL, J. L. & JACOBSON, N. L. 1966 Influence of diet and age on bacterial counts of ileal digesta and feces obtained from young calves. *Applied Microbiology* 14, 70-73.

HAYASHI, T. & KITAHARA, K. 1976 Effect of carbon dioxide and oxygen on the growth of orange-colored *Streptococcus bovis* isolated from bovine rumen. *Journal of General and Applied Microbiology* 22, 301-310.

HEALD, P. J., KROGH, N., MANN, S. O., APPLEBY, J. C., MASSON, F. M. & OXFORD, A. E. 1953 A method for direct viable counts of the facultatively anaerobic microflora in the rumen of sheep on a hay diet. *Journal of General Microbiology* 9, 207-215.

HIGGINBOTTOM, C. & WHEATER, D. W. F. 1954 The incidence of *Streptococcus bovis* in cattle. *Journal of Agricultural Science* 44, 434-442.

HOBSON, P. N. 1971 In *Progress in Industrial Microbiology*, eds Hockenhull, D. J. D. & Churchill, J. R. A. London: Churchill, pp. 41-77.

HOBSON, P. N. & MacPHERSON, M. J. 1952 Amylases of *Clostridium butyricum* and a streptococcus isolated from the rumen of the sheep. *Biochemical Journal* 52, 671-679.

HOBSON, P. N. & MacPHERSON, M. J. 1954 Some serological and chemical studies on materials extracted from an amylolytic *Streptococcus* from the rumen of sheep. *Biochemical Journal* 57, 145-151.

HOBSON, P. N. & MANN, S. O. 1955 Some factors affecting the formation of iodophilic polysaccharide in Group D streptococci from the rumen. *Journal of General Microbiology* 13, 420-435.

HOBSON, P. N., MANN, S. O. & OXFORD, A. E. 1958 Some studies on the occurrence and properties of a large Gram-negative coccus from the rumen. *Journal of General Microbiology* 19, 462-472.

HOBSON, P. N., MANN, S. O. & SUMMERS, R. 1976 Rumen microorganisms in red deer, hill sheep and reindeer in the Scottish highlands. *Proceedings of the Royal Society (Edinburgh)* B 75, 171-180.

HOOGENRAAD, N. J. & HIRD, F. J. R. 1970 Electron-microscopic investigation of the flora of sheep alimentary tract. *Australian Journal of Biological Sciences* 23, 793-808.

HOOGENRAAD, N. J., HIRD, F. J. R., HOLMES, I. & MILLIS, N. F. 1967 Bacteriophages in the rumen contents of sheep. *Journal of General Virology* 1, 575-576.

HORACEK, G. L., FINA, L. R., TILLINGHAST, H. S. & GETTINGS, R. L. 1977 Agglutinating immunoglobulins to encapsulated *Streptococcus bovis* in bovine serum and saliva and a possible relation to feedlot bloat. *Canadian Journal of Microbiology* 23, 100-106.

HUNGATE, R. E. 1957 Microorganisms in the rumen of cattle fed a constant ration. *Canadian Journal of Microbiology* 3, 289-311.

HUNGATE, R. E. 1966 *The Rumen and its Microbes*. London & New York: Academic Press.

HUNGATE, R. E. 1970 Interrelationships in the rumen microbiota. In *Physiology of Digestion and Metabolism in the Ruminant*, ed. Phillipson, A. T. Newcastle, England: Oriel Press.

HUNGATE, R. E., DOUGHERTY, R. W., BRYANT, M. P. & CELLO, R. M. 1952 Microbiological and physiological changes associated with acute indigestion in sheep. *Cornell Veterinarian* 42, 423-449.

HUNGATE, R. E., FLETCHER, D. W., DOUGHERTY, R. W. & BERRENTINE, B. F. 1955 Microbial activity in the bovine rumen: its measurement and relation to bloat. *Applied Microbiology* 3, 161-173.

HUNGATE, R. E., PHILLIPS, G. D., McGREGOR, A., HUNGATE, D. P. & BUECHNER, H. K. 1959 Microbial fermentation in certain mammals. *Science, New York* 130, 1192-1194.

IFKOVITS, R. W. & RAGHEB, H. S. 1968 Cellular fatty acid composition and identification of rumen bacteria. *Applied Microbiology* 16, 1406-1413.

IVERSON, W. G. & MILLIS, N. F. 1976a Characterisation of *Streptococcus bovis* bacteriophages. *Canadian Journal of Microbiology* 22, 847-852.

IVERSON, W. G. & MILLIS, N. F. 1976b Lysogeny in *Streptococcus bovis*. *Canadian Journal of Microbiology* **22**, 853–857.

IVERSON, W. G. & MILLIS, N. F. 1976c Bacteriocins of *Streptococcus bovis*. *Canadian Journal of Microbiology* **22**, 1040–1047.

JACOBSON, D. R., LINDAHL, I. L., McNEILL, J. J., SHAW, J. C., DOETSCH, R. N. & DAVIS, R. E. 1957 Feedlot bloat studies. 2. Physical factors involved in the etiology of frothy bloat. *Journal of Animal Science* **16**, 515–524.

JAYASURIYA, G. C. N. & HUNGATE, R. E. 1959 Lactate conversions in the bovine rumen. *Archives of Biochemistry and Biophysics* **82**, 274–287.

JAYNE-WILLIAMS, D. J. 1976 The application of miniaturized methods for the characterization of various organisms isolated from the animal gut. *Journal of Applied Bacteriology* **40**, 189–200.

JENSEN, R., DEANE, H. M., COOPER, L. J., MILLER, V. A. & GRAHAM, W. R. 1954 The rumenitis-liver abscess complex in beef cattle. *American Journal of Veterinary Research* **15**, 202–216.

JOHN, A., ISAACSON, H. R. & BRYANT, M. P. 1974 Isolation and characteristics of a ureolytic strain of *Selenomonas ruminantium*. *Journal of Dairy Science* **57**, 1003–1014.

JONES, G. A., MacLEOD, R. A. & BLACKWOOD, A. C. 1964 Ureolytic rumen bacteria. I. Characteristics of the microflora from a urea-fed sheep. *Canadian Journal of Microbiology* **10**, 371–378.

JOYNER, A. E. & BALDWIN, R. L. 1966 Enzymatic studies of pure cultures of rumen microorganisms. *Journal of Bacteriology* **92**, 1321–1330.

KANE, J. A. & KARAKAWA, W. W. 1971 Immunochemistry of capsular glucans isolated from *Streptococcus bovis*. *Journal of Immunology* **106**, 103–109.

KANE, J. A., LACKLAND, H., KARAKAWA, W. W. & KRAUSE, R. M. 1969 Chemical studies on the structure of mucopeptide isolated from *Streptococcus bovis*. *Journal of Bacteriology* **99**, 175–179.

KANE, J. A., KARAKAWA, W. W. & PAZUR, H. H. 1972 Glycans from streptococcal cell walls: structural features of a diheteroglycan isolated from the cell wall of *Streptococcus bovis*. *Journal of Immunology* **108**, 1218–1226.

KERN, D. L., SLYTER, L. L., LEFFEL, E. C., WEAVER, J. M. & OLTJEN, R. R. 1974 Ponies vs steers: microbial and chemical characteristics of intestinal ingesta. *Journal of Animal Science* **38**, 559–564.

KISTNER, A., GOUWS, L. & GILCHRIST, F. M. C. 1962 Bacteria of the ovine rumen. II. The functional groups fermenting carbohydrates and lactate on a diet of lucerne (*Medicago sativa*) hay. *Journal of Agricultural Science* **59**, 85–91.

KJELLANDER, J. 1960 *Enteric Streptococci as Indicators of Fecal Contamination of Water*. Copenhagen: Ejnar Munksgaard.

KROGH, N. 1959 Studies on alterations in the rumen fluid of sheep, especially concerning the microbial composition when readily available carbohydrates are added to the food. I. Sucrose. *Acta Veterinaria Scandinavica* **1**, 74–97.

KROGH, N. 1960 Studies on alterations in the rumen fluid of sheep, especially concerning the microbial composition when readily available carbohydrates are added to the food. II. Lactose. *Acta Veterinaria Scandinavica* **1**, 383–410.

KROGH, N. 1961 Studies on alterations in the rumen fluid of sheep, especially concerning microbial composition when readily available carbohydrates are added to the food. III. Starch. *Acta Veterinaria Scandinavica* **2**, 103–119.

KROGH, N. 1963 Clinical and microbiological studies of spontaneous cases of acute indigestion in ruminants. *Acta Veterinaria Scandinavica* **4**, 27–40.

LATHAM, M. J., SHARPE, M. E. & SUTTON, J. D. 1971 The microbial flora of the rumen of cows fed hay and high cereal rations and its relationship to the rumen fermentation. *Journal of Applied Bacteriology* **34**, 425–434.

LATHAM, M. J., SUTTON, J. D. & SHARPE, M. E. 1974 Fermentation and microorganisms in the rumen and the content of fat in the milk of cows given low roughage rations. *Journal of Dairy Science* **57**, 803-810.

LINDSAY, D. B. 1970 Carbohydrate metabolism in ruminants. In *Physiology of Digestion and Metabolism in the Ruminant*, ed. Phillipson, A. T. Newcastle, England: Oriel Press.

LYSONS, R. J., ALEXANDER, T. J. L., HOBSON, P. N., MANN, S. O. & STEWART, C. S. 1971 Establishment of a limited rumen microflora in gnotobiotic lambs. *Research in Veterinary Science* **12**, 486-487.

LYSONS, R. J., ALEXANDER, T. J. L., WELLSTEAD, P. D. & JENNINGS, I. W. 1976 Observations on the alimentary tract of gnotobiotic lambs. *Research in Veterinary Science* **20**, 70-76.

MacPHERSON, M. J. 1953 Isolation and identification of amylolytic streptococci from the rumen of sheep. *Journal of Pathology and Bacteriology* **66**, 95-102.

McBEE, R. H., JOHNSON, J. L. & BRYANT, M. P. 1969 Ruminal microorganisms from elk. *Journal of Wildlife Management* **33**, 181-186.

McWETHY, S. J. & HARTMAN, P. A. 1977 Purification and some properties of an extracellular α-amylase from *Bacteroides amylophilus*. *Journal of Bacteriology* **129**, 1537-1544.

MAKI, L. R. & FOSTER, E. M. 1957 Effect of roughage in the bovine ration on types of bacteria in the rumen. *Journal of Dairy Science* **40**, 905-913.

MANN, S. O. & ØRSKOV, E. R. 1973 The effect of rumen and post-rumen feeding of carbohydrates on the caecal microflora of sheep. *Journal of Applied Bacteriology* **36**, 475-484.

MANN, S. O. & OXFORD, A. E. 1955 Relationships between viable saccharolytic bacteria in the rumen and abomasum of the young calf and kid. *Journal of General Microbiology* **12**, 140-146.

MANN, S. O., MASSON, F. M. & OXFORD, A. E. 1954 Facultative anaerobic bacteria from the sheep's rumen. *Journal of General Microbiology* **10**, 142-149.

MEDREK, T. F. & BARNES, E. M. 1962a The distribution of Group D streptococci in cattle and sheep. *Journal of Applied Bacteriology* **25**, 159-168.

MEDREK, T. F. & BARNES, E. M. 1962b Biochemical and serological properties of strains of *Streptococcus bovis* and related organisms. *Journal of Applied Bacteriology* **25**, 169-179.

MEYER, R. M., BARTLEY, E. E., MORILL, J. L. & STEWART, W. E. 1964 Salivation in cattle. I. Feed and animal factors affecting salivation and its relation to bloat. *Journal of Dairy Science* **47**, 1339-1345.

MIETH, H. 1962 Untersuchungen über das Vorkommen von Enterokokken bei Tieren und Menschen. III. Mitteilung die Enterokokken in den Faeces von Rindern. *Zentralblatt für Bacteriologie, Parasitenkunde, Infektionskrankheiten und Hygiene* Abt. I. Originale **185**, 47-52.

MIKX, F. H. M. & VAN DER HOEVEN, J. S. 1975 Symbiosis of *Streptococcus mutans* and *Veillonella alcalescens* in mixed continuous culture. *Archives of Oral Biology* **20**, 407-410.

MISHRA, B. D., FINA, L. R., BARTLEY, E. E. & CLAYDON, T. J. 1967 Bloat in cattle. 9. The role of rumen aerobic (facultative) mucinolytic bacteria. *Journal of Animal Science* **26**, 606-612.

MISHRA, B. D., BARTLEY, E. E., FINA, L. R. & BRYANT, M. P. 1968 Bloat in cattle. 14. Mucinolytic activity of several anaerobic rumen bacteria. *Journal of Animal Science* **27**, 1651-1656.

MUNDT, J. O., COGGIN, J. H. & JOHNSON, L. F. 1962 Growth of *Streptococcus faecalis* var *liquefaciens* on plants. *Applied Microbiology* **10**, 552-555.

MYLREA, P. J. 1969 The bacterial content of the small intestine of young calves. *Research in Veterinary Science* **10**, 394-395.

NAGA, M. A. & EL-SHAZLY, K. 1968 The metabolic characterisation of the ciliate protozoon *Endiplodinium medium* from the rumen of buffalo. *Journal of General Microbiology* **53**, 305-315.

NEWHOOK, J. C. & TITCHEN, D. A. 1976 Cineradiography of the reticular groove mechanism. *Australian Veterinary Journal* **52**, 132-135.
NIVEN, C. F. Jr., WASHBURN, M. R. & WHITE, J. C. 1948 Nutrition of *Streptococcus bovis*. *Journal of Bacteriology* **55**, 601-606.
OGIMOTO, K. & GIESECKE, D. 1974 The genesis and biochemistry of rumen acidosis. 2. Microorganisms and metabolism of lactic acid isomers. *Zentralblatt für Veterinärmedizin* A. **21**, 532-538.
ORLA-JENSEN, S. 1919 *The Lactic Acid Bacteria*. Copenhagen: Adr. Fred Host & Son.
ØRSKOV, E. R., BENZIE, D. & KAY, R. N. B. 1970 The effects of feeding procedure on closure of the oesophageal groove in young sheep. *British Journal of Nutrition* **24**, 785-795.
PANJARATHINAM, R. & LAXMINARAYANA, H. 1975 Studies on rumen microflora in cows and buffaloes under different feeding regimes. 4, Distribution of bacteria and *in vitro* studies. *Indian Journal of Animal Science* **45**, 173-182.
PARKER, M. T. & BALL, L. C. 1975 Streptococci and aerococci associated with systemic infection in man. *Journal of Medical Microbiology* **9**, 275-302.
PAUL, J. I. 1961 The nitrogen requirements of some members of the viridans group of streptococci. *Australian Journal of Biological Sciences* **14**, 567-579.
PAYNTER, M. J. B., EWERT, D. L. & CHALUPA, W. 1969 Some morphological types of bacteriophages in rumen contents. *Applied Microbiology* **18**, 942-943.
PAZUR, J. H., DROPKIN, D. J., DREKER, K. L., FORSBERG, L. S. & LOWMAN, C. S. 1976 Glycans from streptococcal cell walls. The molecular structure and immunological properties of an antigenic glycan from *Streptococcus bovis*. *Archives of Biochemistry and Biophysics* **176**, 257-266.
PEARSON, H. A. 1965 Rumen organisms in white-tailed deer from South Texas. *Journal of Wildlife Management* **29**, 493-496.
PEARSON, H. A. 1967 Rumen microorganisms in buffalo from Southern Utah. *Applied Microbiology* **15**, 1450-1451.
PEARSON, H. A. 1969 Rumen microbial ecology in mule deer. *Applied Microbiology* **17**, 819-824.
PERRY, K. D., WILSON, M. K., NEWLAND, L. G. M. & BRIGGS, C. A. E. 1955 The normal flora of the bovine rumen. III. Quantitative and qualitative studies of rumen streptococci. *Journal of Applied Bacteriology* **18**, 436-442.
PLEÇEAS, P. 1970 Bactériocines des streptocoques groupe D. 1. Leur incidence. *Archives Roumaines de Pathologie Expérimentale et de Microbiologie* **29**, 229-232.
PLEÇEAS, P., BOGDAN, C. & VEREANU, A. 1971 The bacteriocins of group D streptococci. II. Their action. *Archives Roumaines de Pathologie Expérimentale et de Microbiologie* **30**, 351-358.
POUNDEN, W. D., FERGUSON, L. C. & HIBBS, J. W. 1950 The digestion of rumen microorganisms by the host animal. *Journal of Dairy Science* **33**, 565-572.
PRESCOTT, J. M. & STUTTS, A. L. 1955 Effects of carbon dioxide on the growth and amino acid metabolism of *Streptococcus bovis*. *Journal of Bacteriology* **70**, 285-288.
PRESCOTT, J. M., WILLIAMS, W. T. & RAGLAND, R. S. 1959 Influence of nitrogen source on growth of *Streptococcus bovis*. *Proceedings of the Society for Experimental Biology and Medicine* **102**, 490-493.
PRINS, R. A. & MULDER, I. 1969 Sugar fermentation characteristics, antibiotic sensitivity and fatty acid composition of *Steptococcus bovis*. *Zentralblatt für Veterinärmedizin* **16** B, 731-737.
PRINS, R. A. & VAN DEN VORSTENBOSCH, C. J. A. H. V. 1975 Interrelationships between rumen microorganisms. *Miscellaneous Papers, Landbouwhogeschool, Wageningen* **11**, 15-24.
PRINS, R. A. & VAN DER MEER, P. 1976 On the contribution of the acrylate pathway to the formation of propionate from lactate in the rumen of cattle. *Antonie van Leeuwenhoek*. **42**, 25-31.

RADISSON, J. J., SMITH, C. K. & WARD, G. M. 1956 The mode of action of antibiotics in the nutrition of the dairy calf. 1. Effect of terramycin administered orally on the performance and intestinal flora of young dairy calves. *Journal of Dairy Science* **39**, 1260-1267.

ROBERTSON, W. van B., ROPES, M. W. & BAUER, W. 1940 Mucinase: a bacterial enzyme which hydrolyses synovial fluid mucin and other mucins. *Journal of Biological Chemistry* **133**, 261-276.

ROBINSON, I. M., ALLISON, M. J. & HARTMAN, P. A. 1975 *Anaeroplasma abactoclasticum* gen. nov. sp. nov.: an obligately anaerobic mycoplasma from the rumen. *International Journal of Systematic Bacteriology* **25**, 173-181.

ROBINSON, J. P. & HUNGATE, R. E. 1973 *Acholeplasma bactoclasticum* sp. nov., an anaerobic mycoplasma from the bovine rumen. *International Journal of Systematic Bacteriology* **23**, 171-181.

ROGOSA, M. 1971 Transfer of *Peptostreptococcus elsdenii* Gutierrez *et al.* to the new genus, *Megasphaera* (M. elsdenii (Gutierrez *et al.*) comb. nov.). *International Journal of Systematic Bacteriology* **21**, 187-189.

RUSOFF, L. L., ALFORD, J. A. & HYDE, C. E. 1953 Effect of type of protein on the response of young dairy calves to aureomycin with data on the intestinal microflora. *Journal of Dairy Science* **36**, 45-51.

RYAN, R. K. 1964 Concentrations of glucose and low molecular weight acids in the rumen of sheep following the addition of large amounts of wheat to the rumen. *American Journal of Veterinary Research* **25**, 646-652.

SABBAJ, J., SUTTER, V. L. & FINEGOLD, S. M. 1971 Comparison of selective media for isolation of presumptive Group D streptococci from human faeces. *Applied Microbiology* **22**, 1008-1011.

SANDINE, W. E., RADICH, P. C. & ELLIKER, P. R. 1972 Ecology of the lactic streptococci. A review. *Journal of Milk and Food Technology* **35**, 176-185.

SATTER, L. D. & ESDALE, W. J. 1968 *In vitro* lactate metabolism by ruminal ingesta. *Applied Microbiology* **16**, 680-688.

SCHEIFINGER, C., RUSSELL, M. & CHALUPA, W. 1976 Degradation of amino acids by pure cultures of rumen bacteria. *Journal of Animal Science* **43**, 821-827.

SCHWARTZ, H. M. & GILCHRIST, F. M. C. 1975 Microbial interactions with diet and the host animal. In *Digestion and Metabolism in the Ruminant*, eds McDonald, I. W. & Warner, A. C. I. Sydney, Australia: The University of New England Publishing Unit.

SEELEY, H. W. & DAIN, J. A. 1960 Starch hydrolysing streptococci. *Journal of Bacteriology* **79**, 230-235.

SHARMA, R. M. & PACKER, R. A. 1970 Occurrence and ecologic features of *Streptococcus uberis* in the dairy cow. *American Journal of Veterinary Research* **31**, 1197-2002.

SHARPE, M. E., LATHAM, M. J. & REITER, B. 1976 The immune response of the host animal to bacteria in the rumen and caecum. In *Digestion and Metabolism in the Ruminant*, eds McDonald, I. W. & Warner, A. C. I. Sydney, Australia: The University of New England Publishing Unit.

SHERMAN, J. M. 1937 The Streptococci. *Bacteriological Reviews* **1**, 3-97.

SLYTER, L. L. 1976 Influence of acidosis on rumen function. *Journal of Animal Science* **43**, 910-929.

SLYTER, L. L., OLTJEN, R. R., KERN, D. L. & WEAVER, J. M. 1968 Microbial species including ureolytic bacteria from the rumen of cattle fed purified diets. *Journal of Nutrition* **94**, 184-192.

SLYTER, L. L., OLTJEN, R. R., KERN, D. L. & BLANK, F. C. 1970 Influence of type and level of grain and diethylstilbestrol on the rumen microbial populations of steers fed all-concentrate diets. *Journal of Animal Science* **31**, 996-1002.

SLYTER, L. L., BOND, J., RUMSEY, T. S. & WEAVER, J. M. 1974 Ruminal bacteria, lactic acid and glucose in heifers after fasting and re-feeding. *Journal of Animal Science* **39**, 252-258.

SMART, W. W. G. Jr., BELL, T. A., MOCHRIE, R. D. & STANLEY, N. W. 1964 Pectic enzymes of rumen fluid. *Journal of Dairy Science* **47**, 1220-1223.

SMITH, H. W. 1961 The development of the bacterial flora of the faeces of animals and man: the changes that occur during ageing. *Journal of Applied Bacteriology* **24**, 235-241.

SMITH, H. W. 1962 Observations on the aetiology of neonatal diarrhoea (scours) in calves. *Journal of Pathology and Bacteriology* **84**, 147-168.

SMITH, H. W. 1965 Observations on the flora of the alimentary tract of animals and factors affecting its composition. *Journal of Pathology and Bacteriology* **89**, 95-122.

SMITH, H. W. 1971 The bacteriology of the alimentary tract of domestic animals suffering from *Escherichia coli* infection. *Annals of the New York Academy of Sciences* **176**, 110-125.

SMITH, H. W. & CRABB, W. E. 1961 The faecal bacterial flora of animals and man: its development in the young. *Journal of Pathology and Bacteriology* **82**, 53-66.

SMITH, R. H. 1959 The development and function of the rumen in milk-fed calves. *Journal of Agricultural Science* **52**, 72-78.

SMITH, R. H. 1961 The development and function of the rumen in milk-fed calves. II. Effect of wood shavings in the diet. *Journal of Agricultural Science* **56**, 105-111.

SOKATCH, J. R. 1969 *Bacterial Physiology and Metabolism*. London & New York: Academic Press.

SRIVASTAVA, R. V. N. & SRIVASTAVA, S. K. 1974 Isolation and characterisation of *Streptococcus* spp from the rumen of sheep and buffalo calves. *Indian Journal of Animal Science* **44**, 257-262.

SWITZER, R. E. & EVANS, J. B. 1974 Evaluation of selective media for enumeration of group D streptococci in bovine feces. *Applied Microbiology* **28**, 1086-1087.

SYNGE, R. L. M. 1953 Note on the occurrence of diamino-pimelic acid in some intestinal microorganisms from farm animals. *Journal of General Microbiology* **9**, 407-409.

TARAKANOV, B. V. 1974 Lysogeny among *Streptococcus bovis* cultures isolated from the rumen of cattle and sheep. *Mikrobiologiya* **43**, 391-320.

THOMSON, D. J., BEEVER, D. E., LATHAM, M. J., SHARPE, M. E. & TERRY, R. A. 1978 The effect of inclusion of mineral salts in the diet on dilution rate, the pattern of rumen fermentation and the composition of the rumen microflora. *Journal of Animal Science* (in press).

THORLEY, C. M., SHARPE, M. E. & BRYANT, M. P. 1968 Modification of the rumen bacterial flora by feeding cattle ground and pelleted roughage as determined with culture media with and without rumen fluid. *Journal of Dairy Science* **51**, 1811-1816.

TOMERSKA, H. 1971 Decomposition of pectin *in vitro* by pure strains of rumen bacteria. *Acta Microbiologica Polonica*, Series B. 3 (20): 107-111.

ULYATT, M. J., DELLOW, D. W., REID, C. S. W. & BAUCHOP, T. 1975 Structure and function of the large intestine of ruminants. In *Digestion and Metabolism in the Ruminant*, eds McDonald, I. W. & Warner, A. C. I. Sydney, Australia: The University of New England Publishing Unit.

VIVIANI, R. 1970 Metabolism of long-chain fatty acids in the rumen. *Advances in Lipid Research* **8**, 267-346.

VAN WYK, L. & STEYN, P. L. 1975 Ureolytic bacteria in sheep rumen. *Journal of General Microbiology* **91**, 225-232.

VON BRÜGGEMANN, J. & GIESECKE, D. 1965 Uber das Wachstum von *Streptococcus bovis* in Gegenwart von Ammonium Sulfat als einige Stickstoffquelle. *Zentralblatt für Bakteriologie, Parasitenkunde, Infektionskrankheiten und Hygiene* Abt. 1. Originale **197**, 347-353.

VON BRÜGGEMANN, J., GIESECKE, D. & WALSER-KARST, K. 1967 Beiträge zur Wildbiologie und vergleichenden Tierphysiologie. II. Mikroorganismen im Pansen von Rothirsch (*Ceruus elaphus*) und Reh (*Capreolus capreolus*). *Zeitschrift für Tierphysiologie Tierernahrung und Futtermittelkunde* **23**, 143-151.

WALKER, G. J. & HOPE, P. M. 1964 Degradation of starch granules by some amylolytic bacteria from the rumen of sheep. *Biochemical Journal* **90**, 398–408.
WALLNÖFER, P., BALDWIN, R. L. & STAGNO, E. 1966 Conversion of C^{14}-labelled substrates to volatile fatty acids by the rumen microbiota. *Applied Microbiology* **14**, 1004–1010.
WELLER, R. A., PILGRIM, A. F. & GREY, F. V. 1962 Digestion of foodstuffs of the sheep and the passage of digesta through its compartments. 3. Progress of nitrogen digestion. *British Journal of Nutrition* **16**, 83–90.
WHITELAW, F. G., EADIE, J. M., MANN, S. O. & REID, R. S. 1972 Some effects of rumen ciliate protozoa in cattle given restricted amounts of a barley diet. *British Journal of Nutrition* **27**, 425–437.
WILLIAMS, P. P. & DOETSCH, R. N. 1960 Microbial dissimilation of galactomannan. *Journal of General Microbiology* **22**, 635–644.
WILLIAMS, V. J. & MACKENZIE, D. D. S. 1965 The absorption of lactic acid from the reticulo-rumen of the sheep. *Australian Journal of Biological Sciences* **18**, 917–934.
WINSLOW, C. R. E. & PALMER, G. T. 1910 A comparative study of the intestinal streptococci from the horse, cow and man. *Journal of Infectious Diseases* **7**, 1–16.
WILSSENS, A. & BUTTIAUX, R. 1958 Les bactéries de la flore fécale de la vache saine. *Annales L'Institut Pasteur* **94**, 332–337.
WISEMAN, R. F., JACOBSON, D. R. & MILLER, M. W. 1960 Persistence of lactobacilli and streptococci in bovine rumen during pencillin administration. *Applied Microbiology* **8**, 76–79.
WOJCIECHOWICZ, M. & TOMERSKA, H. 1971 Pectic enzymes in some pectinolytic rumen bacteria. *Acta Microbiologica Polonica* Series A **3 (20)**, 57–62.
WOLIN, M. J. 1964 Fructose-1,6-diphosphate requirement of streptococcal lactic dehydrogenases. *Science, New York* **146**, 775–777.
WOLIN, M. J. 1975 Interactions between the bacterial species of the rumen. In *Digestion and Metabolism in the Ruminant*, eds McDonald, I. W. & Warner, A. C. I. Sydney, Australia: The University of New England Publishing Unit.
WOLIN, M. J., MANNING, G. B. & NELSON, W. O. 1959 Ammonium salts as a sole source of nitrogen for the growth of *Streptococcus bovis*. *Journal of Bacteriology* **78**, 147.
WOLSTRUP, J. JENSEN, V. & JENSEN, K. 1974 The microflora and concentrations of volatile fatty acids in the rumen of cattle fed on single component rations. *Acta Veterinaria Scandinavica* **15**, 244–255.
WRIGHT, D. E. 1960 Pectic enzymes in rumen protozoa. *Archives for Biochemistry and Biophysics* **86**, 251–254.
ZEIDLER, H., TOLLE, A. & HEESCHEN, W. 1968 Zur Beurteilung zytologischbakteriologischer Untersuchungsbefunde im Rahmen der Mastitisdiagnostik. *Milchwissenschaft* **23**, 674–677.
ZIOLECKI, A. & BRIGGS, C. A. E. 1961 The microflora of the rumen of the young calf. II. Source, nature and development. *Journal of Applied Bacteriology* **24**, 148–163.
ZIOLECKI, A., TOMERSKA, H. & WOJCIECHOWICZ, M. 1972 Pectinolytic activity of rumen streptococci. *Acta Microbiological Polonica* Series A **4 (21)**, 183–188.

Streptococci in the Intestinal Flora of Man and other Non-ruminant Animals

G. C. MEAD

ARC Food Research Institute, Norwich, Norfolk, England

CONTENTS

1. Introduction 245
2. Occurrence and distribution in different regions of the intestine 246
3. Factors affecting the streptococcal flora 248
4. Development of the streptococcal flora 248
5. Influence of diet 250
6. Species distribution in various hosts 250
7. Some possible interactions between the streptococcal flora and the host . . . 254
8. Summary 257
9. References 258

1. Introduction

SINCE THE END of the last century there have been numerous studies on the occurrence and properties of streptococci in the intestine of man and other homothermic animals, particularly in relation to the faecal flora. Much of this work has had a significant influence on the classification of streptococci found in the intestine and has led to the development of improved tests for differentiating some of the species present. Difficulties still arise, however, in attempting to classify many of the strains encountered and these will need to be resolved before a full account of the streptococcal flora can be given for any individual host.

There are three main problems concerning classification. First, there is the confusion arising from a lack of precision in some studies when using the terms 'enterococci' or 'faecal streptococci' to describe the organisms being isolated (Deibel 1964). Secondly, many of the isolates loosely referred to within these two categories appear to have a low affinity with established species, while even identifiable strains sometimes show considerable variation in their properties. For example, Raibaud *et al.* (1961) isolated 100 strains of *Streptococcus bovis* from the intestine of the pig and found that the strains could be divided into 38 different types on the basis of carbohydrate fermentation reactions. Experience suggests, however, that less variation occurs in the properties associated with human isolates (Mundt 1973).

The third problem concerns the recognition of 'anaerobic streptococci' and it is unfortunate that Gram positive anaerobic cocci occurring in pairs and chains and forming lactic acid as a major fermentation product are excluded from the genus *Streptococcus* in the current edition of *Bergey's Manual of Determinative Bacteriology* (Buchanan & Gibbons 1974), despite increasing support for their inclusion (Rogosa 1974; Holdeman & Moore 1974; Barnes *et al*. 1977). Reference to anaerobic streptococci has often been made in studies on the inestinal flora without precise definition and without distinguishing the strains from morphologically similar organisms now included in genera such as *Peptococcus* or *Peptostreptococcus* which do not form lactate as a major end product.

Considerable attention has been given in the past to the occurrence of streptococci belonging to Lancefield group D and these organisms usually predominate among the facultatively anaerobic streptococci found in the intestine. Although streptococci other than group D are known to occur, there is much less information on their incidence in different hosts, partly because suitable selective and differential isolation media are unavailable in most cases.

Recent interest in intestinal micro-organisms has been directed towards their possible influence on host physiology and nutrition and, in the case of man, with the role of specific bacteria as direct or indirect agents of disease (reviewed by Drasar & Hill 1974). There are indications that various types of streptococci may be implicated in all these processes, but disease aspects are not covered by the present review.

The term 'enterococci' is used here to include only *S. avium*, *S. faecalis*, *S. faecium* and their respective variants.

2. Occurrence and Distribution in Different Regions of the Intestine

One of the most intriguing differences between man and other host animals with respect to the incidence and distribution of streptococci in the alimentary tract is the nature of the flora, or the lack of one, in the contents of the stomach and proximal part of the small intestine. With humans there is some difficulty in obtaining uncontaminated samples from these regions in live subjects but evidence suggests that the acid contents of the stomach are usually sterile due to the presence of free HCl, while the proximal region of the small intestine normally contains only low numbers of bacteria (see Drasar & Hill 1974). This flora appears to be a transient one, derived mainly from the mouth and respiratory tract and is gradually eliminated from the stomach and proximal part of the small intestine by peristalsis. The presence of viridans streptococci among the predominant flora is attributed mainly to their capacity for prolonged survival under unfavourable conditions (Bauchop 1971).

By contrast with man, it has been shown for a wide range of non-ruminant animal species that the stomach and small intestine can support an extensive microflora including significant numbers of facultatively anaerobic streptococci

Table 1
Numbers of facultatively anaerobic streptococci in different regions of the alimentary tract in various adult animals

Animal species	No. examined	Stomach		Small intestine		Caecum	Faeces
		ant.	post.	ant.	post.		
		Log_{10} median organisms g^{-1} and median pH value ( ) of contents					
Horse	3	6.5 (5.4)	5.0 (3.3)	5.0 (6.7)	6.1 (7.9)	6.0 (7.0)	5.9 (7.5)
Pig	20	6.0 (4.3)	4.4 (2.2)	4.2 (6.0)	6.5 (7.5)	7.0 (6.3)	7.2 (7.1)
Rabbit	11	N (1.9)	N (1.9)	N (6.0)	N (8.0)	4.0 (6.6)	4.7 (7.2)
Rat	7	5.0 (5.0)	4.2 (3.8)	3.0 (6.5)	5.0 (7.1)	6.0 (6.8)	6.2 (6.9)
Mouse	3	6.0 (4.5)	5.3 (3.1)	5.0 (NT)	6.7 (NT)	7.0 (NT)	6.7 (NT)
Chicken	6	*4.0 (4.9)	†3.7 (4.2)	4.0 (5.8)	4.2 (7.8)	6.7 (7.0)	6.5 (7.6)

N, Not found; NT, not tested.
* Crop.
† Gizzard.
Smith (1965a).

(Smith 1965a). In the rhesus monkey, *Macaca mulatta*, the stomach even supports a limited bacterial fermentation and there is evidence (Bauchop 1971) that this part of the intestine has a stable and characteristic bacterial flora. *Streptococcus bovis* occurs among the predominant organisms, which are mainly lactic acid bacteria, and concentrations of lactic acid ranging from 1.2 to 10.6 mg g^{-1} of gastric contents have been demonstrated.

The mouse is an example of an animal which has an extensive flora in the stomach but in this case it has been established that the predominant flora is intimately associated with the mucosal epithelium (Dubos *et al.* 1965). Similar associations between indigenous micro-organisms and gastrointestinal epithelium have been shown with several animal species involving different parts of the intestine (Savage 1972).

Among the predominant organisms found in the murine stomach were anaerobic streptococci (Lancefield group N), occurring at *ca.* $10^8 g^{-1}$ of tissue, with 10^6-$10^8 g^{-1}$ in both small and large intestines. Enterococci were not found by Dubos *et al.* (1965) in any sample examined from the small intestine, although low numbers of these organisms (10^3-$10^4 g^{-1}$) occurred in the large intestine.

Anaerobic streptococci have also been demonstrated in the human jejunum where they colonize the investing mucus, reaching levels up to $10^6 g^{-1}$ (Nelson & Mata 1970) but there is no evidence that either the contents or the mucosal epithelium of the human stomach supports an indigenous streptococcal flora of any kind.

The incidence of facultatively anaerobic streptococci in the human intestine has been reviewed by Drasar & Hill (1974). The organisms are more abundant in the lower regions of the intestine and become established at levels up to *ca.* 10^9 g^{-1} or ml^{-1} of contents in the appendix and in the faeces. A similar situation in the lower intestine was demonstrated for a variety of homothermic animals by Smith (1965*a*); some examples are given in Table 1. The lowest incidence of streptococci occurred in the rabbit which has an intestinal flora comprising very few types of bacteria. The observed incidence of streptococci in this and other animals is related to the varying pH values at different sites along the intestine (Table 1) but is not explained entirely on this basis.

3. Factors Affecting the Streptococcal Flora

The complexity of factors affecting the composition of the gut microflora is well known but poorly understood. Among the more important factors which are likely to affect the streptococcal flora are the age and diet of the host, the structure and activity of the intestine, the possession of immune systems and interactions between different intestinal organisms. For example, gnotobiotic chickens were shown to develop relatively high titres of agglutinins when monocontaminated with a strain of *S. faecalis* subsp. *liquefaciens* previously isolated from a conventional (i.e. not germ-free) chicken (Morishita & Mitsuoka 1973) but the same strain became established in much larger numbers in germ-free birds than in conventional ones (Morishita *et al.* 1972), thus indicating that other intestinal organisms are likely to play a part in limiting the colonization of the intestine by *S. faecalis* under normal circumstances.

Streptococci are among the organisms which gain access to the intestine from the environment of the host at an early stage in development (see below). The nature of the environment can also affect the balance of the intestinal flora by causing various kinds of stress. Tannock & Savage (1974) found that some strains of mice which had been stressed by depriving them of bedding material, food and water for 48 h showed 10- to 100-fold increase in levels of enterococci in the intestine, thereby demonstrating the influence of host factors in controlling the incidence of these organisms.

Factors affecting the apparent composition of the streptococcal flora and relating to methodology have been discussed by Deibel (1964). Difficulties arise in attempting to compare the results of different studies concerning the numbers and types of streptococci in various host animals. These are due to differences in approach to the classification of the isolates obtained and the diversity of isolation media used. Deibel (1964) refers to evidence that some media may favour the isolation of one type of streptococcus at the expense of another.

4. Development of the Streptococcal Flora

For both man and animals there is general agreement that the alimentary tract is either sterile at birth or contains very few micro-organisms. Soon afterwards,

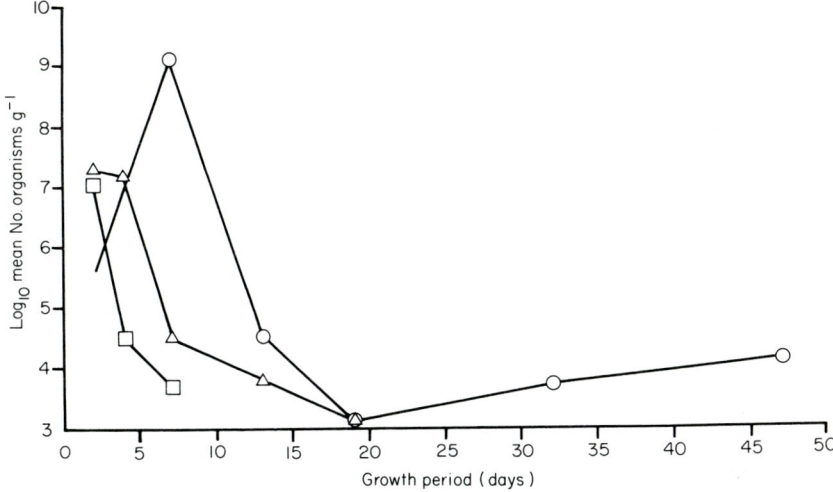

Fig. 1. Changes with age in the incidence of facultatively anaerobic streptococci in the intestine of the chick. □, duodenum; △, caecum; ○, rectum (Ochi et al. 1964).

however, a microflora begins to develop and this follows a similar pattern during the first few weeks of life in most of the animals studied (Smith & Crabb 1961; Smith 1965b). Streptococci were shown to be among the predominant flora of the intestine in the first 2-3 days after birth and the organisms tended to persist in high numbers for several days before declining to markedly lower levels over a period of two or more weeks. Similar results have been obtained by others with pigs (Wilbur et al. 1960) and chickens (Shapiro & Sarles 1949; Ochi et al. 1964; Huhtanen & Pensack 1965; Mead & Adams 1975). Results obtained by Ochi et al. (1964) for different regions of the chick intestine are shown in Fig. 1. The early predominance of streptococci in the faeces of human infants was confirmed by Mitsuoka & Hayakawa (1972). In this case, the organisms eventually reached a level of 10^7-10^8 g^{-1} which then remained relatively stable throughout life. Smith & Crabb (1961) attributed the similarities they observed between different animals in the early behaviour of the faecal flora to the milk diet given in each case and the lack of diversity in gut structure and function in the young animal.

The streptococci showed two different patterns of behaviour in the mouse (Schaedler et al. 1965). While the enterococci followed the usual pattern of decline with age, in the large intestine, anaerobic streptococci became established soon after birth, in the stomach as well as the small and large intestines. Maximum numbers of the anaerobes were reached by about the twelfth day of life and remained thereafter at 10^8-10^9 g^{-1}. These observations are consistent with the view that the enterococci should be regarded as chance invaders of the intestine which are gradually suppressed to a low level during development of the adult flora. Dubos *et al.* (1965) considered the anaerobic streptococci in the mouse to be an autochthonous flora.

5. Influence of Diet

For many years, the nature of the diet has been regarded as a critical factor in determining the composition of the intestinal microflora including the numbers and types of streptococci present, but consistent effects have been difficult to demonstrate, as shown by the following examples of studies made because of a predicted association between the intestinal flora and human cancer. In one instance (Hill *et al.* 1971) it was found that enterococci were consistently more numerous in the faeces of individuals from India, Japan and Uganda who were living on a mainly vegetarian diet than with subjects from England, Scotland and the USA, living on a mixed Western diet. Moreover, *S. faecium* predominated among the streptococci in the former group and *S. faecalis* biotypes in the latter. Different results were obtained, however, by Finegold *et al.* (1974), comparing the faecal flora of US subjects of Japanese origin which had been fed on either US or Japanese diets. In this case the incidence of *S. faecalis* subsp. *faecalis* was *ca.* 10-fold higher in subjects fed on the Japanese diet.

Studies made on the influence of chemically defined diets on the human faecal flora have shown a marked effect in reducing the levels of enterococci (Crowther *et al.* 1973; Bounous & Devroede 1974), but, with another type of primate, the baboon, the effect of feeding a synthetic diet was to increase the numbers of facultatively anaerobic streptococci in all regions of the intestine (Uphill *et al.* 1974).

Another dietary factor is the presence of naturally occurring antimicrobial compounds. Sieburth (1959) examined the intestinal flora of various Antarctic birds and failed to isolate any enterococci. It was found by Sieburth (1961) that these organisms were inhibited by acrylic acid which occurs in a marine alga, *Phaeocystis pouchetii*, a component of the birds' diet. The effect on the streptococcal flora of adding antibiotics to commercial animal feeds will be discussed below.

6. Species Distribution in Various Hosts

Organisms which have been classified as anaerobic streptococci are usually predominant among the total streptococcal flora of human faeces and include *S. hansenii*, *S. intermedius* and *S. morbillorum* (Holdeman & Moore 1974; Finegold *et al.* 1974; Holdeman *et al.* 1976). Anaerobic streptococci, including *S. intermedius*, have also been isolated at $> 10^8 \text{ g}^{-1}$ from the caecal contents of

Table 2
Incidence of facultatively anaerobic streptococci in human faeces

Enterococcus group	No. of positive samples (out of 20)	Log_{10} organisms g^{-1} Range	Mean
S. faecalis subsp. faecalis	14	4–10	9.8
S. faecalis subsp. liquefaciens	5	3–8	7.5
S. faecalis subsp. zymogenes	3	7–10	9.1
S. faecium	5	6–10	9.2
S. faecium subsp. cassiflavus?	1	9	7.8
S. durans	2	5–8	6.9
Other group D streptococci			
S. bovis	5	4–11	9.7
S. equinus?	1	8	6.7
Other streptococci			
S. agalactiae	1	7	7.8
S. cremoris*	4	6–10	9.5
S. equisimilis	1	4	3.7
S. lactis	2	8–10	7.0
S. mitis	3	6–7	6.1
S. MG	2	6–8	7.5
S. pyogenes	1	8	7.8
S. salivarius*	1	4	2.8
S. sanguis	1	4	2.9
S. thermophilus?	1	8	6.8
S. uberis	1	9	8.2
S. zooepidemicus?	1	7	6.6
Unidentified	13	4–11	9.9
Total			10.9

* Isolated from individuals on Japanese diet only.
Finegold et al. (1974).

poultry (Barnes & Impey 1970, 1974) and recently a new species, *S. pleomorphus*, was described following isolation from the same source (Barnes et al. 1977). *Streptococcus pleomorphus* closely resembled the group VIIa of Hare (1967), a type found previously in human faeces but considered to be uncommon.

A detailed analysis of the human faecal flora was made by Finegold et al. (1974) and facultatively anaerobic streptococci belonging to Lancefield groups A, B, C, D, F, H, K and N were among the types isolated (Table 2). Some other types were noted in earlier studies: for example group G (Hare & Maxted 1935; Smith & Sherman 1938), *S. acidominimus* (Seelemann 1954) and *S. avium* (Nowlan & Deibel 1967).

Although many of the streptococci isolated by Finegold et al. (1974) could not be identified, this study confirmed that group D species usually predominate among the classifiable types present. The incidence of strains belonging to groups other than D appears to be sporadic (Table 2). In reviewing the carriage of group B streptococci, Patterson & Hafeez (1976) reported the presence of these organisms in only 5.5-29.1% of human stool samples examined.

Results obtained for the incidence of individual group D species in human faeces have varied with the geographical location of the individuals. Studies made in Egypt, the UK and USA have shown that *S. faecalis* biotypes tend to predominate, with *S. faecium* being found less frequently or at lower levels (Cooper & Ramadan 1955: Moussa 1965; Finegold *et al.* 1974). However, reports on subjects living in continental Europe, India, Japan and Uganda have indicated a reverse trend with the incidence of *S. faecium* being equal to or greater than that of *S. faecalis* (Guthof 1957; Buttiaux 1958; Kjellander 1960; Hill *et al.* 1971). The likely influence of diet and other factors on these results has been discussed above.

Similar variation in the incidence of group D species has been reported for animals, not only between different hosts but with the same animal species examined by different investigators. Some examples are given in Table 3.

Among non-ruminants, a predominance of *S. bovis* or *S. equinus* is usually associated with the horse or pig (Bartley & Slanetz 1960; Fuller *et al.* 1960; Raibaud *et al.* 1961). With other animals, the species found most frequently have been *S. faecalis*, *S. faecium* and intermediate strains (Table 3). The wild animals referred to in Table 3 include at least 25 species of mammals, 17 species of birds and 18 species of reptiles (Mundt 1963).

Table 3 indicates that the most widespread variant of *S. faecalis* is the subspecies *liquefaciens* and, in addition to results given in the table, this organism comprised one-quarter of the enterococci isolated from poultry faeces by Moussa (1965). With human faeces, however, the commonest variant was the subspecies *faecalis* which was isolated from 14/20 samples by Finegold *et al.* (1974) at levels up to 10^{10} g^{-1}. It was not detected in a variety of ruminant and non-ruminant farm animals studied by Cooper & Ramadan (1955) and Ramadan & Sabir (1963). In addition to the relatively uncommon isolation of *S. faecalis* subsp. *faecalis* from chickens and pigs (Table 3), Bartley & Slanetz (1960) found only one strain in animal faeces; in this case the source was a dog.

Further distinctions have been demonstrated between human and animal strains of streptococci isolated from faeces by studying properties other than those used in speciation. Cooper & Ramadan (1955) found that 76.4% of their human isolates could withstand heating at 63°C for 30 min and then grow in a medium containing 0.04% of potassium tellurite. None of the animal strains isolated by either Cooper & Ramadan (1955) or Ramadan & Sabir (1963) survived this combined test regardless of their resistance to tellurite. However, the physiological basis of the test has not been elucidated.

Although *S. bovis* rarely predominates among the group D streptococci in

Table 3

Incidence of group D species in the intestinal contents of man and various non-ruminant animals

Source	Author(s)	No. and type ( ) of sample	% containing group D spp.	S. faecalis subsp. faecalis	S. faecalis subsp. liquefaciens	S. faecalis subsp. zymogenes	S. faecium	S. bovis	Others
Man	Buttiaux (1958)	38 (faeces)	100	57.9	5.0	10.5	92.1	10.5	NF
	Kjellander (1960)	73 (faeces)	?	38.4	NF	15.1	42.5	15.1	12.3
	Finegold et al. (1974)	40 (faeces)	?	72.5	20.0	17.5	35.0	15.0	?
Chicken	Barnes (1958)	50 (caecum)	100	1.7*	58.3*	0.9*	33.9*	NF	5.2*
Turkey	Harrison & Hansen (1950)	12 (caecum)	?	NF	31.3*	NF	NF	NF	68.8*
Pig	Barnes & Ingram (1955)	21 (colon)	76.2	NF	NF	NF	57.1	NF	42.9
	Buttiaux (1958)	52 (faeces)	100	19.2	34.6	NF	92.3	5.7	NF
	Raibaud et al. (1961)	10 (caecum, faeces)	100	NF	NF	NF	NF	100	NF
Wild Mammals	Mundt (1963)	216 (faeces)	71.5	10.7	22.4	1.9	36.0	NF	6.1
Wild Reptiles		70 (faeces)	84.7	19.7	60.5	NF	13.1	NF	6.6
Wild Birds		22 (faeces)	31.8	9.1	13.6	4.5	9.0	NF	4.5

NF, Not found.
* Percentage of strains isolated.

human faeces, those strains isolated from man usually ferment mannitol and fail to form slime in a sucrose medium whereas the reverse is true of animal strains. The human strains of *S. bovis* form a homogeneous group (Kiel & Skadhauge 1973) which appears to be synonymous with the *S. inulinaceus* isolated in large numbers from human faeces by Bartley & Slanetz (1960). Another biotype of *S. bovis* isolated from pigs by Raibaud *et al.* (1961) characteristically produced urease.

These distinctions between biotypes of different origin do not appear to extend to the serological properties of the organisms. Barnes (1964) studied the occurrence of 19 serological types of *S. faecium* and related strains in different hosts and found that several of these types occurred in man as well as cattle, chickens, pigs and sheep; none was found to be host specific.

Almost all animals have ample opportunity to ingest enterococci (Mundt 1963) but the trends cited above indicate that some selection occurs in the gut of the host (Mieth 1962). In Mundt's survey the occurrence of enterococci was least consistent among rodents which do not appear to be natural hosts for these organisms (Ostrolenk & Hunter 1946). Another factor which may be important in relation to the carriage of enterococci is the degree of contact with man. Whereas enterococci and other faecal 'indicator' organisms are commonly present in the intestinal flora of domestic poultry, as they are also in sea gulls which scavenge in highly polluted environments (Wood & Trust 1972), they are less common in birds whose existence is more remote from human influence (Mushin & Ashburner 1962; Mundt 1963; Mead 1965).

7. Some Possible Interactions between the Streptococcal Flora and the Host

During the early stages of development in young, healthy animals such as pigs and poultry there is a brief period during which the growth rate is depressed, an effect which is usually attributed to the activities of the intestinal flora. The condition can be alleviated by supplementing the diet with antibiotics and is not observed with germ-free animals (Jayne-Williams & Fuller 1971). It appears that growth depression is due to a transitory malabsorption syndrome and hence a reduced efficiency in feed utilization (Eyssen & de Somer 1963). From the economic point of view this is particularly important in the rearing of pigs and poultry as food animals but the precise nature of the effect and the possible role of micro-organisms has yet to be elucidated.

A common property of almost all antibiotics known to be effective as growth promoters is their ability to inhibit Gram positive bacteria (Eyssen & de Somer 1963). Of these organisms it is usually the enterococci which predominate in the intestinal microflora of the very young animal, as discussed previously. Moreover, enterococci have been implicated in various studies of growth

Table 4
Properties of enterococci of possible significance in relation to intestinal function and nutrition in the chick

Property	S. faecalis			S. faecium and variants
	subsp. faecalis	subsp. liquefaciens	subsp. zymogenes	
Proteolysis	−	+	±	−
NH$_3$ from arginine	+	+	+	+
Tyramine from tyrosine	+	+	+	±
Deconjugation of bile acids	±	±	±	±
Anaerobic breakdown of uric acid	+	+	+	−

+, positive; −, negative; ±, variable.
Modified from Barnes (1975).

depression and a list of their properties which may be relevant in this respect is given in Table 4. Proteolytic activity is included in the table since it is a possible means of providing substrates suitable for microbial production of ammonia which is recognized as a potentially harmful substance in the small intestine. The formation of tyramine by tyrosine decarboxylation is of interest because tyramine is a vasoconstricting agent which could reduce blood flow in the intestinal mucosa and hence lower the rate of nutrient absorption (Huhtanen & Pensack 1965).

Many intestinal organisms are known to be capable of deconjugating bile acids, including at least some strains of *S. faecalis* biotypes and *S. faecium* subsp. *durans* (Shimada *et al*. 1969). The resultant loss of detergency causes a malabsorption of fats in the small intestine and may interfere even with the absorption of other nutrients, as discussed by Jayne-Williams & Fuller (1971) and Burke *et al.* (1974).

Evidence of a specific role for *S. faecalis* in the malabsorption syndrome was obtained by Eyssen & de Somer (1965, 1967) and by Huhtanen & Pensack (1965) using germ-free chickens. Growth depression was observed following mono-contamination of chicks with a strain of *S. faecalis* but this was enhanced in the presence of a filterable agent, presumed to be a virus, which had been obtained from the gut contents of a conventional chick. Harrison & Coates (1972) found that neither a strain of *S. faecalis* subsp. *liquefaciens* nor a bacteria-free filtrate from chicken faeces depressed growth significantly when given alone to germ-free chicks but a marked effect was produced when both agents were given together. It was thought that the filterable agent might be responsible for damaging the intestinal mucosa, thus facilitating invasion by the streptococcus. Alternatively, the streptococcus may have been the predisposing

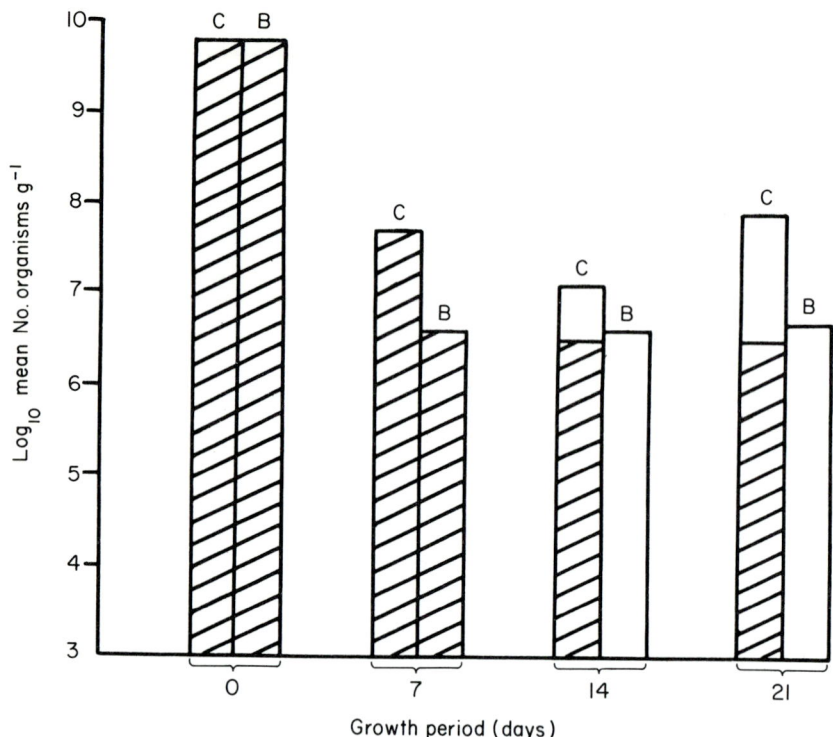

Fig. 2. Effect of dietary bacitracin on the incidence of group D streptococci in the caecum of the developing chick. □, 'total' streptococci; ▨ *S. faecalis* subsp. *liquefaciens*; C, control birds; B, bacitracin-fed birds (six birds in each group) (Barnes & Mead, unpublished).

agent. Moore & Gross (1968) isolated a strain of *S. faecalis* subsp. *liquefaciens* which actively desquamated the intestinal epithelium of young turkeys thus permitting penetration by a strain of *Catenabacterium* sp. and the subsequent development of liver granulomas. Hence, there is the possibility that a specific biotype of *S. faecalis* with a unique property or combination of properties is at least one of the organisms which may be actively involved in growth depression. The effect of dietary antibiotics is not, apparently, to eliminate all streptococci from the intestine but to change the composition of the streptococcal flora to one which is less deleterious to the host. Studies on chickens (Barnes 1958; Elliott & Barnes 1959) showed that the addition of 25 mg l^{-1} of chlortetracycline (CTC) to the diet did not change significantly the numbers of streptococci in the caeca but it resulted in CTC-sensitive *S. faecalis* subsp. *liquefaciens*, which was present prior to treatment, being replaced by a resistant strain of *S. faecalis* subsp. *faecalis*. A recent study of the intestinal flora of the chick up to 3 weeks of age (Barnes & Mead, unpublished) showed a similar effect with 50 mg l^{-1} of dietary bacitracin: *S. faecalis* subsp. *liquefaciens* was eliminated from both the small intestine and caeca within 14 days, whilst the 'total' population of streptococci in the caeca remained relatively high in numbers (Fig. 2). It has been shown that feeding chicks on a diet containing benzylpenicillin results in an increased incidence of *S. faecium* in the small intestine at the expense of *S. faecalis* (Jeffries *et al.* 1977). In this case, however, the effect appears to be indirect, due to the rapid disappearance of penicillin from the intestine, and is probably due to changes in the microflora of the crop and the interaction of this flora with that of the small intestine (Smith 1965*a*; Fuller 1973).

Recently it was shown by Mead (1974) and Mead & Adams (1975) that *S. faecalis* (but not *S. faecium*) can decompose uric acid under anaerobic conditions with the production of ammonia. In view of the predominance of *S. faecalis* in the young bird and the fact that urine containing > 1% uric acid flows back from the large intestine into the caecum, as discussed by Barnes (1975), it is possible that the ammonia produced in the anaerobic environment of the caecum due to uric acid breakdown could be utilized by the bird. Barnes (1975) has suggested that in natural circumstances, when food and water are scarce, the ability of *S. faecalis* and other intestinal bacteria to decompose uric acid may be of benefit to the host since it provides a means of recycling both nitrogen and water derived from the urine, without a parallel accumulation in the intestine of the potentially irritant uric acid.

8. Summary

Much information is available on the occurrence and properties of streptococci in the intestinal tract of different host animals, but problems in classifying many of the strains encountered and conflicting views concerning the recognition of anaerobic streptococci continue to hamper progress in describing more fully the streptococcal flora of the intestine.

The incidence and distribution of streptococci in different regions of the gut, and in some cases their association with intestinal epithelium, varies with the host species. In man the acid contents of the stomach are usually sterile, and few organisms are normally found in the proximal part of the small intestine. By contrast, the same regions in the mouse harbour large numbers of organisms including anaerobic streptococci (Lancefield group N) which rapidly become established in the young animal and are intimately associated with the gastric mucosa. These organisms are considered to be an autochthonous flora (Dubos *et al.* 1965) whereas the enterococci present in the lower bowel are regarded as 'accidental invaders' of the intestine and their behaviour tends to follow a similar pattern in all the host animals that have been studied. A brief phase of predominance shortly after birth is followed by a period of marked decline before the enterococci stabilize at relatively low levels, limited by host factors and probable interactions with other intestinal micro-organisms.

Possible effects of the streptococcal flora on the physiology and nutrition of the host are considered. These include the apparent role of *S.faecalis* subsp. *liquefaciens* in the depression of growth which is observed for a short period in young animals such as pigs and poultry. Evidence is discussed suggesting that dietary antibiotics alleviate growth depression caused by this organism by changing the streptococcal flora of the intestine to one which does not have a deleterious effect on the host.

9. References

BARNES, E. M. 1958 The effect of antibiotic supplements on the faecal streptococci (Lancefield Group D) of poultry. *British Veterinary Journal* **114**, 333-344.

BARNES, E. M. 1964 Distribution and properties of serological types of *Streptococcus faecium, Streptococcus durans* and related strains. *Journal of Applied Bacteriology* **27**, 461-470.

BARNES, E. M. 1975 Development and ecological significance of the avian intestinal flora. In *Proceedings of the First Intersectional Congress of IAMS Vol. 2. Developmental Microbiology, Ecology*, ed. Hasegawa, T. Tokyo: Science Council of Japan.

BARNES, E. M. & INGRAM, M. 1955 The identity and origin of faecal streptococci in canned hams. *Annales de l'Institut Pasteur de Lille* **7**, 115-119.

BARNES, E. M. & IMPEY, C. S. 1970 The isolation and properties of the predominant anaerobic bacteria in the caeca of chickens and turkeys. *British Poultry Science* **11**, 467-481.

BARNES, E. M. & IMPEY, C. S. 1974 The occurrence and properties of uric acid decomposing anaerobic bacteria in the avian caecum. *Journal of Applied Bacteriology* **37**, 393-409.

BARNES, E. M., IMPEY, C. S., STEVENS, B. J. H. & PEEL, J. L. 1977 *Streptococcus pleomorphus* sp. nov., an anaerobic *Streptococcus* isolated mainly from the caeca of birds. *Journal of General Microbiology* **102**, 45-53

BARTLEY, C. H. & SLANETZ, L. W. 1960 Types and sanitary significance of fecal streptococci isolated from feces, sewage and water. *American Journal of Public Health* **50**, 1545-1552.

BAUCHOP, T. 1971 Stomach microbiology of primates. *Annual Review of Microbiology* **25**, 429-435.

BOUNOUS, G. & DEVROEDE, G. J. 1974 Effect of an elemental diet on human fecal flora. *Gastroenterology* **66**, 210-214.

BUCHANAN, R. E. & GIBBONS, N. E. 1974 *Bergey's Manual of Determinative Bacteriology* 8th edn. Baltimore: Williams & Wilkins Co.

BURKE, V., GRACEY, M. & THOMAS, J. 1974 Another untoward effect of enteric microorganisms on intestinal functions. *Australian and New Zealand Journal of Medicine* **4**, 211.

BUTTIAUX, R. 1958 Les streptocoques fécaux des intestines humains et animaux. *Annales de l'Institut Pasteur* **94**, 778-782.

COOPER, K. E. & RAMADAN, F. M. 1955 Studies in the differentiation between human and animal pollution by means of faecal streptococci. *Journal of General Microbiology* **12**, 180-190.

CROWTHER, J. S., DRASAR, B. S., GODDARD, P., HILL, M. J. & JOHNSON, K. 1973 The effect of a chemically defined diet on the faecal flora and faecal steroid concentration. *Gut* **14**, 790-793.

DEIBEL, R. H. 1964 The Group D streptococci. *Bacteriological Reviews* **28**, 330-366.

DRASAR, B. S. & HILL, M. J. 1974 *Human Intestinal Flora.* London & New York: Academic Press.

DUBOS, R., SCHAEDLER, R. W., COSTELLO, R. & HOET, P. 1965 Indigenous, normal and autochthonous flora of the gastrointestinal tract. *Journal of Experimental Medicine* **122**, 67-76.

ELLIOTT, S. D. & BARNES, E. M. 1959 Changes in serological type and antibiotic resistance of Lancefield Group D streptococci in chickens receiving dietary chlortetracycline. *Journal of General Microbiology* **20**, 426-433.

EYSSEN, H. & DE SOMER, P. 1963 The mode of action of antibiotics in stimulating growth of chicks. *Journal of Experimental Medicine* **117**, 127-138.

EYSSEN, H. & DE SOMER, P. 1965 Studies on gnotobiotic chicks: effects of controlled intestinal floras on growth and nutrient absorption. *Ernährungsforschung* **10**, 264-273.

EYSSEN, H. & DE SOMER, P. 1967 Effects of *Streptococcus faecalis* and a filterable agent on growth and nutrient absorption in gnotobiotic chicks. *Poultry Science* **46**, 323-333.

FINEGOLD, S. M., ATTEBERY, H. R. & SUTTER, V. L. 1974 Effect of diet on human fecal flora: comparison of Japanese and American diets. *American Journal of Clinical Nutrition* **27**, 1456-1469.

FULLER, R. 1973 Ecological studies on the lactobacillus flora associated with the crop epithelium of the fowl. *Journal of Applied Bacteriology* **36**, 131-139.

FULLER, R., NEWLAND, L. G. M., BRIGGS, C. A. E., BRAUDE, R. & MITCHELL, K. G. 1960 The normal intestinal flora of the pig. IV. The effect of dietary supplements of penicillin, chlortetracycline or copper sulphate on the faecal flora. *Journal of Applied Bacteriology* **23**, 195-205.

GUTHOF, O. 1957 Streptokokken und Dysbakterie Problem. *Zentralblatt für Bakteriologie, Parasitenkunde, Infektionskrankheiten und Hygiene.* Abt. I. Orig. **170**, 327-333.

HARE, R. 1967 The anaerobic cocci. In *Recent Advances in Medical Microbiology,* ed. Waterson, H. P. Boston: Little Brown & Co.

HARE, R. & MAXTED, W. R. 1935 The classification of haemolytic streptococci from the stools of normal pregnant women and of cases of scarlet fever by means of precipitin and biochemical tests. *Journal of Pathology and Bacteriology* **41**, 513-520.

HARRISON, G. F. & COATES, M. E. 1972 Interrelationship between the growth-promoting effect of fish solubles and the gut flora of the chick. *British Journal of Nutrition* **28**, 213-221.

HARRISON, A. P. & HANSEN, P. A. 1950 The bacterial flora of the cecal feces of healthy turkeys. *Journal of Bacteriology* **50**, 197-210.

HILL, M. J., DRASAR, B. S., ARIES, V., CROWTHER, J. S., HAWKSWORTH, G. & WILLIAMS, R. E. O. 1971 Bacteria and aetiology of cancer of large bowel. *Lancet* **1**, 95-100.

HOLDEMAN, L. V. & MOORE, W. E. C. 1974 New genus *Coprococcus.* Twelve new species and emended descriptions of four previously described species of bacteria from human feces. *International Journal of Systematic Bacteriology* **24**, 260-277.

HOLDEMAN, L. V., GOOD, I. J. & MOORE, W. E. C. 1976 Human fecal flora: variation in bacterial composition within individuals and a possible effect of emotional stress. *Applied and Environmental Microbiology* **31**, 359-375.

HUHTANEN, C. N. & PENSACK, J. M. 1965 The role of *Streptococcus faecalis* in the antibiotic growth effect in chickens. *Poultry Science* **44**, 830-844.

JAYNE-WILLIAMS, D. J. & FULLER, R. 1971 The influence of the intestinal microflora on nutrition. In *Physiology and Biochemistry of the Domestic Fowl.* Vol. 1, eds Bell, D. J. & Freeman, B. M. London & New York: Academic Press.

JEFFRIES, L., COLEMAN, K. & BUNYAN, J. 1977 Antimicrobial substances and chick growth promotion: comparative studies on selected compounds *in vitro* and *in vivo. British Poultry Science* **18**, 295-308.

KIEL, P. & SKADHAUGE, K. 1973 Studies on mannitol-fermenting strains of *Streptococcus bovis*. *Acta Pathologica et Microbiologica Scandinavica Section B* **81**, 10-14.
KJELLANDER, J. 1960 Enteric streptococci as indicators of faecal contamination of water. *Acta Pathologica et Microbiologica Scandinavica* Suppl. 136, **48**, 1-124.
MEAD, G. C. 1965 The distribution of *Streptococcus faecalis* and related biotypes. Ph.D. Thesis, University of London.
MEAD, G. C. 1974 Anaerobic utilization of uric acid by some Group D streptococci. *Journal of General Microbiology* **82**, 421-423.
MEAD, G. C. & ADAMS, B. W. 1975 Some observations on the caecal microflora of the chick during the first two weeks of life. *British Poultry Science* **16**, 169-176.
MIETH, H. 1962 Untersuchungen über das Vorkommen von Enterokokken bei Tieren und Menschen. IV. Mitteilung die Streptokokkenflora in den Faeces von Pferden. *Zentralblatt für Bakteriologie, Parasitenkunde, Infektionskrankheiten und Hygiene.* Abt. I. Orig. **185**, 166-174.
MITSUOKA, T. & HAYAKAWA, K. 1972 Die Faecalflora bie Menschen I. Mitteilung: die Zusammensetzung der Faecalflora der verschiedenen Altersgruppen. *Zentralblatt für Bakteriologie, Parasitenkunde, Infektionskrankheiten und Hygiene.* Abt. I. Orig. **A223**, 333-342.
MOORE, W. E. C. & GROSS, W. B. 1968 Liver granulomas of turkeys—causative agents and mechanism of infection. *Avian Diseases* **12**, 417-422.
MORISHITA, Y. & MITSUOKA, T. 1973 Antibody responses in germ-free chickens to bacteria isolated from various sources. *Japanese Journal of Microbiology* **17**, 181-187.
MORISHITA, Y., MITSUOKA, T., KANEUCHI, C., YAMAMOTO, T., YAMAMOTO, S. & OGATA, M. 1972 Establishment of microorganisms isolated from chickens in the digestive tract of germ-free chickens. *Japanese Journal of Microbiology* **16**, 27-33.
MOUSSA, R. S. 1965 Species differentiation of faecal and nonfaecal enterococci. *Journal of Applied Bacteriology* **28**, 466-472.
MUNDT, J. O. 1963 Occurrence of enterococci in animals in a wild environment. *Applied Microbiology* **11**, 136-140.
MUNDT, J. O. 1973 Litmus milk reaction as a distinguishing feature between *Streptococcus faecalis* of human and non-human origins. *Journal of Milk and Food Technology* **36**, 364-367.
MUSHIN, R. & ASHBURNER, F. M. 1962 Gastrointestinal microflora of mutton birds (*Puffinus tenuirostris*) in relation to 'limy' disease. *Journal of Bacteriology* **83**, 1260-1267.
NELSON, D. P. & MATA, L. J. 1970 Bacterial flora associated with the human gastrointestinal mucosa. *Gastroenterology* **58**, 56-61.
NOWLAN, S. S. & DEIBEL, R. H. 1967 Group Q streptococci I. Ecology, serology, physiology and relationship to established enterococci. *Journal of Bacteriology* **94**, 291-296.
OCHI, Y., MITSUOKA, T. & SEGA, T. 1964 Untersuchungen über die Darmflora des Huhnes III. Mitteilung: die Entwicklung der Darmflora von Küken bis zum Huhn. *Zentralblatt für Bakteriologie, Parasitenkunde, Infectionskrankheiten und Hygiene.* Abt. I. Orig. **193**, 80-95.
OSTROLENK, M. & HUNTER, A. C. 1946 The distribution of enteric streptococci. *Journal of Bacteriology* **51**, 735-741.
PATTERSON, M. J. & HAFEEZ, A. E. 1976 Group B streptococci in human disease. *Bacteriological Reviews* **40**, 774-792.
RAIBAUD, P., CAULET, M., GALPIN, J. V. & MOCQUOT, G. 1961 Studies on the bacterial flora of the alimentary tract of pigs. II. Streptococci: selective enumeration and differentiation of the dominant group. *Journal of Applied Bacteriology* **24**, 285-306.
RAMADAN, F. M. & SABIR, M. S. 1963 Differentiation studies of fecal streptococci from farm animals. *Canadian Journal of Microbiology* **9**, 443-450.
ROGOSA, M. 1974 Family III. Peptococcaceae. In *Bergey's Manual of Determinative Bacteriology* 8th edn, eds Buchanan, R. E. & GIBBONS, N. E. Baltimore: Williams & Wilkins Co.

SAVAGE, D. C. 1972 Associations and physiological interactions of indigenous microorganisms and gastrointestinal epithelia. *American Journal of Clinical Nutrition* **25**, 1372-1379.

SCHAEDLER, R. W., DUBOS, R. & COSTELLO, R. 1965. The development of the bacterial flora in the gastrointestinal tract of mice. *Journal of Experimental Medicine* **122**, 59-66.

SEELEMANN, M. 1954 *Biologie der Streptokokken* 2nd edn. Nürnberg: Verlag Hans Carl., p. 467.

SHAPIRO, S. K. & SARLES, W. B. 1949 Microorganisms in the intestinal tract of normal chickens. *Journal of Bacteriology* **58**, 531-544.

SHIMIDA, K., BRICKNELL, K. S. & FINEGOLD, S. M. 1969 Deconjugation of bile acids by intestinal bacteria: review of literature and additional studies. *Journal of Infectious Diseases* **119**, 273-281.

SIEBURTH, J. McN. 1959 Gastrointestinal microflora of antarctic birds. *Journal of Bacteriology* **77**, 521-531.

SIEBURTH, J. McN. 1961 Antibiotic properties of acrylic acid, a factor in the gastrointestinal antibiosis of polar marine animals. *Journal of Bacteriology* **82**, 72-79.

SMITH, H. W. 1965a Observations on the flora of the alimentary tract of animals and factors affecting its composition. *Journal of Pathology and Bacteriology* **89**, 95-122.

SMITH, H. W. 1965b The development of the flora of the alimentary tract in young animals. *Journal of Pathology and Bacteriology* **90**, 495-513.

SMITH, H. W. & CRABB, W. E. 1961 The faecal bacterial flora of animals and man: its development in the young. *Journal of Pathology and Bacteriology* **82**, 53-66.

SMITH, F. R. & SHERMAN, J. M. 1938 The haemolytic streptococci of human feces. *Journal of Infectious Diseases* **62**, 186-189.

TANNOCK, G. W. & SAVAGE, D. C. 1974 Influence of dietary and environmental stress on microbial populations in the murine gastrointestinal tract. *Infection and Immunity* **9**, 591-598.

UPHILL, P. F., WILDE, J. K. H. & BERGER, J. 1974 Repeated examinations, using the laparotomy sampling technique, of the gastrointestinal microflora of baboons fed a natural or a synthetic diet. *Journal of Applied Bacteriology* **37**, 309-317.

WILBUR, R. D., CATRON, D. V., QUINN, L. Y., SPEER, V. C. & HAYS, V. W. 1960 Intestinal flora of the pig as influenced by diet and age. *Journal of Nutrition* **71**, 168-175.

WOOD, A. J. & TRUST, T. J. 1972 Some qualitative and quantitative aspects of the intestinal microflora of the glaucous-winged gull (*Larus glaucescens*). *Canadian Journal of Microbiology* **18**, 1577-1583.

Streptococci in the Dairy Industry

B. A. Law and M. Elisabeth Sharpe

*National Institute for Research in Dairying,
Shinfield, Reading, Berkshire, England*

CONTENTS

1. Introduction 263
2. Lactic streptococci used in starter cultures in the dairy industry . . . 264
3. Utilization of milk proteins by lactic streptococci 267
4. Lactose utilization by lactic streptococci 268
5. Alternative pathways of carbohydrate metabolism and their effect on the properties of short-life fermented products 270
6. The influence of lactic streptococci on cheese ripening 271
7. The formation of toxic amines in cheese by group D streptococci . . 273
8. Causes of slow growth of lactic streptococci in milk 274
9. Factors stimulating the growth of lactic streptococci 275
10. References 276

1. Introduction

THE STREPTOCOCCI found in man and animals are either associated with pathological symptoms or are present as part of the natural microflora of the host. The types and numbers of streptococci present are governed by the species of animal, the special conditions of the particular habitat in or on the animal, and the presence of other microbial groups. Chance contamination also plays an important role, especially in the case of pathological infections. In contrast to this situation, the streptococci encountered in the dairy industry are, with few exceptions, added deliberately as starter cultures in the manufacture of fermented milk products, and their presence is beneficial. The production of fermented milk products is a major food industry and their value (in 1976) represented 21.6% of the total value of fermented foods throughout the world (Table 1) when over 9,000,000 tonnes of cheese were produced. The commercial production of most of these milk-based foods depends on the fermentation of lactose (milk sugar) to lactic acid by the lactic streptococci. These organisms also produce aroma compounds and their proteolytic action aids the maturation of ripened dairy products. A knowledge of the characteristics and metabolism of the lactic streptococci is therefore of vital importance in the dairy industry.

In pre-industrial cheesemaking, lactic streptococci were allowed to contaminate milk at random from natural sources and it was hoped that suitable

Table 1
The value of the world output of fermented products in 1976

Product	Value (£m)	% of total
Alcoholic drinks: beer, wine, spirits, saké	25,200	72.6
Cheese and fermented milk products	7500	21.6
Other fermented foods: pickles, vinegar, soy sauce etc.	2000	5.8

Data from Bronn, (1976).

strains would grow. Product quality and uniformity were of no great economic consequence in cheesemaking on this scale, but now it is a capital-intensive industry involving factories with capacities to process up to 800,000 l of milk each day into cheese. Under these circumstances the manufacturer requires the provision of cultures of lactic streptococci whose quality is strictly controlled and whose acid- and aroma-producing properties are predictable and matched to a particular product.

2. Lactic Streptococci used as Starter Cultures in the Dairy Industry

The lactic streptococci were originally isolated from souring milk or cream. Lister (1878) isolated an organism from sour milk which he called *Bacterium lactis* and was later designated *Streptococcus lactis* by Orla-Jensen (1919) in his classical monograph based on an intensive investigation of the morphology and biochemical characteristics of the principal bacteria which produced acid in milk and cream. The second organism was named *S. cremoris* (from cream) and a further species, *S. diacetilactis*, was recognized as representing citrate-fermenting streptococci which were distinct from *Leuconostoc* (Matuszewski *et al.* 1936). The 8th edition of *Bergey's Manual of Determinative Bacteriology* (Buchanan & Gibbons 1974) designates *S. diacetilactis* as a subspecies (*S. lactis* subsp. *diacetylactis*, referred to hereafter as *S. diacetylactis*). These lactic streptococci were distinguished from *S. faecalis* (group D streptococci; Lancefield 1933) by Sherman (1938) and assigned to the serological group N by Shattock & Mattick (1943; *S. lactis*), Briggs & Newland (1952; *S. cremoris*) and Briggs (1952; *S. diacetylactis*). The ecology of lactic streptococci has been reviewed recently (Sandine *et al.* 1972) and the available literature suggests that the natural habitat of *S. lactis* is plant material, although strains have been isolated from raw milk. *Streptococcus lactis* has also been isolated at levels of *ca.* $10^4 \, g^{-1}$ from the adult bovine rumen in animals fed a high carbohydrate diet (Sharpe, unpublished) suggesting that it may be part of the minority flora of the rumen. Wild strains of *S. cremoris* rarely occur and its natural habitat is unknown, although it may be present in the same environment as *S. lactis*,

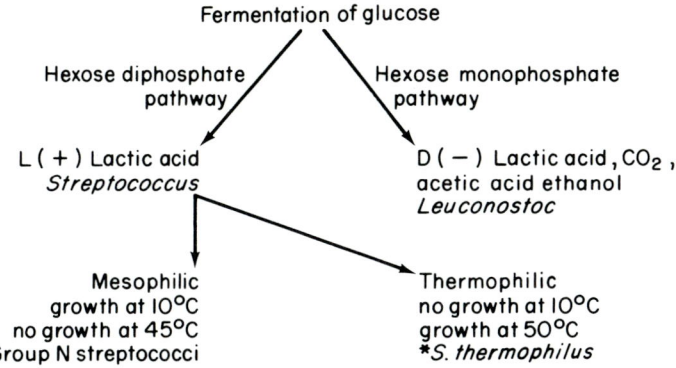

Fig. 1. Relationship between the lactic streptococci within the family Streptococcaceae.
*Some strains cannot ferment glucose on first isolation from milk but require lactose.

but in lower numbers, and new methods of recovery are necessary to confirm this. The natural habitat of *S. thermophilus* is not known but it can be isolated from heat-treated milk (Sherman 1937). Its growth temperature (optimum 37°C, able to grow at 50°C) suggests that it might be an intestinal organism.

The interrelationships between the starter streptococci are shown in Fig. 1. The mesophilic group N streptococci include *S. cremoris*, *S. lactis* and *S. diacetylactis* which, with *S. thermophilus* comprise the homofermentative starters producing L(+) lactic acid. The heterofermentative leuconostocs are included because although they are not streptococci, they are often used in starter cultures for their aroma-producing properties. The homofermentative

Table 2
Some differentiating characteristics of lactic streptococci

	S. lactis	S. diacetylactis	S. cremoris	S. thermophilus
Growth at 39.5°C	+	+	−	+
Growth at 50°C	−	−	−	+
NH_3 from arginine	+	+	−	−
Acid from maltose	+	+	−	−
CO_2 and diacetyl produced from citrate	−	+	−	−
% NaCl inhibiting growth	4.0–6.5	4.0–6.5	2.0–4.0	<2.0
Group N antigen	+	+	+	−

+, Growth or positive reaction; −, no growth or negative reaction.

Table 3
The use of lactic streptococci in fermented milk products

Culture	Product
S. cremoris* S. lactis S. diacetylactis	Most hard cheese, including English territorial (e.g. Cheddar)
S. cremoris S. lactis S. diacetylactis† Leuconostoc spp.†	Mould-ripened cheese (e.g. Stilton) Dutch-type cheese (e.g. Gouda)
S. cremoris S. diacetylactis†† Leuconostoc spp.††	Cottage cheese, cream cheese, cultured cream
S. thermophilus‡	Yoghurt, cooked cheese (e.g. Emmental, Gruyère)

* *S. cremoris* consistently gives cheeses free from off-flavours but pairing with *S. lactis* gives a shorter manufacturing time.
† Produce CO_2 giving desired open texture for mould growth or 'eye' formation in Dutch cheese.
†† Produce characteristic aroma in products.
‡ In associative growth with *Lactobacillus bulgaricus*.

lactic streptococci are differentiated from each other by growth and biochemical characteristics (Table 2). The ability of *S. lactis* and *S. diacetylactis* to grow at 39.5°C, ferment maltose and liberate NH_3 from arginine are the most useful characteristics to differentiate them from *S. cremoris*; *S. diacetylactis* differs from *S. lactis* in being able to ferment citrate. *Streptococcus thermophilus* does not contain group N antigen, and is able to grow at 50°C.

In commercial practice the lactic streptococci are often used as combined cultures as well as singly, depending on the product required (Table 3). For example, cottage cheese is usually made with *S. cremoris* to provide acid and *S. diacetylactis* is added separately to the curd as a cultured cream dressing to provide aroma. *Streptococcus thermophilus* is used with *Lactobacillus bulgaricus* in yoghurt manufacture because more acid is produced in combined culture than would be produced by each organism growing alone, and together they produce relatively larger amounts of acetaldehyde to give the product its characteristic flavour.

The cooked cheese varieties require the use of high curd temperatures ($\sim 45°C$) which only *S. thermophilus* (together with thermophilic lactobacilli) can withstand, whereas Edam or Cheddar cheese can be made with mesophilic group N streptococci since curd temperatures rarely exceed 40°C. However, the choice of starter streptococci for various products involves dairy product technology and as such will be dealt with more fully in another paper in this

Symposium (Cox et al., this volume). The present paper is concerned with the way in which the growth and metabolism of lactic streptococci in milk contribute to the properties of fermented milk products. The lactic streptococci grow in milk using lactose as an energy source and milk proteins as a N source. The efficiency with which these two components are utilized has important consequences on the usefulness of different streptococci as starter cultures.

3. Utilization of Milk Proteins by Lactic Streptococci

The lactic streptococci are nutritionally fastidious and require for growth an exogenous supply of preformed leucine, isoleucine, valine, methionine, arginine, histidine, glutamic acid and, in some cases, phenylalanine, proline and cystine (Law et al. 1976c). Free amino acid concentrations in milk are too low to support the starter growth required in the manufacture of fermented milk products (Lawrence et al. 1976) but the group N streptococci have cell-bound extracellular proteinases which degrade whole casein to peptides (Thomas et al. 1974; Exterkate 1975). These can then be taken up by the cells and hydrolysed intracellularly to their constituent free amino acids by a wide variety of peptidases (e.g. Law et al. 1974; Mou et al. 1975) (Fig. 2). The size exclusion limit for uptake in S. cremoris is reached with peptides containing between four and seven amino acid residues (Law et al. 1976c). Uptake is energy dependent and distinct dipeptide and oligopeptide uptake systems operate (Law 1978). Streptococcus lactis transports intact peptides but can also hydrolyse them

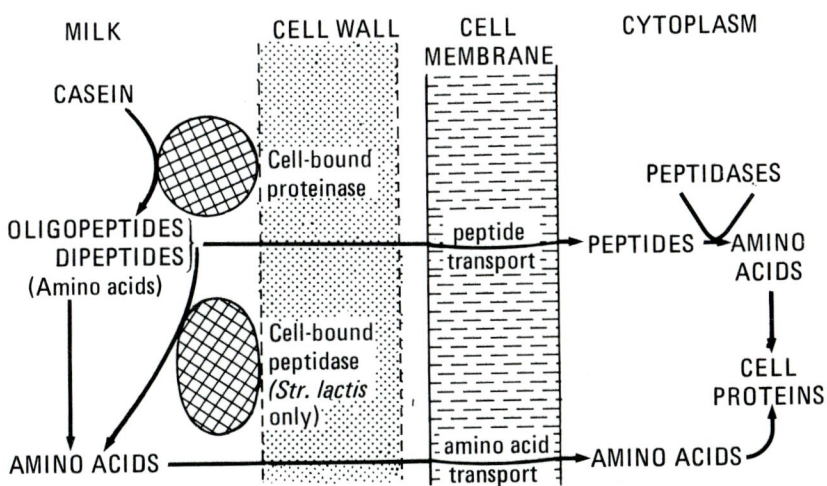

Fig. 2. Role of proteinases, peptidases and peptide transport in the growth of group N streptococci in milk.

extracellularly and is not, therefore, entirely dependent on peptide uptake for growth (Law 1977).

Cultures of group N streptococci often contain small numbers of variants which grow only slowly in milk, but normally in media containing amino acids or peptides. These are designated prt⁻ variants and one such strain has been characterized by the loss of an acid and a neutral proteinase present in the cell wall of the parent strain (Exterkate 1976). Recent evidence that loss of proteolytic activity in *S. lactis* coincides with loss of plasmid DNA of a particular mol. wt suggests that these proteinases may be plasmid linked (Efstathiou & McKay 1976). Starter streptococci also vary widely in their ability to utilize peptides (Law 1977) but the genetic basis for these variations is not known. Since many starter cultures are mixtures of strains and species of lactic streptococci it is probable that interactions occur during associative growth. For example, strains of *S. lactis* would be expected to stimulate the growth of certain strains of *S. cremoris* which grow poorly on peptides, by producing free amino acids as alternative sources of nitrogen.

Growth and acid production by *S. thermophilus* is faster in milk when it is grown with *Lactobacillus bulgaricus*. The lactobacillus has a more active cell-bound proteinase than the streptococcus and presumably aids the growth of *S. thermophilus* by releasing stimulatory peptides. Stimulation by *L. bulgaricus* can be simulated by adding mixtures of valine, histidine, methionine, glutamic acid, leucine and tryptophane (or peptides containing these amino acids) to *S. thermophilus* growing in milk (Shankar & Davies 1977). The mechanism of the stimulation of growth and/or acid production in lactic streptococci by peptides is not clear but the simultaneous supply of several essential amino acids in peptide form probably requires less energy expenditure than the transport of individual amino acids, and is less likely to cause imbalance to the amino acid pool which could interfere with protein synthesis.

4. Lactose Utilization by Lactic Streptococci

The group N streptococci take up lactose into the cell as glucosyl-β-(1, 4)-galactoside-6-phosphate (lactose-P) via a phosphoenolpyruvate-dependent phosphotransferase system (PEP:PTS; McKay et al. 1970; Molskness et al. 1973). A β-D-phosphogalactosidase (β-Pgal) then hydrolyses the lactose-P to glucose and galactose-6-P (Fig. 3). The glucose is metabolized to pyruvate via hexokinase and the normal glycolytic pathway but galactose-P is converted via the fructose epimer, tagatose-P (Bissett & Anderson 1974) to the triose-P stage of glycolysis. The generation of ATP by glycolysis produces reduced pyridine nucleotides ($NADH_2$) which are re-oxidized during the reduction of pyruvate to lactate catalysed by lactate dehydrogenase (LDH). Pyruvate kinase (the enzyme catalysing the conversion of PEP to pyruvate) and LDH are key

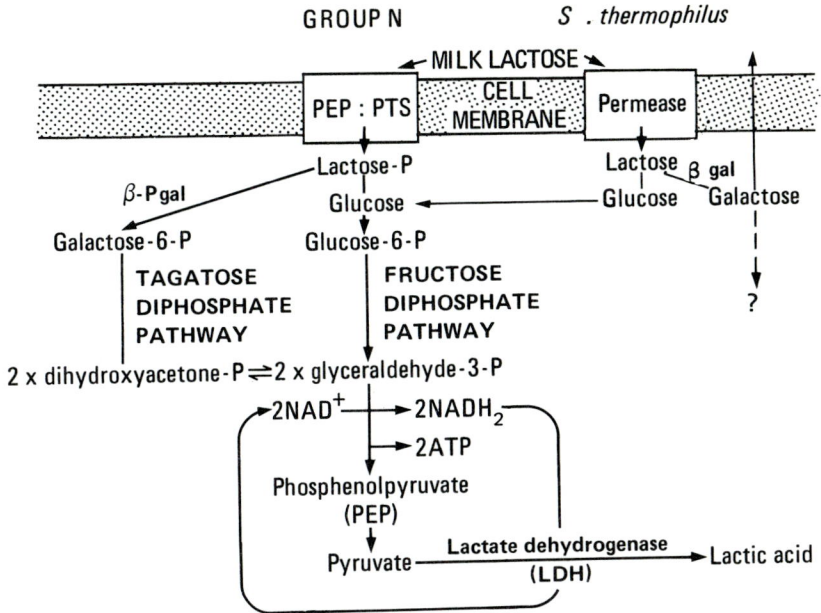

Fig. 3 Major pathways of lactose utilization by the lactic streptococci.

enzymes in the control of lactose metabolism. Both are allosteric; pyruvate kinase has a broad activator specificity (hexose-P and triose-P) but LDH is only activated by fructose diphosphate (FDP) or tagatose diphosphate (TaDP) (Thomas 1976). Further control of lactose utilization is exerted by the dependence of its uptake on PEP.

Group N streptococci produce spontaneous mutants unable to utilize lactose, although they grow normally on glucose. These lac^- mutants cannot synthesize two protein components of the PEP:PTS system and the β-Pgal enzyme (McKay et al. 1970; Cords & McKay 1974). They do not transport thiomethyl-β-D-galactoside (TMG) and grow poorly on lactose or galactose. The relative instability of lactose utilizing ability in a proportion of streptococcal cells has led to the suggestion that, as with the proteinases, the genes coding for the PEP:PTS proteins are carried on plasmid DNA. This is supported by evidence of specific plasmid loss in lac^- mutants of S. lactis (Efstathiou & McKay 1976) and of plasmid DNA tranduction from lac^+ to lac^- strains (McKay et al. 1976). However, in the absence of an established genetic marker for chromosomal DNA in lactic streptococci it is impossible to prove that lac is not genetically linked to a chromosome.

Although lactose utilization by group N streptococci has been studied intensively, much less is known about S. thermophilus. Reddy et al. (1973)

showed that lactose was transported by an energy-dependent, inducible system. There are no reports of a PEP:PTS or β-Pgal in *S. thermophilus* and the lactose appears to be hydrolysed to glucose and galactose by β-gal. The galactose is not metabolized and is excreted into the growth medium (O'Leary & Woychick 1976) but the glucose is converted to lactic acid as in the group N streptococci.

Enzymes for the metabolism of pentoses have been demonstrated in the lactic streptococci (Oram & Reiter 1966; Lees & Jago 1976), but their specific activities are very low (Demko *et al.* 1972) and they probably function to supply pentoses for nucleotide synthesis and NADPH for fatty acid synthesis.

5. Alternative Pathways of Carbohydrate Metabolism and their Effect on the Properties of Short-life Fermented Products

Group N streptococci growing under anaerobic conditions in milk with lactose as their energy source exert tight control over carbohydrate metabolism. The activation of LDH by FDP and TaDP ensures a low intracellular pyruvate concentration so that $NADH_2$ formed during triose-P oxidation exactly balances $NADH_2$ oxidized during pyruvate reduction. In these circumstances, alternative pathways of pyruvate metabolism are not expressed and only lactic acid is produced. Any disturbance of the balance of $NADH_2$ oxidation and intracellular pyruvate concentration may lead to other products being formed from the pyruvate. For example, under aerobic growth conditions (eg. excessive agitation during manufacture), $NADH_2$ oxidase is induced (Bruhn & Collins 1970) and molecular oxygen serves as an alternative to pyruvate as a hydrogen acceptor (Fig. 4). The NAD^+ can then take part in the oxidative decarboxylation of

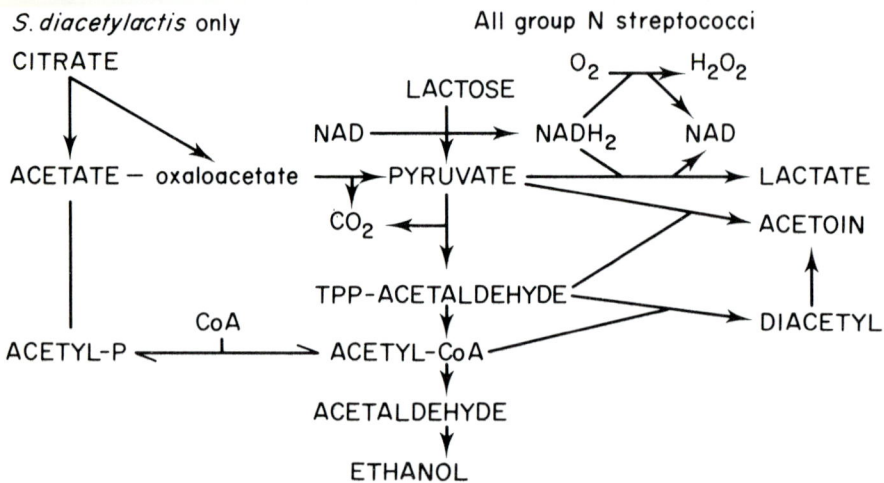

Fig. 4. Alternative pathways of carbohydrate metabolism in lactic streptococci.

pyruvate to acetyl-coenzyme A (acetyl-CoA) and CO_2. Acetyl-CoA undergoes a variety of conversions to acetate, acetaldehyde, ethanol, diacetyl or acetoin (Collins & Bruhn 1970; Anders et al. 1970; Cogan 1976). Pyruvate in excess of that required by LDH for the re-oxidation of $NADH_2$ under anaerobic conditions can be produced by *S. diacetylactis* from the citrate present in milk. The citrate is taken up by an inducible permease system and converted to pyruvate via oxaloacetate with the concomitant production of CO_2 and acetate (Fig. 4). The pyruvate is converted to hydroxyethylthiamine pyrophosphate (TPP-acetaldehyde complex) and to acetyl-CoA (also produced from acetate) (Cogan 1976). The most important reaction in *S. diacetylactis* is that between acetyl-CoA and TPP-acetaldehyde to produce diacetyl (Fig. 4), the compound which imparts the characteristic aroma to cottage cheese and fermented creams.

The products of carbohydrate metabolism generated during the manufacture of fermented milk products play a vital role in determining their properties. Lactic acid lowers the pH of the product and is itself an inhibitor of pathogens such as *Staphylococcus aureus* and spoilage organisms (Daly et al. 1972; Haines & Harman 1973; Babel 1977). The developing acidity also aids clotting by rennet, and texture formation. This also applies to hard cheese where the acid is essential for whey expulsion (syneresis) from curds before pressing. In short-life products the alternative metabolites to lactic acid play a direct role in determining their flavour and quality; diacetyl production from citrate by *S. diacetylactis* in cottage cheese has already been cited. However, the CO_2 produced at the same time (Fig. 4) can cause the cheese curd to float in the whey and spoil its texture, so cottage cheese curd is normally made with *S. cremoris* as an acid producer, then the washed curd is treated with a dressing of cream previously cultured with *S. diacetylactis* and containing diacetyl. Obviously starters which produce diacetyl reductase (catalysing the formation of odourless acetoin from diacetyl) are not used in cottage cheese since low diacetyl concentrations result in a flavourless product.

Acetaldehyde has already been cited as the compound which gives yoghurt its characteristic flavour but although some lactic streptococci can reduce acetyl-CoA to acetaldehyde (Lees & Jago 1976) it is probably produced from the cleavage by *S. thermophilus* of threonine present in milk (Shankar & Davies 1977). Overproduction of acetaldehyde by group N streptococci in cultured buttermilk and sour cream starters leads to a harsh off-flavour (Lindsay et al. 1965).

6. The Influence of Lactic Streptococci on Cheese Ripening

The end products of the fermentation of milk carbohydrates by lactic streptococci play a less obvious role in stored products such as hard cheese, compared with their role in the short-life products. Diacetyl may be involved in Cheddar cheese aroma by acting synergically with compounds produced during maturation (Manning 1974) but overproduction of diacetyl produces an atypical flavour. Carbon dioxide may be beneficial in giving mould-ripened cheese an open texture to aid mould growth or in producing 'eyes' in Dutch cheese, but in Cheddar cheese open texture is a defect. However most of the end products are reduced compounds and, as such, confer a relatively low redox potential on fermented products (Galesloot 1960; Law et al. 1976b). This is particularly important in ripened products such as Cheddar cheese which depends to some extent on reduced compounds such as H_2S and methanethiol (Manning 1974; Manning et al. 1976) for its characteristic flavour. Such compounds would quickly become oxidized if the redox potential of the product were positive. Kristoffersen (1967) has also suggested that a negative redox potential is necessary for balanced flavour development in Cheddar cheese.

During maturation some cheese varieties rely on the activities of particular micro-organisms, such as propionibacteria in Emmenthal cheese or *Penicillium* moulds in Roquefort, to produce characteristic flavour compounds. However in Cheddar cheese, which is economically the most important variety in the Western world, the non-starter microflora has very little effect on the development of typical flavour (Law et al. 1976a). Reiter et al. (1967) showed that Cheddar cheese produced under controlled bacteriological conditions and containing only starter streptococci, developed typical flavour. The enzymes of the starter cells remain active in the cheese after most of the bacteria have died out within the first month of ripening (Law et al. 1974) and they liberate increasing concentrations of free amino acids which can be regarded as flavour precursors (Law et al. 1976b). Also lactic streptococci can release free fatty acids from mono- and diglycerides in milk fat (Stadhouders & Veringa 1973). However, there is no relationship between the concentration of the enzymes of starter streptococci in Cheddar cheese and the rate of typical flavour develop-

Table 4
Off-flavour production by starter streptococci in Cheddar cheese

	Off-flavour	Cause
S. lactis (e.g. NCDO 763) S. lactis subsp. diacetylactis	Fruity	Ethyl butyrate Ethyl hexanoate
S. lactis (e.g. NCDO 604)	Malty	3-Methylbutanal
'Fast' starter strains (most strains of S. lactis but also S. cremoris (NCDO 607)	Bitter	Hydrophobic oligopeptides

Table 5
*Factors affecting growth and acid production
by lactic streptococci in milk*

Intrinsic factors	Loss of proteolytic enzymes
	Loss of lactose metabolism
Extrinsic factors	Lactoperoxidase/thiocyanate/H_2O_2
	Agglutinating antibodies
	Antibiotics
	Bacteriophage infections

ment (Law *et al.* 1976*b*). Indeed, one of the few compounds whose concentration in Cheddar cheese correlates with flavour intensity is methanethiol, and no lactic streptococci so far examined produce this compound in significant quantities (Law & Sharpe 1977). This suggests that the role of the lactic streptococci in cheese maturation is confined to that of supplying the correct environment of pH and redox potential, and to supplying flavour precursors (e.g. free methionine for methanethiol) which are then transformed by non-enzymic reactions to flavour compounds.

In contrast to the apparent lack of positive involvement of lactic streptococci in flavour development in hard cheese, they can produce certain well-defined defects (Table 4). Cheddar cheese made with starter bacteria which continue growing at normal curd cooking temperatures have high starter populations in the finished curd and tend to develop off-flavours on ripening. Lowrie & Lawrence (1972) suggested that such starters contribute excessive amounts of proteolytic enzymes to the cheese and this results in the formation from casein of bitter peptides in concentrations exceeding their flavour threshold. Also, starters which not only reach high populations in curd but also survive in the cheese (mainly strains of *S. lactis*) tend to produce 'fruity' off-flavours associated with the production of ethanol (from acetaldehyde) which then combines with butyric or hexanoic acid (Bills *et al.* 1965) to form fruity flavoured esters (Table 4). Some strains of *S. lactis* contain transaminases and decarboxylases which convert certain amino acids to aldehydes; in particular leucine can be converted to 3-methylbutanal, the primary flavour compound which can impart malty off-flavour to Cheddar cheese (Sheldon *et al.* 1971).

7. The Formation of Toxic Amines in Cheese by Group D Streptococci

The only streptococci known to occur in cheese, other than the deliberately-added starter bacteria, are the group D streptococci which can be present in some Cheddar cheese curds in numbers ranging from 10^4-10^6 g^{-1} (Law *et al.*

1973). Tyramine, histamine and tryptamine are found in many varieties of ripened cheese (Silverman & Kosikowski 1956; Vorgt *et al.* 1974; de Vuyst *et al.* 1976) and although they are normally present in only low concentrations in Cheddar cheese, Dahlberg & Kosikowski (1949) showed that the amines were produced in higher concentrations in cheeses containing *S. faecalis*. Tyramine in cheese may be involved in the onset of migraine attacks in susceptible subjects (Hannington 1967) and also produces hypersensitive crises in patients treated with monoamine oxidase inhibitors (eg. Blackwell & Mabbitt 1965). For this reason proposals to use *S. faecalis* in cheese starters are unlikely to find wide acceptance despite the ability of some strains to acidify milk at a rate compatible with their use in cheesemaking.

8. Causes of Slow Growth of Lactic Streptococci in Milk

Some of the intrinsic factors affecting the rate of growth and acid production by lactic streptococci in milk have already been described (loss of proteinases, loss of lactose metabolism—Table 5). However there are also a number of extrinsic factors affecting starter performance and if they slow down or prevent growth and lactic acid formation they can cause serious product defects and complete loss of a production batch. For example, hard cheese in which insufficient acid has developed tends to have a high moisture content; this results in poor body and texture, rapid proteolysis and off-flavour formation. Also there is a greater possibility of pathogenic bacteria, such as staphylococci, multiplying and/or producing endotoxins if the pH of the cheese is abnormally high (Reiter *et al.* 1964).

Raw milk contains natural inhibitors such as the lactoperoxidase/thiocyanate/hydrogen peroxide system (LP/SCN$^-$/H$_2$O$_2$) and the agglutinins (Table 5). Hydrogen peroxide itself can reach inhibitory levels in aerated starter cultures (Anders *et al.* 1970) but even at non-inhibitory levels it can oxidise milk SCN$^-$, in a reaction catalysed by LP, to an unknown inhibitor (Reiter 1978). However, not all starter strains are sensitive to this system and heating milk to $> 70°$C for 20 min destroys it.

The agglutinins are antibodies produced in the cow and transferred to colostrum and milk. They are inactivated by homogenization, heat treatments given to the milks used for starter manufacture, and by the formation of a rennet coagulum. However they can agglutinate and thus precipitate starters in cottage cheese, which is made without rennet (Emmons *et al.* 1966) and cause slow acid development, sludge and texture defects.

Occasionally, antibiotics used to treat bovine mastitis may be carried over into manufacturing milk supplies and cause slow growth of starter streptococci. Some strains produce antibiotics themselves and inhibit other starters in mixed cultures, causing strain imbalance (Collins 1961). The best known antibiotic produced by *S. lactis* is the polypeptide nisin, which has a very broad spectrum of activity—it inhibits not only other group N streptococci but also groups A, B, E, F, G, K and M as well as staphylococci and clostridia (Mattick & Hirsch 1947; Hurst, this volume).

Another common cause of starter slowness or even failure in milk product manufacture is bacteriophage infection. The lactic acid bacteria phages are all of the DNA type with either prolate polyhedral or isometric heads and usually non-contractile tails. Some phages are species specific but a recent study (Chopin *et al.* 1976) suggests that there are large groups of non-species specific phages. Knowledge of phage relationships between strains of lactic streptococci is necessary so that starter cultures can be varied by rotation from day to day in cheese factories to avoid the build up of any particular phage.

Phage infection in lactic streptococci, as in other bacteria, involves adsorption, digestion of the cell wall, and injection of phage DNA. The nature of the phage receptors on the bacterial cells is uncertain (Reiter 1973) but it is well established that Ca^{2+} is required for adsorption and the use of Ca^{2+}-free media for starter propagation is widespread, particularly in the USA. Latent period and burst size during infection vary according to the temperature (e.g. 2-113 particles, released in 23-56 min at $30°C$; up to 139 particles released in 16-44 min at $37°C$; Lawrence *et al.* 1976). Phage multiplication during cheesemaking is limited by physical factors such as the size of the zone through which expelled whey diffuses from curd particles, carrying the phage with it to infect other cells. Not all phage infections of lactic streptococci result in lysis and many examples of lysogenic strains carrying temperate phages have now been reported (e.g. Lowrie 1974; Reiter & Kirikova 1976). Recent evidence suggests that most mixed strain cheese starters contain lysogenic strains which act as a reservoir of phages that can be lytic for other strains in the mixture (Huggins 1977).

9. Factors Stimulating the Growth of Lactic Streptococci

The stimulatory effect on lactic streptococci of adding yeast extract to milk indicates that milk is not a 'complete' medium. This effect has been attributed to the presence in yeast extract of peroxidase, free amino acids, peptides, purines and pyrimidines and inorganic constituents (Lawrence *et al.* 1976). Group N streptococci require or are stimulated by a number of vitamins, notably niacin, pantothenic acid and biotin (Reiter & Oram 1962) although carbon dioxide had a sparing effect on biotin. Carbon dioxide is essential for the growth of lactic streptococci and is involved in the biosynthesis of aspartic acid and fatty acids via biotin-mediated fixation (Reiter & Oram 1962). This may explain the stimulatory effect of carbon dioxide in some media.

10. References

ANDERS, R. F., HOGG, D. M. & JAGO, G. R. 1970 Formation of hydrogen peroxide by Group N streptococci and its effect on their metabolism. *Applied Microbiology* **19**, 608–612.

BABEL, F. J. 1977 Antibiosis by lactic culture bacteria. *Journal of Dairy Science* **60**, 815–821.

BILLS, D. D., MORGAN, M. E., LIBBEY, L. M. & DAY, E. A. 1965 Identification of compounds responsible for fruity flavour defect of experimental Cheddar cheeses. *Journal of Dairy Science* **48**, 1168–1173.

BISSETT, D. L. & ANDERSON, R. L. 1974 Lactose and D-galactose metabolism in Group N streptococci: Presence of enzymes for both D-galactose 1-phosphate and D-tagatose 6-phosphate pathways. *Journal of Bacteriology* **117**, 318–320.

BLACKWELL, B. & MABBITT, L. A. 1965 Tyramine in cheese related to hypertensive crises after monoamine-oxidase inhibition. *Lancet* **1**, 938–940.

BRIGGS, C. A. E. 1952 A note on the serological classification of *Streptococcus diacetilactis* (Matuszewski et al.) *Journal of Dairy Research* **19**, 167–178.

BRIGGS, C. A. E. & NEWLAND, L. G. M. 1952 The serological classification of *Streptococcus cremoris*. *Journal of Dairy Research* **19**, 160–166.

BRONN, W. K. 1976 *Die Branntwein Wirschaft* **12**, 216.

BRUHN, J. C. & COLLINS, E. B. 1970 Reduced nicotinamide adenine dinucleotide oxidase of *Streptococcus diacetylactis*. *Journal of Dairy Science* **53**, 857–860.

BUCHANAN, R. E. & GIBBONS, N. E. (eds) 1974 *Bergey's Manual of Determinative Bacteriology*, 8th edn. Baltimore: Williams & Wilkins.

CHOPIN, M.–C., CHOPIN, A. & ROUX, C. 1976 Definition of bacteriophage groups according to their lytic action on mesophilic lactic streptococci. *Applied and Environmental Microbiology* **32**, 741–746.

COGAN, T. M. 1976 The utilization of citrate by lactic acid bacteria in milk and cheese. *Dairy Industries International* **41**, 12–16.

COLLINS, E. B. 1961 Domination among strains of lactic streptococci with attention to antibiotic production. *Applied Microbiology* **9**, 200–205.

COLLINS, E. B. & BRUHN, J. C. 1970 Roles of acetate and pyruvate in the metabolism of *Streptococcus diacetylactis*. *Journal of Bacteriology* **103**, 541–546.

CORDS, B. R., & MCKAY, L. L. 1974 Characterization of lactose-fermenting revertants from lactose-negative *Streptococcus lactis* C2 mutants. *Journal of Bacteriology* **119**, 830–839.

DAHLBERG, A. C. & KOSIKOWSKI, F. V. 1948 The relationship of the amounts of tyramine and the numbers of *Streptococcus faecalis* on the intensity of flavour in American Cheddar cheese. *Journal of Dairy Science* **31**, 305–314.

DALY, C., SANDINE, W. E. & ELLIKER, P. R. 1972 Interactions of food starter cultures and food-borne pathogens: *Streptococcus diacetylactis* versus food pathogens. *Journal of Milk and Food Technology* **35**, 349–357.

DE VUYST, A., VERVACK, W. & FOULON, M. 1976 Detection of non-volatile amines in several cheese types. *Le Lait* **56**, 414–422.

DEMKO, G. M., BLANTON, S. J. B. & BENOIT, R. E. 1972 Heterofermentative carbohydrate metabolism of lactose-impaired mutants of *Streptococcus lactis*. *Journal of Bacteriology* **112**, 1335–1345.

EFSTATHIOU, J. D. & MCKAY, L. L. 1976 Plasmids in *Streptococcus lactis*: Evidence that lactose metabolism and proteinase activity are plasmid linked. *Applied and Environmental Microbiology* **32**, 38–44.

EMMONS, D. B., ELLIOT, J. A. & BECKETT, D. C. 1963 Agglutination of starter bacteria, sludge formation and slow acid development in Cottage cheese manufacture. *Journal of Dairy Science* **46**, 600.

EXTERKATE, F. A. 1975 An introductory study of the proteolytic system of *Streptococcus cremoris* strain HP. *Netherlands Milk and Dairy Journal* **29**, 303–318.

EXTERKATE, F. A. 1976 The proteolytic system of a slow lactic acid-producing variant of *Streptococcus cremoris* HP. *Netherlands Milk and Dairy Journal* **30**, 3–8.

GALESLOOT, T. 1960 Effect of oxidising salts upon the oxidation- reduction potential of milk inoculated with starter. *Netherlands Milk and Dairy Journal* **14**, 176-214.
HAINES, W. C. & HARMON, L. G. 1973 Effect of variations in conditions of incubation upon inhibition of *Staphylococcus aureus* by *Pediococcus cerevisiae* and *Streptococcus lactis*. *Applied Microbiology* **25**, 169-172.
HANNINGTON, E. 1967 Preliminary report on tyramine headache. *British Medical Journal* **2**, 550-551.
HUGGINS, A. R. 1977 Incidence and properties of temperate bacteriophages induced from lactic streptococci. *Applied and Environmental Microbiology* **33**, 184-191.
KRISTOFFERSEN, T. 1967 Interrelationships of flavour and chemical changes in cheese. *Journal of Dairy Science* **50**, 279-284.
LANCEFIELD, R. C. 1933 A serological differentiation of human and other groups of haemolytic streptococci. *Journal of Experimental Medicine* **57**, 571-595.
LAW, B. A. 1977 Dipeptide utilization by starter streptococci. *Journal of Dairy Research* **44**, 309-317.
LAW, B. A. 1978 Peptide utilization by Group N streptococci. *Journal of General Microbiology* **105**, 113-118.
LAW, B. A. & SHARPE, M. E. 1977 The formation of methanethiol by bacteria isolated from Cheddar cheese. *Journal of Dairy Research* (in press).
LAW, B. A., CASTANON, M. & SHARPE, M. E. 1976a The effect of non-starter bacteria on the chemical composition and the flavour of Cheddar cheese. *Journal of Dairy Research* **43**, 117-125.
LAW, B. A., CASTANON, M. & SHARPE, M. E. 1976b The contribution of starter streptococci to flavour development in Cheddar cheese. *Journal of Dairy Research* **43**, 301-311.
LAW, B. A., SEZGIN, E. & SHARPE, M. E. 1976c Amino acid nutrition of some commercial cheese starters in relation to their growth in peptone-supplemented whey media. *Journal of Dairy Research* **43**, 291-300.
LAW, B. A., SHARPE, M. E., MABBITT, L. A. & COLE, C. B. 1973 *Sampling & Microbiological Monitoring of Environments*, eds Board, R. G. & Lovelock, D. W. London & New York: Academic Press.
LAW, B. A., SHARPE, M. E. & REITER, B. 1974 The release of intracellular dipeptidase from starter streptococci during Cheddar cheese ripening. *Journal of Dairy Research* **41**, 137-146.
LAWRENCE, R. C., THOMAS, T. D. & TERZAGHI, B. E. 1976 Reviews in the Progress of Dairy Science: Cheese starters. *Journal of Dairy Research* **43**, 141-193.
LEES, G. J. & JAGO, G. R. 1976 Acetaldehyde: an intermediate in the formation of ethanol from glucose in lactic acid bacteria. *Journal of Dairy Research* **43**, 63-73.
LINDSAY, R. C., DAY, E. A. & SANDINE, W. E. 1965 Green flavour defect in lactic starter cultures. *Journal of Dairy Science* **48**, 863-869.
LISTER, J. (1878). On the lactic fermentation and its bearing on pathology. *Transactions of the Pathology Society* **29**, 425-467.
LOWRIE, R. J. 1974 Lysogenic strains of Group N streptococci. *Applied Microbiology* **27**, 210-217.
LOWRIE, R. J. & LAWRENCE, R. C. 1972 Cheddar cheese flavour. IV. A new hypothesis to account for the development of bitterness. *New Zealand Journal of Dairy Science and Technology* **7**, 51-53.
MCKAY, L. L., MILLER, A., SANDINE, W. E. & ELLIKER, P. R. 1970 Mechanisms of lactose utilization by lactic acid streptococci: enzymatic and genetic analyses. *Journal of Bacteriology* **102**, 804-809.
MCKAY, L. L., BALDWIN, K. A. & EFSTATHIOU, J. D. 1976 Transductional evidence for plasmid linkage of lactose metabolism in *Streptococcus lactis* C2. *Applied and Environmental Microbiology* **32**, 45-52.
MANNING, D. J. 1974 Sulphur compounds in relation to Cheddar cheese flavour. *Journal of Dairy Research* **41**, 81-87.

MANNING, D. J., CHAPMAN, H. R. & HOSKING, Z. D. 1976 The production of sulphur compounds in Cheddar cheese and their significance in flavour development. *Journal of Dairy Research* **43**, 313–320.

MATTICK, A. T. R. & HIRSCH, A. 1947 Further observations on an inhibitory substance (nisin) from lactic streptococci. *Lancet* **253**, 5.

MATUSZEWSKI, T. E., PŸENOWSKI, E. & SUPINSKA, J. 1936 *Streptococcus diacetilactis*, a new species. *Polish Agriculture and Forestry Annals* **36**, 1–28.

MOLSKNESS, T. A., LEE, D. R., SANDINE, W. E. & ELLIKER, P. R. 1973 β-D-phosphogalactoside galactohydrolase of lactic streptococci. *Applied Microbiology* **25**, 373–380.

MOU, L., SULLIVAN, J. J. & JAGO, G. R. 1975 Peptidase activities in Group N streptococci. *Journal of Dairy Research* **42**, 147–155.

O'LEARY, V. S. & WOYCHIK, J. H. 1976 Utilization of lactose, glucose and galactose by a mixed culture of *Streptococcus thermophilus* and *Lactobacillus bulgaricus* in milk treated with lactase enzyme. *Applied and Environmental Microbiology* **32**, 89–94.

ORAM, J. D. & REITER, B. 1966 Inhibition of streptococci by lactoperoxidase, thiocyanate and hydrogen peroxide. *Biochemical Journal* **100**, 373–381.

ORLA-JENSEN, S. 1919 *The Lactic Acid Bacteria*, 2nd edn., 1942. Copenhagen.

REDDY, M. S., WILLIAMS, F. D. & REINBOLD, G. W. (1973). Lactose transport in *Streptococcus thermophilus*. *Journal of Dairy Science* **56**, 634–635

REITER, B. 1973 Some thoughts on cheese starters. *Journal of the Society of Dairy Technology* **26**, 3–15.

REITER, B. 1978 Antimicrobial systems in milk. *Journal of Dairy Research* **45**, 131–147.

REITER, B. & KIRIKOVA, M. 1976 The isolation of a lysogenic strain from a multiple strain starter culture. *Journal of the Society of Dairy Technology* **29**, 221–225.

REITER, B., FEWINS, G. F., FRYER, T. F. & SHARPE, M. E. 1964 Factors affecting the multiplication and survival of coagulase-positive staphylococci in Cheddar cheese. *Journal of Dairy Research* **31**, 261–272.

REITER, B., FRYER, T. F., PICKERING, A., CHAPMAN, H. R., LAWRENCE, R. C. & SHARPE, M. E. 1967 The effect of microbial flora on the flavour and free fatty acid composition of Cheddar cheese. *Journal of Dairy Research* **34**, 257–272.

SANDINE, W. E., RADICH, P. C. & ELLIKER, P. R. 1972 Ecology of the lactic streptococci. A review. *Journal of Milk and Food Technology* **35**, 176–184.

SHANKAR, P. A. & DAVIES, F. L. 1977 *Journal of Applied Bacteriology* **43**, Abstract.

SHATTOCK, P. M. F. & MATTICK, A. T. R. 1943 The serological grouping of *Streptococcus lactis* (Group N) and its relationship to *Streptococcus faecalis*. *Journal of Hygiene* **43**, 173–188.

SHELDON, R. M., LINDSAY, R. C., LIBBEY, L. M. & MORGAN, M. E. 1971 Chemical nature of malty flavour and aroma produced by *Streptococcus lactis* var. *maltigenes*. *Applied Microbiology* **22**, 263–266.

SHERMAN, J. M. 1937 The streptococci. *Bacteriological Reviews* **1**, 3–97.

SHERMAN, J. M. 1938 The enterococci and related streptococci. *Journal of Bacteriology* **35**, 81–93.

SILVERMAN, G. J. & KOSIKOWSKI, F. V. 1956 Amines in Cheddar cheese. *Journal of Dairy Science* **39**, 1134–1146.

STADHOUDERS, J. & VERINGA, H. A. 1973 Fat hydrolysis by lactic acid bacteria. *Netherlands Milk and Dairy Journal* **27**, 77–91.

THOMAS, T. D. 1976 Regulation of lactose fermentation in group N streptococci. *Applied and Environmental Microbiology* **32**, 474–478.

THOMAS, T. D., JARVIS, B. D. W. & SKIPPER, N. A. 1974 Localization of proteinases near the cell surface of *Streptococcus lactis*. *Journal of Bacteriology* **118**, 329–333.

VOIGT, M. N., EITERMILLER, R. R., KOEHLER, P. E. & HAMDY, M. K. 1974 Tyramine, histamine and tryptamine content of cheese. *Journal of Milk and Food Technology* **37**, 377–381.

Starters: Purpose, Production and Problems

W. A. Cox and G. Stanley

*Unigate Technical Centre, Bradford-on-Avon,
Wiltshire, England*

AND

J. E. Lewis

*Unigate Foods Ltd, Sturminster Marshall,
Wimborne, Dorset, England*

CONTENTS

1. Introduction	279
2. The composition and purpose of starter cultures	279
3. Production of starters in the dairy industry	281
(a) Conventional handling methods	282
(b) Concentrated starter cultures	284
4. Problems associated with production and use of starters	291
5. References	293

1. Introduction

THE PRODUCTION of milk, and from it such products as cheese, yoghurt, buttermilk, butter and milk powder, is of worldwide economic importance and technological developments have led to the large scale mechanization of many parts of the dairy industry.

The importance of cheese production can be ascertained from the quantity of milk used for its manufacture. In the period April 1975 to March 1976, 7875 million litres of milk were sold in the UK by the Milk Marketing Board; 2228 million litres were used for cheese production—approximately 70% being Cheddar (*Anon.* 1976b). Cheesemaking has grown from traditional origins, and over the years has become acknowledged as a rapidly developing scientifically based industry. However, technology has not yet entirely replaced the artistic skills of the cheesemaker.

2. The Composition and Purpose of Starter Cultures

For the manufacture of fermented milk products, starter cultures are grown in heat-treated milk or milk-based media. The types of cultures in use for hard cheese and cottage cheese are the mesophilic group N streptococci and *Leuconostoc* species; for yoghurt the thermophilic starters *Lactobacillus*

Table 1
Composition of starter cultures used for hard cheese

Type	*Streptococcus* species	Method of use and characteristics	Cheese variety/location
Single strain starters	*S. cremoris* *S. lactis* *S. lactis* var. *diacetylactis*	Single or paired	Cheddar in New Zealand and Australia and selected UK creameries
Multiple strain starters	*S. cremoris* *S. lactis* *S. lactis* var. *diacetylactis* *Leuconostoc* species	Defined mixtures of two or more strains (may be used in pairs)	Hard cheese varieties in United States and certain UK creameries
Mixed strain starters	*S. cremoris* *S. lactis* *S. lactis* var. *diacetylactis* *Leuconostoc* species	Unknown proportions of different strains which can vary on subculture (may be used in pairs)	Most hard cheese varieties in UK and Europe

bulgaricus and *Streptococcus thermophilus*, and for Continental cheeses both the group N streptococci and thermophilic lactobacilli strains are used. In addition, certain penicillium strains (e.g. *Penicillium roqueforti*) are included in mould-ripened cheese (e.g. Stilton), and lactose-fermenting yeasts are used in fermented milk of the koumiss and kefir variety. The bacteriological classification and taxonomy of starters has been discussed by Law & Sharpe (this volume). However, for technological purposes this classification must be extended, particularly with respect to cheese starters. Lawrence *et al.* (1976) suggest a division into single-strain, multiple-strain and mixed-strain starters (see Table 1). Further division of the latter two groups into those starters composed of strains of one species, and those of more than one species, can be contemplated. It should be noted that cottage cheese cultures, although subject to a similar classification scheme do not normally contain *S. lactis* var. *diacetylactis* since these metabolize citrate to give CO_2 which has undesirable consequences in cottage cheese curd production. Yoghurt cultures are usually composed of a mixture of one strain of *S. thermophilus* and one of *Lact. bulgaricus*. However, with the increasing probability of culture inactivation by bacteriophage, the use of mixed or multiple strains may be advantageous.

Starter cultures are added to the milk medium at a predetermined rate so as to allow the conventional stages of manufacture to be followed. Thus, for large scale production of a commodity such as cheese, a bank of suitable starters is required that will consistently produce, either directly or indirectly, the required product. The essential characteristics required of a starter have been described by Cox (1977): the culture should contain the desirable micro-organisms capable of forming lactic acid in the milk medium at a suitable rate; it should effect the desired changes, either directly or indirectly, during manufacture and ripening or maturation of the cheese; and it should be free from contamination by yeasts, moulds, extraneous bacterial species and bacteriophage (except for bacteriophage-carrying cultures). Also it is desirable that the culture be resistant to inhibitory substances (naturally occurring or otherwise) that may be present in the milk medium.

At a molecular level little is known about the interactions of the metabolically produced lactic acid in milk. Foster *et al.* (1958) list the important functions of the acid as: promoting curd formation by rennet, causing the curd to shrink and so promote drainage of whey, reducing the growth of undesirable micro-organisms during cheesemaking and ripening, affecting the elasticity of the finished curd so promoting its fusion into a solid mass, and affecting the nature and extent of enzymic changes during ripening, thus helping to determine the characteristics of the cheese.

3. Production of Starters in the Dairy Industry

Methods for the production of cheese starters, as applicable to the UK, will be discussed, noting that these will vary with the individual creamery. Historically, cheesemakers relied on the naturally occurring streptococcal flora of the milk to act as starters. Often a portion of the whey from the previous day's cheese would be reserved for addition to the vat milk (Cheke 1959). The understanding of lactose fermentation owes much to the detailed studies of Orla-Jensen (1919). Upon this sound bacteriological basis it has been possible to develop starter technology to the level of the present day: a large modern creamery handling up to 700,000 l of milk a day requires the controlled preparation of about 18,000 l of starter.

(a) *Conventional handling methods*

Conventional techniques require the creamery laboratory to produce starter from a master culture (freeze-dried or liquid culture). Several subcultures must be made to ensure the complete revival of the starter, and to produce sufficient inoculum to add to the cheese vat—in hard cheese production this inoculum is 1-2% (v/v). A suitable selection of cultures can be maintained either by regular subculture or by storing at deep-freeze or refrigeration temperatures and drawing on them as required (Lewis 1977).

An alternative conventional method which has been widely adopted within the UK is the Lewis protected system (Lewis 1956; Cox & Lewis 1972). This is based on the centralized production of frozen starters and their handling by aseptic techniques from the master culture stage through to, and including, the bulk starter stage. The 'mother' culture, usually of freeze-dried origin, is maintained in re-usable polythene bottles (114 g capacity) by regular transfer using a sterile double-ended needle (stainless steel, 12-14 gauge) and injecting the inoculum through a disposable rubber seal containing sodium hypochlorite solution (250 mg l^{-1} of available chlorine). This 'mother' culture is used to inoculate 'working' cultures in 850 g capacity polythene bottles, and from these are produced further batches of 'working' cultures. It is these cultures that are added as an inoculum for bulk-starter preparation.

Antibiotic-free milk powder is used for the culture medium throughout starter preparation, and is normally reconstituted to 12% (w/v). This medium is heat-treated inside the polythene bottles. Cheese cultures are incubated at 22°C for 16-18 h, and yoghurt cultures at 37°C for the same time period. This protected system can be used in individual creameries or at a centralized site. In the latter case the freezing of the inoculated but unincubated working cultures at -20°C in the 850 g bottles, is a convenient method of storing and distributing the cultures to individual creameries. Insulated containers are used for such distribution. Frozen cultures maintain their required characteristics for 3-6 months. The preparation of the working cultures for bulk starter inoculation is effected by thawing the cultures for 2 h at 30°C followed by incubation at

22°C for 16–18 h. These clotted cultures are inoculated, via a sterile double-ended needle and rubber seal, at 0.1–0.2% (v/v) into the bulk-starter vessel.

Bulk-starter vessels can be of various types. Those offering most protection are the pressurized vessels: the medium is heat treated inside the sealed vessel without air being expelled or drawn in during the heating and cooling process. Such vessels can be of 40–2000 l capacity. Other vessels in use include those that are pre-sterilized, and fed with medium which has been heat treated and cooled in separate units. The air is expelled during the filling process through a water seal. Alternatively, the medium can be heat treated inside non-pressurized vessels. During heating and cooling, the air is expelled and drawn in through bacteriological filters. These last two systems rely on adequate sterile conditions being maintained throughout the stages of bulk starter preparation and this is not always achieved. The medium for bulk-starters is usually 12% (w/v) reconstituted antibiotic-free skim milk.

It is essential that effective quality control of the cultures be carried out at all stages of culture preparation. A test designed to simulate hard cheese production (vitality test) based on the work of Whitehead & Cox (1932) and Pulay & Zsmko (1969) has been described (Cox & Lewis 1972). This, together with a simple activity test (Anderson & Meanwell 1942) determines the acid-producing activity of the starter. Other control tests used are for the detection of yeasts and moulds, *coli-aerogenes* bacterial strains, and bacteriophage. This latter assessment is either a simple tube test (Anderson & Meanwell 1942) or a double-layer plaque assay (Meanwell & Thompson 1959; Terzaghi & Sandine 1975). Differential media for the various components of a mixed or multiple starter can be employed in laboratory culture control. Of practical importance is the identification of strains capable of metabolizing citrate to CO_2 and diacetyl, and methods for this have been described by Sandine *et al.* (1962; 1972) and Pack *et al.* (1968). Further, an agar medium for distinguishing between *S. lactis*, *S. cremoris* and *S. lactis* var. *diacetylactis* is available (Reddy *et al.* 1972).

The use of heat-treated 12% (w/v) antibiotic-free reconstituted skim milk powder in a protected system has been widely adopted in the UK for bulk-starter preparation and this should reduce bacteriophage problems to a minimum. However, when such a system is not available the bulk-starter can be partially protected from phage by the chemical sequestering of the calcium in the medium (Doull & Meanwell 1953), since phages for lactic streptococci and lactobacilli generally require calcium for multiplication; phosphate-treated milk has been developed for this purpose. (Reiter 1956; Babel 1958; Zoltola & Marth 1966). Its use has not been widely adopted in the UK, partly because of the success of the protected system and cost considerations. Several other factors should also be considered: (1) Sozzi (1972) has demonstrated the ability of certain starter phages to lyse cultures in the absence of available calcium; (2) the absence of available calcium normally only prevents phage multiplication, not its inactivation; (3) trials of the medium have shown that it does not always support satisfactory growth (Czulak & Keogh 1957; Tybeck 1959).

(b) *Concentrated starter cultures*

A proportion of the UK creameries are producing cheese using commercially available concentrated cultures. The production of such cultures is an extension of the centralized operation scheme offering freeze-dried liquid and frozen cultures. The availability and use of these has been discussed (Wigley 1977; Tofte Jesperson 1977). Concentrated cultures are used for the inoculation of bulk-starter vats, thus alleviating the requirement for incubation and testing of working cultures (as is the case with the Lewis system). This responsibility, therefore, is removed from the creamery laboratory. There is also interest in the use of these starters for inoculation directly into the cheese vat (DV1) so eliminating the need for bulk-starter preparation (Baumann & Reinbold 1964, 1966; Accolas & Auclair 1967; Lattey 1968; Rymaszewski *et al.* 1971; Wigley 1977). A further advantage of the concentrate system is that its preparation can be carefully monitored and a uniform, contaminant-free product can be prepared. There are disadvantages to the system: all concentrates do not respond well to storage in a conventional deep-freeze at $-20°C$ (Moss & Speck 1963; Birkjaer *et al.* 1974), so liquid nitrogen facilities, low temperature deep-freezers and solid carbon dioxide are usually made available for storage and transport of cultures (Baumann & Reinbold 1964; Birkjaer *et al.* 1974). The cost of this equipment can be high, as can the cost of the cultures. A further disadvantage is that not all cultures can be concentrated successfully. The production details of the commercially concentrated cultures are not generally available although they have been the source of several patent specifications in this country (*Anon.* 1968*a,b*, 1970, 1972, 1976*a*). Literature reviews on starter concentrates have been published (Lloyd 1971; Gilliland & Speck 1974; Tofte Jesperson 1974; Lawrence *et al.* 1976).

Growth of starters in a milk medium is arrested by metabolically produced lactic acid at *ca.* 10^9 colony-forming units (cfu) ml^{-1} (Sellars 1967). However, with the pH controlled at 6.0–6.5 (Harvey 1965; Cogan *et al.* 1971) and with a suitable nutrient medium such as tryptone, yeast extract and lactose-based broths, starter cell levels of 10^{10} cfu ml^{-1} can be achieved (Lamprech & Foster 1963; Bergère & Hermier 1968; Peebles *et al.* 1969). The addition of such supplements as yeast extract was also shown to reduce the division time of the starter (Pont & Holloway 1968). From our own experience (Stanley 1977) broth-based media have been replaced by enzymic digests of a low lactose milk medium, fortified where necessary, with suitable preparations such as yeast extract. Aseptically prepared cultures were grown in this medium for 16–18 h at 22–23°C and pH 6.5, cooled, concentrated and resuspended automatically to a level of *ca.* 10^{11} cells ml^{-1} using a Westfalia separator. The cell concentrates were aseptically packaged into syringes and stored in liquid nitrogen until required. The complete system must be capable of being adequately cleaned and steam sterilized and this has been achieved with a system of strategically placed steam and Zephyr valves (to control direction of flow). Initially cell yields at the end of fermentation were 10^{10} total cells ml^{-1} for certain single strain cultures (*S. lactis* ML$_8$, *S. cremoris* E$_8$) but others exhibited yields of $\leqslant 10^9$ total cells ml^{-1} (*S. cremoris* Am$_1$, Am$_2$, SK$_{11}$). Nutrient deficiency was found not to be responsible for this, and the inhibition could be effectively removed by the addition of low levels of catalase. In the presence of catalase the cell population of *S. lactis* ML$_8$ reached 1.80×10^{10} cells ml^{-1} and populations of six strains of *S. cremoris* (E$_8$, Sk$_{11}$, Am$_1$, Am$_2$, P$_2$, 166) ranged from 10^{10} to 1.70×10^{10} ml^{-1}. It was assumed, therefore, that hydrogen peroxide, produced metabolically from entrained oxygen (caused by the gentle agitation required for accurate pH control), was responsible for growth inhibition (Anders *et al.* 1970; Gilliland & Speck 1969).

The methodology for assessing accurately the numbers of chain-producing lactic streptococci presents a problem, as does the relating of these numbers to the acid-producing activities of the cultures. Martley (1972) blended cultures to reduce the chain length to a more uniform size; however, this technique can, under certain growth conditions, cause cell damage (Stanley 1974). Accolas & Auclair (1970) graphically correlated various inoculation levels of a culture with the acidity developed in an acid-producing activity test. This technique has been used, together with haemocytometer measurements and plate counts, to provide information on starter acid-producing activity (Stanley 1977). Thus 'activity curves' similar to those of Accolas & Auclair (1970) were constructed to compare the activity of fermented, concentrated and conventional cultures. Such comparisons can be based on: (1) total cells ml^{-1}; (2) total chains ml^{-1}; (3) cfu ml^{-1}; (4) cfu × average chain length ml^{-1} (to compensate for chain length but using a viable count basis).

Using this method it has been shown that for certain single strains (e.g. *S. cremoris* Am$_2$) there was a gross loss of acid-producing activity after fermentation in a high-lactose-based medium (PDSM/Whey) and a further loss on concentration (Fig. 1). The former loss of activity was thought to be due to continued metabolism of excess lactose after stationary phase had been reached (Thomas & Batt 1968, 1969), since the same culture grown in a low lactose medium (MDSM) did not exhibit the same loss. Other cultures such as *S. cremoris* E$_8$ only lost acid-producing activity after concentration and frozen storage (Fig. 2). This was found to be due to the partial inactivation of the culture during concentration and automatic resuspension: the long chains of certain cultures (e.g. *S. cremoris* Am$_1$ and Am$_2$) were broken down (Table 2) and the percentage recovery on agar plates reduced (Table 3). The use of

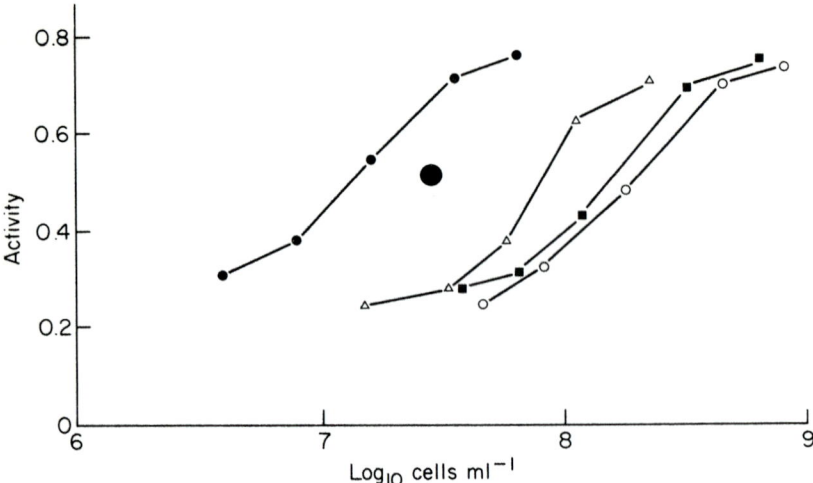

Fig. 1. Activity of cultures of *S. cremoris* Am_2 at different cell concentrations. ●, Conventional culture; ○, fresh concentrate; △, fermenter culture (PDSM/whey); ⬤, fermenter culture (MDSM); ■, frozen and stored concentrate. Activity is expressed as per cent titratable acidity, determined by incubation of the appropriate culture for 6 h at 30°C in pasteurized 10% reconstituted antibiotic-free skim milk powder, followed by titration of 10 ml portions of culture against 0.111 mol l^{-1} sodium hydroxide using 1 ml of 0.5% phenolphthalein as indicator.

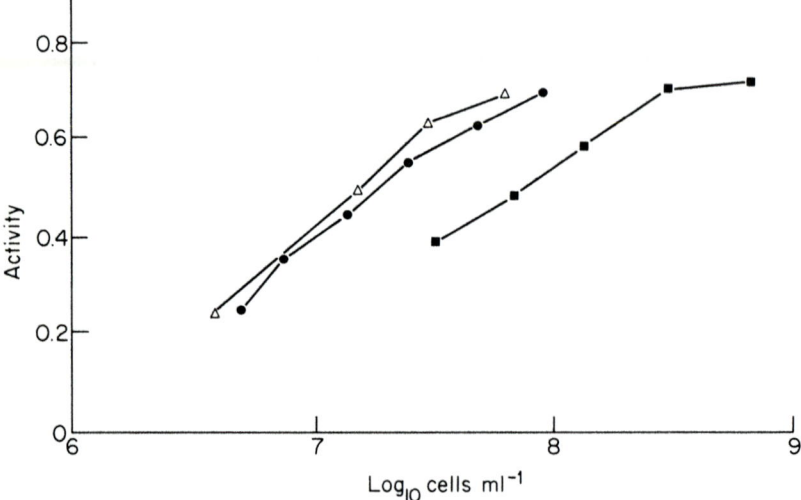

Fig. 2. Activity of cultures of *S. cremoris* E_8 at different cell concentrations. ●, Conventional culture; △, fermenter culture; ■, frozen and stored concentrate.

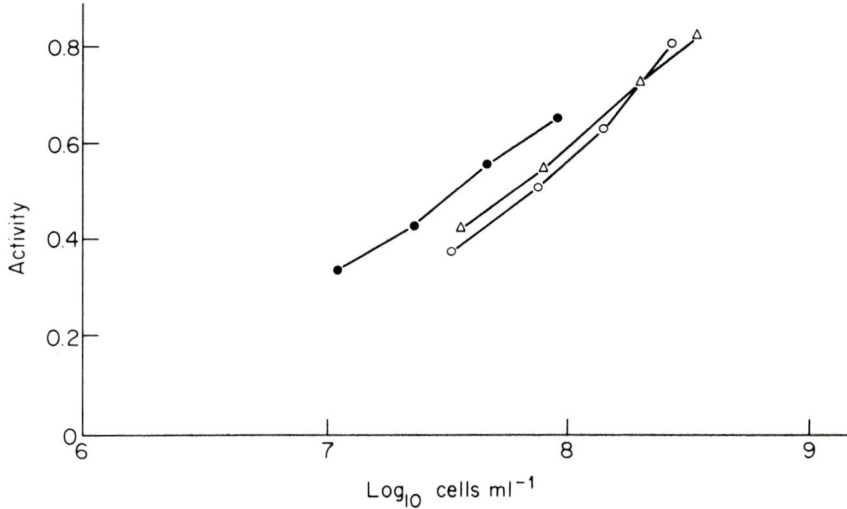

Fig. 3. Activity of cultures of mixed strain 1051 at different cell concentrations. ●, Conventional culture; △, fermenter culture; ○, fresh concentrate.

selected mixed strain cultures has partially overcome this problem in that they appear more resistant to cell damage during concentrate production, as is found with Unigate mixed culture No. 1051 (Fig. 3).

In a medium containing sufficient nutrients, pH-controlled starter growth is limited by the accumulation of metabolic end products such as lactate ions. It should be possible to increase yields by the removal of these metabolites. Gerhardt & Gallup (1963) and Bergère & Hermier (1968), using dialysis techniques, were unable to increase starter yields, although the former successfully produced concentrates of bacteria other than lactic streptococci. Osborne

Table 2
Average chain length during starter concentrate production*

Starter strain	Conventional culture	Fermenter culture	Fresh concentrated culture	Frozen concentrated culture
E_8 (*S. cremoris*)	2.0	2.2	1.3	1.3
P_2 (*S. cremoris*)	1.7	2.1	1.5	1.7
Am_2 (*S. cremoris*)	4.5	8.3	1.6	1.7
Am_1 (*S. cremoris*)	3.0	7.2	1.7	2.0

* Number of streptococcal cells per chain.

Table 3
Viability of starters during concentrate production*

Starter strain	Conventional culture (%)	Fermenter culture (%)	Fresh concentrated culture (%)	Stored concentrated culture (%)
Am_2 (*S. cremoris*)	54	47	8	11
E_8 (*S. cremoris*)	100	100	63	63
P_2 (*S. cremoris*)	100	80	–	50

* Viability expressed as: (average chain length $\times$ colony-forming units $ml^{-1} \times 100$) total cells ml^{-1}.

(1977) using autoclavable 0.2 μm pore size cellulose acetate membranes produced concentrated starter cultures of 10^{11} total cells ml^{-1}. The major requirement of this system was a sufficient membrane area to permit adequate diffusion of the rapidly accumulating lactate; 5 cm^2 of membrane ml^{-1} of fermenter medium was necessary to achieve this, together with a fermenter: reservoir volume ratio of at least 1 to 15. The advantages of such a system over normal concentrate procedures are obvious since the yield is not dependent on a centrifugation stage and the efficiencies of such a process. However, there are likely to be scale-up limitations imposed by the diffusion unit size and at present the volume of the fermenter circuit is held at 500 ml. The potential of the unit, therefore, is seen as its operation on a continuous or semicontinuous basis. Adequate quality control is essential for starter concentrates, both during and after production. This has been described (Osborne 1977) and is similar to that set out by Cox & Lewis (1972). Concentrates are used primarily for bulk-starter preparation, and are added to the bulk medium at a predetermined dosage rate (*ca.* 0.002–0.007% v/v).

As mentioned previously, there is interest in concentrates for DVI, and the system has been assessed experimentally. Under satisfactory growth, concentration and storage conditions, selected cultures can produce acid at the required rate in the cheese vat. Cheddar trials have been carried out (Stanley 1974) using concentrates of *S. lactis* (NCDO 510) which were (1) freshly prepared and cooled, (2) frozen or (3) frozen in liquid nitrogen in droplet form. The starter growth and acidity development in the vat were compared with control cultures prepared by conventional techniques and inoculated into the vat at similar viable cell levels (Figs 4 and 5). The results show the concentrates to be slightly slower in acid production than the control culture but, nevertheless, gave acceptable cheeses. However, the use of DVI on a large scale may be governed by financial not technological considerations.

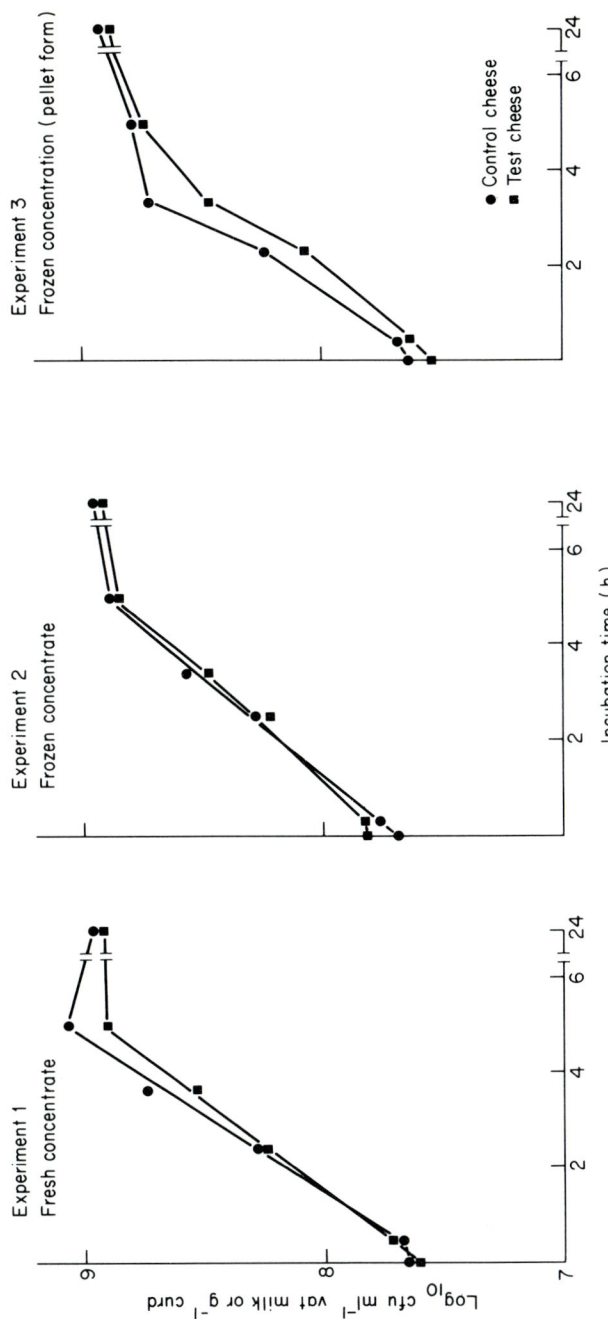

Fig. 4. Use of concentrates for Cheddar cheese manufacture. Growth of *S. lactis* NDCO 510. Experiment 1, fresh concentrate; Experiment 2, frozen concentrate; Experiment 3, frozen concentrate (pellet form). ●, Control cheese; ■, test cheese. *Ordinate scale*; $\log_{10}$ colony forming units (cfu) ml^{-1} of vat milk for samples taken before 1 h incubation time; $\log_{10}$ cfu g^{-1} of curd for samples taken after 1 h incubation time.

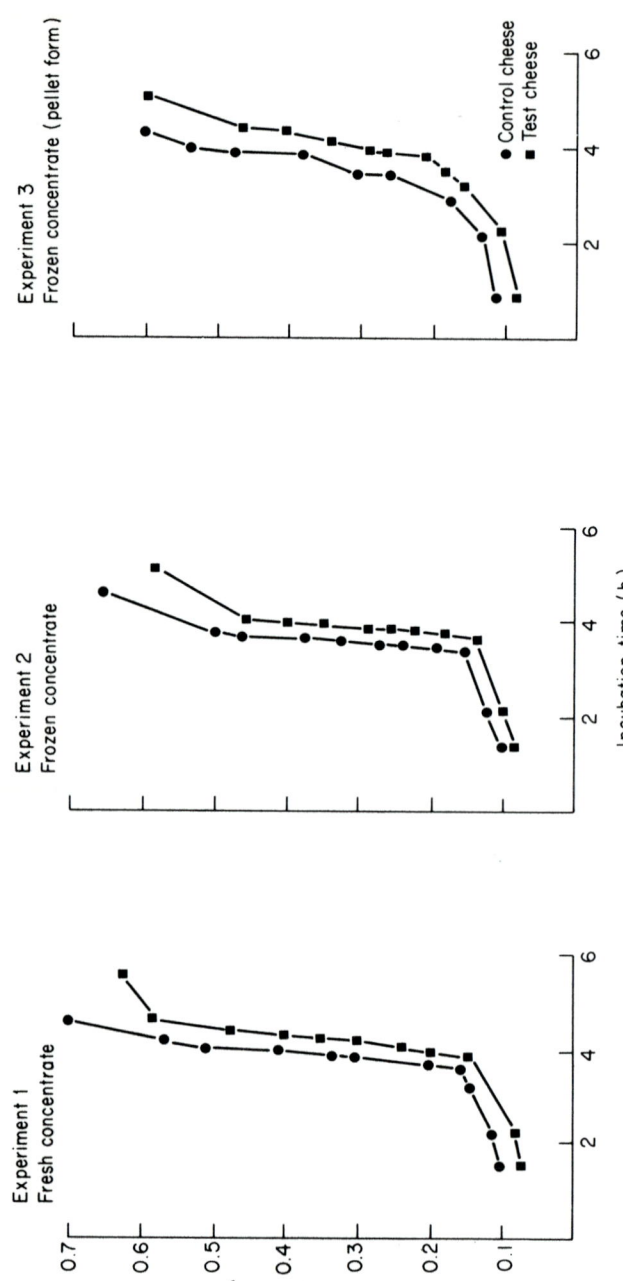

Fig. 5. Use of concentrates for Cheddar cheese manufacture. Titratable acidities of whey samples. Experiment 1, fresh concentrate; Experiment 2, frozen concentrate; Experiment 3, frozen concentrate (pellet form). ●, Control cheese; ■, test cheese.

The preparation of highly active freeze-dried starters would appear to be the next development in starter technology and has been attempted on a small scale with varying degrees of success (Stadhouders *et al.* 1969; Speckman *et al.* 1974). However, freeze-dried starters for DVI have been developed recently in France by Vitex (Europe) in collaboration with L'Air Liquide (France) and their high degree of acid-producing activity has been demonstrated by Cheddar trials at NIRD (Chapman 1978). The advantages of such starters include ease of transport and inoculation into the cheese vat, no requirement for refrigeration storage temperatures and elimination of all starter preparation in the creamery.

4. Problems Associated With Production and Use of Starters

Where appropriate, problems associated with starter production have already been discussed. The prevention of starter inactivation by phage has been partially overcome by the use of protected bulk systems and phage-resisting medium. A further precaution is the use of phage-unrelated 'rotations' (Anderson & Meanwell 1942; Whitehead & Hunter 1947; Lawrence & Pearce 1972). In New Zealand rotations of phage-unrelated pairs of single strains are used, each pair consisting of a fast and slow strain (this denoting the rate of acid production at scald temperatures). Continued monitoring of the whey is required to detect the presence of phage races that might cross-relate previously unrelated cultures (Lawrence *et al.* 1976). A reduction of the ripening period (the time between starter and rennet additions) is a useful precaution against phage inactivation of the starter (Hunter 1944; Meanwell & Thompson 1959). It is common practice in the UK to use mixed starters in pairs, and, where information is available, as phage-unrelated rotations. Effective cleaning/disinfection systems and steam sterilization of the cheese vats should be employed. Providing all the recommended procedures are adopted, then phage is not a major problem.

Agglutinins (antibodies) present in raw milk, particularly during the lactation period, are capable of agglutinating sensitive starter bacteria. The mechanism of this interaction is not well understood (Reiter 1973). For hard cheese processes the inhibitory effect is minimal, since the action of agglutinins is prevented by the formation of a rennet coagulum (Stadhouders & Hassing 1974). Their presence can be detrimental, however, in cottage cheese manufacture (Emmons *et al.* 1966; Emmons & Elliott 1967). Where, by interacting with sensitive starters, a sediment or 'sludge' forms at the bottom of the cheese vat. Emmons & Tuckey (1967) recommend a simple laboratory test for detecting sensitive organisms and suggest that sediment formation is associated with the use of high gas-producing cultures, the use of aged milk and milk containing added calcium. At present, suitable selection and screening of cultures would appear the most practical method of preventing sediment formation.

Another naturally occurring inhibitor in milk is the lactoperoxidase/thiocyanate/hydrogen peroxide system. Wright & Tramer (1957, 1958) began work in this area, with further identification of the participating moieties of the system by Jago & Morrison (1962) and Reiter, Pickering & Oram (1964). From a practical point of view many starters are resistant to the peroxidase system and a simple tube test can be used to determine sensitivity. Thus, again, continual starter selection and screening is essential, noting that, resistant cultures can give rise to sensitive mutants (Auclair & Vassal 1963) and the peroxidase in the heat-treated reconstituted skim milk used for starter propagation will be inactivated, so allowing any sensitive mutant to proliferate (Reiter 1973).

Antibiotics occur in milk as a result of their use in the therapeutic treatment of bovine mastitis, and the subject has been discussed in detail by Mol (1975). The common causative organisms of mastitis are *S. agalactiae, S. uberis, S. dysgalactiae* and *Staphylococcus pyogenes*, with *Staphylococcus pyogenes* being considered the most important (Wilson 1964). The antibiotic preparations available in the UK for intramammary infusion include: penicillin G, erythromycin, ampicillin, cloxacillin, streptomycin, aureomycin, neomycin and novobiocin. These are formulated to give quick or slow release of the antibiotic for treatment during lactation or for dry cow therapy. Withholding times, during which the milk must not be offered for retail, are given by the manufacturers. However, antibiotics are often detected in milk (Tramer 1964) and a testing scheme has been established by the Joint Committee of the Milk Marketing Board and the Dairy Trade Federation whereby farm milk is screened for antibiotics and other inhibitory substances, and a penalty scheme operated. The original test for this screening, developed by Wright & Tramer (1961), has been superseded by the 'Intertest' based on work by Jacobs *et al.* (1972). This test is designed to fail samples that contain the equivalent of 0.02 international units (iu) of penicillin ml^{-1}, and depends on lactic acid production by the test organism *S. thermophilus* and its detection by a colour change of the dye bromocresol purple. The presence of inhibitory substances is determined by a slowness in, or absence of, colour change of the dye.

Antibiotics in milk are undesirable on three grounds: medical, legal, and economic.

1. Individuals can be sensitive or develop sensitivity to antibiotics; their presence in milk can lead to the selection of antibiotic resistant strains or the transfer of resistance to pathogenic strains; the natural flora of the intestinal tract could be modified with possible ill effects.
2. The adulteration of milk with antibiotics is illegal (Section 2, Food and Drugs Act 1955), and successful prosecutions have been brought for milk "not of the nature, quality or substance required".
3. Antibiotics in milk can lead to the inactivation or partial inactivation of cheese and yoghurt starters. Yoghurt cultures (in particular *S. thermo-*

philus) are affected by 0.01-0.2 iu of penicillin ml^{-1}, 0.02-0.03 iu of penicillin ml^{-1} being sufficient to cause a yoghurt culture to fail. Cheese starters are variable in sensitivity but more resistant than yoghurt cultures. Levels of 0.05-0.3 have been shown to be detrimental (Dearden & Meanwell 1962, Meanwell 1962, Tramer 1964). In certain circumstances, milk is treated with penicillinase before adding the appropriate starter culture, and this has been adopted as common practice by many yoghurt producers.

As a final point, variations in the composition of mixed (and possibly multiple) strain starters should be considered; this is of consequence to the UK dairy industry. Strain dominance by one of the strains in a mixture can occur during daily subculturing (Reddy *et al.* 1972), thus leaving the culture open to attack by a single phage. In practice mixed starters are selected for their ability to withstand such subculturing, but nevertheless changes in milk composition can affect starter balance. Differences in manganese content, for instance, have been shown to alter the ratio of leuconostocs and streptococci (DeMan 1961). Thus it is recommended that milk powder, to be used for culture preparation, be produced only under the best seasonal conditions (Cox & Lewis 1972). The relationships of strains within a mixed starter are not well understood and Cox (1977) reported that attempts to recombine isolates of mixed cultures were of only limited success.

The whole field of mixed cultures requires further investigation and, in particular, the relation of experimental results to the starter production situation for conventional and concentrated cultures, and performance in the cheese vat. Further, classification studies should be carried out on these mixed strain starters, based not only on conventional biochemical/genetic tests, but on the applied characteristics which are of direct importance in the preparation of fermented milk products.

5. References

ACCOLAS, J. P. & AUCLAIR, J. 1967 Conservation à l'état congelé de suspensions de bactéries lactiques concentrées sous faible volume. *Le Lait* **47**, 253-260.

ACCOLAS, J. P. & AUCLAIR, J. 1970 Détermination de l'activité acidifiante des suspensions concentrées congelées de bactéries lactiques. *Le Lait* **50**, 609-626.

ACCOLAS, J. P. & VASSAL, Y. 1963 Occurrence of variants sensitive to agglutinins and to lactoperoxidase in a lactenin-resistant strain of *Streptococcus lactis*. *Journal of Dairy Research* **30**, 345-349.

ANDERS, R. F., HOGG, D. M., & JAGO, G. R. 1970. Formation of hydrogen peroxide by Group N streptococci and its effect on their growth and metabolism. *Applied Microbiology* **19**, 608-612.

ANDERSON, E. B. & MEANWELL, L. J. 1942 The problem of bacteriophage in cheese making. *Journal of Dairy Research* **13**, 58-72.

ANON. 1968a Mixed bacterial concentrates for the fermentation of milk. *British Patent* 1,110,977.

ANON. 1968b Mixed bacterial concentrates for the fermentation of milk. *British Patent* 1,110,978.
ANON. 1970 Cell culture concentrates of lactic acid-producing bacteria. *British Patent* 1,205,733.
ANON. 1972 Production of cell culture concentrates. *British Patent* 1,260,247.
ANON. 1976a A method of producing a composition for use as an inoculant material in the preparation of starters and fermented milk products. *British Patent* 1,421,226.
ANON. 1976b *United Kingdom Dairy Facts and Figures*. The Federation of United Kingdom Milk Marketing Boards, pp. 101 & 110.
BABEL, F. J. 1958 New developments in the propagation of lactic cultures: culture media and bacteriophage inhibition. *Journal of Dairy Science* **41**, 697–698.
BAUMANN, D. P., & REINBOLD, G. W. 1964 Preservation of lactic cultures. *Journal of Dairy Science* **47**, 674.
BAUMANN, D. P., & REINBOLD, G. W. 1966 Freezing of lactic cultures. *Journal of Dairy Science* **49**, 259–264.
BERGÈRE, J. L. & HERMIER, J. 1968 La production massive de cellules de streptocoques lactiques. II. Croissance de *Streptococcus lactis* dans un milieu à pH constant. *Le Lait* **48**, 13–30.
BIRKJAER, H. E., ANDERSON, A. K., & LARSEN, H. 1974 Freezing at $-40°C$ of starters for cheese manufacture. *19th International Dairy Congress* 1E, 434.
CHAPMAN, H. R. 1978 Direct vat inoculation of milk with freeze-dried starters for making cheddar cheese. *Journal of the Society of Dairy Technology* **31**, 99–101.
CHEKE, V. 1959 *The Story of Cheese-making in Britain*. London: Routledge & Kegan Paul.
COGAN, T. M., BUCKLEY, D. J., & CONDON, S. 1971 Optimum growth parameters of lactic streptococci used for the production of concentrated starter cultures. *Journal of Applied Bacteriology* **34**, 403–409.
COX, W. A. 1977 Characteristics and use of starter cultures in the manufacture of hard pressed cheese. *Journal of the Society of Dairy Technology* **30**, 5–15.
COX, W. A. & LEWIS, J. E. 1972 Methods of handling and testing starter cultures. In *Safety in Microbiology*, eds Shapton, D. A. & Board, R. G. Society for Applied Bacteriology Technical Series No. 6. London & New York: Academic Press.
CZULAK, J., & KEOGH, B. 1957 Trials of phage-resisting medium. *Australian Journal of Dairy Technology* **12**, 54.
DEARDEN, K. & MEANWELL, L. J. 1962 The influence of penicillin on acid development and the bacterial flora of cheddar cheese. *16th International Dairy Congress* **B**, 737–746.
DEMAN, J. C. 1961. De invloed van mangaan op zuursels. *Netherlands Milk & Dairy Journal* **15**, 202–203.
DOULL, D. P., & MEANWELL, L. J. 1953 Phage suppression and the cultivation of lactic streptococci. *13th International Dairy Congress* **3**, 1114–1120.
EMMONS, D. B., & ELLIOTT, J. A. 1967 Effect of homogenisation of skimmilk on rate of acid development, sediment formation, and quality of cottage cheese made with agglutinating cultures. *Journal of Dairy Science* **50**, 957.
EMMONS, D. B. & TUCKEY, S. L. 1967 *Cottage Cheese and Other Cultured Milk Products*. New York: Chas. Pfizer & Co. Inc.
EMMONS, D. B., ELLIOTT, J. A. & BECKETT, D. C. 1966 Effect of lactic streptococcal agglutinins in milk on curd formation and manufacture of cottage cheese. *Journal of Dairy Science* **49**, 1357–1366.
FOSTER, E. M., NELSON, F. E., SPECK, M. L., DOETSCH, R. N. & OLSON, J. E. 1958 *Dairy Microbiology*. London: MacMillan & Co. Ltd.
GERHARDT, P. & GALLUP, D. M. 1963 Dialysis flask for concentrate culture of microorganisms. *Journal of Bacteriology* **86**, 919–929.
GILLILAND, S. E. & SPECK, M. L. 1969 Biological response of lactic streptococci and lactobacilli to catalase. *Applied Microbiology* **17**, 797–800.
GILLILAND, S. E. & SPECK, M. L. 1974 Frozen concentrated cultures of lactic starter bacteria. A review. *Journal of Milk & Food Technology* **37**, 107–111.

HARVEY, R. J. 1965 Damage to *Streptococcus lactis* resulting from growth at low pH. *Journal of Bacteriology* **90**, 1330-1336.
HUNTER, G. J. E. 1944 The influence of bacteriophage on the cheese-making process. *Journal of Dairy Research* **13**, 294-301.
JACOBS, J., KLASENS, MARION & PENNINGS, A. 1972 A simple rapid test for the detection of antibiotics residues in milk. *Tijdschrift voor Diergneeskunde* **97**, 548-550.
JAGO, G. R. & MORRISON, M. 1962 Anti-streptococcal activity of lactoperoxidase III. *Proceedings of the Society of Experimental Biology, New York* **III**, 585-588.
LAMPRECH, E. D. & FOSTER, E. M. 1963 The survival of starter organisms in concentrated suspensions. *Journal of Applied Bacteriology* **26**, 359-369.
LATTEY, JANET M. 1968 Studies on ultra-deep frozen cheese starters. *New Zealand Journal of Dairy Technology* **3**, 35-41.
LAWRENCE, R. C. & PEARCE, L. E. 1972 Cheese starters under control. *Dairy Industries* **37**, 73-78.
LAWRENCE, R., THOMAS, T. D. & TERZAGHI, B. E. 1976 Reviews in the progress of dairy science: cheese starters. *Journal of Dairy Research* **43**, 141-193.
LEWIS, J. E. 1956 A new approach to the problem of phage control during the production of commercial cheese-starters. *Journal of the Society of Dairy Technology* **9**, 123-128.
LEWIS, J. E. 1977 Starter manufacture at individual cheese factories. *Journal of the Society of Dairy Technology* **30**, 32-35.
LLOYD, G. T. 1971 New developments in starter technology. *Dairy Science Abstracts* **33**, 411-416.
MARTLEY, F. G. 1972 The effect of cell numbers in streptococcal chains on plate-counting. *New Zealand Journal of Dairy Science and Technology* **7**, 7-11.
MEANWELL, L. J. 1962 The influence of penicillin on cheese starter activity. *Journal of Applied Bacteriology* **25**, 128-136.
MEANWELL, L. J. & THOMPSON, N. 1959 The influence of rennet on bacteriophage infection in the cheese vat. *Journal of Applied Bacteriology* **22**, 281-286.
MOL, H. 1975 *Antibiotics and Milk*. Rotterdam: Balkema.
MOSS, C. W. & SPECK, M. L. 1963 Injury and death of *Streptococcus lactis* due to freezing and frozen storage. *Applied Microbiology* **11**, 326-329.
ORLA-JENSEN, S. 1919 *The Lactic Acid Bacteria*. Copenhagen: Høst & Son.
OSBORNE, R. J. W. 1977 Production of frozen concentrated cheese starters by diffusion culture. *Journal of the Society of Dairy Technology* **30**, 40-44.
PACK, M. Y., VEDAMUTHU, E. R., SANDINE, W. E., ELLIKER, P. R. & LEESMENT, H. 1968 Effect of temperature on growth and diacetyl production by aroma bacteria in single and mixed strain lactic cultures. *Journal of Dairy Science* **51**, 339.
PEEBLES, M. M., GILLILAND, S. E. & SPECK, M. L. 1969 Preparation of concentrated lactic streptococci starters. *Applied Microbiology* **17**, 805-810.
PONT, E. G. & HOLLOWAY, G. L. 1968 A new approach to the production of cheese starter. *Australian Journal Dairy Technology* **23**, 22-29.
PULAY, G. & ZSMKO, M. 1969 Activity test of mesophilic starters. *Tejepar* **18**, 25.
REDDY, M. S., VEDAMUTHU, E. R., WASHAM, C. J. & REINBOLD, G. W. 1972 Agar medium for differential enumeration of lactic streptococci. *Applied Microbiology* **24**, 947-952.
REITER, B. 1956 Inhibition of lactic streptococcus bacteriophage. *Dairy Industries* **21**, 877-879.
REITER, B. 1973 Some thoughts on cheese starters. *Journal of the Society of Dairy Technology* **26**, 3-15.
REITER, B., PICKERING, A. & ORAM, J. D. 1964 In *Microbial Inhibitors in Food*, ed. Molin, N. Stockholm: Almqvist & Wiksell, p. 297.
RYMASZEWSKI, J. POZNANSKI, S. & MAGINSKA, Cz. 1971 Utilisation de la biomasse des bactéries de la fermentation lactique pour la fabrication des fromages. *Le Lait* **51**, 23.

SANDINE, W. E., ELLIKER, P. R. & HAYS, H. 1962 Cultural Studies on *Streptococcus diacetilactis* and other members of the lactic streptococcus group. *Canadian Journal of Microbiology* **8**, 161-174.

SANDINE, W. E., RADICH, P. C. & ELLIKER, P. R. 1972 Ecology of the lactic streptococci. A review. *Journal of Milk & Food Technology* **35**, 176-184.

SELLARS, R. L. 1967 In *Microbial Technology*, ed. Peppler, H. J. New York: Reinhold, p. 34.

SOZZI, T. 1972 Etude sur l'exigence en calcium des phages des ferments lactiques. *Milchwissenschaft* **27**, 503-507.

SPECKMAN, C. A., SANDINE, W. E. & ELLIKER, P. R. 1974 Lyophilized lactic acid starter culture concentrates: preparation and use in inoculation of vat milk for cheddar and cottage cheese. *Journal of Dairy Science* **57**, 165-173.

STADHOUDERS, J. & HASSING, F. 1974 Enhancement of the acid production of some lactic streptococci by milk-peroxidase. *19th International Dairy Congress* 1E, 369-370.

STADHOUDERS, J., JANSEN, L. A. & HUP, G. 1969 Preservation of starters and mass production of starter bacteria. *Netherlands Milk and Dairy Journal* **23**, 182-199.

STANLEY, G. 1974 Factors affecting the growth of lactic streptococci. Ph.D. Thesis, University of Reading.

STANLEY, G. 1977 The manufacture of starters by batch fermentation and centrifugation to produce concentrates. *Journal of the Society of Dairy Technology* **30**, 36-39.

TERZAGHI, B. E. & SANDINE, W. E. 1975 Improved medium for lactic streptococci and their bacteriophages. *Applied Microbiology* **29**, 807-813.

THOMAS, T. D. & BATT, R. D. 1968 Survival of *Streptococcus lactis* in starvation conditions. *Journal of General Microbiology* **50**, 367-382.

THOMAS, T. D. & BATT, R. D. 1969 Degradation of cell constituents by starved *Streptococcus lactis* in relation to survival. *Journal of General Microbiology* **58**, 347-362.

TOFTE JESPERSON, N. J. 1974 Recent developments in dairy starter cultures. *South African Journal of Dairy Technology* **6**, 63-68.

TOFTE JESPERSON, N. J. 1977 The use of commercially available concentrated starters. *Journal of the Society of Dairy Technology* **30**, 47-51.

TRAMER, J. 1964 Antibiotics in milk. *Journal of the Society of Dairy Technology* **17**, 95-100.

TYBECK, E. 1959 Growth experiments in "Cockade" PRM with bacteria used in the manufacture of Emmental cheese. *15th International Dairy Congress* **2**, 611-615.

WILSON, C. D. 1964 The mastitis problem. *Journal of the Society of Dairy Technology* **17**, 142-148.

WHITEHEAD, H. R. & COX, G. A. 1932 A method for the determination of vitality in starters. *New Zealand Journal of Science and Technology* **13**, 304-309.

WHITEHEAD, H. R. & HUNTER, G. J. E. 1947 Bacteriophage in cheese manufacture. Contamination from farm equipment. *Journal of Dairy Research* **15**, 112-120.

WIGLEY, R. C. 1977 The use of commercially available concentrated starters. *Journal of the Society of Dairy Technology* **30**, 45-47.

WRIGHT, R. C. & TRAMER, J. 1957 The influence of cream rising upon the activity of bacteria in heat-treated milk. *Journal of Dairy Research* **24**, 174-183.

WRIGHT, R. C. & TRAMER, J. 1958 Factors influencing the activity of cheese starters. The role of milk peroxidase. *Journal of Dairy Research* **25**, 104-118.

WRIGHT, R. C. & TRAMER, J. 1961 The estimation of penicillin in milk. *Journal of the Society of Dairy Technology* **14**, 85-87.

ZOLTOLA, E. A. & MARTH, E. H. 1966 Dry-blended phosphate-treated milk media for inhibition of bacteriophages active against lactic streptococci. *Journal Dairy Science* **49**, 1343-1349.

Nisin: Its Preservative Effect and Function in the Growth Cycle of the Producer Organism

A. HURST

*Health Protection Branch, Health and Welfare Canada
Ottawa, Canada*

CONTENTS

1. General Introduction	297
2. Nisin as a food preservative	298
(a) Introduction	298
(b) Some properties of nisin	299
(c) Use of nisin in food preservation	299
3. Function of nisin in the growth cycle of the producer organism	301
(a) Synthesis of nisin	301
(b) Function	306
(c) Conclusions	310
4. References	311

1. General Introduction

NISIN is a polypeptide antibiotic synthesized by strains of *Streptococcus lactis*. The producer organisms belong to serological group N, and the name was derived from the letters group N Inhibitory Substance (Mattick & Hirsch 1947). A related antibiotic, diplococcin, is produced by another member of the lactic streptococci, *S. cremoris* (Oxford 1944). Reports appear from time to time suggesting that other streptococci also produce antibiotics. Thompson & Shibuya (1946) isolated enterococci capable of producing antibiotics. Murray & Loeb (1950) reported that three strains of β-haemolytic streptococci grown on blood agar produced antibiotics. Using cellophane sacs they obtained cell-free antibiotic preparations. Kafel & Ayres (1969) noted antagonism between enterococci and bacilli in canned hams but did not isolate the inhibitor. Iandolo *et al.* (1965) observed the repression of *Staphylococcus aureus* by *S. diacetylactis* but ascribed the inhibition to competition for nutrients. Daly *et al.* (1972) described practical food preservative applications of *S. diacetylactis*. Pinheiro *et al.* (1968) isolated an inhibitory substance from *S. diacetylactis* which turned out to be acetic acid. Hirsch & Wheater (1951) examined 10,000 isolates of various streptococci and only observed antibiotic effects with the lactic streptococci. It appears that antibiotic production among streptococci other than the lactic ones is likely to exist but is rare. A false impression of antibiotic synthesis can be easily obtained with this genus because they compete successfully with many groups of organisms, due to their rapid growth rate and to their acid and peroxide formation.

'Inhibitory' streptococci which were probably nisin producers, were first observed by Rogers in 1928. They were rediscovered by Whitehead and collaborators in New Zealand in 1933 (Whitehead 1933; Whitehead & Riddet 1938). The New Zealand workers were seeking to explain slow acid development in cheesemaking, the principal cause of which is phage attack rather than inhibitory streptococci. In consequence these organisms were again forgotten until 1943 when Meanwell described them as a cause of slowness in cheesemaking in Britain. The inhibitory substance was clearly recognized as an antibiotic in 1944 by Mattick & Hirsch, who used the strains originally isolated by Meanwell. The antibiotic has found application as a food preservative, but because this aspect has been covered in several previous reviews, it will be dealt with only briefly in this chapter. The biosynthesis and possible growth regulatory function of nisin has not been previously reviewed and this aspect will be discussed in greater detail.

2. Nisin as a Food Preservative

(a) *Introduction*

Hirsch *et al.* (1951) first used nisin as a food preservative. They made Swiss type cheese with a nisin-producing starter culture which effectively prevented 'blowing' faults caused by *Clostridium butyricum* and *Cl. tyrobutyricum*. Subsequently they used a similar technique for the preservation of processed cheese (McClintock *et al.* 1952). These promising results stimulated the manufacture of nisin and a food preservation grade antibiotic came on the market. In this form nisin is sold mixed with skim milk powder. In many European countries it then became an accepted and legal food additive. Its use in food preservation was described by Berridge (1953) and by Hawley (1957, 1962) while Gillespy (1957) described the use of nisin for certain canned vegetables; its application in canned green peas may be particularly important (Kiss *et al.* 1968). More recently, East-European countries have been active in applying nisin to canned vegetables, and nisin is known to be manufactured in Poland and the USSR (Gudkov *et al.* 1973). Nisin in food preservation was reviewed by Marth (1966) and Jarvis & Morisetti (1969). Additional applications of nisin not listed here are given by Jarvis and Morisetti and include evaporated milk, sterilized milk, tomatoes, chocolate-flavoured milk, carrot purée, gelatin and mushrooms. The interaction of nisin-producing streptococci and sporeforming microbes (Hurst 1972), and the effect of nisin on the microecology of foods (Hurst 1973) have been reviewed also.

(b) *Some properties of nisin*

Nisin is a polypeptide synthesized by certain strains of *S. lactis*. Maximal amounts are formed in rich complex media and in milk; there is a positive correlation between the number of cells ml^{-1} and the amount of nisin synthesized (Hirsch 1951; Egorov et al. 1971). This synthesis appears to be episomally controlled (Kozak et al. 1974).

Numerous methods have been proposed for the assay of nisin, the most convenient being either turbidimetric (Berridge & Barrett 1952; Hurst 1966a) or agar diffusion methods (Tramer & Fowler 1964). Nisin diffuses through agar only slowly, but Tramer and Fowler used the detergent Tween 20 to increase the diffusion rate. They also defined a standard unit (Reading Unit) for its biological activity, and this standard has been accepted by WHO (1969). The approximate activity of 1 μg of pure nisin is 40 Reading Units (RU).

Nisin is most soluble and stable in aqueous solutions at pH 2. Its stability and solubility decrease at neutrality, and alkali inactivates it. Chymotrypsin and nisinase also inactivate it (Alifax & Chevalier 1962; Jarvis 1967; Jarvis & Mahoney 1969; Jarvis & Farr 1971).

Nisin is a narrow spectrum antibiotic effective only against Gram positive organisms. It is bactericidal rather than bacteristatic (Hirsch 1950). Early investigations held out promise of its use for treatment of streptococcal infections, tuberculosis and staphylococcal mastitis in cows (Mattick & Hirsch 1947; Hirsch & Mattick 1949). However, therapeutic application was largely unsuccessful because of the insolubility of nisin at physiological pH and its consequent local irritant effect (Taylor et al. 1949; Gowans et al. 1952).

(c) *Use of nisin in food preservation*

(i) *Technical considerations*

Considering the properties of nisin already described, successful application requires that: (1) the food be acid to ensure stability during both processing and storage; (2) the spoilage organisms to be controlled should be Gram positive, i.e. nisin sensitive. These conditions are satisfied in processed Gruyere cheese, in which nisin was first used successfully. This product is at about pH 5.5-6.0, and the spoilage organisms are anaerobic sporeformers (*Cl. tyrobutyricum*), the heating and the microecology of the food having eliminated other organisms. For this application nisin is used at levels of 100-400 $RU\,g^{-1}$ added to the melted cheese at the same stage as the melting salts (Fowler & McCann 1972). Gillespy (1957) carried out nisin trials with canned beans in tomato sauce inoculated with *Cl. thermosaccharolyticum* and found no spoilage when nisin was used at a concentration of 200 $RU\,ml^{-1}$. Wajd & Kalra (1976) used 100 $RU\,ml^{-1}$ of milk for the preparation of 'sterilized milk'. Nisin extended the

Table 1
Spoilage of lentil soup after 6 weeks storage

Temperature of storage °C	Number of control cans				Number of cans with nisin			
	Initially stored	Hard swells	Flat sour	Not spoiled	Initially stored	Hard swells	Flat sour	Not spoiled
Ambient	42	0	0	42	28	0	0	28
37	81	46	34	1	79	0	5	74
55	20	10	10	0	19	1	3	15

Gibbs & Hurst (1964).
Nisin was added at 400 RU ml^{-1} before heating.

shelf-life of this product for up to 60 days, whereas corresponding samples without nisin spoiled in 3 to 7 days.

On the other hand, even 400 RU ml^{-1} offers no guarantee of complete protection if bacteriologically low quality starting material is used. Thus use of nisin does not enable lowering of the hygienic quality of products. For example Gibbs & Hurst (1964) (see Table 1) failed to obtain complete preservation of canned lentil soup made from highly contaminated starting material.

It is important to note that canned low acid foods (pH of 4.6 or higher) in which nisin is used as an adjunct to processing, must receive a minimum heat treatment to ensure the destruction of *Cl. botulinum*. Strains of this organism are more nisin resistant than other clostridia. Because some spoilage organisms are more heat resistant than *Cl. botulinum*, cans are often heated beyond the minimum required to destroy the pathogen. However, spoilage organisms are nisin sensitive; use of nisin in low acid foods permits this extra heating to be saved with consequent possible improvements in organoleptic quality of the food and savings of cost and fuel. In high acid foods (pH $<$ 4.6) *Cl. botulinum* does not develop and use of nisin may offer even greater advantages (Boone 1966).

(ii) *Effect of nisin on spores*

Ramsier (1960) showed that nisin was sporicidal to *Cl. butyricum*, confirming the earlier findings of Hirsch & Grimsted (1954). Nisin was adsorbed from solution by spores causing release of 260-nm-absorbing substances (Gould 1964). In this respect, as well as in some others, nisin acted like a cationic detergent. Early reports suggested that nisin decreased the heat resistance of spores (Lewis *et al.* 1954; O'Brien *et al.* 1956) but this was later shown to be due to carryover of the antibiotic into the culturing medium (Campbell & Sniff 1959; Campbell *et al.* 1959; Denny *et al.* 1961). Also, an early suggestion that heat-damaged spores have increased nisin sensitivity was not confirmed by Tramer (1966). The carryover of nisin occurred because of its adsorption to the spores; after treatment with trypsin, the spores could be germinated and grown normally (Campbell & Sniff 1959; Thorpe 1960).

The effect of nisin on bacillus spores was investigated by Hitchins et al. (1963) and by Gould (1964). They showed that nisin allowed germination, i.e. darkening under phase-contrast microscopy and loss of heat resistance, but prevented post-germinative swelling, opening of the spore coat and outgrowth (formation of the first vegetative cell). Gould (1962), Gould & Hurst (1962) and Jarvis (1967) divided bacilli into two groups according to the way in which spore coats opened. The small-celled species (e.g. *Bacillus subtilis*) appeared to open their coats by mechanical pressure and outgrowth was prevented by 2-10 RU of nisin ml^{-1} of culture. Large-celled species (e.g. *B. cereus*) appeared to open their coats by lysis. The lytic spores were much more resistant to nisin, more than 100 RU ml^{-1} of culture were needed to prevent outgrowth.

(iii) *Public health considerations*

This facet of the nisin problem is clearly discussed by Jarvis & Morisetti (1969). Wild, nisin-producing organisms occur in milk and dairy products so that nisin is a natural constituent of dairy products, and it has probably been consumed since the time of domestication of milk-producing animals. The use of nisin in food preservation seems reasonable because nisin is not used clinically, it is non-toxic (Fraser et al. 1962) and it does not induce resistant strains in the consumer since it is inactivated by digestive enzymes (Heineman & Williams 1966; Jarvis & Mahoney 1969; Cowell et al. 1971). In addition, nisin is not used as a feed additive in animal husbandry. It seems unlikely, therefore, that nisin could induce multiple antibiotic-resistant enterobacteriae and its use in food preservation appears to be quite safe.

3. Function of Nisin in the Growth Cycle of the Producer Organism

(a) *Synthesis of nisin*

(i) *Chemistry*

The polypeptide can be prepared from culture fluids or the cells of the producer organism. Methods for its concentration and purification have been described (Berridge et al. 1952; Bailey & Hurst 1971). Nisin is available commercially and can be used as a starting material for the isolation of pure nisin Willimowska-Pelc et al. 1976).

Originally the molecular weight of nisin was thought to be 7000 (Cheeseman & Berridge 1959); however, it was shown later that the molecular weight was

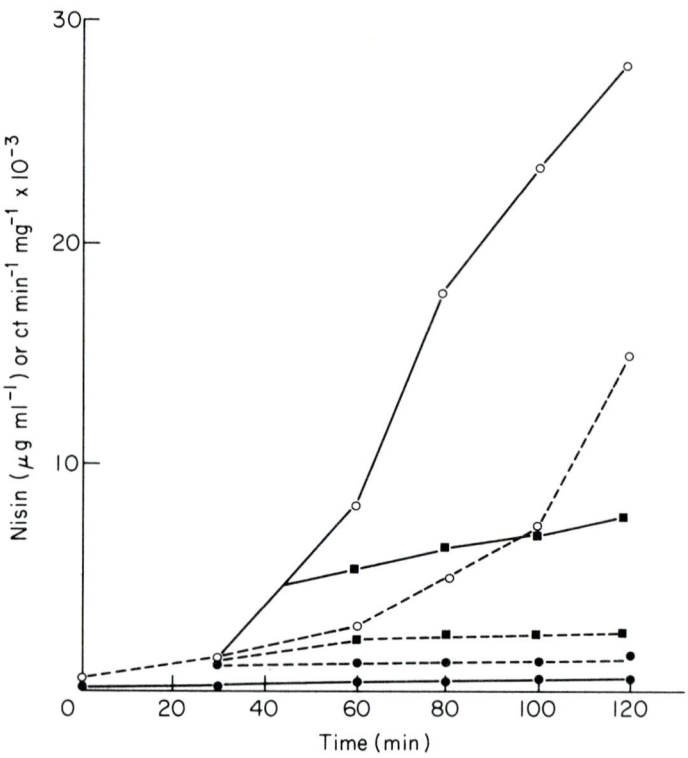

Fig. 1. The effect of chloramphenicol (20 μg ml^{-1}) on nisin synthesis and incorporation of [U-^{14}C]-L-glutamic acid by *S. lactis*. ——, radioactivity; - - -, nisin; ○, control; ●, chloramphenicol added at time zero; ■, chloramphenicol added at 50 min (after Hurst 1966a).

3500 and that nisin contained dehydroalanine (Gross & Morell 1967). Normally, nisin occurs as a dimer with the molecular weight of 7000 (Jarvis *et al.* 1968). It contains L-amino acids and the unusual S-amino acids lanthionine and β-methyl-lanthionine (Newton *et al.* 1953). These unusual amino acids are common to nisin and subtilin (Gross *et al.* 1969). The configuration of the atoms in the nisin molecule is unusual and very complex (Gross & Morell 1970: Morell & Gross 1973; Knox & Keck 1975).

(ii) *Synthesis by whole cells*

Hurst (1966a) devised a reaction mixture of amino acids, salts, growth factors, glucose and buffer in which *S. lactis* incorporated radioactive tracers and synthesized nisin. Penicillin and mitomycin, to which the organism is sensitive, had no effect on nisin synthesis. Actinomycin D inhibited RNA synthesis immediately but nisin synthesis was inhibited after a delay of 60 min.

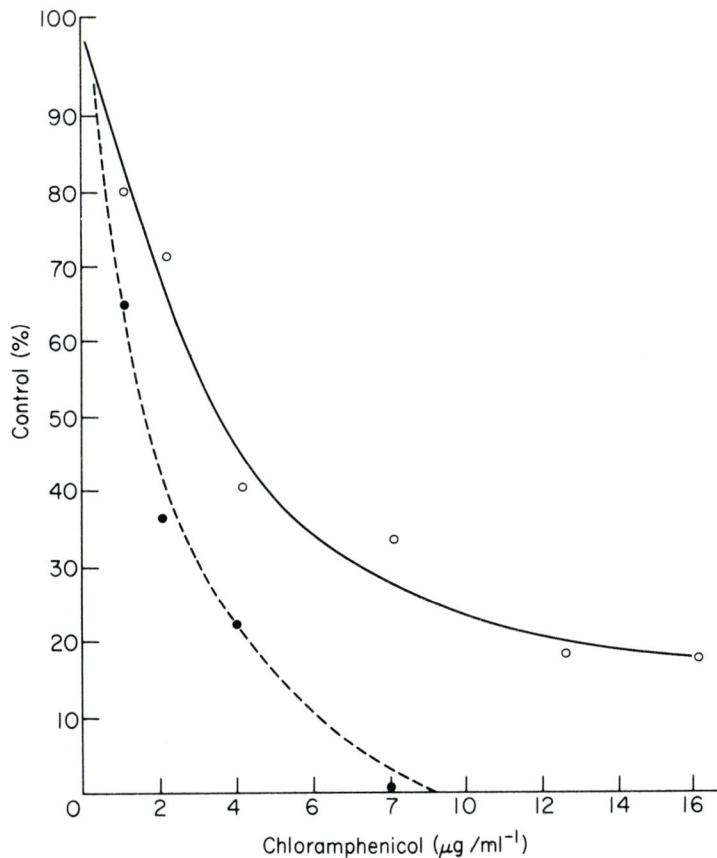

Fig. 2. Protein and nisin synthesis by *S. lactis* in the presence of various concentrations of chloramphenicol, as % of control without antibiotic. ○——○, protein; ●- - -●, nisin (after Hurst 1966a).

The lack of inhibition by mitomycin suggests that nisin synthesis was independent of newly formed DNA but the actinomycin D effect suggested dependence on messenger RNA which had an unusually long half-life.

Antibiotics which interfered with protein synthesis e.g. chloramphenicol, puromycin and tetracycline, also interfered with nisin synthesis. The inhibition was immediate and occurred irrespective of whether the antibiotic was added at the beginning of an experiment or after nisin synthesis was vigorously underway (Fig. 1). Nisin synthesis was more sensitive than protein synthesis as shown in Fig. 2.

These observations were extended by Ingram (1969, 1970). He used the reaction mixture of Hurst (1966a) but instead of measuring nisin by bioassay, he determined incorporation of radioactive amino acids into nisin to follow *de novo* synthesis. The nisin was then isolated and purified, and the properties of this material examined by chromatography and polyacrylamide gel electrophoresis. Ingram confirmed that inhibitors of protein synthesis inhibited nisin synthesis preferentially and that cysteine was a precursor of both lanthionine and β-methyllanthione. These two thioether amino acids were thought to be formed after cysteine was incorporated into a polypeptide chain. Serine and threonine are the other precursor amino acids of lanthionine and methyllanthione, respectively. Ingram (1970) proposed that the latter two precursors are dehydrated to the dehydroamino acids which then condense with cysteine to give the two unusual S-amino acids of nisin. Radioactive serine and threonine were incorporated into nisin which contained neither of these amino acids thus providing additional evidence for his scheme.

One difficulty of this work lay in the conflicting observations that nisin contains non-protein amino acids, yet its synthesis was sensitive to chloramphenicol and puromycin. These inhibitors affect protein synthesis at the ribosomal level. An invariant genetic code does not permit ribosomal synthesis of substances containing unusual amino acids. According to the suggestion of Ingram (1969, 1970) polypeptide chains containing the usual amino acids cysteine, serine and threonine would be first synthesized by a ribosomal mechanism. Subsequent modification of these residues might form nisin.

This notion was further confirmed and extended by the work of Hurst & Paterson (1971). Hurst (1967) had earlier isolated strains of *S. lactis* which did not produce nisin antibiotic but produced instead low molecular weight proteins which, by Sephadex chromatography and polyacrylamide electrophoresis, had the properties of the antibiotic. Preparations of the non-antibiotic protein incubated with cell extract of the nisin-producing strain generated an antibiotic which resembled nisin (Hurst & Paterson 1971).

Nisin is synthesized after the exponential phase of growth (see below) and the extract, possessing the enzyme activity converting precursor protein to nisin, was obtained just before the start of detectable nisin synthesis (at 300 min in Fig. 3). The highest specific activity was obtained later, but such cell extracts were not used because they were contaminated with nisin. The specific activity then rapidly declined and was undetectable at 8 h (Fig. 3). The activity was heat labile and decreased when incubated for longer than 30 min at 30°C.

The enzyme activity was readily obtained by blending cells with micro-beads for 45 s which affected neither the phase-contrast appearance nor the Gram reaction of the cells. Because the nisin-generating ability could be solubilized long before other proteins, the location of the enzyme and the site of conversion of pronisin into nisin was likely to be at the cell surface. These results and those of Ingram (1969, 1970) suggest that nisin may be synthesized in at least two

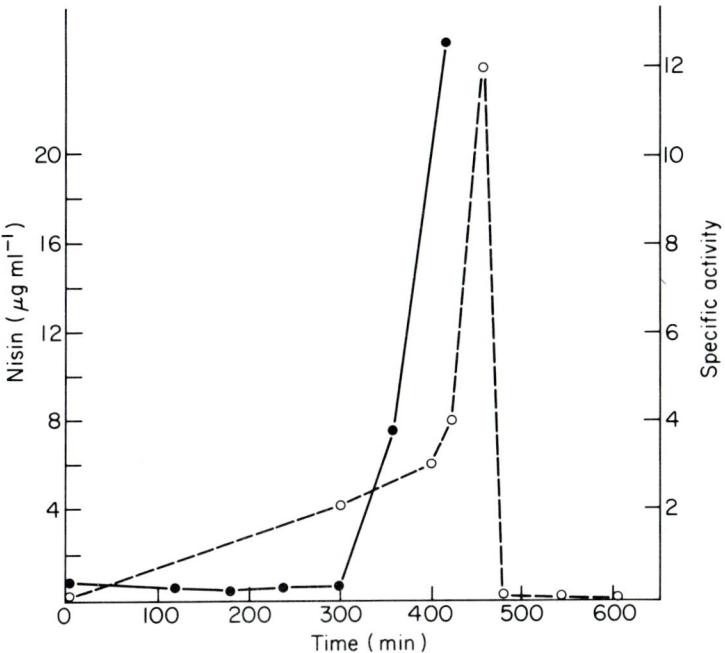

Fig. 3. Nisin synthesis and specific activity of nisin-generating cell extracts of *S. lactis*. (Specific activity = μg of nisin generated mg^{-1} of protein in 30 min at 30°C.) ●——●, nisin synthesis; ○- - -○, specific activity of cell extracts (after Hurst & Paterson 1971).

stages: first, nisin precursor is made by the accepted protein synthetic mechanisms involving transcription and translation; secondly, post-translational modifications are necessary, probably in the envelope of the cell, to obtain an active antibiotic.

(iii) *Location of nisin in the producer organism*

Cells of the nisin-producing organism grown in different ways were disrupted chemically and physically. The cells were treated chemically with hot and cold 5% trichloroacetic acid and other solvents. For the physical treatments different methods of cell disintegration were used followed by differential centrifugation (White & Hurst 1968). Both methods showed that nisin was located in the outer layers of the producer organism, which is consistent with the results described above for the location of nisin-generating enzyme (Table 2).

(iv) *Conclusions*

The evidence reviewed suggests that nisin is synthesized in two stages: by a ribosomal mechanism in the cell interior when pronisin is formed. This is

Table 2
Distribution of nisin in cellular fractions of the producer organism

Descriptive terms*	% of total recovered following fractionation	
	Chemical	Physical
'Walls'	36	43
'Membranes'	57	4
'Cell contents'	3	7
'Cells saps'		5
	96	59

* These are descriptive terms only. To take them too literally would be misleading. For example the 'wall' is the residue after cold and hot TCA extraction, extraction with aqueous ethanol and digestion with trypsin. In the physical fractionation, wall is the 10,000 g pellet; it may contain membrane particles as well.
Composed from Tables 2 and 3 of White & Hurst 1968.

followed by an enzymic conversion of the pronisin to nisin, in the envelope of the producer organism. However, neither pronisin nor the nisin-generating enzyme(s) have been isolated and studied. Until such time the scheme of nisin synthesis remains suggestive rather than definitive. The technical difficulties in such further studies should not be underestimated; pronisin may be formed early in the growth cycle when cell densities are low requiring large volumes of media to be handled rapidly. A way out of this difficulty might be to use mutants of the nisin-producing streptococcus which are deficient in the nisin-generating enzyme(s).

(b) *Function*

(i) *Nisin and the growth cycle of the producer organism*

Nisin is a secondary metabolite. Its biosynthesis starts after active protein, DNA and RNA synthesis and after an increase of about 50% in the dry weight of the organism (Hurst 1966*b*). Nisin inhibits and lyses the producer organism when added after commencement of growth but before its active synthesis (Hurst & Kruse 1972).

When nisin is added to fresh broth medium and the producer organism is inoculated afterwards, growth is slightly delayed. However, after this short delay the organism grows at the same rate as the nisin-free control. In contrast, if the organism is first grown and nisin added subsequently (up to 2 h after inoculation), the producer organism is highly sensitive to nisin and growth is completely inhibited (Hurst & Kruse 1972). These results show that at the start of growth either nisin is inactivated or the organism changes in its nisin sensitivity. The evidence to be presented discusses both these aspects. The suggestion will be made that nisin and other similar basic peptides have a regulatory function for the producer organisms and that antibiotic activity may be incidental.

(ii) *The disappearance of nisin*

It was repeatedly observed that after subculture the amount of bioassayable nisin per unit dry wt of cells decreased (Fig. 3) (Hurst 1966b). Was the disappearance of nisin apparent or real? To answer this a culture (pH controlled to 6.8) was grown to stationary phase, and the cell-associated nisin was measured at intervals. The cellular nisin mg dry wt^{-1} increased from 2 to 17 h and changed from about zero to 1.3% of the dry wt. At this stage the culture was inoculated into fresh medium (1% v/v) and the nisin content of the cells was again determined immediately. The concentration had fallen dramatically to 1/10 its immediate previous level (0.13% of the dry wt) and then continued to decline slowly to about zero at 2 h after subculture (Hurst 1968). The disappearance of nisin could not be explained enzymically but the results were compatible with the suggestion that nisin was bound to living cells in the form of Ca complex. Hurst & Lazarus (1968) demonstrated rapid calcium uptake by cells immediately after inoculation into fresh medium. This calcium was firmly bound and could not be removed by washing with pH 4.2 buffer. The cellular Ca concentration decreased in the course of the growth cycle and showed a complex inverse relationship to cellular nisin content (Fig. 4).

(iii) *Effect of cations on growth initiation*

The biochemical events that occurred when new growth was initiated on subculture thus include the disappearance of nisin, apparently connected with the uptake of calcium. Since uptake of calcium could be expected to alter the optical properties of cell suspensions, this was studied (Hurst 1969). Concomitant electron microscopic observations were also made (Hurst & Stubbs 1969).

On subculture the optical density of cell suspensions declined; when the complex medium was replaced by a solution of buffered salts plus glucose the optical phenomenon was still observed. A solution containing only tris buffer, Ca^{2+} and glucose also gave an optical reaction. This reaction required living cells and glucose and was temperature dependent (Hurst 1969). The reaction was poisoned by p-mercuribenzoate and iodoacetate but insensitive to ouabain, dinitrophenol and n-ethylmaleimide (Hurst & Kruse 1970). It occurred rapidly at 30°C and reached its maximum within 10 min. On subculture, the permeability properties of the cells had altered within 4 min: stationary phase

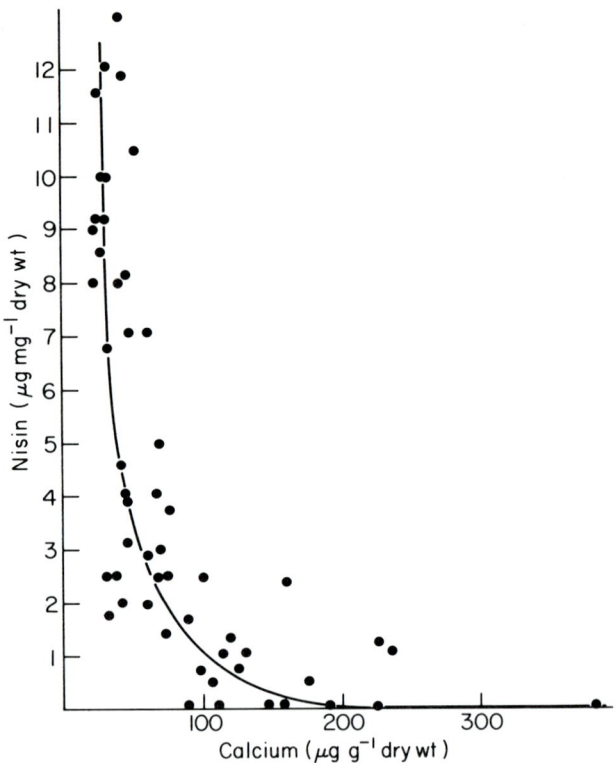

Fig. 4. Relationship between cellular calcium and nisin in *S. lactis* (after Hurst & Lazarus 1968).

cells suspended in ammonium molybdate showed clear structures of wall, membranes, adhesions of membranes to wall, septae and mesosomes. Four minutes after subculture, structures could no longer be observed, presumably because the cells had become permeable to ammonium molybdate. The cell interiors were stained solidly black (Hurst & Stubbs 1969).

During the optical reaction, cells incorporated $^{45}Ca^{2+}$ and lost nisin bioassay activity. Omission of glucose or addition of iodoacetate stopped the optical reaction and the $^{45}Ca^{2+}$ uptake, and nisin was not inactivated. Unsuccessful attempts were made to find some interaction between Ca^{2+} and isolated, purified nisin. The only explanation which fitted the facts was that, in the presence of glucose, living cells incorporated calcium which then led to a configurational change of the cells which was measured by the optical reaction, and that this configurational change was associated with nisin inactivation (Hurst & Kruse 1970).

The inactivation of nisin before the commencement of growth appears to be connected with the length of the lag phase of growth. There is a correlation between lag phase and cellular nisin content.

(iv) *Correlation between the length of the lag phase and cellular nisin level*

The disappearance of cellular nisin upon inoculation, described earlier, is followed by a gradual further decline in nisin content. In a similar way, the length of the lag phase after inoculation appears to be related to the nisin content of the inoculated cells and diminishes with diminishing nisin content. The end of the lag phase coincides with the beginning of new nisin synthesis. Then the length of the lag and the concentration of nisin both increase to a maximum in late stationary phase (Hurst 1966b).

It has already been mentioned that nisin synthesis is more sensitive to chloramphenicol than protein synthesis. Hurst & Dring (1968) used low levels of chloramphenicol to grow cells which were nisin deficient and which had a shorter lag than the control organisms grown without chloramphenicol. Figure 5 shows this surprising result: control cells in stationary phase contained about 25 μg nisin mg dry wt^{-1} and when subcultured into fresh medium, grew after a lag of 190 min. Cells grown in medium containing chloramphenicol (1 μg ml^{-1}) contained only 10 μg nisin mg dry wt^{-1} and after subculture grew after a shorter lag than control cells (130 min). The correlation coefficient between the length of lag phase and the nisin content of control or chloramphenicol grown cells was highly significant ($P = 0.884$).

A high correlation coefficient does not signify a causal relation; it can not be deduced that nisin controls the length of the lag. This point is especially important when considering this species as a whole: all organisms have a growth cycle with lag phase, exponential phase and stationary phase, but only few strains produce nisin. Thus it was difficult to see how nisin could have a regulatory function. Hurst (1967) examined this point in a small survey of wild *S. lactis* strains. Thirty-five of forty cultures examined produced an inhibitor which was nisin like. Thus, antibiotic production, albeit in small amounts, is a common property in this species. Of the five strains which produced no antibiotic, three were further studied. The basic proteins of these strains were extracted and examined by polyacrylamide gel electrophoresis using methods developed for nisin studies. All three strains produced peptides which resembled nisin by molecular weight and charge; the peptides from these strains could not be separated from nisin by Sephadex chromatography, yet they had no antibiotic activity. The basic peptides from one of these strains were later used by Hurst & Paterson (1971) to generate nisin. This is described above in Section 3(a)(ii).

Thus strains which do not produce nisin produced nisin-like basic proteins which might serve the same function in cell regulation as the antibiotic. It is

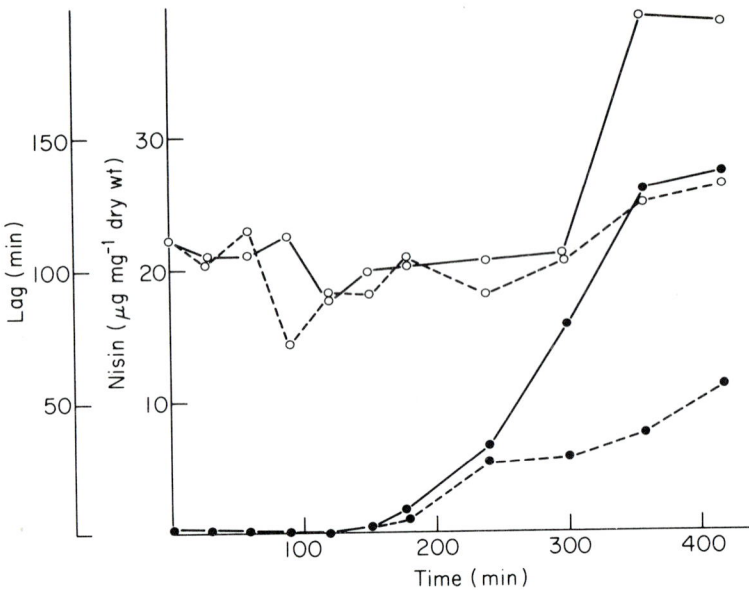

Fig. 5. The effect of chloramphenicol (1 µg ml^{-1}) on nisin synthesis and length of lag phase of growth. ●——●, cellular nisin (µg mg^{-1} dry wt) in original control culture; ○——○, period of lag when samples taken at intervals from original control culture were subcultured into fresh medium; ●---●, cellular nisin (µg mg^{-1} dry wt) in culture containing chloramphenicol; ○---○, period of lag when samples taken at intervals from the chloramphenicol containing culture were subcultured into fresh medium (after Hurst & Dring 1968).

possible that antibiotic activity is not required for the regulatory function. It is suggested that this function could be connected with the initiation and termination of the growth cycle.

(c) Conclusions

Section 3 of this chapter reviews chiefly the author's own work. It is fairly well established that *S. lactis* synthesizes nisin in at least two stages: a pronisin is produced by a protein synthetic mechanism and the final conversion to the antibiotic occurs enzymically at the cell surface. There is speculative evidence on the function of the antibiotic to the producer organism: nisin and other basic peptides may be concerned in the initiation and halting of growth because nisin disappears at the start of growth to be newly synthesized in stationary phase. This disappearance coincides with Ca^{2+} uptake, increased cellular permeability and diminished lag. Most strains of *S. lactis* make nisin. The few that do not, produce basic peptides which have a composition close enough to nisin to be accepted by cell extracts for the generation of nisin.

The growth regulatory hypothesis for the function of nisin would be strengthened if a component of the producer cell could be shown to be nisin sensitive and that this inhibition occurred in late exponential phase. Alternatively, nisin and other basic peptides may form structural units of the resting cell, and a direct 'switch' function of nisin may not exist.

Information on these points is important if we are to understand why cells make such apparently useless substances. In the case of the antibiotic, one may argue that its primary function is the advantage it confers to the producer organism in a strongly competitive situation (Hirsch 1952). But what could be, for example, the function of enterotoxin B to the producer organism? There are a number of points in common between the synthesis of nisin by *S. lactis* and enterotoxin B by *Staphylococcus aureus*, but we have no idea why the latter substance should be produced (Hurst & Kruse 1972).

4. References

ALIFAX, R. & CHEVALIER, R. 1962 Etude de la nisinase produite par *Streptococcus thermophilus*. *Journal of Dairy Research* **29**, 233-240.
BAILEY, F. J. & HURST, A. 1971 Preparation of a highly active form of nisin from *Streptococcus lactis*. *Canadian Journal of Microbiology* **17**, 61-67.
BERRIDGE, N. J. 1953 The antibiotic nisin and its use in the making and processing of cheese. *Chemistry and Industry* **1953**, 1158-1161.
BERRIDGE, N. J. & BARRETT, J. 1952 A rapid method for the turbidimetric assay of antibiotics. *Journal of General Microbiology* **6**, 14-20.
BERRIDGE, N. J., NEWTON, G. G. F. & ABRAHAM, E. P. 1952 Purification and nature of the antibiotic nisin. *Biochemical Journal* **52**, 529-535.
BOONE, P. 1966 Mode of action and applications of nisin. *Food Manufacture* **41**, 49-51.
CAMPBELL, L. L. & SNIFF, E. E. 1959 Effect of subtilin and nisin on spores of *Bacillus coagulans*. *Journal of Bacteriology* **77**, 766-770.
CAMPBELL, L. L., SNIFF, E. E. & O'BRIEN, R. T. 1959 Subtilin and nisin as additives that lower the heat-process required of canned foods. *Food Technology* **13**, 462-464.
CHEESEMAN, G. C. & BERRIDGE, N. J. 1959 Observations on the molecular weight and chemical composition of nisin A. *Biochemical Journal* **71**, 185-194.
COWELL, N. D., ALLEN, A. R. & JARVIS, B. 1971 The *in vivo* effect of nisin on the microflora of the oral cavity. *Journal of Applied Bacteriology* **34**, 787-791.
DALY, C., SANDINE, W. E. & ELLIKER, P. E. 1972 Interaction of food-starter cultures and food-borne pathogens: *Streptococcus diacetylactis* versus food pathogens. *Journal of Milk and Food Technology* **35**, 349-357.
DENNY, C. B., SHARPE, L. E. & BOHRER, C. W. 1961 Effects of tylosin and nisin on canned-food-spoilage bacteria. *Applied Microbiology* **9**, 108-110.
EGOROV, N. S., BARANOVA, I. P. & KOZLOVA, Y. I. 1971 Optimization of nutrient medium composition for the production of the antibiotic nisin. *Mikrobiologia* **40**, 860-864.
FOWLER, G. G. & McCANN, B. 1972 The use of nisin in the food industry. *Food Industry, South Africa* **25**, 49-55.
FRASER, A. C., SHARRATT, M. & HICKMAN, J. R. 1962 The biological effects of food additives. I. Nisin. *Journal of the Science of Food and Agriculture* **13**, 32-42.
GIBBS, B. M. & HURST, A. 1964 Limitations of nisin as a preservative in non-dairy foods. In *Microbial Inhibitors in Foods*, ed. Molin, J. Stockholm: Almqvist & Wiksell.
GILLESPY, T. G. 1957 Nisin trials. *Fruit and Vegetable Canning and Quick Freezing Research Association*, Leaflet No. 3, Chipping Campden, Glos, UK

GOULD, G. W. 1962 Microscopical observations on the emergence of cells of *Bacillus* from spores under different cultural conditions. *Journal of Applied Bacteriology* **25**, 35–41.

GOULD, G. W. 1964 Effect of food preservatives on the growth of bacterial spores. In *Microbial Inhibitors in Foods*, ed. Molin, N. Stockholm: Almqvist & Wiksell.

GOULD, G. W. & HURST, A. 1962 Inhibition of bacillus spore development by nisin and subtilin. *8th International Congress of Microbiology*, Abstract A2-11.

GOWANS, J. L., SMITH, N. & FLOREY, H. W. 1952 Some properties of nisin. *British Journal of Pharmacology* **7**, 438–449.

GROSS, E. & MORELL, J. L. 1967 The presence of dehydroalanine in the antibiotic nisin and its relationship to activity. *Journal of the American Chemical Society* **89**, 2791-2793.

GROSS, E. & MORELL, J. L. 1970 Nisin. The assignment of sulfide bridges of β-methyl-lanthionine to a novel bicyclic structure of identical ring size. *Journal of the American Chemical Society* **92**, 2919-2920.

GROSS, E., MORELL, J. L. & CRAIG, L. C. 1969 Dehydroalanyllysine: Identical COOH-terminal structures in the peptide antibiotics nisin and subtilin. *Proceedings of the National Academy of Sciences USA* **62**, 952-956.

GUDKOV, A. V., TROFIMOVA, T. I., DOLIDZE, G. G., LYNBIMOVA, L. A., SILEVA, M. N. & BLAGUSHNIA, R. F. 1973 Comparison of nisin preparations of English, Polish and Soviet manufacture (transl.). *Antibiotiki* **18**, 162-165.

HAWLEY, H. B. 1957 Nisin in food technology. *Food Manufacture* **32**, 370 & 430.

HAWLEY, H. B. 1962 The uses of antibiotics in canning. In *Antibiotics in Agriculture*, ed Woodbine, M. London: Butterworth.

HEINEMAN, B. & WILLIAMS, R. 1966 Inactivation of nisin by pancreatin. *Journal of Dairy Science* **49**, 312-313.

HIRSCH, A. 1950 The assay of the antibiotic nisin. *Journal of General Microbiology* **4**, 70-83.

HIRSCH, A. 1951 Growth and nisin production of a strain of *Streptococcus lactis*. *Journal of General Microbiology* **5**, 208-221.

HIRSCH, A. 1952 The evolution of the lactic streptococci. *Journal of Dairy Research* **19**, 290-293.

HIRSCH, A. & GRIMSTED, E. 1954 Methods for the enumeration of anaerobic spore formers from cheese, with observations on the effect of nisin. *Journal of Dairy Research* **21**, 101-110.

HIRSCH, A. & MATTICK, A. T. R. 1949 Some recent applications of nisin. *Lancet* **2**, 190-197.

HIRSCH, A. & WHEATER, D. M. 1951 The production of antibiotics by streptococci. *Journal of Dairy Research* **18**, 193-197.

HIRSCH, A., GRIMSTED, E., CHAPMAN, H. R. & MATTICK, A. T. R. 1951 A note on the inhibition of an anaerobic sporeformer in swiss-type cheese by a nisin producing streptococcus. *Journal of Dairy Research* **18**, 205-206.

HITCHINS, A. D., GOULD, G. W. & HURST, A. 1963 The swelling of bacterial spores during germination and outgrowth. *Journal of General Microbiology* **30**, 445–453.

HURST, A. 1966a Biosynthesis of the antibiotic nisin by whole *Streptococcus lactis* organisms. *Journal of General Microbiology* **44**, 209-220.

HURST, A. 1966b Biosynthesis of the antibiotic nisin and other basic peptides by *Streptococcus lactis* grown in batch culture. *Journal of General Microbiology* **45**, 503-513.

HURST, A. 1967 Function of nisin and nisin-like basic proteins in the growth cycle of *Streptococcus lactis*. *Nature, London* **214**, 1232-1234.

HURST, A. 1968 Apparent destruction of nisin by the producer organism before initiation of growth in *Streptococcus lactis*. *Nature, London* **219**, 403-404.

HURST, A. 1969 Change in the absorbancy of bacterial suspensions before initiation of growth. *Journal of Bacteriology* **97**, 1062-1068.

HURST, A. 1972 Interactions of food-starter cultures and food-borne pathogens: the antagonism between *Streptococcus lactis* and sporeforming microbes. *Journal of Milk and Food Technology* **35**, 418–423.

HURST, A. 1973 Microbial antagonism in foods. In *Microbial Food-borne Infections and Intoxications*, eds Hurst, A. & de Man, J. M. Ottawa, Canada: Health Protection Branch, Health and Welfare Canada.

HURST, A. & DRING, G. J. 1968 The relation of the lag phase of growth to the synthesis of nisin and other basic proteins by *Streptococcus lactis* grown under different cultural conditions. *Journal of General Microbiology* 50, 383-390.

HURST, A. & KRUSE, H. 1970 The correlation between change in absorbancy, calcium uptake, and cell-bound nisin activity in *Streptococcus lactis*. *Canadian Journal of Microbiology* 16, 1205-1211.

HURST, A. & KRUSE, H. 1972 Effect of secondary metabolites on the organisms producing them: Effect of nisin on *Streptococcus lactis* and enterotoxin B on *Staphylococcus aureus*. *Antimicrobial Agents and Chemotherapy* 1, 277-279.

HURST, A. & LAZARUS, W. 1968 Calcium uptake during growth of *Streptococcus lactis*. *Nature, London* 219, 404-405.

HURST, A. & PATERSON, G. M. 1971 Observations on the conversion of an inactive precursor protein to the antibiotic nisin. *Canadian Journal of Microbiology* 17, 1379-1384.

HURST, A. & STUBBS, J. M. 1969 Electron microscopic study of membranes and walls of bacteria and changes occurring during growth initiation. *Journal of Bacteriology* 97, 1466-1479.

IANDOLO, J. J., CLARK, C. W., BLUM, L. & ORDAL, Z. J. 1965 Repression of *Staphylococcus aureus* in associative culture. *Applied Microbiology* 13, 646-649.

INGRAM, L. 1969 Synthesis of the antibiotic nisin: formation of lanthionine and β-methyllanthionine. *Biochimica et biophysica acta* 184, 216-219.

INGRAM, L. 1970 A ribosomal mechanism for synthesis of peptides related to nisin. *Biochimica et biophysica acta* 224, 263-265.

JARVIS, B. 1967 Resistance to nisin and production of nisin-inactivating enzymes by several *Bacillus* species. *Journal of General Microbiology* 47, 33-48.

JARVIS, B. & FARR, J. 1971 Partial purification, specificity and mechanism of action of the nisin-inactivating enzyme from *Bacillus cereus*. *Biochimica et biophysica acta* 227, 232-240.

JARVIS, B. & MAHONEY, R. R. 1969 Inactivation of nisin by alpha-chymotrypsin. *Journal of Dairy Science* 52, 1448-1450.

JARVIS, B. & MORISETTI, M. D. 1969 The use of antibiotics in food preservation. *International Biodeterioration Bulletin* 5, 39-61.

JARVIS, B., JEFFCOAT, J. & CHEESEMAN, G. C. 1968 Molecular weight distribution of nisin. *Biochimica et biophysica acta* 168, 153-155.

KAFEL, S. & AYRES, J. C. 1969 The antagonism of enterococci on other bacteria in canned hams. *Journal of Applied Bacteriology* 32, 217-232.

KISS, I., KISS, K. N., FARKAS, J., FABRI, I. & VAS, K. 1968 Further data on the application of nisin in pea preservation. *Ellelmiszentudomany*, Budapest 2, 51-57.

KNOX, J. R. & KECK, P. C. 1975 β-methyllanthionine: a sulfur amino acid in subtilin and nisin antibiotics. *Acta crystallographia* B31, 2698-2700.

KOZAK, W., RAJCHERT-TRZPIL, M. & DOBRZANSKI, W. T. 1974 The effect of proflavin, ethidium bromide and an elevated temperature on the appearance of nisin-negative clones in nisin-producing strains of *Streptococcus lactis*. *Journal of General Microbiology* 83, 295-302.

LEWIS, J. C., MICHENER, H. D., STUMBO, C. R. & TITUS, D. S. 1954 Antibiotics in food processing: Additives accelerating death of spores by moist heat. *Journal of Agriculture and Food Chemistry* 2, 298-302.

McCLINTOCK, M., SERRES, L., MARZOLF, J. J., HIRSCH, A. & MOCQUOT, G. 1952 Action inhibitrice des streptocoques producteurs de nisine sur le development des sporules anaerobies dans le fromage de Gruyère fondu. *Journal of Dairy Research* 19, 187-193.

MARTH, E. H. 1966 Antibiotics in foods—naturally occurring, developed and added. *Residue Reviews* 12, 65-161.

MATTICK, A. T. R. & HIRSCH, A. 1944 A powerful inhibitory substance produced by group N Streptococci. *Nature, London* **154**, 551

MATTICK, A. T. R. & HIRSCH, A. 1947 Further observations on an inhibitory substance (nisin) from lactic streptococci. *Lancet* **2**, 5-12.

MEANWELL, L. J. 1943 The influence of raw milk quality on "slowness' in cheesemaking. *Proceedings of the Society of Agricultural Bacteriologists* **19**, Abstract.

MORELL, J. L. & GROSS, E. 1973 Configuration of the beta carbon atoms of β-methyllanthionine residue in nisin. *Journal of the American Chemical Society* **95**, 6480-6481.

MURRAY, R. G. E. & LOEB, L. J. 1950 Antibiotics produced by micrococci and streptococci that show selective inhibition with the genus *Streptococcus*. *Canadian Journal of Research* **28E**, 177-185.

NEWTON, G. G. F., ABRAHAM, E. P. & BERRIDGE, N. J. 1953 Sulphur-containing amino-acids of nisin. *Nature, London* **171**, 606.

O'BRIEN, R. T., TITUS, D. S., DEVLIN, K. A., STUMBO, C. R. & LEWIS, J. C. 1956 Antibiotics in food preservation. II. Studies on the influence of subtilin and nisin on the thermal resistance of food spoilage organisms. *Food Technology* **10**, 352-355.

OXFORD, A. E. 1944 Diplococcin, an antibacterial protein elaborated by certain milk streptococci. *Biochemical Journal* **38**, 178-182.

PINHEIRO, A. J. R., LISKA, B. J. & PARMELEE, C. E. 1968 Properties of substances inhibitory to *Pseudomonas fragi* produced by *Streptococcus citrovorus* and *Streptococcus diacetilactis*. *Journal of Dairy Science* **51**, 183-187.

RAMSIER, H. R. 1960 The action of nisin on *Clostridium butyricum*. *Archiv für Mikrobiologie* **37**, 57-94.

ROGERS, L. A. 1928 The inhibiting effect of *Streptococcus lactis* on *Lactobacillus bulgaricus*. *Journal of Bacteriology* **16**, 321-325.

TAYLOR, J. I., HIRSCH, A. & MATTICK, A. T. R. 1949 The treatment of bovine streptococcal and staphylococcal mastitis with nisin. *Veterinary Record* **61**, 197-198.

THOMPSOM, R. & SHIBUYA, M. 1946 The inhibitory action of saliva on the diphtheria bacillus: The antibiotic effect of salivary streptococci. *Journal of Bacteriology* **51**, 671-684.

THORPE, R. H. 1960 The action of nisin on spoilage bacteria I. The effect of nisin on the heat resistance of *Bacillus stearothermophilus*. *Journal of Applied Bacteriology* **23**, 136-143.

TRAMER, J. & FOWLER, G. G. 1964 Estimation of nisin in foods. *Journal of the Science of Food and Agriculture* **15**, 522-528.

WAJD, H. R. A. & KALRA, M. S. 1976 Nisin as an aid for extending shelf life of sterilized milk. *Journal of Food Science and Technology, Mysore* **13**, 6-8.

WHITE, R. J. & HURST, A. 1968 The location of nisin in the producer organism, *Streptococcus lactis*. *Journal of General Microbiology* **53**, 171-179.

WHITEHEAD, H. R. 1933 A substance inhibiting bacterial growth produced by certain strains of lactic streptococci. *Biochemical Journal* **27**, 1793-1800.

WHITEHEAD, H. R. & RIDDET, W. 1938 Slow development of acidity in cheese manufacture. *New Zealand Journal of Agriculture* **46**, 225-229.

WILLIMOWSKA-PELC, A., OLICHWEIR, Z., MALICKA-BLASZKIEWICZ, M. & MEJBAUM-KATNELLENBOGEN, W. 1976 The use of gel-filtration for the isolation of pure nisin from commercial products. *Acta Microbiologica Polonica* **25**, 71-77.

WHO 1969 Specifications for identity and purity of some antibiotics. World Health Organization/Food Add./69.34, pp. 53-67.

Streptococci of Lancefield Groups A, B and D and Those of Buccal Origin in Foods: Their Public Health Significance, Monitoring and Control

D. A. A. Mossel, P. G. H. Bijker and I. Eelderink

*Department of the Science of Food of Animal Origin,
Faculty of Veterinary Medicine
University of Utrecht, The Netherlands*

CONTENTS

1. Ecological principles 315
    (a) Group A streptococci 315
    (b) Group B streptococci 316
    (c) Group D streptococci 316
    (d) Organisms of oral origin 318
2. Monitoring of foods for streptococci 318
    (a) Principles 318
    (b) Group A streptococci 320
    (c) Group B streptococci 320
    (d) Group D streptococci 321
    (e) Streptococci of human oral origin 325
3. Control of contamination by streptococci 328
4. References 329

1. Ecological Principles

(a) *Group A streptococci*

THE STREPTOCOCCI of the Lancefield group A are certainly pathogenic to man, when consumed with food. In the past, when the disease was almost endemic, their absorption could lead to scarlet fever (Godfrey 1929), but nowadays infection with group A streptococci transmitted by food rarely causes more than tonsillitis ('septic sore throat'), a less serious disease (Andrews & Fuchs 1948; Savage 1949; Boissard & Fry 1955; Farber & Korf 1958; Taylor & McDonald 1959; Otte & Ritzerfeld 1960; Dudding et al. 1969; Hill et al. 1969; McCormick et al. 1976). However, in some instances serious sequelae may occur, probably due to the immunological condition of the patient; these comprise acute glomerulonephritis and rheumatic fever, which are serious diseases indeed (see Parker and Maxted, this volume).

The foods mostly involved in outbreaks of foodborne tonsillitis e.g. salads, cereal puddings and similar commodities, are almost always intensively manipulated before distribution. This situation is similar to that of hepatitis A: it seems that exposure of foods for a long period of time to human contamination, from respiratory or enteric sources, creates a risk of the foods becoming dangerously contaminated with pathogenic agents characterized by low minimal infectious doses.

(b) *Group B streptococci*

Streptococci of the Lancefield group B are also pathogenic to man, although it seems that only neonates, compromised people, i.e. those predisposed to infection (Bayer et al. 1976), and the very old are really at risk. The infections occurring in such populations are, however, serious, the most frequently described diseases being endocarditis, pneumonia and other respiratory distress, pyelonephritis, rheumatoid arthritis, meningitis, urogenital and veneral diseases and endometritis (Erbsloh & Grün 1949; Kahler & Aicher 1952; Kexel 1965; Kexel & Schönbohm 1965; Butter & de Moor 1967; McKnight et al. 1969; Steinitz et al. 1971; Svartz 1972; Franciosi et al. 1973; Anthony & Conception 1975; Allow et al. 1976; Bayer et al. 1976 ; Hemming et al. 1976; Mhalu 1976; Patterson & Hafeez 1976; Wallin & Forsgren 1976).

Whereas the role of food is beyond doubt when group A streptococci are concerned, the situation is much less clear in the case of group B organisms (Hahn et al. 1974). First of all, the question whether group B streptococci of animal and human origin are identical is far from resolved. Some workers think they are (Obiger 1975), others point to essential differences in characteristics, such as dissimilation of lactose and pigment formation (Butter & de Moor 1967; Jensen 1976; Mhalu 1976). But even if the two groups are identical, or have the same phylogenetic origin, it has been demonstrated only occasionally that oral absorption of group B streptococci leads to clinical disease. The low tolerance of these bacteria to acid and bile salts makes it doubtful whether they can colonize the lower areas of the alimentary canal at all (Obiger 1975).

(c) *Group D streptococci*

Group D streptococci can cause systemic disease in man (Bayer et al. 1976b). Ingram & Barnes (1955) and later investigators (Lang 1961; Luond & Gasser 1964; Sedova 1970; Pusztai et al. 1972) thought that group D streptococci, when consumed with food, could also cause febrile gastroenteritis. The evidence for this opinion is only circumstantial. It is also noteworthy that with improved diagnostic techniques for the assessment of the causes of foodborne diseases such reports have become very rare (Ingram & Barnes 1955). This supports the earlier opinion of the Wisconsin School that group D streptococci are not implicated as a cause of food-transmitted gastroenteritis (Deibel & Silliker 1963). On the other hand, there is no doubt that prolific growth of group D streptococci in foods may lead to the formation of clinically significant levels of pressor amines; these may cause a particular type of disease spread by foods heavily populated with group D streptococci (Legroux & Levaditi 1947; van Veen & Latusan 1950; Parrot & Nicot 1965; Doeglas et al. 1967; Mossel 1968; Bulajic 1973; Merson et al. 1974; Gellman et al. 1975; Foo 1976; Rice et al. 1976). It is noteworthy that pressor amines are very thermostable and will therefore remain active after application of modes of heat processing which eliminate all the viable streptococci. Consequently, any monitoring for the presence of streptococci must be carried out before final processing.

The major significance of Lancefield group D streptococci in foods of animal origin is their function as index and indicator organisms. The general principle of the use of index organisms (Mossel 1978) is that the probability that foods processed for safety will be contaminated with pathogens can be assessed by counting more numerous and usually harmless groups of bacteria such as group D streptococci (Drion & Mossel 1977). It is not, of course, essential for pathogens to be actually found. Their absence in the presence of group D streptococci should never be considered as a 'false positive' result, but rather, following the definition of indicator organisms (Mossel 1978), as a demonstration that the procedure of processing for safety has failed. While it may not have led to the presence of pathogens in a particular instance, it may well do so on a future occasion.

Lancefield group D streptococci have a thermal resistance (defined as the time/temperature/exposure required to achieve about six overall decimal reductions of colony-forming units, cfu, at pH $ca.$ 7 and a_w $ca.$ 0.98) which far exceeds that of most non-sporing pathogenic bacteria (Park 1928; Webster & Esselen 1956; Nevot et al. 1958; Obiger 1976). This also applies to chemical resistance (Shannon et al. 1964) and to resistance to drying (Ostrolenk et al. 1947; Clark et al. 1966) and freezing (Fanelli & Ayres 1959; Kereluk & Gunderson 1959a, b; Raj et al. 1961) similarly defined. Resistance of group D streptococci seems to be of the same order as that of hepatitis A virus (Provost et al. 1973), with the reservation that, despite recently accrued detailed knowledge of that virus, the numbers of virions inactivated by such treatments are not yet exactly known. Hence, if Lancefield group D streptococci have been eliminated (defined as $\log_{10}$ cfu $< 0{-}2$ g^{-1}) from a given food or food environment by the processing treatment, hepatitis A virus, if present initially, will also have been inactivated. Consequently, quantitative examination, of foods and food preparation areas susceptible to faecal contamination by human manipulation, for Lancefield group D streptococci can yield extremely useful information both for the Food Processing Industry and for the Public Health Authorities. However, the detection of streptococci in numbers of cfu exceeding those laid down in Reference Values (Mossel 1978), should never be interpreted as indicating that in such ecosystems hepatitis A virus will invariably, or even frequently, occur.

Provided the errors in the sense of microbial ecology stressed above are avoided, the use of index and indicator organisms in general and that of Lancefield group D streptococci in particular, can be of great value in monitoring the processing of foods and the disinfection of the food environment, according to previously specified 'good practices'.

(d) *Organisms of oral origin*

Investigations on the aetiology of dental caries have provided useful information on the occurrence of streptococci in the human oral cavity. From the point of view of food contamination, the most relevant types are those which predominate in mixed oral secretions, colloquially known as saliva. The most frequently encountered types are *Streptococcus mitior*, *S. salivarius* and *S. sanguis*, with *S. milleri* and *S. mutans* in the minority (Hardie & Marsh this volume). These streptococcal types do not belong to a well-defined Lancefield serotype: mostly, but relatively infrequently, the K antigen occurs in the three former groups, but the F and H antigen are also occasionally encountered (Parker & Ball 1976).

Given their regular occurrence in the human mouth, it would be pointless to attribute any health significance to the presence of the buccal types of streptococci in foods. By contrast, their occurrence in foods processed for safety (*vide supra*) indicates re-contamination from the human respiratory tract. This may have introduced pathogenic agents as well. Every effort must be made to prevent such recontamination.

2. Monitoring of Foods for Streptococci

(a) *Principles*

Medical food microbiology is one of the youngest branches of laboratory medicine. It is therefore not surprising, that on the one hand too much is expected from it, whereas on the other hand methods available are often not yet entirely satisfactory.

The most relevant aspect of the former is that bacteriological monitoring of foods is *per se* totally inadequate to assure good microbiological quality of the products when eaten. Quality assurance should rather rely on the application of Codes of Good Manufacturing and Distribution Practices (Wilson 1955). Once that has been achieved, the level and fluctuations of quality of the product reaching the consumer will be such that spot checks make sense (Mossel 1977*a*).

The latter should always be applied with discretion, so that available laboratory facilities are used optimally. One of the elements of dependable monitoring is that the fewest possible criteria will be used to assess the bacteriological condition of a given type of food or meal. These have to be chosen after careful study of the microbial ecology of the product under consideration along the lines of Section 1 of this paper. In no circumstances should the food microbiologist yield to the obsession of examining all foods for all those organisms that may possibly present a hazard. Rather should reliance be placed on the use of suitably chosen index or indicator organisms (Mossel 1978) to assess the products' overall microbiological condition (Drion & Mossel 1977).

In this context it may be useful to differentiate between three ecologically different categories of foods. Group 1 consists of commodities processed for safety: heated foods and meals, fermented sausages, chlorinated water supplies and biologically purified fresh seafoods, for example. Group 2 comprises those foods that have not been exposed to any treatment leading to decontamination, henceforth to be indicated as 'raw'. Finally, group 3 comprises food samples that are suspected of having been incriminated in outbreaks of diseases transmitted by foods. We will follow this system in reviewing the examination procedures which seem to be rational for given situations.

In doing so, we will refrain from an evaluation of the functioning of given media, or procedures. There are at least two reasons for doing so. First, a given method often functions well only for a limited group of commodities, because their microbial association is such that only trifling numbers of organisms that may challenge the selectivity of a given medium will occur on the plates. For foods of a different category, particularly those with a considerable interfering 'background' flora, a more selective medium may have to be used. This may be more toxic to the organisms sought and therefore not to be recommended for use with foods in which few interfering contaminants occur. A second point is, that only rather recently have effective steps been built into procedures for the microbiological examination of foods that also allow recovery of microbial cells, sublethally damaged by processing. These cells are, as a rule, incompetent to develop freely on the selective media that are fully suitable for non-stressed populations of the same taxon and, in fact, require a 'resuscitation' treatment to recover competence. Pending collaborative testing and evaluation of optimal methodology for resuscitation (Mossel & Corry 1977) every laboratory should apply the techniques which it finds effective in daily practice; quibbling about the performance of different procedures used in different laboratories is pointless.

Finally it goes without saying that monitoring of foods is fruitless unless reference values for 'wholesomeness' have been established. For the following it will be assumed that such values are available and have been determined by the only valid procedure: assessment of the highest attainable standards when the best possible manufacturing, storage and distribution practices are faithfully followed (Mossel 1977a).

(b) *Group A streptococci*

These bacteria would certainly not be determined routinely in foods of groups 1 and 2 above (Lee & Koburger 1970). However in investigations on outbreaks of septic sore throat, where epidemiological evidence suggests that food may be involved, their detection is obviously required. In addition their determination is sometimes advisable in monitoring the cleanliness of the food environment.

Various selective media have been suggested for the detection of group A streptococci in clinical specimens, and we have found that these are usually also reasonably effective for estimating the numbers of colony-forming units of group A streptococci in foods. Blood agar is invariably the basal medium and various selective antimicrobial agents such as sodium azide, crystal violet, fusidic acid, polymyxin, neomycin, gentamicin and nalidixic acid can be used as inhibitors of other groups of bacteria occurring in foods (Lowbury et al. 1964; Vincent et al. 1971; Black & van Buskirk 1971 and pp. 372-395). It is advisable to incubate such blood agar plates under anaerobic conditions to suppress interfering organisms even more. To facilitate the later selection of colonies for further study, spread plates are generally used for the enumeration of group A streptococci in foods and the food environment. A preliminary resuscitation treatment, aimed at the restoration of sublethally impaired cells may be required in many instances.

Identification of suspect isolates relies on the usual criteria: haemolysis, mode of attack on aesculin, arginine and lactose, growth at $45°C$ and tolerance of bile (Parker & Ball 1976).

(c) *Group B streptococci*

These bacteria are only sought infrequently in any foods of the ecologically defined groups 1-3 above. However, the veterinary profession is interested in their presence in raw milk.

The selective media in current use for the detection of group A streptococci are suitable for the detection of group B as well (Mason et al. 1976). The addition of aesculin to such media as recommended by Wilson and Salt (this volume) can be helpful, group B streptococci being virtually never able to dissimilate aesculin under these conditions (Jokipii & Jokipii 1976), whereas of group A organisms some 20% of cultures can use it (Facklam et al. 1974; Parker & Ball 1976).

The use of such selective media can also be of help in epidemiological research on the possible aetiological role of orally absorbed group B streptococci in human disease (Hahn et al. 1974), studies which certainly should be pursued.

(d) *Group D streptococci*

Recently, a reviewer counted no less than about 80 different selective media recommended for the enumeration of Lancefield group D streptococci in foods (Barnes 1976). This illustrates well how little this area of food microbiology is settled. Reasons for this, are not only the general limitations of selective media outlined previously, but also a taxonomic difficulty. Most authors seem to disagree on whether to include the non-enterococcal group D streptococci, *S. bovis* and *S. equinus*, in the enumerations. One of the arguments is that these types are much less robust than the enterococci, which makes their inclusion in counts somewhat superfluous (Feachem 1974). In addition, including these types in counts considerably increases the requirements for the performance of the medium: the *bovis/equinus* group is much more readily inhibited by selective agents than the enterococcal types (*vide infra*), which impairs the productive properties of a medium thus composed. Yet, in principle, a good medium for Lancefield group D streptococci should allow fair recovery of the *bovis/equinus* groups as well.

The present authors originally used the blood agar medium of Packer (1943), relying on the use of crystal violet and sodium azide as selective agents together with an elevated incubation temperature. This procedure had a quite acceptable productivity as well as selectivity (Mossel *et al.* 1957; Mossel 1963). However, the use of blood in this medium is not always convenient for non-clinical laboratories. For general purpose use a different medium is advantageous. Various authors have recommended selective, indicating media for this purpose, and particularly the indicator system aesculin/iron (Rochaix 1924; Colobert & Morélis 1958) in a mixture with azide, alone, or with bile salts (Meyer & Schönfeld 1926; Swan 1954; Pavlova *et al.* 1972; Switzer & Evans 1974; Brodsky & Schiemann 1976) or antibiotics (Levin *et al.* 1975; Mossel *et al.* 1975; Mossel 1977).

We have studied the performance of aesculin-azide media with different concentrations of kanamycin (Heeschen *et al.* 1968; van der Wiel-Korstanje & Winkler 1975) and one concentration of amikacin (Finland *et al.* 1976; Linzenmeier *et al.* 1976). In addition we have again investigated whether an increased incubation temperature (Mossel *et al.* 1957) could improve the selectivity of the procedure. The results obtained with pure cultures and with raw materials of animal origin with high 'total' counts are shown in Tables 1-7. They well illustrate the points discussed above. When 'background flora' is no problem, the less selective medium containing $10\ \mu g\ ml^{-1}$ of kanamycin functions quite satisfactorily (Table 4). However, the examination of materials with a considerable non-streptococcal microflora requires either increased selectivity obtained by incubation at *ca.* 42°C (Tables 6 and 7) or an increase in the kanamycin concentration to $20\ \mu g\ ml^{-1}$, both unfortunately at the

Table 1
Recovery on kanamycin-aesculin-azide agar of viable cells of Lancefield group D streptococci at 37°C

Strain		Colonies (duplicates) obtained from same dilution on medium		Type of halo of colony on KAA
		TSBA*	KAA20†	
S. bovis	BRD	50–65	33–37	Black
	UK	124–134	0–0	–
S. durans	9	186–213	161–183	Black
	13	247–268	245–252	Black
	17	177–208	167–181	Black
S. equinus	BRD	229–258	283–310	Black
	UK	153–186	191–221	Grey
S. faecalis	32	174–182	136–153	Black
	55	219–234	195–235	Black
	82	295–308	291–304	Black
	84	156–163	142–164	Black
	118	121–157	113–148	Black
	L1	37–62	65–74	Black
	L2	79–103	82–101	Black
S. faecium	15	335–395	365–401	Black

* TSBA, tryptone-soya-peptone agar, containing (g l^{-1} of distilled water): tryptone, 15; soya peptone, 5; NaCl, 5; agar, 15. Sterilization, 15 min at 121°C; final pH, 7.3. Spread plates, incubated ≤ 3 days at 37°C.
† KAA20, kanamycin-aesculin-azide agar, containing (g l^{-1} of distilled water): tryptone, 20; yeast extract, 5; kanamycin sulphate, 0.020; NaCl, 5; sodium citrate, 1; aesculin, 1; ferric ammonium citrate, 0.5; sodium azide, 0.15; agar, 15. Sterilization, 15 min at 120°C; final pH at 25°C, 7.1. Spread plates, incubated 18–20 h at 37°C.

expense of productivity (Tables 1 and 2). The latter is well illustrated by the last line of Table 5. Whereas increasing the kanamycin concentration from 10 to 20 µg ml^{-1} gave rise to a most satisfactory confirmation rate when minced meat was examined, it resulted in worse results with offal (gullet) because the background was less affected than the streptococci.

Luckily, only infrequently have severely contaminated materials to be examined, namely when production line studies to assess the efficiency of processing for safety are to be carried out. In daily routine only a limited number of commodities have to be examined for group D streptococci. These are all processed specimens of group 1 above: drinking water (Mossel 1977), the disinfected food environment (Mossel et al. 1976), biologically purified seafoods (Slanetz et al. 1968), dried milk products and other weaning foods (Mattick et al. 1945; Kjellander & Nygren 1962; Mossel et al. 1973), packaged, sliced,

Table 2
Recovery of bacteria other than Lancefield group D streptococci on kanamycin-aesculin-azide agar at 37°C

Organism	Colonies (duplicates) obtained from same dilution on medium		Type of halo of colony on KAA
	TSBA*	KAA20*	
Aerococcus 8	50–63	60–69	Light black
Aerococcus 2	72–87	68–83	do.
Aerococcus 3	127–130	69–84	do.
Aerococcus 5	126–141	104–156	do.
Arthrobacter	ca. 10^2	$<1 - <1$	
Bacillus cereus	ca. 10^3	$<1 - <1$	
Candida sp.	ca. 10^5	$<1 - <1$	
Citrobacter freundii	10^4	$<1 - <1$	
Enterobacter aerogenes	2×10^4	$<1 - <1$	
Escherichia coli	ca. 10^4	$<1 - <1$	
Lactobacillus casei	ca. 10^4	$<1 - <1$	
Lactobacillus sp.	ca. 10^3	$<1 - <1$	
Proteus vulgaris	2×10^3	$<1 - <1$	
Pseudomonas aeruginosa 002	ca. 10^5	$<1 - <1$	
Ps. aeruginosa 025	ca. 10^5	$<1 - <1$	
Salmonella typhimurium	ca. 10^4	$<1 - <1$	
Staphylococcus aureus 8043	ca. 10^4	$<1 - <1$	
Staph. aureus 7886	ca. 10^4	$<1 - <1$	
Staph. aureus 7767	ca. 10^4	$<1 - <1$	
Staph. aureus 7760	ca. 10^4	$<1 - <1$	
Staph. aureus 7867	5×10^3	$<1 - <1$	
Staph. aureus 3813	5×10^3	$<1 - <1$	
Staph. aureus Leiden	ca. 10^4	$<1 - <1$	
Staph. aureus Dijkmann	ca. 10^4	$<1 - <1$	
Streptococcus lactis	ca. 2×10^5	19–20	
S. salivarius	245–250	142–157	Black
Yersinia enterocolitica	2×10^3	$<1 - <1$	

* For composition, preparation, inoculation and incubation of media, see footnotes to Table 1.

cooked meat products (Mossel 1962; Mantel & Beck 1977), large size canned ham (Ingram & Barnes 1955) and starter preparations used in the manufacture of fermented sausage (Favre & Ramet 1977). These specimens as a rule contain very few organisms of the groups that, according to literature reports (Burman 1967; Barnes & Corry 1969; Oblinger 1975; Mundt 1976) and our own observations, interfere most frequently with the use of kanamycin-azide-aesculin agars: namely aerococci, staphylococci and lactobacilli. They can therefore be plated on the less inhibitory kanamycin agar—obviously after a preceding resuscitation treatment (cf. Table 3).

Table 3
Recovery of cell populations of Lancefield group D streptococci heated for 30 min at 60°C, pH 7

Strain	Colonies obtained from same dilution on medium at 37°C		
	TSBA*	KAA20*	BAA†
Streptococcus bovis	0.6×10^4	<1	6
S. faecalis L1	0.8×10^4	2	0.5×10^4
S. durans	10^4	<1	0.9×10^3
S. faecium 15	0.6×10^4	$0.1 - 10^3$	0.1×10^4
S. faecalis 84	0.4×10^4	1	0.5×10^4
S. equinus Kiel	0.5×10^2	<1	<1

* For composition, preparation, inoculation and incubation, see footnotes to Table 1.
† BAA, bile-aesculin-azide agar, containing (g l^{-1} of distilled water): tryptone, 20; yeast extract, 5; ox bile, 10; aesculin, 1; ferric ammonium citrate, 0.5; NaCl, 5; sodium azide, 0.15; agar, 15. Sterilization, 15 min at 120°C; final pH at 25°C, 7.1. Inoculation with 0.1 ml amounts as spread plates, incubated 18–20 h at 37°C.

In all instances confirmation of the identity of a representative proportion of typical isolates is required. This relies on the Sherman criteria valid for all group D streptococci: growth at 45 ± 0.1°C, typical morphology, negative catalase reaction, growth in the presence of 40% bile and fermentative anaerogenic attack

Table 4
Recovery of Lancefield group D streptococci from fresh proteinaceous staple foods on various types of aesculin–azide agars, with reference to the inhibition of 'background' organisms

Commodity	Proportion of 'total confirmed' to 'total examined' typical colonies (Σc), and background (b)* at 37°C on			
	BAA† c/e b*	KAA10†† c/e b*	KAA20† c/e b*	AAA10‡ c/e b*
Offal (chicken liver)	364/577 +	312/315 θ	254/254 θ	321/336 θ
Minced beef and pork	148/178 +++	139/149 +++	85/85 +	40/40 +
Poultry meat	59/64 +	43/43 θ	36/36 θ	28/33 θ
Mean	190/273 ++	165/169 +	125/125 θ	130/136 θ

* Background semi-quantitatively assessed as: +++, strong interference; ++, definite interference; +, almost equal numbers of background and sought colonies, i.e. with black halos; θ, virtually only colonies with black halos.
† Composition, preparation, inoculation and incubation, see footnotes to Table 3.
†† As KAA20, but with only 10 μg ml^{-1} kanamycin.
‡ As KAA10, but kanamycin replaced by 10 μg ml^{-1} amikacin.

Table 5

Confirmation rates, i.e. proportion between numbers of colonies with black halos, confirmed as Lancefield group D streptococci and numbers of such colonies examined, on kanamycin-aesculin-azide agar at 37°C

Product	Number of samples examined	Confirmation rate on KAA10*	Confirmation rate on KAA20*
Minced beef and pork	66	0.71	0.93
Offal (gullet)	30	0.80	0.59

* For composition, preparation, inoculation and incubation of media, see footnotes to Table 4.

on glucose (Mossel 1963; Barnes 1976). Strains may be typed, if required, using the remaining Sherman criteria.

Reliable media for the enumeration of Lancefield group D streptococci may also render service in the assessment of the still equivocal role of these bacteria in foodborne diseases, other than pressor amine syndromes. As we have indicated earlier the evidence that Lancefield group D streptococci can give rise to enteric syndromes is most contradictory, but some evidence (e.g. Sedova 1970) cannot be disregarded. Factors like particular attributes of strains or influences exerted by properties of the food wherein these cocci proliferated are worthy of further investigation. Retrospective examination of foods aetiologically involved in outbreaks can be helpful and in this approach reliable enumeration techniques are essential.

(e) *Streptococci of human oral origin*

The monitoring of foods processed for safety relies mostly on an examination for Enterobacteriaceae (Drion & Mossel 1977) and, in some instances discussed above, for Lancefield group D streptococci. Hence there is, in general, no need for a routine search for oral streptococci in foods etc. of groups 1 and 2 above. As these organisms are without pathogenic significance they will not be included in epidemiological investigations on commodities of group 3 either, unless a virus, originating from the oro-pharynx, is suspected, on clinical data.

We found the oral streptococci of some use, though, in investigations on the efficacy of manual and machine dish-washing, particularly of glasses in times of outbreaks of Pfeiffer's glandular fever. Two media are available for this purpose, both containing sucrose as the carbon source (pp. 380-383). Chapman's (1944) formula relies on the use of crystal violet and trypan blue as inhibitors and

Table 6

Recovery ($cfu\ ml^{-1}$) of Lancefield group D streptococci on various azide–aesculin agars incubated at elevated temperature

Medium and incubation temperature*		S. faecalis				S. faecium					S. durans				S. bovis Kiel	S. equinus Kiel
		9	82	118	1	7	8	K1	K17	K22	K23	K50			
TSBA†	37°C	164	34	96	147	96	173	102	129	58	162	31		31	70
	42°C	147	40	84	165	90	169	96	131	54	155	34		25	78
KAA10†	37°C	142	22	57	145	96	156	101	133	66	176	31††		<1	<1
	42°C	159	30	65	168	99	145	95	94††	57	142††	38		<1	<1
KAA20†	37°C	124	35	66	143	81	189	103††	<1	80††	115††	42††		<1	<1
	42°C	119	36	68	132	83††	147	82	<1	<1	<1	26††		<1	<1
BAA†	37°C	174	35	78	152	45	152	100	157	61	180	29††		30††	67††
	42°C	154	35	70	130††	14	170	98	133	64	166	27††		<1	69††

* In air incubator, with temperature variation of $ca.\ \pm 1°C$ from indicated value.
† For composition, preparation, inoculation and incubation of media see footnotes to Tables 1 and 4.
†† Halos less black than usual or even colourless.

Table 7
Recovery of Lancefield group D streptococci from offal on various aesculin-azide media at 42°C

Total specific (t) and confirmed (c) counts (cfu/g) on medium incubated for 18–20 h in an air incubator at 42°C

Product	BAA*			KAA10*			KAA20*		
	t	c	background†	t	c	background†	t	c	background†
Udder	122	67	+	112	102	θ	48	48	θ
Udder	13	13	+	2	2	θ	<1	<1	θ
Udder	54	54	θ	84	84	+	10	10	θ
Chicken liver	56	56	+++	159	159	θ	140	140	θ
Chicken liver	46	46	++	68	60	+	90	90	θ
Gullet	11	11	++	11	11	θ	9	9	θ
Gullet	19	19	++	130	130	θ	120	120	θ
Mean	46	38	++	81	78	θ	59	59	θ

* For composition, preparation and sterilization of media, see footnotes to Table 4.
† For quantification of interfering background, see footnote to Table 4.

tellurite as index chemical, and de Stoppelaar & de Moor's (1967) medium contains sulphite, acetate and a sulphonamide. In studies on foods severely contaminated with other bacteria, increasing the selective properties of these media by the addition of the newer antibiotics, referred to in connection with the media to be used for the enumeration of group A streptococci, is often required.

Colonies obtained on suitable media have always to be further examined for the rather heterogeneous characters of the bacteria sought, principally *S. mitior, salivarius* and *sanguis* (Parker & Ball 1976; Facklam 1977).

3. Control of Contamination by Streptococci

Whenever streptococci of any of the above groups are detected in production line samples of products of group 1 in greater numbers than those attainable by good practice, obviously the matter should be examined. Merely reporting the results without recommending ways to correct defects is of little value: it is a guiding principle of medicine that diagnosis is no more than the first stage of therapy.

When defects are detected in samples of foods or water, processed for safety, it is obvious that the handling has been inadequate. The standards for adequate processing are summarized in the mnemonic expression LIPS = Longitudinal Integrity of Processing for Safety, a concept introduced many years ago by Dack (1965). LIPS includes the following three essential elements (Mossel 1977a).

1. Adequate bactericidal treatment leading to the elimination of pathogens initially present: heating of foods and meals, acid fermentation of dairy and meat products, biological purification of seafoods, chlorination of water, etc.
2. Absence of recontamination which would nullify the effect of the treatment referred to under 1, in the phase between processing for safety and sealed packaging.
3. Subsequent storage under conditions which prevent proliferation of the very low numbers of indicator organisms surviving processing for safety, i.e. the order of 0 to $-1 \log_{10}$ cfu g^{-1} or cfu ml^{-1}.

Each of these three steps has to be monitored separately, leading to an overall evaluation of the processing and thus to the elucidation of the faults. These have to be remedied on the basis of guidelines which can be found in the literature (Mossel 1977a). It is essential that every attempt at improvement is subsequently verified by the examination of production line samples. The methodology indicated in the previous sections has to be used for this purpose. It is necessary that adequate resuscitation of sublethally stressed cells of Lancefield group D streptococci prior to plating on any selective medium is never ignored (Allen et al. 1953; Windle Taylor & Burman 1964; Payne & Morley 1976).

Once the optimal LIPS method has been established, monitoring can henceforth rely on checking of the parameters of the process: F_0 values in case of heat treatments, ultimate pH and velocity of pH drop in the case of fermented products, bacterial condition of effluents from the biological purification of shellfish, chlorine levels in water purification, etc. Full bacteriological spot checks can thus be limited to a minimum within reach of the available laboratory facilities.

4. References

ALLEN, L. A., PIERCE, M. A. F. & SMITH, H. M. 1953 Enumeration of *Streptococcus faecalis* with particular reference to polluted waters. *Journal of Hygiene* 51, 458–467.

ALLOW, R. C., DRISCOLL, S. G., EFFMANN, E. L., GROSS, I., JOLLES, C. J., UANY, R. & WARSHAW, J. B. 1976 A comparison of early-onset group B streptococcal neonatal infection and the respiratory-distress syndrome of the newborn. *New England Journal of Medicine* 294, 65–70.

ANDREWS, J. & FUCHS, A. W. 1948 Pasteurization and its relation to health. *Journal of the American Medical Association* 138, 128–131.

ANTHONY, B. F. & CONCEPCION, N. F. 1975 Group B streptococcus in a general hospital. *Journal of Infectious Diseases* 132, 561–567.

BARNES, E. M. 1976 Methods for the isolation of faecal streptococci. *Laboratory Practice* 25, 145–147.

BARNES, E. M. & CORRY, J. E. L. 1969 Microbial flora of raw and pasteurized egg albumen. *Journal of Applied Bacteriology* 32, 193–205.

BAYER, A. S., CHOW, A. W., ANTHONY, B. F. & GUZE, L. B. 1976a Serious infections in adults due to group B streptococci. Clinical and serotypic characterization. *American Journal of Medicine* 61, 498–503.

BAYER, A. S., SEIDEL, J. S., YOSHIKAWA, T. T., ANTHONY, B. F. & GUZE, L. B. 1976b Group D enterococcal meningitis. *Archives of Internal Medicine* 136, 883–886.

BLACK, W. A. & BUSKIRK, F. VAN 1973 Gentamicin as a selective agent for the isolation of beta haemolytic streptococci. *Journal of Clinical Pathology* 26, 154–156.

BOISSARD, J. M. & FRY, R. M. 1955 A food-borne outbreak of infection due to *Streptococcus pyogenes*. *Journal of Applied Bacteriology* 18, 478–483.

BRODSKY, M. H. & SCHIEMANN, D. A. 1976 Evaluation of Pfizer selective enterococcus and KF media for recovery of fecal streptococci from water by membrane filtration. *Applied and Environmental Microbiology* 31, 695–699.

BULAJIC, Z. 1973 Histaminsko trovanje u jednoj bolnici. *Hrana i Ishrana* 14, 203–207.

BURMAN, N. P. 1967 Recent advances in the bacteriological examination of water. In *Progress of Microbiological Techniques*. London, pp. 185–212.

BUTTER, M. N. W. & MOOR, C. E. DE 1967 *Streptococcus agalactiae* as a cause of meningitis in the newborn, and of bacteraemia in adults. *Antonie van Leeuwenhoek* 33, 439–450.

CHAPMAN, G. H. 1944 The isolation of streptococci from mixed cultures. *Journal of Bacteriology* 48, 113–114.

CLARK, W. S., REINBOLD, G. W. & RAMBO, R. S. 1966 Enterococci and coliforms in dehydrated vegetables. *Food Technology* 20, 1353–1356.

COLOBERT, L. & MORELIS, P. 1958 Nouveau milieu sélectif pour l'isolement des entérocoques, d'emploi pratique pour leur recherche et leur numération dans l'eau. *Annales de l'Institut Pasteur, Paris* 94, 120–122.

DACK, G. M. 1956 Evaluation of microbiological standards for foods. *Food Technology* 10, 507–509.

DEIBEL, R. H. & SILLIKER, J. H. 1963 Food-poisoning potential of the enterococci. *Journal of Bacteriology* **85**, 827-832.

DOEGLAS, H. M. G., HUISMAN, J. & NATER, J. P. 1967 Histamine intoxication after cheese. *Lancet* **2**, 1361-1362.

DRION, E. F. & MOSSEL, D. A. A. 1977 The reliability of the examination of foods, processed for safety, for enteric pathogens and Enterobacteriaceae: a mathematical and ecological study. *Journal of Hygiene* **78**, 301-324.

DUDDING, B. A., DILLON, H. C., WANNAMAKER, L. W., KILTON, R. M., CHAPMAN, S. S. & ANTHONY, B. F. 1969 Postepidemic surveillance studies of a food-borne epidemic of streptococcal pharyngitis at the United States Air Force Academy. *Journal of Infectious Diseases* **120**, 225-236.

ERBSLOH, F. & GRÜN, L. 1949 Galtstreptokokken der serologischen Gruppe B als Erreger der Endokarditis lenta. *Deutsche medizinische Rundschau* **3**, 508.

FACKLAM, R. R. 1977 Physiological differentiation of viridans streptococci. *Journal of Clinical Microbiology* **5**, 184-201.

FACKLAM, R. R., PADULA, J. F., THACKER, L. G., WORTHAM, E. C. & SCOYERS, B. J. 1974 Presumptive identification of group A, B and D streptococci. *Applied Microbiology* **27**, 107-113.

FANELLI, M. J. & AYRES, J. C. 1959 Methods of detection and effect of freezing on the microflora of chicken pies. *Food Technology* **13**, 294-300.

FARBER, R. E. & KORFF, F. A. 1958 Foodborne epidemic of group A beta hemolytic streptococcus. *Public Health Reports Washington* **73**, 203-209.

FAVRE, C. & RAMET, F. 1977 Les ferments lactiques utilisés dans l'industrie des produits carnés. *Revue technique Vétérinaire des Aliments d'Origine Animale* **130**, 19-21.

FEACHEM, R. 1974 Faecal coliforms and faecal streptococci in streams in the New Guinea highlands. *Water Research* **8**, 367-374.

FINLAND, M., GARNER, C., WILCOX, C. & SABATH, L. D. 1976 Susceptibility of recently isolated bacteria to amikacin in vitro: comparisons with four other aminoglycoside antibiotics. *Journal of Infectious Diseases* **134**, S297-S307.

FOO, L. Y. 1976 Scombroid poisoning. Isolation and identification of "saurine". *Journal of the Science of Food and Agriculture* **27**, 807-810.

FRANCIOSI, R. A., KNOSTMAN, J. D. & ZIMMERMAN, R. A. 1973 Group B streptococcal neonatal and infant infections. *Journal of Pediatrics* **82**, 707-718.

GELLMAN, M., MANSDORF, W., BEIN, M., SHAHIDI, S., MARR, J. S. & MUNSON, L. 1975 Scombroid poisoning—New York City. *Morbidity and Mortality Weekly Report* **24**, 342-347.

GODFREY, E. S. 1929 Age distribution in milk-borne outbreaks of scarlet fever and diphtheria. *American Journal of Public Health* **19**, 257-264.

HAHN, G., HEESCHEN, W., REICHMUTH, J. & TOLLE, A. 1974 Interrelations entre les streptocoques du groupe B qui provoquent des infections chez l'homme et les bovins. *Le Lait* **54**, 252-268.

HEESCHEN, W., TOLLE, A. & ZEIDLER, H. 1968 Die selektive Züchtung und Anreicherung Koagulase-positiver Staphylokokken und enteropathogener Streptokokken in flüssigen Nährmedien. *Archiv für Lebensmittelhygiene* **19**, 184-189.

HEMMING, V. G., McCLOSKEY, D. W. & HILL, H. R. 1976 Pneumonia in the neonate associated with group B streptococcal septicemia. *American Journal for Diseases of Children* **130**, 1231-1233.

HILL, H. R., ZIMMERMAN, R. A., REID, G. V. K., WILSON, E. & KILTON, R. M. 1969 Food-borne epidemic of streptococcal pharyngitis at the United States Air Force Academy. *New England Journal of Medicine* **280**, 917-921.

INGRAM, M. & BARNES, E. 1955 Streptococci in pasteurized canned hams. *Annales de l'Institut Pasteur Lille* **7**, 101-111.

JENSEN, N. E. 1976 Forgaeringstyper af gruppe B streptokokker. *Nordisk Veterinar Medicine* **28**, 434-443.

JOKIPII, A. M. M. & JOKIPII, L. 1976 Presumptive identification and antibiotic susceptibility of group B streptococci. *Journal of Clinical Pathology* **29**, 736-739.

KAHLER, J. & AICHER, J. 1952 Der Galtstreptokokkus als Erreger einer ulzero-polypösen Endokarditis beim Menschen. *Zeitschrift für allgemeine Pathologie* **88**, 312.

KERELUK, K. & GUNDERSON, M. F. 1959a Studies on the bacteriological quality of frozen meat pies. I. Bacteriological survey of some commercially frozen meat pies. *Applied Microbiology* **7**, 320-323.

KERELUK, K. & GUNDERSON, M. F. 1959b Studies on the bacteriological quality of frozen meat pies. IV. Longevity studies on the coliform bacteria and enterococci at low temperature. *Applied Microbiology* **7**, 327-328.

KEXEL, G. 1965 Ueber das Vorkommen der B-Streptokokken beim Menschen. *Zeitschrift für Hygiene und Infektionskrankheiten* **151**, 336-348.

KEXEL, G. & SCHÖNBOHM, S. 1965 *Streptococcus agalactiae* als Erreger von Säuglingsmeningitiden. *Deutsche medizinische Wochenschrift* **90**, 258-261.

KJELLANDER, J. & NYGREN, B. 1962 On the occurrence of faecal streptococci in industrial food products. *Acta Pathologica Microbiologica Scandinavica* Suppl. **154**, 323-326.

LANG, K. 1961 Ueber Lebensmittelschädigungen durch Streptokokken. *Archiv für Lebensmittelhygiene* **12**, 125-127.

LEE, I. & KOBURGER, J. A. 1970 Incidence and identification of some beta-hemolytic streptococci in foods. *Journal of Milk and Food Technology* **33**, 323-325.

LEGROUX, R. & LEVADITI, J. 1947 Origine de l'histamine présente dans la chair de thons responsables d'intoxications collectives. *Comptes Rendues Société de Biologie* **141**, 998-1000.

LEVIN, M. A., FISCHER, J. R. & CABELLI, V. J. 1975 Membrane filter technique for enumeration of enterococci in marine waters. *Applied Microbiology* **30**, 66-71.

LINZENMEIER, G., NAUMANN, P., NEUSSEL, H. & ROSIN, H. 1976 In vitro susceptibility of clinically important bacteria to amikacin: correlation of results of broth dilution and disk sensitivity tests and effect of medium composition. *Journal of Infectious Diseases* **134**, S262-S270.

LOWBURY, E. J. L., KIDSON, A. & LILLY, H. A. 1964 A new selective blood agar medium for *Streptococcus pyogenes* and other haemolytic streptococci. *Journal of Clinical Pathology* **17**, 231-235.

LÜOND, H. & GASSER, H. 1964 Faekalstreptokokken als Lebensmittelvergifter. *Mitteilungen auf dem Gebiete der Lebensmitteluntersuchung und Hygiene* **55**, 144-149.

McCORMICK, J. B., KAY, D. M., HAYES, P. & FELDMAN, R. 1976 Epidemic streptococcal sore throat following a community picnic. *Journal of the American Medical Association* **236**, 1039-1041.

McKNIGHT, J. F., ELLIS, J., JENSEN, K. A. & FRANZ, B. 1969 Group B streptococci in neonatal deaths. *Applied Microbiology* **17**, 926.

MANTEL, Th. & BECK, G. 1977 Zur mikrobiologischen Situation vorverpackter Brühwurst. *Fleischwirtschaft* **57**, 245-247.

MASON, E. O., WONG, P. & BARRETT, F. F. 1976 Evaluation of four methods for detection of group B streptococcal colonization. *Journal of clinical Microbiology* **4**, 429-431.

MATTICK, A. T. R., HISCOX, E. R. & CROSSBY, E. L. 1945 The effect of temperature of pre-heating, of clarification and of bacteriological quality of the raw milk on the keeping properties of whole-milk powder dried by the Kestner spray-process. Part II. The effect of the various factors upon the bacterial (plate) count of the intermediate products and of the final powder. *Journal of Dairy Research* **14**, 135-144.

MERSON, M. H., BAINE, W. B. & GANGAROSA, E. J. 1974 Scombroid fish poisoning: outbreak traced to commercially canned tuna fish. *Journal of the American Medical Association* **228**, 1268-1269.

MEYER, K. & SCHÖNFELD, H. 1926 Ueber die Unterscheidung des Enterococcus vom *Streptococcus viridans* und die Beziehungen beider zum *Streptococcus lactis*. *Zentralblatt für Bakteriologie und Parasitenkunde, Abteilung I, Originale* **99**, 402-416.

MHALU, F. S. 1976 Infection with *Streptococcus agalactiae* in a London hospital. *Journal of Clinical Pathology* **29**, 309-312.

MOSSEL, D. A. A. 1962 Significance of micro-organisms in foods. *In: Chemical and Biological Hazards in Food*, eds Ayres, J. C., Kraft, A. A., Snyder, H. E. & Walker, H. W., Ames, Iowa State University Press, pp. 157-201.

MOSSEL, D. A. A. 1963 The suitability of streptococci of Lancefield's group D for the estimation of the wholesomeness of milk products. *Netherlands Milk and Dairy Journal* 17, 404-415.

MOSSEL, D. A. A. 1968 Bacterial toxins of uncertain oral pathogenicity. In *The Safety of Foods*, eds Ayres, J. C. *et al.*, Westport, Conn., USA, Avi Publishing Company, pp. 168-182.

MOSSEL, D. A. A. 1977*a Microbiology of Foods. Occurrence, Prevention and Monitoring of Hazards and Deterioration*. Utrecht. The University.

MOSSEL, D. A. A. 1977*b* Microbiological quality assurance of water in relation to food hygiene. *Archiv für Lebensmittelhygiene* 28, 1-2.

MOSSEL, D. A. A. 1978 Index and indicator organisms—a current assessment of their usefulness and significance. *Food Technology Australia* 30, accepted for publication.

MOSSEL, D. A. A. & CORRY, J. E. L. 1977 Detection and enumeration of sublethally injured pathogenic and index bacteria in foods and water processed for safety. *Alimenta* 16, Sondernummer Mikrobiologie, 19-34.

MOSSEL, D. A. A., DIEPEN, H. M. J. VAN & BRUIN, A. S. DE 1957 The enumeration of faecal streptococci in foods, using Packer's crystal violet sodium azide blood agar. *Journal of applied Bacteriology* 20, 265-272.

MOSSEL, D. A. A., EELDERINK, I., VOR, H. DE & KEIZER, E. D. 1976 Use of agar immersion, plating and contact (AIPC) slides for the bacteriological monitoring of food, meals and the food environment. *Laboratory Practice* 25, 393-395.

MOSSEL, D. A. A., HARREWIJN, G. A. & SPRANG, F. J. VAN 1973 Microbiological quality assurance for weaning formulae. *In The Microbiological Safety of Food*, eds Hobbs, B. C. & Christian, J. H. B. London & New York. Academic Press, pp. 77-88.

MOSSEL, D. A. A., HARREWIJN, G. A., SPREEKENS, K. VAN, BRIDSON, E. Y., BIJKER, P. G. H. & BOER, H. B. DE 1975 Experience with the use of Colobert's modified medium (bile aesculin azide agar) for the enumeration of Lancefield group D streptococci in various foods. *Journal of Applied Bacteriology* 39, vi.

MUNDT, J. O. 1976 Streptococci in dried and frozen foods. *Journal of Milk and Food Technology* 39, 413-416.

NEVOT, A., LAFONT, Ph. & LAFONT, J. 1958 De la destruction des bactéries par la chaleur. Etude de l'efficacité de la pasteurisation du lait. *Monographie No. 18, Institut National d'Hygiène*, Paris. 140p.

OBIGER, G. 1975 Gruppe B Streptokokken im Genitale der Frau. *Archiv für Gynakologie* 218, 65-84.

OBIGER, G. 1976 Untersuchungen über die Thermostabilität bedeutsamer Infektionserreger unter den Bedingungen der Milchpasteurisation. *Archiv für Lebensmittelhygiene* 27, 137-144.

OBLINGER, J. L.1975 Recovery of streptococci from a variety of foods: a comparison of several media. *Journal of Milk and Food Technology* 38, 323-326.

OSTROLENK, M., KRAMER, N. & CLEVERDON, R. C. 1947 Comparative studies of enterococci and *Escherichia coli* as indices of pollution. *Journal of Bacteriology* 53, 197-203.

OTTE, H. J. & RITZERFELD, W. 1960 Massenerkrankung an Angina durch Streptokokken in Lebensmitteln. *Deutsche medizinische Wochenschrift* 85, 1625-1628.

PACKER, R. A. 1943 The use of sodium azide (NaN_3) and crystal violet in a selective medium for Streptococci and *Erysipelothrix rhusiopathiae*. *Journal of Bacteriology* 46, 343-349.

PARK, W. H. 1928 Thermal death points of streptococci. *American Journal of Public Health* 18, 710-714.

PARKER, M. T. & BALL, L. C. 1976 Streptococci and Aerococci associated with systemic infection in man. *Journal of Medical Microbiology* 9, 275-302.

PARROT, J. L. & NICOT, G. 1965 Le rôle de l'histamine dans l'intoxication alimentaire par le poisson. *Bulletin de la Société Scientifique d'Hygiëne Alimentaire* **53**, 76–82.

PATTERSON, M. J. & HAFEEZ, A. E. B. 1976 Group B streptococci in human disease. *Bacteriological Reviews* **40**, 774–792.

PAVLOVA, M. T., BREZENSKI, F. T. & LITSKY, W. 1972 Evaluation of various media for isolation, enumeration and identification of fecal streptococci from natural sources. *Health Laboratory Science* **9**, 289–298.

PAYNE, J. & MORLEY, J. S. 1976 Recovery of tellurite resistance by heat injured *Streptococcus faecalis*. *Journal of General Microbiology* **94**, 421–424.

PROVOST, P. J., ITTENSOHN, O. L., VILLAREJOS, V. M., ARGUEDAS, G. J. A. & HILLEMAN, M. R. 1973 Etiologic relationship of marmoset-propagated CR 326 hepatitis A virus to hepatitis in man. *Proceedings of the Society for Experimental Biology and Medicine* **142**, 1257–1267.

PUSZTAI, S., VETESI, F. & HOCH, V. 1972 Toxicity testing of enterococcus strains responsible for food poisoning. *Acta Veterinaria Academia Scientia Hungarica* **22**, 299–306.

RAJ, H., WIEBE, W. J. & LISTON, J. 1961 Detection and enumeration of fecal indicator organisms in frozen sea foods. II. Enterococci. *Applied Microbiology* **9**, 295–303.

RICE, S. L., EITENMILLER, R. R. & KOEHLER, P. E. 1976 Biologically active amines in food: A review. *Journal of Milk and Food Technology* **39**, 353–358.

ROCHAIX, A. 1924 Milieu à l'esculine pour le diagnostic différential des bactéries du groupe strepto-entéropneumocoque. *Comptes Rendus Société de Biologie* **90**, 771–772.

SAVAGE, W. 1949 Milk-borne infections in Great Britain. *British Journal of Social Medicine* **3**, 45–55.

SEDOVA, N. N. 1970 A study of the role of enterococci in the aetiology of food poisoning. *Voprosy Pitaniya* **29**, 82–87. *Abstracts of Hygiene* **45**, 1145.

SHANNON, E. L., REINBOLD, G. W. & CLARK, W. S. 1964 Heat and chlorine resistance of enterococci. *Journal of Dairy Science* **47**, 666.

SLANETZ, L. W., BARTLEY, C. H. & STANLEY, K. W. 1968 Coliforms, fecal streptococci and *Salmonella* in seawater and shellfish. *Health Laboratory Science* **5**, 66–78.

STEINITZ, H., SCHUCHMANN, L. & WEGNER, G. 1971 Leberzirrhose, Meningitis und Endokarditis ulceropolyposa bei einer Neugeborenensepsis durch B-Streptokokken (*Strept. agalactiae*). *Archiv für Kinderheilkunde* **183**, 382.

STOPPELAAR, J. D. DE, HOUTE, J. VAN & MOOR, C. E. DE 1967 The presence of dextran-forming bacteria, resembling *Streptococcus bovis* and *Streptococcus sanguis*, in human dental plaque. *Archives of Oral Biology* **12**, 1199–1201.

SVARTZ, N. 1972 The primary cause of rheumatoid arthritis is an infection—the infectious agent exists in milk. *Acta Medica Scandinavica* **192**, 231–239.

SWAN, A. 1954 The use of bile-esculin medium and of Maxted's technique of Lancefield grouping in the identification of enterococci (group D streptococci). *Journal of Clinical Pathology* **7**, 160–163.

SWITZER, R. E. & EVANS, J. B. 1974 Evaluation of selective media for enumeration of group D streptococci in bovine feces. *Applied Microbiology* **28**, 1086–1087.

TAYLOR, P. J. & McDONALD, M. A. 1959 Milkborne streptococcal sore throat. *Lancet* **1**, 330–333.

VEEN, A. G. VAN & LATUASAN, H. E. 1950 Fish poisoning caused by histamine in Indonesia. *Documenta Neerkandica Indonesia Morbis Tropicis* **2**, 18–20.

VINCENT, W. F., GIBBONS, E. W. & GAAFAR, H. A. 1971 Selective medium for the isolation of streptococci from clinical specimens. *Applied Microbiology* **22**, 942–943.

VON KURNATOWSKY, H. A., SIERRA-CALLEJAS, J. L., HENKEL, W. & DIEDERICK, K. W. 1977 Foudroyant tödlich verlaufende Myokarditis durch Streptokokken der Gruppe B. *Deutsche medizinische Wochenschrift* **102**, 439–441.

WALLIN, J. & FORSGREN, A. 1976 Group B streptococci in venereal disease clinic patients. *British Journal of Venereal Diseases* **51**, 401–404.

WEBSTER, R. C. & ESSELEN, W. B. 1956 Thermal resistance of food poisoning organisms in poultry stuffing. *Journal of Milk and Food Technology* **19**, 209–212.

WIEL-KORSTANJE, J. A. A. VAN DER & WINKLER, K. C. 1975 The faecal flora in ulcerative colitis. *Journal of Medical Microbiology* **8**, 491–501.
WILSON, G. S. 1955 Symposium on food microbiology and public health: general conclusions. *Journal of Applied Bacteriology* **18**, 629–630.
WINDLE TAYLOR, E. & BURMAN, N. P. 1964 The application of membrane filtration techniques to the bacteriological examination of water. *Journal of Applied Bacteriology* **27**, 294–303.

Streptococci as Indicators in Water Supplies

N. P. Burman, Janet K. Stevens and A. W. Evans

Thames Water Authority,
New River Head Laboratories
London, England

CONTENTS

1. Introduction 335
2. Routine application of faecal streptococci counts 336
   (a) Faecal streptococci as supplementary indicator organisms . . . 336
   (b) Methodology in the United Kingdom 336
   (c) Application to river and sewage samples 337
3. The use of faecal coliform: faecal streptococci ratios to indicate the source of faecal pollution 340
   (a) Theoretical basis 340
   (b) Application to river and sewage samples 341
4. Acknowledgements 346
5. References 346

1. Introduction

ESCHERICHIA COLI is the organism most widely used as an indicator of faecal pollution of water in this country, but the significance of other members of the coliform group is debatable. Outside the body, coliform organisms other than *E. coli* frequently show greater powers of survival than *E. coli* and may multiply on decaying vegetation or other materials that satisfy their nutritional requirements. When the faecal nature of pollution is doubtful, as in situations where large numbers of coliform organisms are isolated in the absence of *E. coli*, it is of value to search for other intestinal bacteria such as faecal streptococci or *Clostridium perfringens*. For the purposes of this paper the term faecal streptococci will be taken to refer to the Lancefield serological group D streptococci; that is the *Streptococcus faecalis-faecium-durans* group (enterococci) and the *S. bovis-equinus* group.

Faecal streptococci, like *E. coli* and *Cl. perfringens*, do not normally multiply in water and are found where there has been contamination by excreta. They are not found in pure waters or in virgin soils and sites remote from human and animal life (Medrek & Litsky 1960; Holden 1970). The occurrence of faecal streptococci in water, therefore, infers either direct or indirect faecal contamination (Deibel 1964).

2. Routine Application of Faecal Streptococci Counts

The most common examples of the use of faecal streptococci counts by the Water Authorities are in the assessment of the sanitary significance of coliform organisms isolated in the absence of *E. coli* from new and repaired water mains and water sources such as springs or wells which may be used in rural areas as sources of potable water after minimal treatment.

(a) *Faecal streptococci as supplementary indicator organisms*

The basic assumption behind the employment of faecal streptococci counts in these cases is that under adverse environmental conditions the enterococci die off more slowly than *E. coli* and therefore, despite an initial inferiority in numbers, the enterococci may after a passage of time become the more numerous organisms (Burman 1961; Geldreich 1966). Thus the enterococci may become the more readily detectable indicator organisms in samples of water subject to remote or intermittent faecal pollution (Burman 1961).

In the case of new and repaired water mains the coliform organisms and faecal streptococci that are occasionally isolated after chlorination usually represent the survivors of organisms associated with soil contamination. Such contamination is difficult to disinfect, particularly when the soil is in pipe joints. The presence of faecal streptococci in water samples from mains in the absence of *E. coli* may be due to the greater resistance of faecal streptococci to chlorination compared with *E. coli* (Kjellander 1960; Buttiaux & Mossel 1961), but can equally well be attributed to the effect of the different die-off rates of these organisms in faecally contaminated soil (van Donsel *et al.* 1967) prior to the introduction of the soil into the pipework. Surviving coliform organisms may multiply on greases, jointing compounds and other materials to give high counts, the sanitary significance of which is ascertained by testing for faecal streptococci.

(b) *Methodology in the United Kingdom*

The standard tests for water supply bacteriology in the United Kingdom have been comprehensively described (*Anon.* 1969). The recommended membrane filtration method for enumeration of faecal streptococci utilizes the membrane enterococcus agar (M-EA) of Slanetz & Bartley (1957), but with incubation for 4 h at 37°C as a 'resuscitation' step, followed by 44 h at 44–45°C, in contrast to the 48 h incubation period at 35°C suggested by Slanetz & Bartley (1957) and in Standard Methods (*Anon.* 1971). The multiple tube method of Hannay & Norton (1947) may be used where filtration of suitable volumes is prevented by excessive amounts of suspended solids.

The technical modification of increasing the incubation temperature for the membrane filtration technique was originally adopted because incubation at 35-37°C was found, with river samples, to lead to the recovery of a large proportion of false positive isolates, including streptococci of indeterminate status that did not belong to the Lancefield serological group D (Burman 1961; Windle Taylor & Burman 1964). Incubation at 45°C eliminated most of these unclassifiable strains, and counts could be directly assessed without recourse to confirmatory procedures (*Anon.* 1966).

(c) *Reliability in practice*

The following experimental work was performed to compare the effect of incubation at 37 and 45°C on the selectivity and sensitivity of the M-EA medium used in the membrane filtration technique for the isolation of faecal streptococci.

River water samples were taken at weekly intervals over a period of five months (February to June) at two intakes on the River Lee in north London. Samples of settled sewage and the final treated effluent were collected on six occasions during the same period from a large sewage treatment works in Middlesex. Suitable volumes or dilutions of the samples were filtered through cellulose acetate membrane filters which were placed on plates of M-EA. Filtrations were performed in duplicate and the membranes were incubated either at 37°C for 48 h or at 37°C for 4 h, followed by 44 h at 45°C.

Colonies appearing on the membranes after 48 h incubation or a representative proportion of them were checked for purity by dilution streaking on a glucose, yeast and meat extract medium as used by Mead (1965) and Gram stained using Lillie's modification as described by Cowan (1974). The original colonial morphology was recorded and colonies categorized by size into three classes; >1 mm, 1-0.5 mm and <0.5 mm. These classes were further divided on the basis of pigment production from the reduction of the 2, 3, 5, -triphenyltetrazolium chloride (TTC) incorporated in M-EA into maroon, red and pink groups. Isolates showing streptococcal morphology (chains of Gram positive cocci) were then inoculated on the glucose medium supplemented with 10% horse blood and 40% ox bile and incubated at 37°C for 3 days. Isolates growing within 3 days on the 40% bile medium were streaked on plates of the tyrosine-sorbitol-thallous acetate agar (TSTA) of Mead (1963) and incubated for 3 days at 45°C. Cultures showing zones of clearing (decarboxylation of tyrosine) and an even maroon colouration (reduction of TTC) were accepted as *S. faecalis sensu strictu* (Mead 1963).

The results are shown in Table 1. With water samples from the River Lee, confirmation of isolates as bile-tolerant streptococci was improved on average from 63 to 85% on increasing the incubation temperature from 37 to 45°C.

Table 1
Selectivity of M-EA with river and sewage samples

	River samples incubated at		Sewage samples incubated at			
	37°C	37–45°C	37°C	37–45°C	37°C	37–45°C
Number of colonies investigated	449	224	259	316	234*	290*
(As % total investigated):						
Bile-tolerant streptococci	63	85	60	76	66	83
Bile-sensitive streptococci	21	3	20	3	22	4
Streptococci not isolated	16	12	20	21	12	13
TSTA positive	1	4	1	4	1	6

* Discounting large maroon colonies of Gram negative rods.

Sewage samples demonstrated a similar increase from 60 to 76% confirmation as bile-tolerant streptococci.

The distribution of the classes of colonial morphology observed on M-EA are shown in Figs 1 and 2 for river waters. A similar distribution was obtained for sewage. With both river and sewage samples incubated at 37°C the majority of isolates produced maroon colonies. Incubation at 45°C increased the proportion of maroon colonies. The bile-sensitive streptococci and those colonies from which streptococci were not isolated were concentrated in the maroon colony classes of < 1 mm diameter, apart from a significant group of Gram negative rods that produced large (> 1 mm) maroon colonies on the membranes from some sewage samples. The presence of these organisms was noted since particularly poor confirmation rates for bile-tolerant streptococci were observed in the samples in which the organisms occurred.

If Gram negative rods producing large maroon colonies (which with experience could be distinguished from those of streptococci) were discounted from the results, then the confirmation rates for bile-tolerant streptococci in sewage samples incubated at 37 and 45°C became 66 and 83%, respectively (Table 1). The results for the sewage and treated effluent samples became comparable with the results from the river samples. Both river and sewage sample results are similar to those found by G. Stanfield (pers. comm.) in recent work with sewage in which confirmation rates for faecal streptococci, based on the criteria of Sherman (1937) were 69% at 35°C and 87% at 44°C.

Although faecal streptococci grow in the presence of 40% bile, (Deibel 1964; Cowan 1974) the recovery rate of these organisms at 37°C on M-EA was no more than 66% overall and 85% at 45°C in our investigation. These results do not agree with the statements by Slanetz & Bartley (1957) and in Standard Methods (*Anon.* 1971) that M-EA medium at 35°C is 100% selective for faecal streptococci.

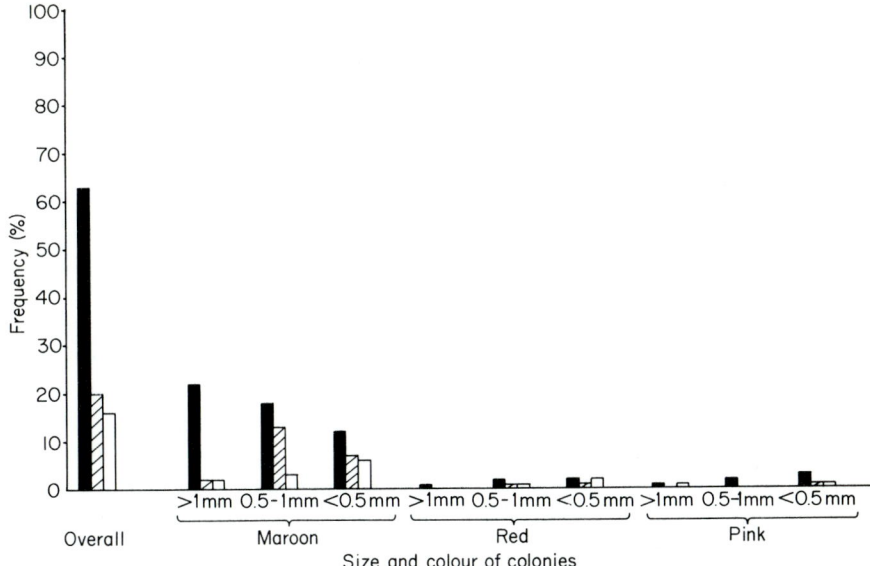

Fig. 1. Distribution of colony types isolated from river water on M-EA. Incubation for 48 h at 37°C. ■, Bile-tolerant streptococci; ▨, bile-sensitive streptococci; ☐, non streptococci.

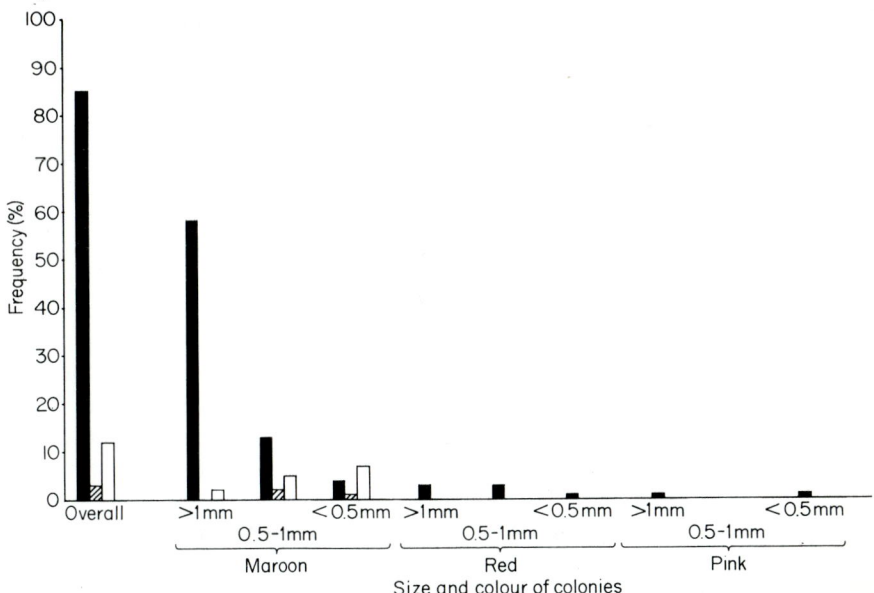

Fig. 2. Distribution of colony types isolated from river water on M-EA. Incubation for 4 h at 37°C, followed by 44 h at 45°C. ■, Bile-tolerant streptococci; ▨, bile-sensitive streptococci; ☐, non streptococci.

Table 2
Recovery of faecal streptococci on M-EA at 37°C and 45°C

	Ratio 37°C count: 45°C count	
	Range	Geometric mean
River samples		
Presumptive counts	1-125:1	5.6:1
Bile-tolerant streptococci	0.8-100:1	3.6:1
Sewage samples		
Presumptive counts	2.5-22:1	7.0:1
Bile-tolerant streptococci	1.3-4.4:1	2.3:1

The distribution of colony types (Figs 1 and 2) suggests that although most of the isolates that were not faecal streptococci could be grouped in two classes (maroon; 1-0.5 mm and < 0.5 mm), modification of counting criteria to exclude these classes would not substantially improve the reliability of direct counts from membranes incubated at 37°C.

The use of incubation at 45°C improved the confirmation rates of counts, but with the samples examined, this improvement in selectivity led to a reduction in sensitivity as demonstrated by the higher recoveries at 37°C (Table 2). Table 2 gives the ratio of 37°C counts to 45°C counts from a set of 20 river samples and 12 sewage samples. Because of the considerable variation in the ratios found in individual samples, the results are presented as the range and geometric mean of the ratios.

3. The Use of Faecal Coliform: Faecal Streptococci Ratios to Indicate the Source of Faecal Pollution

(a) *Theoretical basis*

It has been suggested (Geldreich & Kenner 1969; Mara 1974; Feachem 1975) that the ratio of faecal coliforms to faecal streptococci may be used to indicate the source of faecal pollution of water. The basis for distinguishing human faecal pollution from animal pollution lies in the initial bacterial composition of the faeces and the different survival rates of faecal coliforms and faecal streptococci in natural waters.

In human faeces the faecal coliforms (FC) outnumber the faecal streptococci (FS), of which the enterococci predominate (Cooper & Ramadan 1955; Kenner *et al.* 1960; Mead 1965). Human faeces and sewage exhibit FC to FS ratios >4 (Geldreich & Kenner 1969). Conversely, in the faeces of animals, particularly those of cattle, the predominant faecal streptococci are non-enterococci of the *S. bovis-equinus* group which outnumber the faecal coliforms to give FC to FS ratios < 0.7 (Kjellander 1960; Raibaud *et al.* 1961; Geldreich 1966).

It is therefore theoretically possible to ascribe a human or animal source to faecal pollution of water on the evidence of FC to FS ratios greater than 4 or less than 0.7, provided that the pollution is known to be of recent origin. Geldreich & Kenner (1969) recommended that initial FC to FS ratios should only be considered if samples are taken in the first 24 h following discharge of the polluting faecal material to a watercourse. This would not take account of the time interval between defecation and discharge and in practice it is usually difficult or impossible to determine the age of faecal pollution of water. In these circumstances the observation of FC to FS ratio changes in a stored sample may give some indication of the source of pollution (Feachem 1976).

The reported die-off rates of faecal streptococci and faecal coliforms in natural waters (Kjellander 1960; Geldreich & Kenner 1969; McFeters *et al.* 1974) can be summarized as: *S. bovis-equinus* group > faecal coliforms > *S. faecalis-faecium* group.

It follows that samples giving an initial FC to FS ratio greater than 4 that decreased with time should indicate faecal pollution from a human source, whilst samples giving initial FC to FS ratios less than 0.7 which subsequently rose should indicate an animal source (Feachem 1975).

(b) *Application to river and sewage samples*

Since the ability to discriminate between human and animal pollution using relatively simple tests would be of advantage to the Water Authorities (in an attempt to determine the source of the pollution), the survival of faecal streptococci and faecal coliforms was investigated in stored water samples.

Samples of settled sewage, treated sewage effluent and river water from the sources described previously were stored in clear glass bottles of 2.8 l capacity, either in the dark or in the light in a north facing room. Temperature fluctuations during the period of the experiments were $20 \pm 3°C$. The bottles were sampled daily after being vigorously shaken for 1 min, and faecal streptococci and *E. coli* were enumerated using the membrane filtration methods for the bacteriological examination of water supplies (*Anon.* 1969). Presumptive *E. coli* counts were confirmed by the subculture of a representative proportion of colonies from the membranes; isolates producing acid and gas from lactose peptone water and indole from tryptone water within 24 h at $44 \pm 0.25°C$ were accepted as *E. coli*. Confirmed isolates were tested for oxidase production by the method of Kovačs as described by Cowan (1974).

The oxidase tests failed to reveal any oxidase positive isolates, thus eliminating *Aeromonas* spp. Since the faecal coliform group (Geldreich 1966) includes organisms which do not produce indole from tryptophan at 44°C, the confirmed test for *E. coli* may exclude some faecal coliforms from the results. However with the samples examined, over 90% (frequently 100%) of the isolates producing acid and gas from lactose at 44°C were confirmed as *E. coli* and therefore *E. coli* counts can be taken as faecal coliform counts.

All the samples demonstrated FC to FS ratios greater than 4. The settled sewage samples represent recent faecal pollution (probably less than 24 h) and therefore, as expected, the FC to FS ratio indicates human pollution, although some animal-derived faecal material may also be present. In the absence of further inputs the treated sewage effluent can be regarded as representing human faecal pollution. The river Lee samples were taken at points where the major source of faecal pollution was considered to be treated sewage effluents. FC to FS ratios > 4 support this view, but could theoretically arise from animal pollution that occurred some considerable time previously. The similar percentage of faecal streptococci counts that were confirmed as *S. faecalis sensu strictu* from both river and sewage samples (Table 1) also suggests that the

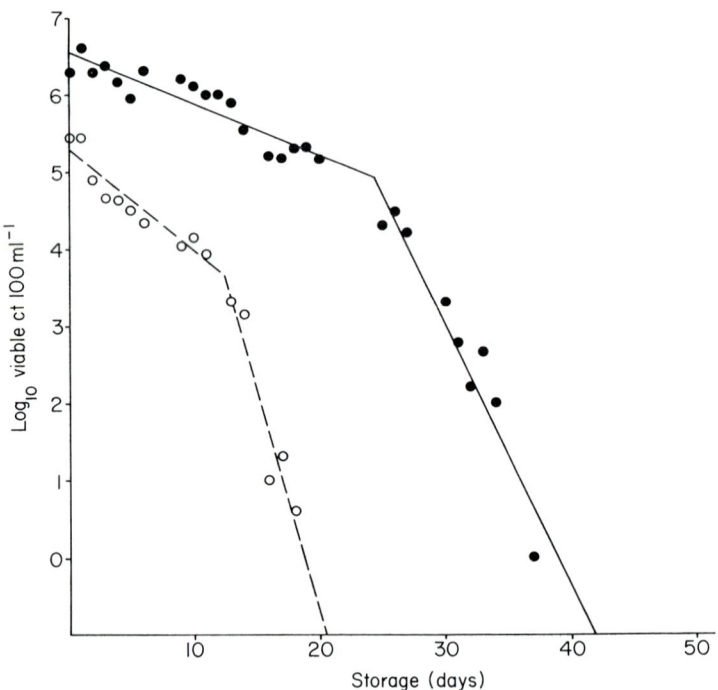

Fig. 3. The survival of *E. coli* and faecal streptococci in stored settled sewage in the light. Storage temperature: 20 ± 3°C. ●, *E. coli*; ○, faecal streptococci.

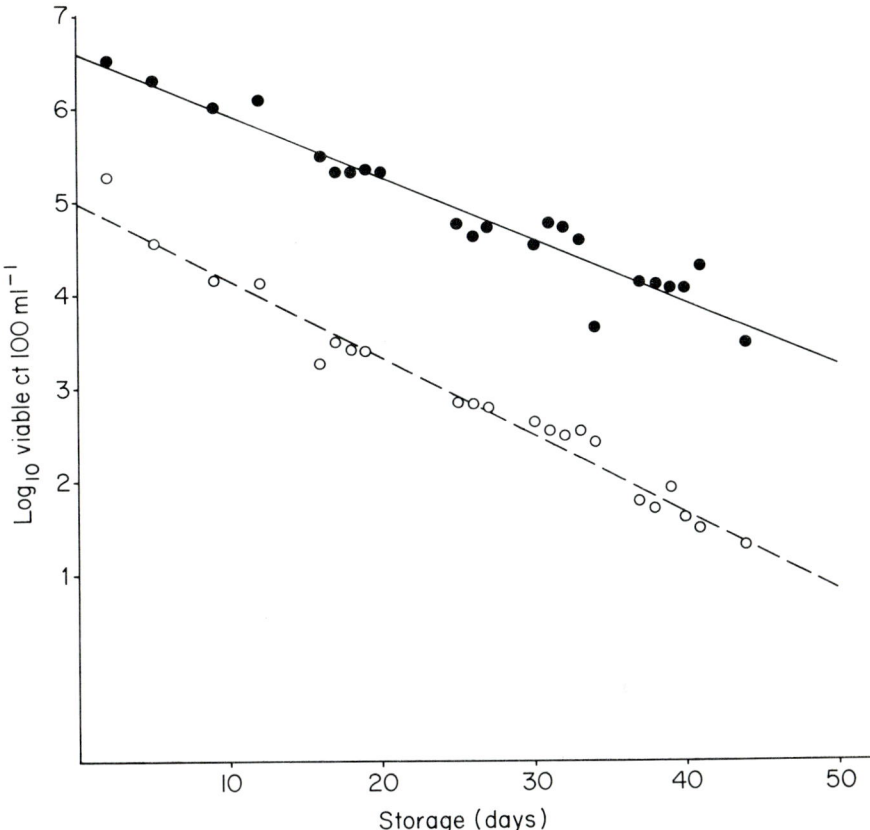

Fig. 4. The survival of *E. coli* and faecal streptococci in stored settled sewage in the dark. Storage temperature: $20 \pm 3°C$. ●, *E. coli*; ○, faecal streptococci.

degree of human faecal contribution to the samples was similar, since this organism is associated mainly with human faecal material in this country (Mead 1965) and the USA (Bartley & Slanetz 1960).

The observed die-off rates of faecal streptococci were similar to or greater than those of *E. coli* in the following stored samples:

1. settled sewage in the light and in the dark (Figs 3 and 4);
2. settled sewage diluted to 10%, 1% and 0.1% with dechlorinated mains water;
3. river water (Fig. 5).

The storage of treated sewage effluent gave variable results (Fig. 6). In two samples the die-off rate for faecal streptococci was similar to or greater than that of *E. coli* (Fig. 6b and c). However in a third sample the *E. coli* died off faster than the faecal streptococci (Fig. 6a).

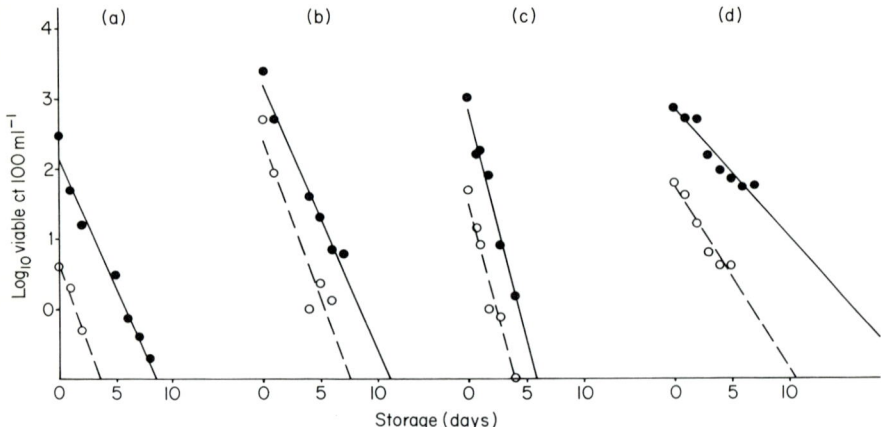

Fig. 5. The survival of *E. coli* and faecal streptococci in four stored samples of river water. Storage temperature: (a), (b), (c), $20 \pm 3°C$; (d), $5-8°C$. ●, *E. coli*; ○, faecal streptococci.

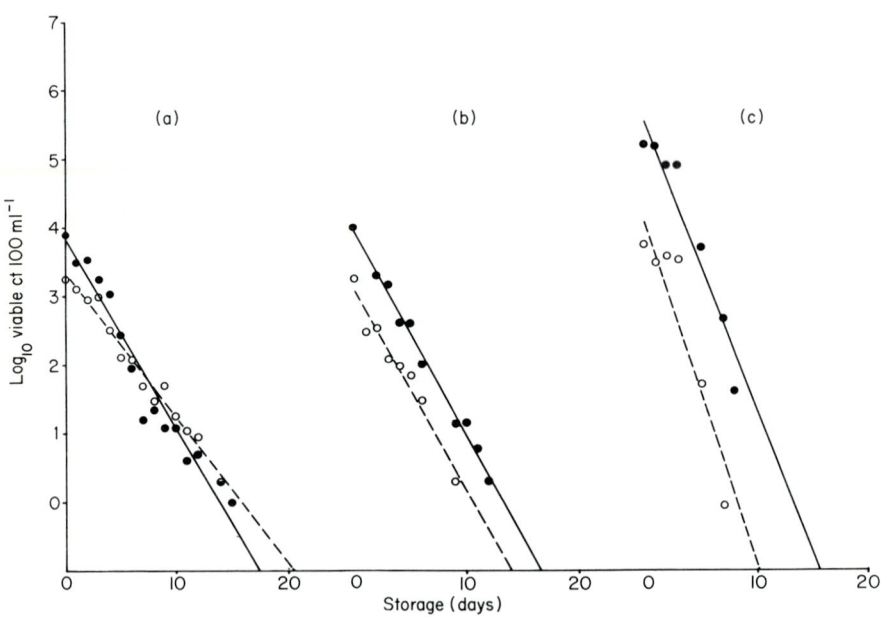

Fig. 6. The survival of *E. coli* and faecal streptococci in three stored samples of treated sewage effluent. Storage temperature: $20 \pm 3°C$. ●, *E. coli*; ○, faecal streptococci.

Therefore, apart from one sample of treated sewage effluent, the FC to FS ratios remained steady or increased in stored samples.

Storage of settled sewage in the light and the dark (Figs 3 and 4) revealed that in the dark the die-off rates for faecal streptococci and *E. coli* remained constant for up to seven weeks. In the light, the levels of the indicator organisms showed the reductions as observed in the dark for an initial period after which the die-off rates increased. The faecal streptococci showed an increase in die-off rate after *ca.* 14 days' storage. The *E. coli* die-off rate was steady for approximately the first 25 days, after which the rate increased. In the dark, pH decreased from 7.2 to 6.8 over the storage period, whilst in the light, pH increased to 8.5, presumably due to the algal growth which developed in settled sewage samples after the first week of storage. Similar algal blooms and an increase in pH from 7.0 to 7.5 were observed in stored samples of treated sewage effluent, but in these samples the indicator organisms had nearly died out by the time that the blooms appeared. No blooms were observed in the river water samples during the period in which the indicator organisms were detected and the pH remained stable at values 7.2 and 7.4. These pH values are within the limits pH 4-9 which were suggested by Geldreich & Kenner (1969) for the application of FC to FS ratios.

It could be postulated that in the case of settled sewage samples the nutrient concentrations would enhance the survival of *E. coli* compared with that of faecal streptococci, since the nutrient concentration required for regrowth of *E. coli* is considerably less than that for faecal streptococci (Evison & James 1975). The similar results obtained with settled sewage, diluted settled sewage and river waters suggest that the greater survival of *E. coli* than faecal streptococci was not due to this factor.

Since faecal streptococci die-off rapidly at higher temperatures (20-30°C) on storage (Evison & James 1975) it is possible that the storage temperature of $20 \pm 3°C$ accentuated the die-off rates of faecal streptococci. However, storage of river water at 5-8°C (Fig. 5d) revealed that faecal streptococci died off faster than *E. coli*, the die-off rates of both indicators being considerably less than those observed at 20°C.

With the techniques used, attenuation of the faecal streptococci on storage may have led to reduced recovery of the viable organisms present, producing an apparent die-off greater than actually occurred. On occasions when additional samples were taken from the stored waters and faecal streptococci enumerated using incubation at 37°C, the initial numerical superiority due to incubation at 37°C over that at 45°C (expressed as ratios in Table 2) diminished as storage continued until similar counts were obtained with either incubation temperature. Incubation at 37°C, which would make these results directly comparable with those of other workers who used more natural conditions of storage (e.g. McFeters *et al.* 1974), would therefore accentuate the die-off rates for faecal streptococci. The use of a non-selective resuscitation step prior to incubation on the M-EA medium has been shown to produce equivalent or reduced counts, compared with the methods used in these studies with samples of river water and sewage (*Anon.* 1966).

The similar counts obtained using either 37 or 45°C incubation temperatures after a period of storage suggest that in the situations where the faecal streptococcus test would normally be applied by the Water Authorities, namely the investigation of remote faecal pollution, the use of either incubation temperature would be acceptable.

It would appear from this work that storage of samples representing human faecal pollution may not produce the changes in FC to FS ratios reported for similar water samples under natural conditions (Kjellander 1960; McFeters *et al.* 1974). According to the criteria suggested by Feachem (1975) the source of pollution in the samples examined would be 'uncertain' with the exception of the treated sewage effluent sample in which FC to FS ratios fell, thus indicating a human source.

The storage of water samples under practical conditions seems unlikely in the case of human pollution to indicate the source of faecal contamination. The literature suggests that more 'natural' (and operationally impractical) storage conditions are necessary to produce useable changes in FC to FS ratios.

4. Acknowledgements

This paper is published by permission of Mr Hugh Fish, Director of Scientific Services, Thames Water Authority. We wish to thank B. G. Foxwell and I. Ferguson for their technical assistance.

5. References

ANON. 1966 Faecal streptococci. In *Forty-first Report on the Results of the Bacteriological, Chemical and Biological Examination of the London Waters for the Years 1963-1964*. London: Metropolitan Water Board.
ANON. 1969 *The Bacteriological Examination of Water Supplies.* Reports on Public Health and Medical Subjects No. 71, 4th edn. London: HMSO
ANON. 1971 *Standard Methods for the Examination of Water and Wastewater*, 13th edn. American Public Health Association.
BARTLEY, C. H. & SLANETZ, L. W. 1960 Types and significance of fecal streptococci isolated from feces, sewage and water. *American Journal of Public Health* **50**, 1545-1552.
BURMAN, N. P. 1961 Some observations on coli-aerogenes bacteria and streptococci in water. *Journal of Applied Bacteriology* **24**, 368-376.
BUTTIAUX, R. & MOSSEL, D. A. A. 1961 The significance of various organisms of faecal origin in foods and drinking water. *Journal of Applied Bacteriology* **24**, 353-364.

COOPER, K. E. & RAMADAN, F. M. 1955 Studies in the differentiation between human and animal pollution by means of faecal streptococci. *Journal of General Microbiology* **12**, 180–190.

COWAN, S. T. 1974 *Manual for the Identification of Medical Bacteria*, 2nd edn. London: Cambridge University Press.

DEIBEL, R. H. 1964 The group D streptococci. *Bacteriological Reviews* **28**, 330–366.

EVISON, L. M. & JAMES, A. 1975 *Bifidobacterium* as an indicator of faecal pollution in water. *Progress in Water Technology* **7**, 57–66.

FEACHEM, R. G. 1975 An improved role for faecal coliform to faecal streptococci ratios in the differentiation between human and non-human pollution sources. *Water Research* **9**, 689–690.

FEACHEM, R. G. 1976 Author's reply. *Water Research* **10**, 569–570.

GELDREICH, E. E. 1966 *Sanitary Significance of Fecal Coliforms in the Environment*. Water Pollution Control Research Series Publication No. WP-20-3. United States Department of the Interior, Federal Water Pollution Control Administration, Cincinnati, Ohio.

GELDREICH, E. E. & KENNER, B. A. 1969 Concepts of fecal streptococci in stream pollution. *Journal of the Water Pollution Control Federation* **41**, R336–R352.

HANNAY, C. L. & NORTON, I. L. 1947 Enumeration, isolation and study of faecal streptococci from river water. *Proceedings of the Society for Applied Bacteriology* **10**, 39–45.

HOLDEN, W. S. (ed) 1970 *Water Treatment and Examination*. London: J. & A. Churchill, p. 234.

KENNER, B. A., CLARK, H. F. & KABLER, P. W. 1960 Fecal streptococci. II. Quantification of streptococci in feces. *American Journal of Public Health* **50**, 1553–1559.

KJELLANDER, J. 1960 Enteric streptococci as indicators of fecal contamination of water. *Acta Pathologica et Microbiologica Scandinavica* Suppl. 136, **48**, 1–133.

McFETERS, G. A., BISSONNETTE, G. K., JEZESKI, J. J., THOMSON, C. A. & STUART, D. G. 1974 Comparative survival of indicator bacteria and enteric pathogens in well water. *Applied Microbiology* **27**, 823–829.

MARA, D. D. 1974 *Bacteriology for Sanitary Engineers*. Edinburgh & London: Churchill-Livingstone.

MEAD, G. C. 1963 A medium for the isolation of *Streptococcus faecalis sensu strictu*. *Nature, London* **197**, 1323–1324.

MEAD, G. C. 1965 The distribution of *Streptococcus faecalis* and related biotypes. Ph.D. Thesis, University of London.

MEDREK, T. F. & LITSKY, W. 1960 Comparative incidence of coliform bacteria and enterococci in undisturbed soil. *Applied Microbiology* **8**, 60–63.

RAIBAUD, P., CAULET, M., GALPHIN, J. V. & MOCQUOT, G. 1961 Studies on the bacterial flora of pigs. *Journal of Applied Bacteriology* **24**, 285–306.

SHERMAN, J. M. 1937 The streptococci. *Bacteriological Reviews* **1**, 1–97.

SLANETZ, L. W. & BARTLEY, C. H. 1957 Numbers of enterococci in water, sewage and feces determined by the membrane filter technique with an improved medium. *Journal of Bacteriology* **74**, 591–595.

VAN DONSEL, D. J., GELDREICH, E. E. & CLARKE, N. A. 1967 Seasonal variations in survival of indicator bacteria in soil and their contribution to storm water pollution. *Applied Microbiology* **15**, 1362–1370.

WINDLE TAYLOR, E. & BURMAN, N. P. 1964 The application of membrane filtration techniques to the bacteriological examination of water. *Journal of Applied Bacteriology* **27**, 294–303.

Damage and Recovery in Streptococci

J. PAYNE

ARC Food Research Institute,
Norwich, Norfolk, England

CONTENTS

1. Introduction 349
2. Damage caused by exposure to cold 351
3. Damage caused by exposure to heat 354
4. Damage caused by other stress conditions 356
5. Recovery from cold-induced damage 358
6. Recovery from heat-induced damage 359
7. General comments 366
8. References 368

1. Introduction

EXPOSURE OF BACTERIA to stress conditions often results in damaged cells that show physiological deficiencies. These cells may be lethally or sublethally damaged. The distinction between the two types of damage is made by experiment and is dependent upon the culture methods employed. With this qualification, lethally damaged cells may be defined as those that are unable to form colonies on any of the media used, and sublethally damaged cells as those that can form colonies on some media but not on others. Having made this distinction between lethal and sublethal damage this review will be concerned with the latter and will refer to it simply as damage.

Damage may be recognized by a requirement for additional nutrients, an increased sensitivity to NaCl or other selective agents, a decreased temperature range allowing growth, a prolonged lag phase prior to cell division, leakage of material absorbing at 260 nm and metabolic damage in the form of reduced enzyme activity (Table 1). The damage is usually reversible; after incubation in a suitable environment damaged cells regain the characteristics of the undamaged organisms. Restoration of the undamaged state is a process that precedes cell division; its progress may be observed as an increase in colony count on a medium on which only undamaged cells grow without any corresponding increase in count on a medium on which both damaged and undamaged cells form colonies.

It has been recognized for many years that the composition of the plating medium can affect the survival of bacteria after exposure to different stress conditions (Harris 1963). Nelson (1943) reviewed the literature reporting

Table 1
Effects of stress conditions on streptococci

Organism	Stress conditions	Effects of stress	Reference
S. lactis	Storage at $-20°C$	Increased nutritional requirement	Moss & Speck (1963)
S. lactis	Storage at $3°C$	Decreased proteinase and acid producing activity	Cowman & Speck (1965a, b)
S. lactis	Starvation	Increased growth lag	Thomas & Batt (1968)
S. faecalis	Storage at $-20°C$	Increased sensitivity to NaCl, azide and a peptide requirement	Morichi & Yano (1972), Morichi & Irie (1973)
S. faecalis	Heated at $60°C$ for 15 and 30 min	Increased sensitivity to NaCl, pH, temperature and increased growth lag	Clark et al. (1968)
S. faecalis	Heated at $60°C$	Increased sensitivity to NaCl, KCl, $MgCl_2$ and incubation temperature	Beuchat & Lechowich (1968a, b)
S. faecalis	Heated at $60°C$ and $76°C$	Increased sensitivity to sodium azide	Ienistea et al. (1970a, b)
S. faecalis	Heated at $60°C$ for 4 min	Increased sensitivity to potassium tellurite, increased growth lag	Payne & Morley (1976)
S. faecium	Heated at $55°C$ for 15 min	Increased sensitivity to NaCl, increased growth lag, leakage of material absorbing at 260 nm	Duitschaever & Jordan (1974)

increased counts of organisms from pasteurized dairy products when improved media were used and, using a variety of bacteria, showed that heated cells were more exacting in their requirements for growth than were unheated cells. In a later study (Nelson 1944) it was shown that the number of survivors could be affected not only by the presence of certain kinds of nutrients but also by the amounts and the stage at which the peptone supplement was added to the basal medium during preparation. Organisms subjected to cold stress were shown to behave in a similar manner (Straka & Stokes 1959; Moss & Speck 1963).

Although there is an extensive literature on the survival of bacteria subjected to a variety of stress conditions, such as heat, cold, chemicals, drying, ultraviolet and ionizing radiations, not all studies have examined damage and recovery (for reviews, see Allwood & Russel 1970; Ray & Speck 1973; Gray & Postgate 1976; Skinner & Hugo 1976). In such cases the organisms have usually been subjected to heat, cold or ultraviolet or ionizing radiation. The majority of the work with streptococci has been concerned with damage caused by heat or cold and this review will deal almost entirely with the sublethal damage and recovery of streptococci exposed to extremes of temperature.

2. Damage Caused by Exposure to Cold

In streptococci most investigations on damage caused by refrigeration and subzero temperatures have been with *S. faecalis* and *S. lactis*. The findings with the latter organism have been related to its use and that of other lactic streptococci in the dairy industry.

Moss & Speck (1963) found that freeze-damaged cells of *S. lactis* were unable to form colonies on a medium made restrictive by reducing the tryptone concentration from 2 to 0.5%, suggesting that the damaged cells required peptides. A similar requirement has been demonstrated in freeze-damaged *S. faecalis* (Morichi & Irie 1973). Undamaged cells grew on a defined chemical medium that was peptide free, and on which only 43% of the cells surviving freezing were able to grow.

Cowman & Speck (1965a, b) observed a decrease in proteinase activity and acid-producing ability following storage of *S. lactis* at either 3 or $-20°C$, but such losses were much less marked after storage at $-196°C$ in liquid nitrogen.

The effect of storage at $3°C$ on the activity and properties of the proteinase enzyme of lactic streptococci has been extensively studied (Cowman & Speck 1967; Cowman *et al*. 1967; Speck & Cowman 1970). Whole cells were shown to contain an intracellular and a membrane-bound proteinase but only the latter lost activity during storage of the purified enzyme at $3°C$, the rate of loss being comparable to that in intact cells held at $3°C$ (Cowman & Speck 1967). More recent work has questioned the criteria used by these authors to determine the location of the membrane-bound enzyme and concluded that it is most probably cell wall bound (Thomas *et al*. 1974). Cowman & Speck (1967) used sedimentation velocity and gel filtration before and after subjecting to cold stress to show gross structural changes of the purified membrane/cell wall-bound enzyme, concomitant with loss of activity, after storage at $3°C$. At $22°C$ the active enzyme existed primarily in the monomer and dimer forms but, after storage at $3°C$ for 24 h, the enzyme polymerized and lost activity (Fig. 1). Since the *p*-hydroxymercuribenzoate derivative of the enzyme maintained its initial activity and did not undergo these structural changes when subjected to cold, Cowman *et al*. (1967) suggested that sulphydryl groups may be important in maintaining the structure of the enzyme. So far there is no evidence to suggest that this sublethal loss of proteinase activity is reversible.

Inability to grow on selective media containing sodium azide and loss of NaCl tolerance by *S. faecalis* after freezing at $-20°C$ have been reported (Morichi & Yano 1972; Morichi & Irie 1973). Of cells surviving freezing, 93%

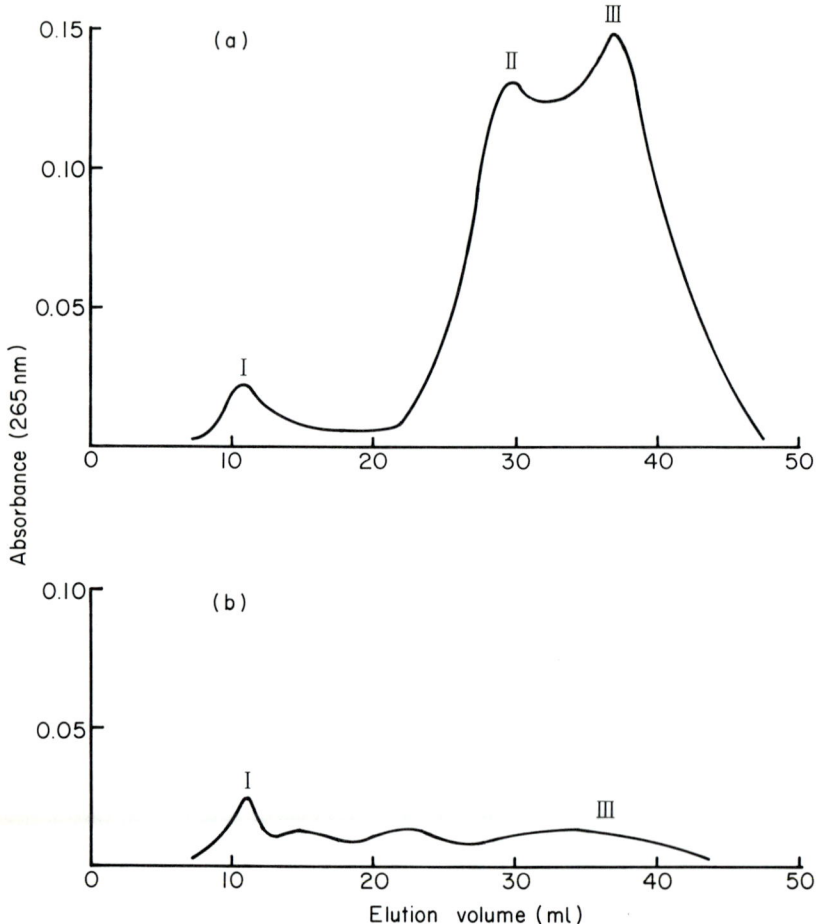

Fig. 1. Effect of storage at 3°C on the elution profile of the membrane proteinase from *S. lactis* subjected to Sephadex G-100 gel filtration. (a) Stored at 22°C; (b) stored at 3°C. Peaks II and III contained activity (from the data of Cowman et al. 1967).

were sensitive to 6% NaCl and 29–44% were sensitive to 0.05% azide although only 4% were sensitive to 0.02% azide.

The extent of damage caused to cells by freezing-and-thawing can vary according to the precise conditions involved. These factors include, amongst others, time of storage at subzero temperature, rate of freezing and possibly subsequent rate of thawing and the composition of the suspending medium.

Streptococcus lactis showed the greatest injury during the early stages of storage at 3 or −20°C (Moss & Speck 1963; Cowman & Speck 1965a, b). During the first 8 days of storage at −20°C, 37% of cells of *S. faecalis* lost viability while

94 and 29% of the survivors were sensitive to NaCl and azide, respectively. Thereafter the number of survivors and the proportion of these that were damaged showed little change (Morichi & Yano 1972).

Moss & Speck (1963) reported that, when *S. lactis* was frozen in buffered distilled water, there was a greater loss of viability when the freezing rate was slow, but that injury was more pronounced when the freezing rate was fast suggesting that the extent of injury was dependent on the resistance of the population to the rate of freezing. However, these authors expressed the numbers of injured, uninjured and surviving cells as a percentage of the initial population prior to freezing. If their data are recalculated so that the injured cells are expressed as a percentage of the survivors, then it appears that rate of freezing, while affecting survival, did not have any great effect on the proportion of the survivors that were injured (Table 2). When non-fat milk was the suspending medium the rate of freezing had little effect on survival or injury. While the effect of freezing rate on injury was not examined in other strains, its effect on survival was found to vary between strains, some being more resistant to fast freezing rates than to slow freezing rates. The effect of freezing rate on the extent of injury therefore appears complex.

Using *S. lactis* Baumann & Reinbold (1966) have investigated the effect of growth temperature on the extent of freeze injury, measured as loss of acid-

Table 2
Effect of storage time and freezing rate on survival and injury of S. lactis

Freezing menstruum	Time at $-20°C$ (days)	Fast freezing*		Slow freezing†	
		Survival (%)	Proportion of survivors injured (%)	Survival (%)	Proportion of survivors injured (%)
Buffered distilled water	1	99	40	44	31
	3	94	30	28	39
	7	62	58	28	39
	14	53	64	28	46
	28	50	66	18	66
Non-fat milk (10%)	1	100	8	98	9
	3	100	9	91	4
	7	98	8	85	1
	14	84	2	80	8
	28	68	0	70	6

* Fast freezing temperature change: 20°C to $-20°C$ in a dry ice-acetone bath at $-35°C$.
† Slow freezing temperature change: 20°C to $-20°C$ in a deep-freeze cabinet.
Recalculated data from Table 1 of Moss & Speck (1963).

producing ability. They found that cells grown at 32°C were less extensively injured than those grown at 15°C when frozen at −20°C. No difference in the extent of injury could be related to growth temperature when cells were frozen at −196°C.

Survival and extent of injury of *S. faecalis* frozen in a variety of different media has been found to vary (Morichi & Yano 1972). When frozen in 1% peptone or 1% soytone, survival was reduced to 4 and 33% respectively but there was no evidence of any sublethal injury with respect to azide sensitivity. However, 53% of the initial population survived freezing in distilled water and of these 25% were sensitive to azide (Table 3).

3. Damage Caused by Exposure to Heat

The most widely reported effects of sublethal heat treatment are those of increased sensitivity to NaCl and selective agents. Heated cells of *S. faecalis* and *S. faecium* have been shown to become sensitive to sodium chloride (Beuchat & Lechowich 1968a; Clark *et al.* 1968; Duitschaever & Jordan 1974), sodium azide (Ienistea *et al.* 1970a, b), potassium tellurite (Payne & Morley 1976) and also to the selective agents in the thallous acetate-tetrazolium agar described by Barnes (1956). Duitschaever & Jordan (1974) also suggested that damage to the cell membrane occurred during heating of *S. faecium* at 55°C since there was a continuous loss of compounds absorbing at 260 nm. Clark *et al.* (1968), as well as using increased sensitivity to NaCl to assess damage, also examined the effect of heat on other characteristics of the organism. They found that heat-damaged cells became more sensitive to pH, incubation temperature and methylene blue (Table 4). Beuchat & Lechowich (1968b) have also noted an increased sensitivity

Table 3
Effect of freezing medium* on survival and injury of S. faecalis

Freezing medium	Survival (%)	Survivors sensitive to azide i.e. injured (%)
Distilled water	53	25
0.9% NaCl	38	28
1% Beef extract (Difco)	69	15
1% Peptone (Difco)	4.7	0
1% Soytone (Difco)	33	0
1% soluble starch	9.8	30
1% gelatin	79	17

* Suspensions were frozen at −20°C for 1 month.
Morichi & Yano (1972).

Table 4

Heat injury* of S. faecalis *R57 measured by different assay systems*

Plate count medium	Incubation temperature (°C)	Injured survivors (%)
TSA pH 7.3	37	0
TSA pH 7.3	45	67.9
TSA pH 7.3	10	99.08
TSA pH 9.6	37	98.5
TSA + 0.01% methylene blue pH 7.3	37	99.94
TSA + 5.5% NaCl pH 7.3	37	99.97

* Cells were heated in phosphate buffer pH 6.0 at 60°C for 30 min.
TSA, Trypticase-soy agar.
Clark *et al.* (1968).

to post-injury incubation temperature with *S. faecalis*, while Payne (unpublished) found that the increased sensitivity to incubation temperature became more pronounced as the heating time increased (Fig. 2).

Heating *S. faecalis* at 60°C produced damaged cells that became sensitive to selective agents in the following order: potassium tellurite > sodium chloride > sodium azide > thallous acetate-tetrazolium agar (Fig. 3). Loss of tellurite resistance occurred in at least 99.99% of the survivors after 4 min at 60°C. To produce this damage, expressed as loss of resistance to NaCl, heating for 15 min at 60°C was required. Most of the published work on thermal damage has used loss of NaCl resistance to determine the proportion of the population damaged after heating times of the order of 15 min. Therefore loss of tellurite resistance as a means of assessing numbers of damaged cells provides a system which allows the investigation of events occurring during the early stages of sublethal heat injury.

Gee & Payne (to be published), using stationary phase cultures of *S. faecalis* exposed to mild heat stress (4 min at 60°C), have shown that the profile of soluble proteins separated by polyacrylamide gel electrophoresis did not alter from that of unheated cells. When the cell protein was labelled by previous growth of the organism with $[U - {}^{14}C]$ L-valine or $[U - {}^{14}C]$ L-aspartic acid, the distribution of radioactivity in the protein peaks was also unaltered. These results suggest that when these cells were exposed to short heat treatments there was no breakdown of cell protein. Damaged cells also failed to lose potassium or magnesium ions in contrast to reports by Hurst *et al.* (1973, 1975) for *Staphylococcus aureus* heated at 52°C for 15 min. It is also interesting to note that 4 min at 60°C did not damage the cells of *S. faecalis* such that their sensitivity to post-injury incubation temperature was increased (Fig. 2).

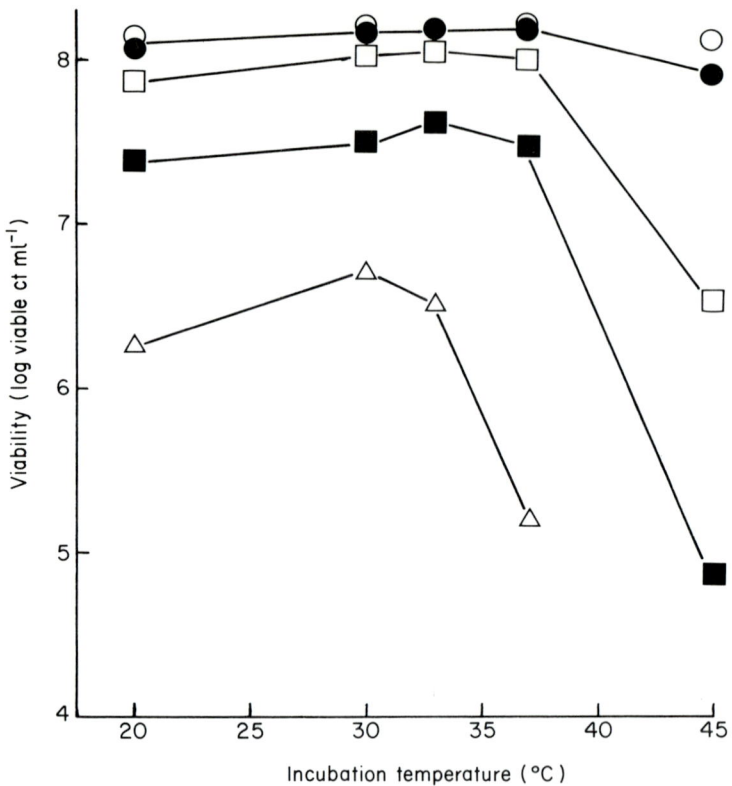

Fig. 2. Effect of post-injury incubation temperature on survival of heated *S. faecalis* EBF 30/39. Cells were heated in heart infusion (HI) broth at 60°C for 0 (○); 4 (●); 10 (□); 21 (■) and 42 min (△), plated on HI agar and incubated at the temperature indicated until the colony count was constant.

Beuchat & Lechowich (1968a) found that the RNA content of *S. faecalis* was reduced when the cells were subjected to heat treatments sufficient to reduce the viable population by 99 and 99.99%, respectively. However, since loss of viability occurred, one cannot distinguish whether loss of RNA resulted in sublethal or lethal damage.

4. Damage Caused by Other Stress Conditions

Scheusner *et al.* (1971) have investigated the effect of sanitizers on *S. faecalis*. When cells were exposed to 100 μg ml^{-1} of the quaternary ammonium compound, methylalkyltrimethyl ammonium chloride for 30 s at 0°C, 77% of the population lost viability on trypticase-soy agar and 56% of the survivors were injured, failing to form colonies on azide-dextrose agar.

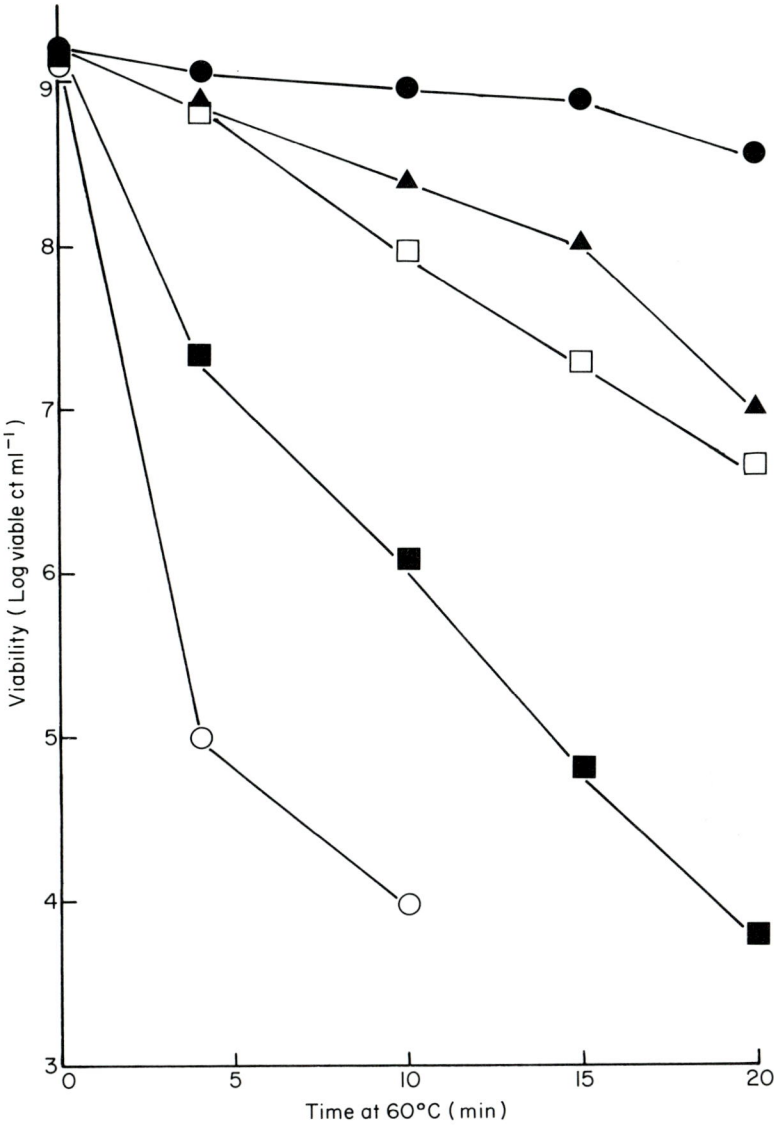

Fig. 3. Loss of resistance to selective agents by heat-injured *S. faecalis* EBF 30/39. Cells were heated in HI broth at 60°C in sealed ampoules, serially diluted and plated on HI agar (●); HI agar + 0.04% potassium tellurite (○); HI agar + 3% NaCl (■); HI agar + 0.04% sodium azide (□); thallous acetate tetrazolium agar (▲). Colonies were counted after 72 h at 33°C.

Metabolic damage in the form of reduced growth rate and reduced enzyme specific activity has been shown to occur in *S. lactis* when shifted to a growth medium with a pH below 5 (Harvey 1965). A period of growth at the new pH was required before the damage was expressed. When the organisms were transferred from a growth medium at pH 6.3 to one at pH 5.5, the cells immediately grew at the rate normal for the new pH. However, when a culture growing at pH 6.3 was shifted to a medium at pH 4.5, there was a transient growth rate higher than normal for pH 4.5 which lasted for about one-half the doubling time. Acetate kinase, hexokinase and glycolytic activity of the cells were shown to decrease rapidly with decreasing pH below 5 but above this the specific activities remained unaltered, irrespective of the pH of the medium. Harvey (1965) concluded that the reduced activity was due either to a lower differential rate of synthesis of the enzyme protein involved, or that a proportion of each enzyme was synthesized in an inactive form or rapidly inactivated after synthesis. Since he showed that the rate of overall protein synthesis was unaltered when the organism was shifted to the low pH the former hypothesis would imply that some proteins were synthesized at an increased differential rate.

Damage as represented by an increased lag prior to cell division on transfer to a growth medium has been shown to occur in *S. lactis* incubated under starvation conditions in phosphate buffer containing 10^{-5} mol l^{-1} EDTA at pH 7.0 (Thomas & Batt 1968). Addition of casamino acids and 10^{-3} mol l^{-1} Mg^{2+} to this starvation medium reduced the extent of damage but addition of 10^{-2} mol l^{-1} lactose enhanced the rate at which cells became damaged.

5. Recovery from Cold-induced Damage

Morichi & Irie (1973) damaged cells of *S. faecalis* by suspending in distilled water and freezing at -20°C for 24 h before thawing. After this treatment 40% of the population had lost viability and 90% of the survivors were sensitive to 6% NaCl. When these damaged cells were incubated in tryptone-yeast extract-glucose recovery broth (TYG) at 37°C the NaCl sensitive cells recovered salt tolerance within 60 min. Addition of chloramphenicol, streptomycin, puromycin, cycloserine or penicillin to the recovery broth did not affect return to salt tolerance indicating that neither protein nor cell wall synthesis were required for recovery of injured cells (Fig. 4a). However, addition of actinomycin D prevented 50% of the injured population from recovering suggesting that RNA biosynthesis was required by some of the damaged cells for recovery (Fig. 4b). Addition of 6% NaCl to recovery broth completely inhibited recovery and RNA biosynthesis in freeze-damaged cells as shown by measuring the incorporation of [2 – ^{14}C] uracil. The authors concluded that this accounted for the failure of freeze-damaged cells to form colonies on media containing high concentrations of NaCl.

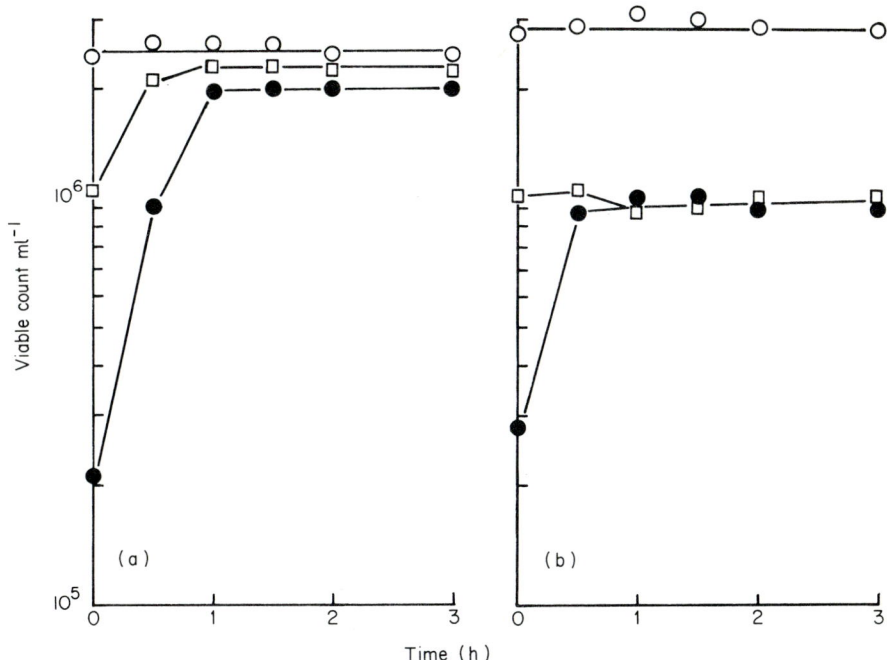

Fig. 4. Recovery of freeze damage in *S. faecalis*. Freeze-damaged cells were incubated at 37°C in (a) TYG broth + 50 μg ml^{-1} chloramphenicol or (b) TYG broth + 20 μg ml^{-1} actinomycin D. Counts were made at intervals on TYG agar (o), TYG agar + 6% NaCl (•) and a defined medium lacking peptides (□) (from Morichi & Irie 1973).

The same workers also investigated recovery when damage was expressed as a peptide requirement. Recovery of this proportion of the damaged population was independent of protein and cell wall synthesis but, unlike recovery of salt tolerance, was entirely dependent on RNA biosynthesis (Fig. 4). Hence the freeze-damaged population of *S. faecalis* contained at least two types of cells, those sensitive to NaCl that could grow on a peptide-free defined medium and those that required peptides for growth. The former cell type could recover in the absence of RNA synthesis whereas the cells requiring peptide could not.

Recovery of sodium azide resistance by freeze-damaged cells has been shown to occur in TYG broth after incubation at 37°C for 1 h (Morichi & Yano 1972). Protein synthesis was not involved and the effect on recovery of inhibiting RNA and cell wall synthesis was not investigated.

6. Recovery from Heat-induced Damage

The post-injury incubation period required for the recovery of heat-damaged streptococci (4 to 5 h) has been found to be longer than that required for the recovery from cold damage (0.5 to 1 h). During the recovery period the injured cells regained their ability to form colonies on selective agar, and it was not until the end of recovery that evidence of cell division was observed (Clark *et al.*

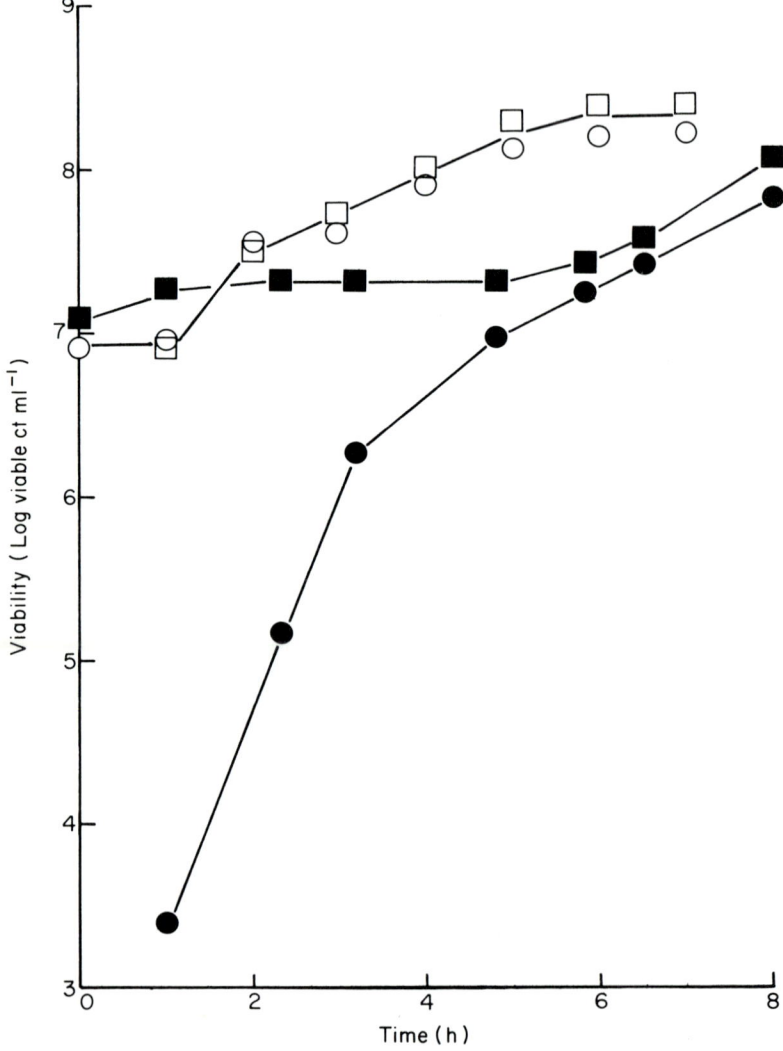

Fig. 5. Recovery of tellurite resistance by heat-injured *S. faecalis* EBF 30/39. Unheated cells (open symbols) or those heated at 60°C for 4 min (closed symbols) were incubated in HI broth at 33°C. Growth and/or recovery were determined by viable counts on HI agar with (○, ●) or without (□, ■) 0.04% potassium tellurite (from Payne & Morley 1976).

1968; Duitschaever & Jordan 1974; Payne & Morley 1976). Unheated cells incubated in a broth recovery medium showed an increase in number on both selective and non-selective agar media at least 3 h before heated cells began to multiply (Fig. 5).

Composition of the recovery medium and the severity of the heat treatment used to damage the cells can affect the recovery time. *S. faecalis* heated at 60°C for 4 min took 4-5 h to recover tellurite resistance but, when exposed to 60°C for 21 min, the injured cells required up to 24 h to recover (Payne, unpublished). When *S. faecalis* was heated at 60°C for 15 min no loss of viability resulted but more than 99% of the population were sensitive to 6% NaCl (Clark *et al.* 1968). The recovery of these cells was examined in a variety of enrichment media commonly used for streptococci such as dextrose broth, tryptose-phosphate broth, azide-dextrose broth, buffered azide-glucose-glycerol broth (BAGG), SF medium and KF medium as well as trypticase-soy broth (TSB). Since the dextrose broth and tryptose-phosphate broth did not contain a selective agent these media proved as good for recovery as TSB. The addition of azide to the dextrose broth, while not preventing recovery, increased the time required to about 8 h. The increased azide concentration and the presence of bromocresol purple in KF medium and BAGG broth extended the recovery time to 8 to 12 h while, with the SF medium, complete recovery took 24 h. Hence, as the medium became more selective, the recovery time increased.

Thus the presence of azide was not lethal to heat-damaged cells and did not prevent recovery. In contrast to this finding addition of 0.04% tellurite to heart infusion broth recovery medium had a lethal effect on a proportion of the injured population of *S. faecalis* damaged at 60°C for 4 min, while the remainder of the injured population were able to recover (Fig. 6) (Payne, unpublished). Hence tellurite incorporated into solid medium prevented recovery of damaged cells but did not prevent recovery of all damaged cells in a liquid recovery medium. This may be because in the liquid recovery medium containing tellurite the ratio of cell number to tellurite concentration was greater than in the solid medium by a factor of about 10^3. Corry (1970) noted that salmonellae showed increased numbers of bizarre cell types on 0.4% selenite agar in comparison to the appearance of cells in 1% selenite broth and concluded that the solid medium was more toxic than its liquid counterpart. Erwin & Haight (1973) have shown that NaCl was lethal to heat-damaged *Staph. aureus* immediately after injury, and that a transition from a lethal to inhibitory effect of salt occurred within 30 min in a medium devoid of NaCl. No such transition from a lethal to inhibitory action of tellurite on heat-damaged *S. faecalis* was observed. If injured cells were allowed to recover for 1 or 2 h prior to the addition of tellurite, this selective agent was still lethal to some of the cells which remained damaged.

Recovery from heat-damage in one strain of *S. faecalis* in defined media has been shown to be relatively poor compared with recovery in a complex medium (Clark et al. 1968). Recovery in a phosphate buffer-amino acid-glucose system was only 1.6% after 6 h and addition of vitamins and nucleic acids to this medium did not improve recovery to any significant extent. A medium containing phosphate buffer, glucose and casein hydrolysate gave 24% recovery but even this was less than the 63% obtained in TSB (Table 5). In contrast, using a different strain of *S. faecalis*, Payne (unpublished) has found that cells heated at 60°C for 4 min or 21 min could recover tellurite resistance in a medium

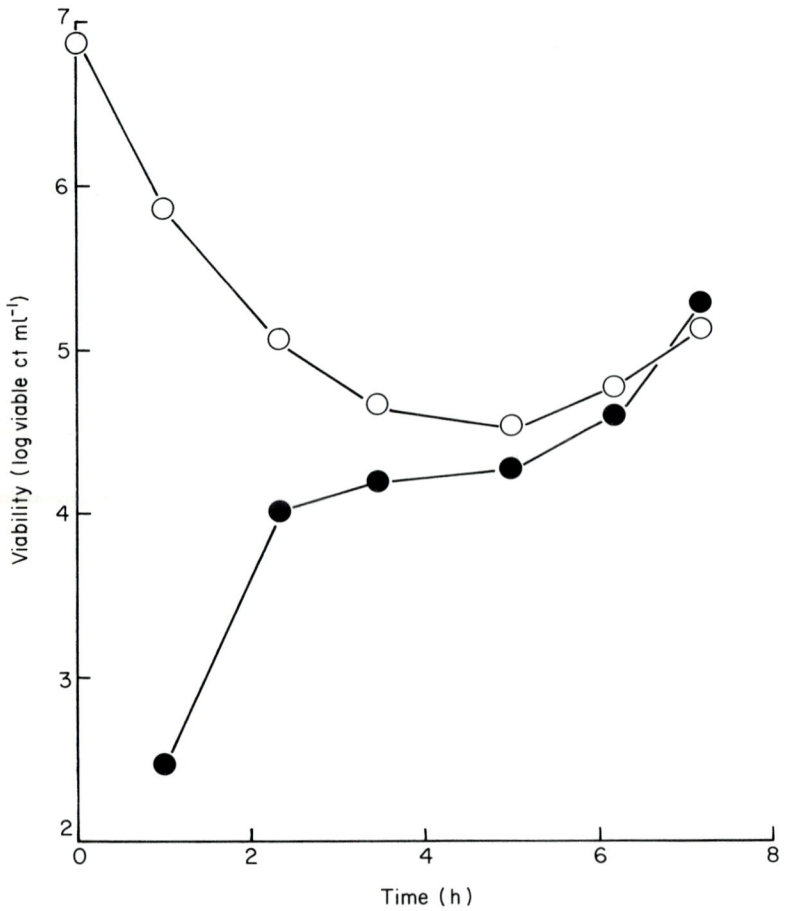

Fig. 6. Effect of tellurite on recovery of tellurite resistance by heat-injured *S. faecalis* EBF 30/39. Cells were heated in HI broth at 60°C for 4 min then incubated at 33°C in HI broth plus 0.04% potassium tellurite. Viable counts were made on HI agar (o) and HI agar + 0.04% tellurite (•).

Table 5
Recovery of heat-injured S. faecalis *R57 in various media*

Recovery medium	Recovery at 6 h (%)
Control	63.0
Modified Fildes (amino acids, glucose, phosphate buffer)	1.6
Modified Fildes plus folic acid, pyridoxine, biotin, uracil, xanthine, adenine, guanine, threonine, isoleucine and tryptophan	3.8
10% Hydrolysed casein, 5% glucose, 50 mmol l^{-1} phosphate buffer	24.0
Complete synthetic medium	30.0

Cells were heat-damaged in 100 mmol l^{-1} phosphate buffer pH 6.8 at 60°C for 15 min and resuspended in recovery medium pH 6.8 to a density 3 × 10^8 cells ml^{-1} and incubated at 37°C.
Clark *et al.* (1968).

containing potassium phosphate buffer, glucose and casein hydrolysate provided that the cell density was less than 1 × 10^8 viable cells ml^{-1} or 1 × 10^7 viable cells ml^{-1}, respectively (Fig. 7). In the absence of glucose or casein hydrolysate, or if sodium phosphate was substituted for potassium phosphate buffer, recovery was limited to about 10%, but addition of 1 mmol l^{-1} KCl to the recovery medium with sodium phosphate buffer gave complete recovery. This recovery system did not support growth of either damaged cells that had recovered or undamaged cells, and hence the process of recovery could be separated from and was independent of growth.

A direct comparison of these findings with those of Clark *et al.* (1968) with defined recovery media is difficult due to the different conditions used. However, the density at which the damaged cells were resuspended in the recovery medium may have had some influence on recovery. The latter authors resuspended injured cells at a density of 3 × 10^8 viable cells ml^{-1} after a 15 min heat treatment whereas Payne found that as the heat treatment was increased from 4 to 21 min the level of viable cells in the recovery medium became more important.

Recovery in the presence of chloramphenicol, penicillin and actinomycin D, to prevent protein, cell wall and RNA biosynthesis respectively, has shown that salt-sensitive injured *S. faecalis* did not require cell wall or protein synthesis for recovery (Clark *et al.* 1968). However, Payne & Morley (1976), while concluding that cell wall synthesis was not necessary for recovery of tellurite resistance, found that protein synthesis was required for all but 10% of the injured population to recover (Fig. 8). Since some of the damaged cells could recover in the absence of protein synthesis, the heated population must have contained at

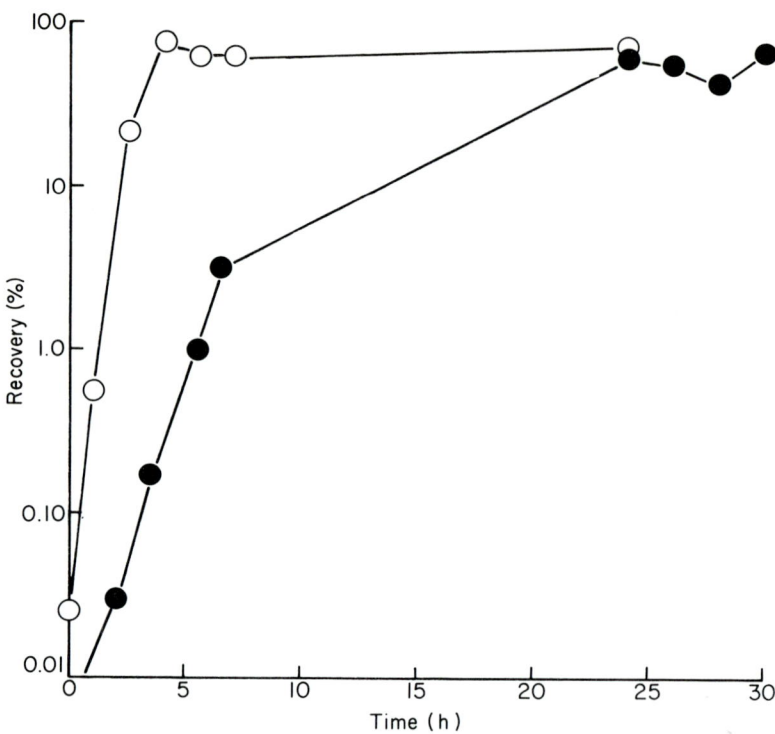

Fig. 7. Recovery of tellurite resistance by heat-injured *S. faecalis* EBF 30/39. Cells were heated in HI broth at 60°C for 4 min (o) or 21 min (•), harvested and washed in 100 mmol l^{-1} phosphate buffer pH 7.1 and resuspended in 100 mmol l^{-1} potassium phosphate buffer pH 7.1, 0.5% glucose, 0.1% casein hydrolysate and incubated at 33°C. Samples were removed at intervals for serial dilution and viable counts on HI agar with and without 0.04% potassium tellurite. Recovery is the HI agar + 0.04% tellurite count as a percentage of the HI agar count.

least two types of injured cell with differing degrees of damage. The requirement for RNA biosynthesis during recovery has been reported in all cases so far examined (Clark et al. 1968; Duitschaever & Jordan 1974; Payne & Morley 1976).

Using an isotope dilution technique the extent of protein synthesis during recovery of tellurite resistance in *S. faecalis* has been determined. Gee & Payne (to be published) have found that, of the total protein content of injured cells allowed to recover for 6 h, 30-40% was synthesized during the recovery period. In the presence of chloramphenicol or the absence of casein hydrolysate, isotope dilution was negligible and less than 20% of the injured cells recovered. Separation of the soluble protein by polyacrylamide gel electrophoresis and

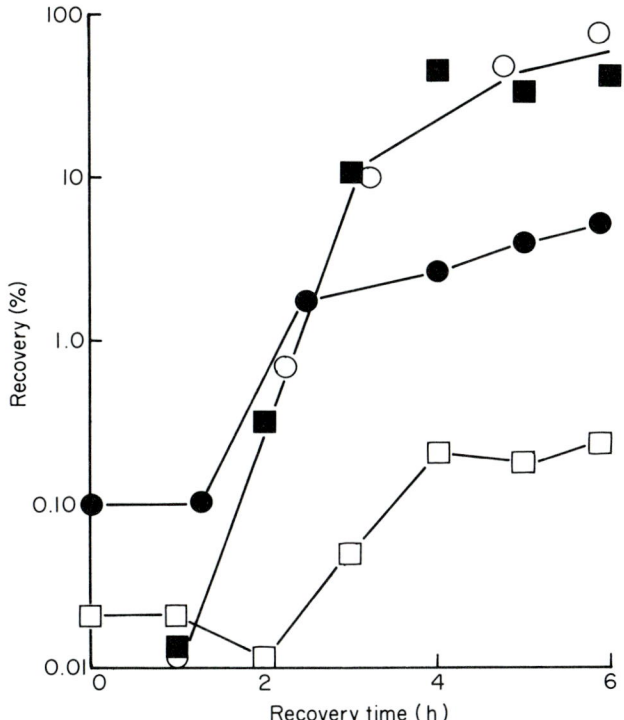

Fig. 8. Recovery of tellurite resistance by heat-damaged *S. faecalis* EBF 30/39. Cells were heated in HI broth at 60°C for 4 min and allowed to recover at 33°C in HI broth containing no addition (○); 100 μg ml^{-1} chloramphenicol (●); 5 μg ml^{-1} actinomycin D (□); 100 μg ml^{-1} penicillin (■). The count on HI agar + 0.04% potassium tellurite is expressed as a percentage of the HI agar count.

analysis of the distribution of radioactivity in such proteins from injured cells recovered in the presence of $[U-^{14}C]$ L-valine has shown that the protein synthesis was of a general nature and not restricted to a few proteins. However, this does not imply that the synthesis of all of these proteins was required for recovery.

Duitschaever & Jordan (1974) have analysed the fatty acid composition of uninjured and recovered *S. faecium* heated at 55°C. During recovery they noted an increase in the saturated fatty acid components, particularly lauric (C_{12}), myristic (C_{14}), palmitic (C_{16}) and stearic (C_{18}) acids. They also found that when recovered cells were exposed to a second heat treatment there was a substantial increase in heat resistance and a concomitant decrease in the number of injured cells amongst the survivors. Such cells maintained the altered fatty acid profile and the increased heat resistance through 12 subcultures. The authors concluded that selection for heat resistance and salt tolerance could occur rapidly in sublethally injured *S. faecium*.

7. General Comments

During the last 10 years there has been an increasing interest in sublethally damaged micro-organisms relative to food (Lewicki & Silverman 1969; Hobbs *et el*. 1971; Busta 1975). It is now well documented that exposure to environmental stress can produce damaged cells which can recover under suitable conditions.

and subsequent recovery in streptococci. Both types of damage result in increased sensitivity of the surviving cells to NaCl and other selective agents and, while cell wall biosynthesis does not appear to be necessary for recovery, RNA biosynthesis is, although this requirement seems to be more important for heat-damaged cells. The requirement for protein synthesis is more variable and may depend on several factors including the organism and even the strain used. Several organisms have been shown to require protein synthesis for recovery from heat damage (Tomlins & Ordal 1971; Hurst *et al*. 1973; Emswiler *et al*. 1976; Gomez *et al*. 1976; Payne & Morley 1976) whereas others do not (Iandolo & Ordal 1966; Sogin & Ordal 1967; Clark *et al*. 1968; Tomlins *et al*. 1971). From the evidence so far available cold damage in streptococci appears to be repaired more rapidly than damage caused by heat.

Since damaged cells show altered characteristics compared with undamaged cells one needs to consider carefully the media and conditions to be used to determine the numbers of these organisms isolated from a variety of environments. If sublethally damaged cells are present, the use of some selective and enrichment media or the wrong choice of incubation temperature may give a false low count. A resuscitation period in non-selective media prior to enumerating organisms, particularly from food products, has been recommended (see review by van Schothorst 1976). However, it is difficult to generalize about recovery times since these seem to vary depending on several factors, including the extent of damage and the stress that was responsible for the damage.

This review has centred largely on the damage caused, and the recovery from reversible damage, in streptococci subjected to cold and heat stress. It is evident that the literature on this topic is far from extensive and more papers have been published on damage and recovery in other organisms (for reviews see Gray & Postgate 1976; Skinner & Hugo 1976). In view of the importance of streptococci it is surprising that this group of organisms has not received more attention from this point of view.

There are many areas of damage and recovery in streptococci about which very little or nothing is known. Ribosome and ribosomal RNA degradation,

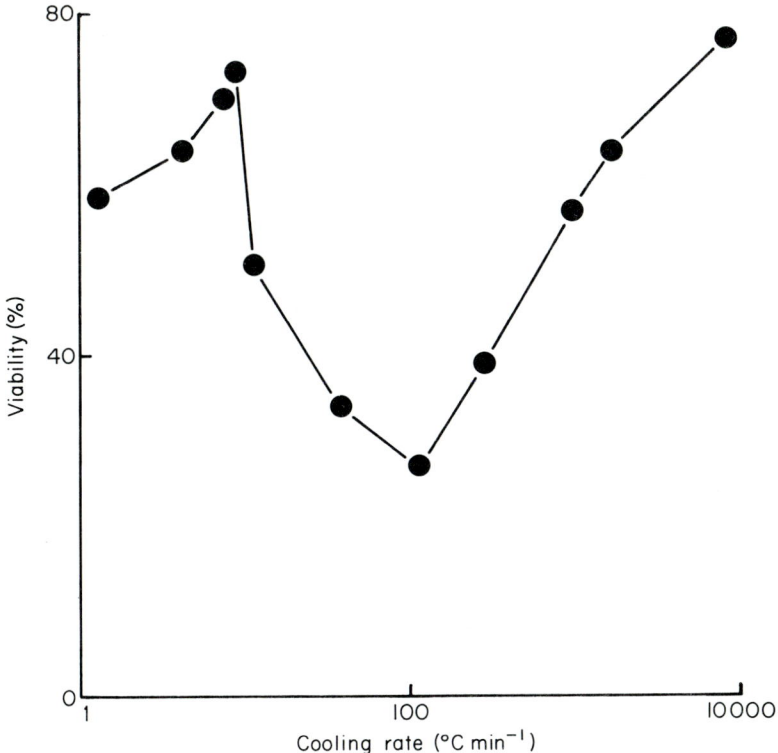

Fig. 9. Effect of cooling rate on survival of *S. faecalis* after freezing in water. Cells were thawed rapidly at $1000°C$ min^{-1} (from Calcott *et al.* 1976).

loss of metabolic activity due to enzyme inactivation and damage to DNA as consequences of sublethal treatments are some areas that need to be investigated more fully. Work by Calcott *et al.* (1976) has shown that *S. faecalis* when subjected to freezing has a maximum and minimum survival at cooling rates of $10°C$ min^{-1} and $100°C$ min^{-1}, respectively (Fig. 9). However no information is available on the relationship between rates of freezing and sublethal damage. More information is needed on how streptococci behave when exposed to chemical stress and whether cells damaged by such treatments can recover. We need to determine precisely the lesions caused by cold, heat, chemicals, ultraviolet and ionizing radiations and by combination of these agents and to determine the physiological processes involved during recovery. Obtaining a proper understanding of the behaviour of damaged cells will lead to an improved approach to the monitoring and control of these organisms.

8. References

ALLWOOD, M. C. & RUSSELL, A. D. 1970 Mechanisms of thermal injury in non-sporulating bacteria. *Advances in Applied Microbiology* **12**, 89-119.
BARNES, E. M. 1956 Methods for the isolation of faecal streptococci (Lancefield Group D) from bacon factories. *Journal of Applied Bacteriology* **19**, 193-203.
BAUMANN, D. P. & REINBOLD, G. W. 1966 Freezing of lactic cultures. *Journal of Dairy Science* **49**, 259-264.
BEUCHAT, L. R. & LECHOWICH, R. V. 1968a Effect of salt concentration in the recovery medium on heat injured *Streptococcus faecalis*. *Applied Microbiology* **16**, 772-776.
BEUCHAT, L. R. & LECHOWICH, R. V. 1968b Survival of heated *Streptococcus faecalis* as affected by phase of growth and incubation temperature after thermal exposure. *Journal of Applied Bacteriology* **31**, 414-419.
BUSTA, F. F. 1976 Practical implications of injured micro-organisms in food. *Journal of Milk and Food Technology* **39**, 138-145.
CALCOTT, P. H., LEE, S. K. & MACLEOD, R. A. 1976 The effect of cooling and warming rates on the survival of a variety of bacteria. *Canadian Journal of Microbiology* **22**, 106-109.
CLARK, C. W., WITTER, L. D. & ORDAL, Z. J. 1968 Thermal injury and recovery of *Streptococcus faecalis*. *Applied Microbiology* **16**, 1764-1769.
CORRY, J. E. L. 1970 Resistance and recovery of Salmonellae. M.Sc. Thesis, University of Bristol.
COWMAN, R. A. & SPECK, M. L. 1965a Activity of lactic streptococci following storage at refrigeration temperatures. *Journal of Dairy Science* **48**, 1441-1444.
COWMAN, R. A. & SPECK, M. L. 1965b Ultra-low temperature storage of lactic streptococci. *Journal of Dairy Science* **48**, 1531-1532.
COWMAN, R. A. & SPECK, M. L. 1967 Proteinase enzyme system of lactic streptococci I. Isolation and partial characterisation. *Applied Microbiology* **15**, 851-856.
COWMAN, R. A., SWAISGOOD, H. E. & SPECK, M. L. 1967 Proteinase enzyme system of lactic streptococci II. Role of membrane proteinase in cellular function. *Journal of Bacteriology* **94**, 942-948.
DUITSCHAEVER, C. L. & JORDAN, D. C. 1974 Development of resistance to heat and sodium chloride in *Streptococcus faecium* recovering from thermal injury. *Journal of Milk and Food Technology* **37**, 382-386.
EMSWILER, B. S., PIERSON, M. D. & SHOEMAKER, S. P. 1976 Sub-lethal heat stress of *Vibrio parahaemolyticus*. *Applied and Environmental Microbiology* **32**, 792-798.
ERWIN, D. G. & HAIGHT, R. D. 1973 Lethal and inhibitory effects of sodium chloride on thermally stressed *Staphylococcus aureus*. *Journal of Bacteriology* **116**, 337-340.
GOMEZ, R. F., BLAIS, K. D., HERRERO, A. & SINSKEY, A. J. 1976 Effects of inhibitors of protein, RNA and DNA synthesis on heat-injured *Salmonella typhimurium* LT2. *Journal of General Microbiology* **97**, 19-27.
GRAY, T. R. G. & POSTGATE, J. R. (eds) 1976 *Survival of Vegetative Microbes*. Society for General Microbiology Symposium No. 26. Cambridge: Cambridge University Press.
HARRIS, N. D. 1963 The influence of the recovery medium and the incubation temperature on the survival of damaged bacteria. *Journal of Applied Bacteriology* **26**, 387-397.
HARVEY, R. J. 1965 Damage to *Streptococcus lactis* resulting from growth at low pH. *Journal of Bacteriology* **90**, 1330-1336.
HOBBS, B. C., OLSON, J. C., SPECK, M. L., COWMAN, R. A., ORDAL, Z. J., NELSON, F. E. & SINSKEY, A. J. 1971 Restoration of sub-lethally impaired bacterial cells in foods. *Journal of Milk and Food Technology* **34**, 548-552.
HURST, A., HUGHES, A., BEARE-ROGERS, J. L. & COLLINS-THOMPSON, D. L. 1973 Physiological studies on the recovery of salt tolerance by *Staphylococcus aureus* after sub-lethal heating. *Journal of Bacteriology* **116**, 901-907.

HURST, A., HUGHES, A., DUCKWORTH, M. & BADDILEY, J. 1975 Loss of D-alanine during sub-lethal heating of *Staphylococcus aureus* S6 and magnesium binding during repair. *Journal of General Microbiology* **89**, 277-284.

IANDOLO, J. J. & ORDAL, Z. J. 1966 Repair of thermal injury in *Staphylococcus aureus*. *Journal of Bacteriology* **91**, 134-142.

IENISTEA, C., CHITU, M. & ROMAN, A. 1970*a* Heat resistance in milk of some strains of group D streptococci from pasteurized milk and the influence exerted on their growth by selective media. *Zentralblatt für Bakteriologie, Parasitenkunde, Infektionskrankheiten und Hygiene* Abt. I Originale **215**, 173-181.

IENISTEA, C., CHITU, M. & ROMAN, A. 1970*b* Influence exerted on the growth of heat-injured faecal streptococci by the addition of blood or milk to sodium azide media. *Zentralblatt für Bakteriologie, Parasitenkunde, Infektionskrankheiten und Hygiene* Abt. I Originale **215**, 182-186.

LEWICKI, P. P. & SILVERMAN, G. J. 1969 Mechanisms concerned in the repair and recovery of sub-lethally impaired cells occurring in foods. In *The Microbiology of Dried Foods*, eds Kampelmacher, E. H., Ingram, M. & Mossel, D. A. A. Haarlem, Netherlands: Grafische Industrie.

MORICHI, T. & IRIE, R. 1973 Factors affecting repair of sub-lethal injury in frozen or freeze-dried bacteria. *Cryobiology* **10**, 393-399.

MORICHI, T. & YANO, N. 1972 Sodium azide sensitivity of *Streptococcus faecalis* cells subjected to freezing. *Journal of the Food Hygienic Society Japan* **13**, 29-35.

MOSS, C. W. & SPECK, M. L. 1963 Injury and death of *Streptococcus lactis* due to freezing and frozen storage. *Applied Microbiology* **11**, 326-329.

NELSON, F. E. 1943 Factors which influence the growth of heat treated bacteria I. Comparison of four agar media. *Journal of Bacteriology* **45**, 395-403.

NELSON, F. E. 1944 Factors which influence the growth of heat treated bacteria II. Further studies on media. *Journal of Bacteriology* **48**, 473-477.

PAYNE, J. & MORLEY, J. S. 1976 Recovery of tellurite resistance by heat-injured *Streptococcus faecalis*. *Journal of General Microbiology* **94**, 421-424.

RAY, B. & SPECK, M. L. 1973 Freeze-injury in bacteria. *CRC Critical Reviews in Clinical Laboratory Sciences* **4**, 161-213.

SCHEUSNER, D. L., BUSTA, F. F. & SPECK, M. L. 1971 Injury of bacteria by sanitizers. *Applied Microbiology* **21**, 41-45.

SKINNER, F. A. & HUGO, W. B. (eds) 1976 *Inhibition and Inactivation of Vegetative Microbes*. Society for Applied Bacteriology Symposium No. 5, London & New York: Academic Press.

SOGIN, S. J. & ORDAL, Z. J. 1967 Regeneration of ribosomes and ribosomal ribonucleic acid during repair of thermal injury to *Staphylococcus aureus*. *Journal of Bacteriology* **94**, 1082-1087.

SPECK, M. L. & COWMAN, R. A. 1970 Preservation of lactic streptococci at low temperatures. In *Culture Collections of Micro-Organisms*, eds Iizuka, H. & Hasegawa, T. Baltimore: University Park Press.

STRAKA, R. P. & STOKES, J. L. 1959 Metabolic injury to bacteria at low temperatures. *Journal of Bacteriology* **78**, 181-185.

THOMAS, T. D. & BATT, R. D. 1968 Survival of *Streptococcus lactis* in starvation conditions. *Journal of General Microbiology* **50**, 367-382.

THOMAS, T. D., JARVIS, B. D. W. & SKIPPER, N. A. 1974 Localisation of proteinase(s) near the cell surface of *Streptococcus lactis*. *Journal of Bacteriology* **118**, 329-333.

TOMLINS, R. I. & ORDAL, Z. J. 1971 Requirements of *Salmonella typhimurium* for recovery from thermal injury. *Journal of Bacteriology* **105**, 512-518.

TOMLINS, R. I., PIERSON, M. D. & ORDAL, Z. J. 1971 Effect of thermal injury on the TCA cycle enzymes of *Staphylococcus aureus* MF31 and *Salmonella typhimurium* 7136. *Canadian Journal of Microbiology* **17**, 759-765.

VAN SCHOTHORST, M. 1976 Resuscitation of injured bacteria in foods. In *Inhibition and Inactivation of Vegetative Microbes*, eds Skinner, F. A. & Hugo, W. B. London & New York: Academic Press.

Isolation Media for Streptococci: Proceedings of a Discussion Meeting

CONTENTS

Principles	372
Ella M. Barnes	
Streptococci of human diseases	375
P. W. Ross	
Streptococci of animal diseases	379
C. D. Wilson	
Oral streptococci	380
J. M. Hardie and P. D. Marsh	
Streptococci of the normal intestinal flora	384
G. C. Mead	
Dairy streptococci	386
M. Elisabeth Sharpe	
Streptococci in food	392
D. A. A. Mossel	
Streptococci in water	393
N. P. Burman and A. W. Evans	
Summary of discussion	394
M. Ingram and Ella M. Barnes	
Appendix	395

Editors' note. In each contribution the information on isolation media is summarized in sections according to the following plan: types of streptococci being isolated, habitat and list of media and their formulations. For each medium are given: incubation conditions, type of colony and/or reaction, contaminants likely to interfere with isolation, and references.

Principles

ELLA M. BARNES

ARC Food Research Institute, Norwich, Norfolk, England

THE DIFFERENT TYPES of isolation media given in the following pages reflect the fact that the genus *Streptococcus* contains many groups and species differing in a large number of properties. There are those that are important pathogens whilst other species are used throughout the dairying industry as starters in the production of cheeses, yoghurts and other products. In some instances, where the organisms form the major population present, the use of selective agents in the media is unnecessary whilst in other environments, such as the alimentary tract of man and animals, or when the streptococci are used as faecal indicator organisms in water or foods, the problem is to isolate the particular organisms required when they are considerably outnumbered by other bacteria.

In the choice of enrichment and selective media one is looking for a set of conditions which permit the growth of the required streptococcus whilst inhibiting the other organisms which predominate in a particular habitat. This session is concerned with the majority of the streptococci which are facultative anaerobes. There is very little information concerning the selective isolation of those anaerobic cocci which are now included in the genus *Streptococcus*. One of the main selective inhibitors used in media is sodium azide which was shown to inhibit the bacterial cytochromes and to reduce catalase activity (Smith 1954). Hence it is not surprising that the streptococci, which are generally catalase negative and lacking in cytochromes, are amongst the most resistant organisms, but one is faced with the problem of resistant lactobacilli and other related groups which are also catalase negative and without cytochromes. The other inhibitor often used is thallous acetate which tends to be active against similar groups of organisms but the mode of action is still not understood. The selective activity of sodium azide and thallous acetate was studied by Richards *et al*. (1945) and their results, shown in Table 1, indicate a considerable variation in the resistance of the various species of streptococci to these compounds.

Their results also demonstrate the resistance of some other important bacteria to these selective agents. Hence the use of additional selective agents such as crystal violet or ethyl violet which are most active against many of the Gram positive bacteria, as can be seen in Table 1.

In addition to these selective agents, antimicrobial compounds such as polymyxin, neomycin, gentamicin, kanamycin or nalidixic acid may also be used. The choice of selective agents, media and incubation conditions will depend on the types of streptococci being isolated and the requirement to inhibit the other types of organisms commonly found in the same environment.

Because of the widely different properties found amongst these streptococci the basal medium itself may vary considerably. In addition the other micro-

Table 1
Inhibitory concentrations of sodium azide, thallous acetate and crystal violet for streptococci and other organisms

Species	No. strains tested	Inhibitory concentration* of		
		Sodium azide	Thallous acetate	Crystal violet
S. faecalis/faecium	5	200	80	200,000
S. salivarius†	1	155	80	200,000
S. thermophilus	1	3300	125	166,000
S. uberis	1	1250	83	250,000
S. agalactiae	10	1000	83	250,000
S. pyogenes	3	1480	143	330,000
S. lactis	3	1000	143	500,000
S. cremoris	3	6600	200	1,700,000
Staph. aureus	5	1700	143	2,000,000
Micrococcus spp.	5	20,000	2500	5,000,000
B. coli	7	5000	500	16,000
B. aerogenes	2	3000	250	16,000
Proteus vulgaris	3	3000	333	25,000
Pseudomonas spp.	3	13,000	2000	12,500
Bacillus spp.	6	6660	1000	2,000,000

* The figures represent the dilution, e.g. 100,000 represents 1 g in 100,000.
† Exact identity of this strain is uncertain.
Basal broth (g l^{-1}): peptone, 10; Lemco, 10; glucose, 5; pH 6.8.
Data of Richards et al. (1945).

organisms will vary according to the environment so that a selective agent which may be best under one set of conditions will not be so useful in another. An increased inhibitory effect on the streptococci themselves may be obtained when the selective agents are used with an incubation temperature nearer the maximum for growth, e.g. 45°C instead of 37°C for *Streptococcus faecalis* or combined with other inhibitory compounds such as bile salts, or a less favourable pH for growth. Similarly, the contaminants are also affected, micrococci being inhibited in thallous acetate-tetrazolium agar at pH 6.0 but much less so at 7.0 (Barnes 1959). Sodium azide is more inhibitory for *E. coli* at pH 6.8 than at 7.4 (Edwards 1938).

On the other hand, if more nutritious media are used, as when blood is incorporated or the complex requirements for the growth of many lactobacilli are met, then the medium may be less inhibitory.

Other substances are incorporated for colony differentiation, e.g. 5% sucrose for the formation of extracellular polysaccharides by the oral streptococci, or aesculin and iron salts to detect aesculin hydrolysis by the blackening of the medium and colonies. Tetrazolium is used in the media in two ways depending on pH and other substances present. At pH 7.0 all of the streptococci reduce it to the insoluble red compound formazan ensuring that small colonies can be more readily detected in the media and particularly on the membrane filters. At pH 6.0 it is used to differentiate between various types of streptococci.

Further complications arise from the need for resuscitation processes for damaged cells, particularly following heating or other unfavourable conditions, prior to isolating on the selective media. This may be achieved in a number of ways but basically by transferring from a less inhibitory environment to the more selective one.

Finally, it should be emphasized that the important question to ask before one chooses a particular medium is which streptococci one is trying to isolate and for what purpose. It is possible that if one is looking for streptococci in general there may be a number of types of streptococci which are not being isolated on any of the selective media used at present. If one is looking for a particular organism, whether it is a pathogen or a faecal indicator organism, there is a much better chance of finding a good selective medium.

References

BARNES, E. M. 1959 Differential and selective media for the faecal streptococci. *Journal of the Science of Food and Agriculture* **12**, 656–662.

EDWARDS, S. J. 1938 The diagnosis of streptococcus mastitis by cultural methods. *Journal of Comparative Pathology and Therapy* **51**, 250–263.

RICHARDS, T., SOULIDES, D. A. & (the late) SOULIDES, E. 1945 The inhibitory effect of crystal violet, sodium azide and thallous acetate upon mastitis streptococci and other common milk flora. *Proceedings of the Society for Applied Bacteriology (Abstracts)* 44–46.

SMITH, L. 1954 Bacterial cytochromes. *Bacteriological Reviews* **18**, 106–130.

Streptococci of Human Diseases
P. W. Ross
Department of Bacteriology, University Medical School, Edinburgh, Scotland

Types

β-Haemolytic streptococci, Lancefield groups A, C and G.

Habitat

Upper respiratory tract and skin.

Media

Non-selective solid media

Any good quality nutrient agar or peptone and a meat extract (Cruickshank *et al.* 1975). Commercial preparations such as Columbia Agar Base (Oxoid CM 331) or Agar Base (Difco 0045) are highly suitable; pH *ca.* 7.3; 5-10% blood (horse, sheep or human) is added. Although sheep blood inhibits growth of haemolytic *Haemophilus influenzae*, zones of β-haemolysis are frequently smaller on this medium. No glucose should be present in a human blood medium as this will prevent haemolysis, nor should human blood have a high titre to antistreptolysin O as this may also interfere with haemolysis (Wahl 1959). Horse blood is satisfactory in that areas of haemolysis are generally clear.

Selective solid media

Addition of crystal violet to blood agar inhibits growth of Gram positive organisms including staphylococci (Pike 1945). This is used at a 1 in 500,000 concentration (0.0002%). Addition of 1 in 16,000 sodium azide limits growth of *Neisseria* spp., some *Haemophilus* spp. and other Gram negative bacilli (Pike 1945). However, as β-haemolysis may be prevented or change into the α-type, care must be exercised in reading cultures: in most clinical situations the use of sodium azide is unnecessary.

Non-selective liquid media

A dehydrated modification of the medium described by Todd & Hewitt (1932) is obtainable commercially (Difco 0492; Oxoid CM 189). The Oxoid product contains (g l^{-1}): infusion from 450 g fat-free minced beef, 10.0; Tryptone (Oxoid L 42), 20.0; glucose, 2.0; $NaHCO_3$, 2.0; NaCl, 2.0; Na_2HPO_4, 0.4; pH *ca.* 7.8. Dissolve 36.4 g in 1 l of distilled water. Mix well; distribute into containers and sterilize by autoclaving at 115°C for 10 min.

Conditions

Optimal temperature for growth is 37°C (range 22-42°C) and optimal duration of incubation is 18-24 h, though neither is critical. Two plates should be incubated, one aerobically (e.g. a plain blood agar plate) and the other anaerobically (e.g. the crystal violet blood agar plate) in 90% H_2; 20% CO_2. Anaerobic incubation enhances the speed of production, amount and completeness of β-haemolysis (Fry 1933), and has an inhibitory effect on *Neisseria, Haemophilus* and *Corynebacterium* spp. (Williams 1958).

Colonies and reactions

Colonies are small after 24 h incubation (0.5-1.0 mm diam), discrete and low convex. Considerable variations in degrees of haemolysis are possible between groups A, C and G, the two latter producing haemolysis in the main larger areas, and also within groups: type of blood may also influence this. Colonies may be matt when freshly isolated and become smooth or glossy on subculture. Mucoid strains may also occur within group A.

Contaminants

The nature of contaminants depends on the area swabbed and will be upper respiratory tract or skin commensals. The use of selective media will minimize difficulties caused by contamination.

References

CRUICKSHANK, R., DUGUID, J. P., MARMION, B. P. & SWAIN, R. H. A. 1975 *Medical Microbiology*, 12 edn, vol. 2. Edinburgh, London & New York: Churchill Livingstone, pp. 104 & 111.
FRY, R. M. 1933 Anaerobic methods for the identification of haemolytic streptococci. *Journal of Pathology and Bacteriology* 37, 337-340.
KRUMWIEDE, E. & KUTTNER, A. G. 1938 A growth-inhibiting substance for the influenza group of organisms in the blood of various species. *Journal of Experimental Medicine* 67, 429-441.
PIKE, R. M. 1945 The isolation of haemolytic streptococci from throat swabs. Experiments with sodium azide and crystal violet in enrichment broth. *American Journal of Hygiene* 41, 211-220.
TODD, E. W. & HEWITT, L. F. 1932 A new culture medium for the production of antigenic streptococcal haemolysin. *Journal of Pathology and Bacteriology* 35, 973-974.
WAHL, R. 1959 In *Rheumatic Fever: Epidemiology and Prevention*, eds Cruickshank, R. & Glynn, A. A. Oxford: Blackwell Scientific Publications, p. 15.
WILLIAMS, R. E. O. 1958 Laboratory diagnosis of streptococcal infections. *Bulletin of the World Health Organization* 19, 153-176.

Types

β-Haemolytic streptococci, Lancefield group B (*S. agalactiae*).

Habitat

Genito-urinary tract, gut and upper respiratory tract of the adult and skin, gut and upper respiratory tract of the newborn.

Media

Non-selective and selective agar media and the Todd-Hewitt broth used for groups A, C and G can also be used but as many of the specimens for culture will be heavily contaminated by Gram negative bacilli and Gram positive organisms, it is essential that selective media are used. Additional selective agar media containing neomycin and nalidixic acid have been described (Vincent et al. 1971) as well as selective broth media containing gentamicin and nalidixic acid (Baker & Barrett 1973). Concentrations recommended for incorporation in Todd-Hewitt broth were 8 μg ml^{-1} of gentamicin and 15 μg ml^{-1} of nalidixic acid, but some recent work has shown that 8 μg ml^{-1} of gentamicin may inhibit some group B strains, and that 4-6 μg ml^{-1} may be a more acceptable concentration. This selective broth medium inhibits staphylococci and many Gram negative enteric organisms. Baker et al. (1976) preferred a selective broth to a selective plate medium. Other aminoglycosides may prove to be as effective as gentamicin for incorporation in a selective medium. Fallon (1975) described the production of pigment by 85% of group B strains on the surface of Columbia agar plates incubated anaerobically and a recent modification (Islam 1977) has also produced encouraging results.

Conditions

As for groups A, C and G. It is again important to incubate at least one plate anaerobically. Swabs may be (a) plated directly on selective media or (b) on plated on ordinary media then immersed overnight in selective broth and plated out on the following day or (c) put immediately into selective broth and plated out after 18-24 h.

Colonies and reactions

Group B colonies are greyer, softer and often slightly larger than those of group A. Haemolysis is generally of the β-type but the zone on blood agar is much narrower and not so clearly demarcated as with groups A, C or G. Some strains produce greenish zones and some may be totally non-haemolytic. Double zones of haemolysis may be produced under special conditions of incubation.

Contaminants

Depending on the area swabbed, there will be skin, vaginal, urethral, gut or upper respiratory-tract contaminants, most of which are eliminated by the use of the selective media described above.

References

BAKER, C. J. & BARRETT, F. F. 1933 Selective broth medium for isolation of group B streptococci. *Applied Microbiology* **26**, 884-885.

BAKER, C. J., GOROFF, D. K., ALPERT, S. L., HAYES, C. & McCORMACK, W. M. 1976 Comparison of bacteriological methods for the isolation of group B *Streptococcus* from vaginal cultures. *Journal of Clinical Microbiology* **4**, 46-48.

FALLON, R. J. 1975 The rapid identification of Lancefield group B streptococci. *Journal of Clinical Pathology* **27**, 902-905.

ISLAM, A. K. M. S. 1977 Rapid recognition of group B streptococci. *Lancet* **1**, 356-357.

VINCENT, W. F., GIBBONS, W. E. & GAAFAR, H. A. 1971 Selective medium for the isolation of streptococci from clinical specimens. *Applied Microbiology* **22**, 942-943.

Streptococci of Animal Diseases

C. D. WILSON

The Central Veterinary Laboratory, Weybridge, Surrey, England

Types

All types of streptococci found in milk with special reference to group B streptococci.

Habitat

Bulk-milk collected from farm tanks.

Medium

Five per cent sheep blood agar containing: aesculin, 0.1%; polymyxin B sulphate, 17 units ml^{-1}; neomycin sulphate, 4.2 µg ml^{-1}; sodium furidate, 0.5 µg ml^{-1}; pH, 7.5 (Lewbury *et al.* 1964).

Conditions

The plates are incubated for 48 h at 37°C.

Colonies and reactions

Streptococcus agalactiae does not split aesculin and is either non-haemolytic or shows a variable amount of β-haemolysis. The colonies are semi-translucent with a raised centre and a crenated edge. Occasionally colonies which are firm to touch with the loop and grow into the medium are seen. They are very difficult to dislodge. *Streptococcus dysgalactiae* is recognized by its production of greening round the colonies which are glistening and can be easily moved by a loop. *Streptococcus uberis* splits the aesculin but has to be differentiated from the other aesculin-splitting streptococci. The β-haemolytic streptococci have to be identified further, serologically and biochemically.

Contaminants

The three antibiotics combine to inhibit the growth of the organisms likely to be present in milk as pathogens or contaminants. These are staphylococci, coliforms and other Gram negative organisms and anthracoids.

Reference

LEWBURY, E. J. L., HUDSON, A. & LILLY, H. A. 1964 A new selective blood agar medium for *Streptococcus pyogenes* and other haemolytic streptococci. *Journal of Clinical Pathology* 17, 231.

Oral Streptococci

J. M. HARDIE AND P. D. MARSH

Oral Microbiology Department, The London Hospital Medical College, London, England

Types

Streptococcus sanguis, S. mitior, S. mutans, S. milleri, S. salivarius (and others).

Habitat

The oral cavity, particularly dental plaque.

Medium (1) Trypticase-yeast extract-cystine (TYC) 5% sucrose agar

Routinely we use the 5% sucrose medium of De Moor & De Stoppelaar. This medium contains (g l^{-1}): Trypticase (BBL)*, 15; Yeast Extract (Difco)*, 5; L-cystine, 0.2; Na_2SO_3, 0.1; NaCl, 1; Na_2HPO_4, 12; H_2O, 2; $NaHCO_3$, 2; sodium acetate (3 H_2O), 20; sucrose, 50; agar, 12; pH, 7.3. The medium is autoclaved at 121°C for 15 min. (*These ingredients can be replaced by Lab M Tryptone and Yeast Extract, respectively.)

Commercial availability

TYC 5% sucrose medium can be obtained from: London Analytical and Bacteriological Media Ltd, (Lab M).

Selectivity

TYC can be made partially selective for *S. mutans* and *S. milleri* by addition of 0.1% sulphadimetine. Add the sulphonamide to molten and cooled agar.

Conditions

Normally 3-5 days anaerobically. Some of the species will also grow well aerobically on this medium.

Colonies and reactions

In common with several of the media devised for oral streptococci the 5% sucrose content of TYC allows the formation of extracellular polysaccharides (glucans and fructans) by some species. Some of the typical colonial forms for the various species are described below, but identification should always be confirmed by other criteria. In our experience TYC is particularly useful for the recognition of differences in colonial morphology. Colony characteristics of oral streptococcal species are given below.

S. sanguis. Grey, white or colourless, 1-3 mm diam (sometimes larger), rough or smooth, hard and rubbery, often adherent and difficult to remove, may distort agar, may produce watery polysaccharide ('drop' or 'puddle').

S. mitior. (a) Hard type. Similar to *S. sanguis.* (b) Soft type. 0.5-2.0 mm diam, grey, white or colourless, round, smooth, soft, not adherent.

S. mutans. White, grey or yellow, 0.5-2.0 mm diam, rough, irregular, heaped, may resemble frosted glass, usually crumbly, sometimes hard and adherent, occasional soft, smooth varieties also occur, may produce 'water drop' or 'puddle' of polysaccharide. Colonies tend to come off whole when picked from plates.

S. salivarius. White or grey, 2-6 mm diam domed, usually smooth, soft and mucoid, sometimes with hard outer crust. Older colonies may become hard.

S. milleri. White or grey, 1-2 mm diam, often rough, dry and crumbly, may be smooth and soft.

S. bovis. Very large, mucoid colonies which tend to drip into lid of Petri dish (isolated from animals' mouths, not reported from humans).

Other streptococci. Enterococci and most other streptococci usually produce small, soft, 'nondescript' colonies on TYC.

Contaminants

Although TYC is not a strictly selective medium, most of the organisms which grow on it are found to be streptococci. Staphylococci and micrococci are isolated occasionally. Many of the other common oral genera (e.g. *Actinomyces, Veillonella, Neisseria, Bacteroides*) either fail to grow or produce very small (0.5 mm diam) colonies.

References

BOWDEN, G. H., HARDIE, J. M. & SLACK, G. L. 1975 Microbial variations in approximal dental plaque. *Caries Research* 9, 253-277.

DE STOPPELAAR, J. D., VAN HOUTE, J. & DE MOOR, C. E. 1967 The presence of dextran-forming bacteria, resembling *Streptococcus bovis* and *Streptococcus sanguis,* in human dental plaque. *Archives of Oral Biology* 12, 1199-1201.

Medium (2) Sodium azide-blood agar

This medium is used in an attempt to estimate the total number of streptococci in oral samples. It does not allow the easy recognition of different colonial types of streptococci. To make the medium: (a) melt 100 ml of Oxoid Blood Agar Base No. 2; (b) cool to 50°C, add 2 ml of 1% sodium azide solution and mix; (c) add 5% of defibrinated horse blood and pour plates.

Medium (3) Mitis-Salivarius Agar

This medium, described originally by Chapman (1944), is probably the most widely used selective medium for oral streptococci and is available from several commercial companies (e.g. BBL, Difco, Oxoid). It contains 5% sucrose, trypan blue, crystal violet and potassium tellurite as selective agents. MSA has been used extensively by dental researchers for isolation of *S. mutans*, *S. sanguis* and *S. salivarius*.

Where problems occur with Gram negative spreaders, MSA can be made more selective by addition of 100 units ml^{-1} of Polymyxin B sulphate (Fitzgerald & Adams 1975).

References

CARLSSON, J. 1967 Presence of various types of non-haemolytic streptococci in dental plaque and other sites of the oral cavity in man. *Odontologist Revy* 18, 55-74.
CHAPMAN, G. H. 1944 The isolation of streptococci from mixed cultures. *Journal of Bacteriology* 48, 113-114.
EDWARDSSON, S. 1970 The caries inducing properties of variants of *Streptococcus mutans*. *Odontologist Revy* 21, 153-157.
FITZGERALD, R. J. & ADAMS, B. D. 1975 Increased selectivity of Mitis-Salivarius agar containing polymyxin. *Journal of Clinical Microbiology* 1, 239-240.
JORDAN, H. V., KRASSE, B. & MOLLER, A. 1968 A method of sampling human dental plaque for certain caries-inducing streptococci. *Archives of Oral Biology* 13, 191-227.
KRASSE, B. 1966 Human streptococci and experimental caries in hamsters. *Archives of Oral Biology* 11, 429-436.
LILJEMARK, W. K., OKRENT, D. H. & BLOOMQUIST, C. G. 1976 Differential recovery of *Streptococcus mutans* from various Mitis-Salivarius agar preparations. *Journal of Clinical Microbiology* 4, 108-109.

Medium (4) MM10 sucrose (Loesche & Syed 1973)

This is a non-selective 5% sucrose agar, containing trypticase, yeast extract and blood (amongst other things), which allows recognition of the distinctive colonies of polysaccharide-producing streptococci without inhibiting the rest of the oral flora.

Reference

LOESCHE, W. H. & SYED, S. A. 1973 The predominant cultivable flora of carious plaque and carious dentine. *Caries Research* 7, 201-216.

Medium (5) BCY (Ikeda & Sandham 1972)

Another non-selective medium, consisting of brain-heart infusion agar plus casitone, yeast extract, cysteine HCl and 5% horse blood. It is reported to allow simultaneous enumeration of *S. mutans* and 'total bacteria'.

Reference

IKEDA, T. & SANDHAM, H. J. 1972 A medium for the recognition and enumeration of *Streptococcus mutans*. *Archives of Oral Biology* 17, 601-604.

Medium (6) MC

This was originally described as a selective medium for *S. mutans*, but has since been shown to support the growth of *S. milleri*. It is essentially similar to MSA, with the addition of 0.1% of sulphadimetine.

References

CARLSSON, J. (1968) A medium for isolation of *Streptococcus mutans*. *Archives of Oral Biology* **12**, 1657-1658.

MEJARE, B. & EDWARDSSON, D. 1975 *Streptococcus milleri* (Guthof). An indigenous organism of the human oral cavity. *Archives of Oral Biology* **20**, 757-762.

Medium (7) MSB

This is another selective medium for *S. mutans*. It is similar to MSA, but contains 0.2 units ml^{-1} of bacitracin and the sucrose content is increased to 20% (Gold *et al.* 1973).

In a comparison of five of the isolation media that have been described for *S. mutans* it was found that certain serotypes (a, d and g) did not grow on some of the media. Counts on the selective media were 10% lower than on non-selective media (Emilson & Bratthall 1976).

References

GOLD, O. G., JORDAN, H. V. & VAN HOUTE, J. 1973 A selective medium for *Streptococcus mutans*. *Archives of Oral Biology* **18**, 1357-1364.

EMILSON, C. G. & BRATTHALL, D. 1976 Growth of *Streptococcus mutans* on various selective media. *Journal of Clinical Microbiology* **4**, 95-98.

Streptococci of the Normal Intestinal Flora

G. C. MEAD

ARC Food Research Institute, Norwich, Norfolk, England

Types

Mainly group D.

Habitat

The intestine.

Medium

Thallous acetate-tetrazolium-glucose agar (TlTG; Barnes 1956). The basal medium contains (g l^{-1}): Bacteriological Peptone (Evans), 10; Lab-Lemco powder (Oxoid), 8; agar (New Zealand), 14.

The peptone and Lab-Lemco powder are dissolved in half the required volume of distilled water, making allowance for the additions (a)-(c) below. The

Table 2

Growth of reference strains of streptococci on the TlTG medium of Barnes (1956)

Strains		Incubation temperature (°C)	
		37	*45
Streptococcus agalactiae	NCTC 8190	+	−
	8542	+	−
S. avium	W32	+	−
	W33	+	−
S. bovis (mannitol +)	TM/C/101	+	−
(mannitol −)	TM/C/33	+	−
S. equinus	N/C32	+	−
S. faecalis	GB 112	+	+
S. faecium	GM/C/5	+	+
S. faecium subsp. *durans*	E/63	+	+
S. lactis	NCIB 6681	+	−
S. mitis	NCTC 10712	−	−
S. salivarius	NCTC 7366	+	−
S. suis	W80 − PM33	−	−
	W81 − PM34	−	−

+, Growth; −, no growth (up to 48 h).
* Preliminary incubation for 4 h at 37°C.

pH value is adjusted to 6.0-6.1. The agar is dissolved separately and the two solutions are mixed and distributed in 92 ml amounts before autoclaving at 121°C for 15 min.

The following solutions are added to each 92 ml of molten basal medium: (a) 5 ml of 20% (w/v) filter-sterilized glucose; (b) 2 ml of 5% (w/v) thallous acetate, autoclaved at 115°C for 15 min; (c) 1 ml of 1% (w/v) 2:3:5 triphenyltetrazolium chloride (filter-sterilized). Agar plates are prepared for surface inoculation and dried before use.

Conditions

Usually 37°C for 20-24 h. Streptococci other than those of group D will grow, especially when the incubation period is extended to 48 h. Selectivity for *S. faecalis/faecium* biotypes is markedly increased by incubating for 4 h at 37°C followed by 24-48 h at 45°C (Hall 1975), as shown in Table 2.

Colonies and reactions

Description of colonies (after 24 h at 37°C):
(1) white or pale pink—*S. faecium* and variants, *S. avium*, *S. bovis*, (some strains), *S. equinus*, some other non-group D spp. and unclassified strains;
(2) maroon or maroon centre with narrow white periphery—*S. faecalis* biotypes;
(3) very small maroon—*S. bovis* (some strains), group N streptococci.

Contaminants

Lactobacilli, *Staphylococcus aureus*, pediococci and streptococci other than group D.

Another selective medium which can be used to isolate a wider range of streptococci is Mitis-Salivarius agar (Chapman 1944).

References

BARNES, E. M. 1956a Tetrazolium reduction as a means of differentiating *Streptococcus faecalis* from *Streptococcus faecium*. *Journal of General Microbiology* **14**, 57-68.

BARNES, E. M. 1956b Methods for the isolation of faecal streptococci (Lancefield Group D) from bacon factories. *Journal of Applied Bacteriology* **19**, 193-203.

BARNES, E. M. 1958 The effect of antibiotic supplements on the faecal streptococci (Lancefield Group D) of poultry. *British Veterinary Journal* **114**, 333-344.

CHAPMAN, G. H. 1944 The isolation of streptococci from mixed cultures. *Journal of Bacteriology* **48**, 113-114.

FINEGOLD, S. M., ATTEBERY, H. R. & SUTTER, V. L. 1974 Effect of diet on human faecal flora; comparison of Japanese and American diets. *American Journal of Clinical Nutrition* **27**, 1456-1469.

HALL, L. P. 1975 *A Manual of Methods for the Bacteriological Examination of Frozen Foods*, 2nd edn. Chipping Campden: Campden Food Preservation R.A., p. 22.

Dairy Streptococci

M. ELISABETH SHARPE

National Institute for Research in Dairying,
Shinfield, Reading, Berkshire, England

Dairy streptococci to be isolated and enumerated include the mesophilic streptococci *Streptococcus lactis*, *S. lactis* subsp. *diacetylactis* and *S. cremoris*, and the thermophilic species *S. thermophilus*. The importance of these organisms is not their chance occurrence in an environment, but their effect, as part of an industrial process, on milk and dairy products. They are deliberately added in large numbers of 10^6-10^8 ml^{-1} of milk, levels which are so much higher than those of the natural microflora that it is unnecessary to take this microflora into account when counting these streptococci initially or during processing of the product.

Other enumerations of the mesophilic starters are to assess production media and methods and to check cell viability in large scale production of concentrated (frozen or lyophilized) starters, which are now supplied to the cheesemaking industry. Once again there is no competition with other types of organisms in making counts.

Thus, media for isolation of these streptococci are not selective but are general purpose media giving good growth of the organism and on which they can be enumerated effectively.

For mesophilic starters the usual medium is yeast-glucose agar (YGA), which has been in use for many years (Sharpe & Naylor 1957). Another more recent medium is the β-glycerophosphate medium of Terzaghi & Sandine (1975), in which the buffering effect of the β-glycerophosphate results in large colonies and more rapid growth than on YGA. This is useful for enumeration studies: on YGA some of these lactic streptococci, especially *S. cremoris*, must be deep plated if they are to be incubated aerobically; if they are surface plated, they must be incubated in a CO_2 atmosphere. In both cases 48 h incubation is necessary. The glycerophosphate medium gives colonies comparable in number and size after aerobic incubation for only 18 h, thus removing the need to use an anaerobic jar for the CO_2, and getting a result 24 h earlier. The well known stimulatory effect of CO_2 on growth of these lactic streptococci is observed on YGA, but not necessarily on the glycerophosphate medium.

For *S. thermophilus*, enumeration of these organisms in connection with yoghurt manufacture, and occasionally in Swiss cheesemaking, is required. In yoghurt, for analysis of the mixed starter consisting of *S. thermophilus* and *Lactobacillus bulgaricus* a medium which suppresses the lactobacillus in order to enumerate the streptococci is needed. A yeast-lactose agar has been used in the past (some strains of *S. thermophilus* will not ferment glucose). But only pin-point streptococcus colonies appear and the lactobacillus also may grow, although very sparsely. Shankar & Davies (1977) have found the glycerophosphate medium to be satisfactory as *S. thermophilus* grows profusely and the lactobacillus is completely suppressed. It is not known why this lactobacillus is inhibited as Terzaghi & Sandine (1975) found that many other lactobacilli grow well on this medium.

If required to isolate these organisms from habitats of mixed organisms, where they are not the major part of the microflora, the following are recommended.

Mesophilic streptococci: use the tetrazolium medium of Barnes (1956), described in the contribution of Mead. Group N streptococci grow on this medium as red colonies. Other streptococci and some lactobacilli will also be present.

Streptococcus thermophilus: heat the sample to 60°C for 30 min. Plate out on β-glycerophosphate medium or on YLA.

The other types of media required for dairy streptococci are diagnostic media. These are to distinguish and enumerate the different types of lactic streptococci present in a multiple strain starter. These starters may contain *S. lactis*, *S. cremoris* and *S. diacetylactis* which, having different properties, may be important for different processes. It is therefore essential to know which of those organisms are present and possibly to monitor their proportions.

In the diagnostic media described below, advantage is taken of different characteristics of these three organisms. Using the M16 medium of Thomas (1973), *S. lactis* and *S. cremoris* are differentiated by the ability of *S. lactis* to hydrolyse arginine. *Streptococcus cremoris* produces lactic acid which turns the BCP indicator yellow, producing yellow colonies. In contrast *S. lactis* which also produces lactic acid, metabolizes the arginine present to produce ammonia. This neutralizes the lactic acid so the colonies remain white. *Streptococcus diacetylactis*, if present, also hydrolyses arginine and therefore produces white colonies. It is essential not to use a concentration of lactose higher than the recommended 0.2%.

To detect the aroma-producing *S. diacetylactis* which breaks down citrate to produce diacetyl, a medium incorporating insoluble Ca citrate is used: *S. diacetylactis* utilizes the citrate and this is demonstrated by a clearing zone around the colonies, in the turbid Ca citrate agar. Citrate-utilizing leuconostocs if present will also produce zones of clearing but these are much smaller and take longer to develop.

Types

Mesophilic lactic streptococci (*S. lactis*, *S. lactis* subsp. *diacetylactis*, *S. cremoris*) and the thermophilic streptococcus *S. thermophilus*.

Habitat

Milk, dairy products, multiple strain starters.

General purpose media

Medium (1) Yeast-glucose agar (YGA)

This is a general purpose medium for mesophilic lactic streptococci. It contains (g l^{-1}): peptone (Oxoid), 10; meat extract, 8; yeast extract, 3; NaCl, 5; glucose, 5 (add after filtration of medium); agar, 20. Adjust pH to 7.0 with 1N NaOH. Filter. Add glucose. Dispense in 100 ml amounts in screwcapped bottles. Autoclave at 121°C for 15 min. Final pH, 6.8 to 7.0. This medium was used traditionally at the NIRD for many years before 1957 but it is difficult to find a reference to a paper giving the formula. YGA without the agar can be used as a broth medium.

Conditions

Use either (i) pour plates of dilutions (1.0 ml sample + *ca*. 12 ml melted agar) and incubate aerobically or (ii) surface plates of dilutions—Miles-Misra drop counts or 0.1 ml spread out on surface of well-dried plates. Incubate in anaerobic jar in atmosphere of H_2 90% + CO_2 10% for 48 h at 30°C.

Colonies

Small, cream coloured, smooth, 0.5 to 1.0 mm diam.

Contaminants

Unlikely to be present (see below) but would probably be lactobacilli.

Reference

NAYLOR, J. & SHARPE, M. E. 1957 Lactobacilli in Cheddar cheese. I. The use of selective media for isolation and of serological typing for identification. *Journal of Dairy Research* 25, 92-103.

Medium (2) M17 (β-glycerophosphate medium)

This is: (a) a general purpose medium for isolation of mesophilic lactic streptococci; (b) a selective medium for isolation of *S. thermophilus* from yoghurt; (c) used for assay of bacteriophages from lactic streptococci (not further considered). The medium contains (l^{-1}): Polypeptone BBL, 5g; Phytone Peptone BBL, 5g; yeast extract, 2.5g; meat extract, 5g; lactose, 5g; ascorbic acid, 0.5g; β-disodium glycerophosphate (Koch Light), 19g; 1.0 mol l^{-1} $MgSO_4.7H_2O$ solution, 1 ml; agar, 15g.

When dispensed in 10 ml amounts, add the lactose to the other ingredients, tube out and autoclave. When dispensed in 100 ml amounts in screwcapped bottles, autoclave at 121°C for 15 min then add previously sterilized (Seitz-filtered) lactose. This medium used as a broth (agar omitted) gives profuse growth of all streptococci mentioned below.

For isolating mesophilic starters the medium is used at pH 7.15

Conditions

Miles-Misra counts or 0.1 ml spread on the surface of well-dried plates. Incubate for 18 h at 30°C, aerobically.

Colonies

White, smooth 1.5 to 2.0 mm diam.

Contaminants

Unlikely to be present (*see below) but would probably be lactobacilli.

Reference

TERZAGHI, B. E. & SANDINE, W. E. 1975 Improved medium for lactic streptococci and their bacteriophages. *Applied Microbiology* **29**, 807–813.

As a selective medium for S. thermophilus *the medium is used at pH 6.8*

Conditions

Use pour plates of dilutions of yoghurt (1.0 ml in 12.0 ml of melted M17 agar). Incubate at 37°C for 48 h aerobically.

Colonies

White, lens shaped.

Contaminants

None.

Reference

SHANKAR, P. A. & DAVIES, F. L. 1977 A note on the suppression of *Lactobacillus bulgaricus* in media containing β-glycerophosphate and application of such media to selective isolation of *Streptococcus thermophilus* from yoghurt. *Journal of the Society of Dairy Technology* **30**, 28–30.

* Media YGA and M17 are general purpose media for isolation, enumeration and cultivation of starter streptococci from cheese-milk and cheese. They are *not* selective, and rely on the fact that the starter has been added at much higher levels than the natural milk microflora present, and by multiplication during cheesemaking continue to dominate the flora. When, in a mature cheese, multiplying lactobacilli reach a higher level than the falling numbers of starter streptococci, the latter cannot readily be enumerated.

Medium (3) Yeast-lactose agar (YLA)

This medium is used to isolate *S. thermophilus* from yoghurt. Prepared as for YGA, but the glucose is replaced by 0.5% lactose.

Conditions

Streaked surface plates are incubated for 48 h at 37°C aerobically.

Colonies

0.5 mm diam.

Contaminants

Lactobacillus bulgaricus from the yoghurt starter may grow sparsely. YLA without the agar can be used as a broth medium.

Diagnostic media for isolation of different mesophilic starters from a multiple strain starter

Medium (1) M16-BCP for S. cremoris

This medium contains (g l^{-1}): meat extract, 5; yeast extract, 2.5; Polypeptone BBL, 5; Phytone BBL, 5; ascorbic acid, 0.5; sodium acetate, 1.8; L-arginine HCl, 4; agar, 15; lactose (add after autoclaving), 2; bromocresol purple (BCP) (add after autoclaving), 0.05.

Adjust pH to 6.8 with 1N NaOH and autoclave at 121°C for 15 min. After cooling to 50°C add autoclaved lactose solution (2%) and autoclaved BCP (0.5%). Final pH, 6.8.

Conditions

Surface plates of dilutions (0.1 ml spread on surface of well-dried plates) incubated for 24-48 h at 30°C in an atmosphere of air, 95% + CO_2, 5%.

Reaction

Strains of *S. cremoris* ferment lactose, producing lactic acid which changes the BCP indicator to yellow, producing yellow colonies. In contrast *S. lactis* which also produces lactic acid, in addition metabolizes arginine to produce ammonia and other products. The ammonia neutralizes the lactic acid and the colonies remain white.

Contaminants

None producing yellow colonies.

Reference

THOMAS, T. D. 1973 Agar medium for differentiation of *Streptococcus cremoris* from other bacteria. *New Zealand Journal of Dairy Science and Technology* **8**, 70-71.

Medium (2) Medium for detecting aroma-producing streptococci.

This contains (g l^{-1}): tryptone, 5; yeast extract, 2.5; casamino acids, 5; K_2HPO_4, 1.25; calcium citrate, 10; carboxymethylcellulose (CMC), 15; agar, 15.

Ten grams of calcium citrate and 15 g of CMC are suspended in 500 ml distilled water (DW) and stirred and warmed (50°C) until a homogeneous turbid suspension is formed. Fifteen grams of agar are dissolved in 500 ml DW added hot, mixed well, and the required amounts of tryptone, yeast extract, casamino acids and K_2HPO_4 are added. Steam for 15 min, adjust pH to 5.6 with 6N HCl. Dispense 100 ml amounts into screwcapped bottles. Sterilize at 121°C for 15 min. Cool to 50°C, add 5 ml sterile reconstituted skim milk and 10 ml of sterile 3% (w/v) $CaCO_3$ in distilled water to 100 ml of basal medium. Final pH, 6.35. Must not be higher.

Conditions

0.1 ml sample or dilutions spread on the surface of well-dried plates. Incubated in anaerobic jar in atmosphere of H_2, 90% + CO_2, 10% for 6 days at 30°C.

Colonies

Streptococci (*S. diacetylactics*) able to utilize Ca citrate show a clear zone free from the turbid Ca citrate round their colonies.

Contaminants

Other citrate utilizing strains present in multiple strain starters such as the leuconostocs, will also produce zones of clearing, but they grow more slowly and are slow citrate fermenting, so that they will be much smaller colonies with smaller zones of clearing.

Acknowledgment

We are grateful to F. L. Davies for helpful discussions on these media.

Reference

REDDY, M. S., VEDAMUTHU, E. R., WASHAM, C. J. & REINBOLD, G. W. 1972 Agar medium for differential enumeration of lactic streptococci. *Applied Microbiology* 24, 947–952.

(*Note.* In the original paper this medium also contains L-arginine HCl and BCP and is designed to differentiate *L. cremoris* from other starter organisms. The medium of Thomas, however, is more satisfactory in that it gives better growth of *S. cremoris*.)

Streptococci in Food

D. A. A. MOSSEL

Faculty of Veterinary Medicine,
University of Utrecht, Utrecht, The Netherlands

Types

Lancefield group D.

Habitat

Raw and processed foods of animal origin; drinking water and bottled mineral waters.

Medium (1)

This contains (g l^{-1}): tryptone, 20; yeast extract (powder), 5; NaCl, 5; sodium citrate, 1; aesculin, 1; ferric ammonium citrate, 0.5; sodium azide, 0.15; agar, 15; kanamycin sulphate, 0.020.

Medium (2)

This is Medium (1), but with only 0.010 g l^{-1} of kanamycin sulphate.
Both media have a pH value of 7.1 ± 0.1 at *ca.* 50°C.

Conditions

Medium (1). Incubated for 16-20 h at 37 ± 1°C.
Medium (2). Incubated for 16-20 h at 42 ± 1°C.

Colonies and reactions

White or transparent, with black or dark brown halo.

Contaminants

Aerococcus, Corynebacterium, Lactobacillus.

References

COLOBERT, L. & MORÉLIS, P. 1958. Nouveau milieu sélectif pour l'isolement des entérocoques, d'emploi pratique pour leur recherche et leur numération dans l'eau. *Annales de l'Institut Pasteur, Paris* **94**, 120-122.

FACKLAM, R. R., PADULA, J. F., THACKER, L. G., WORTHAM, E. C. & SCONYERS, B. J. 1974 Presumptive identification of group A, B and D streptococci. *Applied Microbiology* **27**, 107-113.

MOSSEL, D. A. A., BIJKER, P. H. G., EELDERINK, I. & SPREEKENS, K. A. VAN 1977 The selective enumeration of Lancefield group D streptococci in foods, water and the food environment, using kanamycin aesculin azide agars at various temperatures. Presented at the 1977 Summer Meeting of the Society for Applied Bacteriology. Journal of Applied Bacteriology, offered for publication.

Streptococci in Water
N. P. Burman and A. W. Evans
Thames Water Authority, New River Head Laboratories, London, England

Types
Group D streptococci.

Habitat
Polluted waters.

Medium
Slanetz & Bartley's medium (Membrane Enterococcus Agar) contains (g l^{-1} of distilled water): tryptose, 20; yeast extract, 5; glucose, 2; K_2HPO_4, 4; sodium azide, 0.4; agar, 12; 2, 3, 5, -triphenyltetrazolium chloride (TTC), 0.1; the first six ingredients are steamed with the distilled water to dissolve. The pH should be 7.2 and no adjustment should be necessary. The TTC is added as 10 ml of a 1% sterile solution and the medium poured directly into Petri dishes without further sterilization. The medium with or without TTC should not be stored and re-melted. When placed in a sealed container to prevent drying, poured plates can be kept at 4°C for at least 6 months. pH, 7.2.

Conditions
37°C for 4 h followed by 44–45°C for 44 h.

Colonies and reactions
Colonies coloured from maroon to pink due to reduction of TTC to the formazan.

Contaminants
Contaminants variable; some Gram positive and Gram negative rods and occasional streptococci other than group D.

Reference
REPORT (1969). The Bacteriological Examination of Water Supplies. Reports on Public Health and Medical Subjects No. 71. London: HMSO.

Summary of Discussion

M. INGRAM

ARC Meat Research Institute, Langford, Bristol, England

AND

ELLA M. BARNES

ARC Food Research Institute, Norfolk, Norwich, England

The discussion, following the papers dealing with isolation of streptococci from various environments, was mainly concerned with selective isolation where the organisms formed a minority of the total numbers of micro-organisms present. No single medium seemed likely to select all streptococci. Even a range of selective media cannot reliably do this, because some species are likely to be missed on the media at present used; e.g. *S. thermophilus* which is difficult to detect unless present as a major part of the population.

In discussing the differential and selective properties of the various media the following points were made.

1. In most streptococcal media the glucose is autoclaved in the medium and may increase its reducing properties. No evidence was presented for any harmful effects, and it was suggested as beneficial with some species, e.g. *S. bovis*.
2. The source and purity of the dyes used affects their selective action. This was particularly the case with trypan blue.
3. There is increasing use of antibiotics, particularly aminoglycosides, in selective media. Care should be taken to test adequately before incorporating a new antibiotic, which may be more inhibitory. For example, gentamicin had been found too inhibitory for β-haemolytic streptococci of group B when used at $8 \mu g \, ml^{-1}$; it was much better at $4 \mu g \, ml^{-1}$.
4. Incubation may be in air or anaerobically under 90% H_2 + 10% CO_2. The latter helps the selectivity, especially useful with less inhibitory media. For routine use in examination of foods (other than milk products) aerobic incubation with a more inhibitory medium is preferred.
5. Incubation temperatures from 30 to 45°C have been suggested. The higher temperatures are likely to be more selective, but may require somewhat lower concentrations of inhibitors.

Attention was drawn to the problem of differentiating *Listeria monocytogenes* from *S. agalactiae* on selective media used for the isolation of these organisms from milk. It was pointed out that *L. monocytogenes* has a similar sensitivity to the selective agents and will also grow in the presence of nalidixic acid at the concentrations used.

In using faecal streptococci as indicator organisms, whether in water or foods, it was important to relate the types to their origins. For example, in using them to monitor faecal contamination in a slaughterhouse for cattle, the selective medium should be able to select for *S. bovis* which is the predominant streptococcus in the intestines of cattle, whilst in poultry processing *S. faecalis* and *S. faecium* can more appropriately be used as the faecal indicator organisms.

Appendix

List of UK suppliers of media and ingredients

Becton, Dickinson (UK) Ltd (BBL products)
 York House, Empire Way, Wembley, Middlesex HA9 0PS

Difco Laboratories
 PO Box 14B, Central Avenue, West Molesey, Surrey KT8 0SE.

Evans Medical Ltd.
 Speke Boulevard 24, Liverpool, Lancashire.

Koch-Light Laboratories Ltd.
 Poyle, Colnbrook, Buckinghamshire SL3 0BZ.

London Analytical & Bacteriological Media Ltd. (Lab M)
 50 Mark Lane, London EC3R 7QJ.

Oxoid Ltd.
 Wade Road, Basingstoke, Hampshire RG24 0PW.

Selected Abstracts Presented at the Summer Conference

Session I

Identification of Streptococci of the Viridans Group
D. B. DRUCKER
Department of Bacteriology & Virology, University of Manchester, Oxford Road, Manchester, Lancs, England

Recently it has been shown (Drucker & Green 1977, *Journal of Dental Research* **56**, A159) that certain oral strains of *Streptococcus milleri* are capable of inducing moderate amounts of dental caries in the gnotobiotic rat test system. Bacteriological examination of the strains involved has shown them to be similar, or identical, to many strains of *S. milleri* derived from pyogenic abscesses or wound infections and which produce virtually no dental caries in gnotobiotic rats. The strains which differed in cariogenicity and site of isolation from the 'medical' *S. milleri* were put through our computerized GLC-chemotaxonomic system and compared with many other strains of viridans-group streptococci and representatives of other groups. Only data for *S. milleri* will be discussed at this juncture.

The GLC (gas-liquid chromatography) system consists of an analysis of cellular fatty acids as their methyl carboxylic esters, formed under nitrogen with methanolborontrifluoride. The analysis is performed on polar and non-polar columns (Drucker & Veazey 1977, *Applied Environmental Microbiology* **33**, 221-226) and retention times and peak areas calculated automatically. The GLC data are computer analysed in three stages. I, peaks are tentatively identified with normalization of data; II, all pairs of data are compared and a matrix of coefficients is formed for 'unknown' and 'known' strains and III, strains are identified using the matrix.

It appears that most *S. milleri* strains can be automatically identified by GLC, producing coefficients of 0.99 when compared with 'library' strains (1.00 = identity). The cariogenic *S. milleri*, however, were not recognized by this means and have rather different fatty acid profiles from the non-cariogenic strains of *S. milleri*. Possibly the Swedish designation *S. turkosis* might be useful for such organisms; this was the name given to certain oral streptococci which were later termed *S. milleri*.

Streptococci Isolated from the Human Bloodstream

I. CRAWFORD

Microbiology Department, Stepping Hill Hospital

AND

C. RUSSELL

*Dental School, University of Manchester,
Oxford Road, Manchester, Lancs, England*

This is a preliminary report of a survey in which 52 cultures of streptococci, isolated from the blood of patients with a variety of clinical conditions, have so far been examined. The sources were hospital bacteriology laboratories, mainly in the north-west region. We requested sweep subcultures from the first plate upon which the organisms were isolated on subculture from the original broth. All except one, which was CO_2 dependent, grew in air. After primary subculture in our laboratory 12 separate colonies were selected from each plate and incubated overnight at 37°C in digest broth. These broth cultures were then examined according to seven morphological, biochemical and physiological criteria. On this basis two of the cultures were found to be mixed (two species in each case). A further set of 27 characteristics were recorded and the 54 isolates identified. The two mixed cultures were *S. mitior* and *S. milleri* in one case and *S. salivarius* and *S. milleri* in the other. The percentage proportions of the total 54 isolates were: *S. mitior*, 35.2; *S. sanguis*, 20.5; *S. milleri*, 9.2; *S. faecalis*, 7.5; *S. salivarius*, 7.5; Lancefield group B streptococci, 5.5; *S. bovis*, 3.7; *S. pyogenes*, 3.7; *S. equisimilis*, 3.7; *S. agalactiae* and a non-groupable β-haemolytic streptococcus, 1.8 each. Each isolate consisted of a single subtype. Using information provided by the completed questionnaires an attempt has been made to correlate the clinical condition of the patients with the organisms isolated. Among the 27 species isolated from the 25 endocarditis cases, the largest single clinical group, the percentage distribution was: *S. mitior*, 41.4; *S. sanguis*, 27.6; *S. salivarius*, 13.8; *S. milleri*, 10.4; *S. faecalis*, 3.4 and *S. equisimilis*, 3.4.

In comparison with work elsewhere, the absence of *S. bovis* and *S. mutans* and the relatively large proportion of cases with *S. salivarius* should be noted.

Streptococci and Lactobacilli associated with the Gastric Epithelium of the Neonatal Pig

P. A. BARROW, R. FULLER AND B. E. BROOKER

National Institute for Research in Dairying, Shinfield, Reading, Berks, England

Samples of stomach tissue were examined for adhering bacteria by light- and electronmicroscopy. Rod and coccal forms were seen attached to the *pars oesophagea* (a patch of squamous epithelium adjacent to the oesophageal opening into the stomach). Viable counts of washed *pars oesophagea* tissue yielded lactobacilli and less frequently, streptococci. The total count per unit area decreased in pigs up to one month old.

The most frequently isolated species were *Lactobacillus fermentum, Lact. salivarius, Lact. acidophilus, S. salivarius* and *S. bovis*. Their frequency of isolation was not directly related to their differing abilities to adhere *in vitro*. Our evidence suggests that successful colonization of the stomach by a bacterium depends on its ability to adhere to the *pars oesophagea* and/or its growth rate in the milk diet.

The attachment is tissue specific in that it occurred only with squamous epithelial cells and not with columnar cells from the small intestine and caecum. It is also host specific; bacteria isolated from dairy sources or animals other than pigs fail to adhere to pig oesophageal cells *in vitro*. The basis of this specificity is being investigated and a preliminary characterization of the adhesion determinants has been made.

It is suggested that attachment of bacteria to the *pars oesophagea* is a mechanism for ensuring the rapid colonization of the piglet stomach by specific strains of lactobacilli and streptococci. The lactic acid produced by these organisms may be important in maintaining a low pH in the stomach at a time when endogeneous HCl secretion is not fully effective.

The Development of a Phage-typing System for Group B Streptococci
JACQUELINE STRINGER
Central Public Health Laboratory, Colindale, London NW9 5HT, England

Lysogeny among group B streptococci isolated from human sources has not been reported previously. A large number of phages active against group B streptococci were obtained by cross-culture after induction by mitomycin C.

The lytic spectra of the phages showed a partial correlation with the distribution of type antigens among strains. Within each serotype, however, a number of distinct patterns of lysis were observed. So far, these appear to be reproducible and consistent among sets of supposedly related isolates.

There is little doubt that early-onset neonatal disease due to group B streptococci is acquired from the mother at childbirth. The epidemiology of late-onset neonatal meningitis is not understood, and the possibility of infection from other human sources in the hospital or outside it requires investigation. The existing serotyping scheme—in which only five types can be recognized—is insufficiently discriminating for this purpose. Phage typing shows considerable promise as a means of making fine divisions within serotypes.

Chemical Methods in the Classification of Lactic Acid and related Bacteria

M. D. COLLINS, M. GOODFELLOW, D. E. MINNIKIN AND D. JONES*

*Departments of Microbiology and Organic Chemistry, The University, Newcastle upon Tyne NE1 7RU and *Department of Microbiology, The University, Leicester LE1 7RH, Leics, England*

Isoprenoid quinones are widely distributed in bacteria and the distribution of some of the various structural types has proved useful in the classification of actinomycete and related taxa (Collins, M. D., Pirouz, T., Goodfellow, M. & Minnikin, D. E. 1977 *Journal of General Microbiology* **100**, 221).

Menaquinones, 2-methyl-3-polyisoprenyl-1,4-naphthoquinones, were isolated from a variety of lactic acid and related bacteria and characterized by UV and mass spectrometry. The predominant menaquinone found in strains of *Brochothrix thermosphacta* (*Microbacterium thermosphactum*), *Listeria grayi*, *L. monocytogenes* and *L. murrayi* contained seven isoprene units, abbreviated as MK-7, whereas *L. denitrificans* possessed MK-9 as the major component. The methyl group in the 2-position was absent from *S. faecalis* strains, a result in agreement with previous studies (Baum, R. H. & Dolin, M. I. 1965 *Journal of Biological Chemistry* **240**, 3425). Dimethyl menaquinones with nine isoprene units, DMK-9, were found to be the major isoprenologue in all of the streptococci examined. Representatives of *Lactobacillus mali* differed from all of the other strains examined by possessing MK-8 as the main component.

These preliminary findings suggest that the distribution of isoprenoid quinones may be of value in clarifying the relationships of lactic acid and related bacteria.

An Experimental Study of Infectivity of Streptococci for the Upper Respiratory Tract

D. N. KURL

Department of Pathology, University of Cambridge, Cambs, England

Experimental infection of mice was carried out with different species of streptococci by direct inoculation of the throat. The course of the infection and its effect on the normal flora was examined until the infection was eliminated completely.

The resistance of recovered mice to homologous and heterologous reinfection was examined. The sera of recovered animals were compared with pre-infection sera; this promised to yield serological evidence of infection, and possibly of protective antibody.

The ability of the streptococcus to adhere to the upper respiratory tract appears to be a prerequisite for infection and pathogenicity. The M protein is alleged to help in the adherence of both *S. pyogenes* and *S. salivarius*. Preliminary studies have been made of the effect of M protein and anti-M antibody on our test system.

The extent of cross-infection of the cage-mates was examined. This was an attempt not only to demonstrate the natural spread of an artificially-seeded infection, but also to provide information of value in the design of the experiments.

Measurement, Synthesis and Distribution of *Streptococcus mutans* Insoluble and Soluble Dextran–Sucrase Activity

T. J. MONTVILLE, C. L. COONEY AND A. J. SINSKEY
Department of Nutrition and Food Science, MIT, Cambridge, Massachusetts, USA

A filter disc assay which measures the incorporation of ^{14}C glucose into water soluble, insoluble and total dextran was used to determine the dextran sucrase activity of 10 strains of oral streptococci. Total dextran sucrase activity ranged from 139 to 2000 DSU cell^{-1} $\times$ 10^{-3}. Insoluble activities ranged from 40 to 375 DSU cell^{-1} $\times$ 10^{-13}. Non-adherent and non-cariogenic strains were observed to have only low levels of insoluble dextran sucrase activity when compared with the cariogenic strains.

The quantitative effect of soluble dextran on dextran sucrase activities was strain variable. Cultures grown in media containing Dextran T-10 showed decreased insoluble dextran sucrase activity and increased soluble activity. This effect was reproducible in culture to which chloromphenicol had been added and was shown to be immediate upon the addition of dextran. No change in the distribution of dextran sucrase was observed. The results suggest a physical rather than metabolic mechanism. Additional experiments showed that the inhibition of insoluble activity by Dextran T-10 was non-competitive with a $K_1 = 4 \times 10^{-2}$ mM. K_m for the supernatant insoluble dextran sucrase activity was determined to be 20 mM sucrose.

Total and insoluble dextran sucrase activities were measured in cell associated and supernatant fractions of *S. mutans* GS-5 grown in several media. While the amount of cell associated and supernatant activity varies greatly as a function of media, the whole culture activity appears constant. The distribution of dextran sucrase could be altered without changing the whole culture activity, indicating that the distribution of the enzyme may be regulated independently of its synthesis. Similar studies with *S. mutans* strains PR-89, LM-7 and 6715 showed that the distribution is both strain and growth media dependent. Strain GS-5 had significant cell-associated activity in a medium devoid of sucrose. In all cases the ratio of insoluble to total dextran sucrase activity was higher in the cell-associated fractions than in the cell-free supernatant.

Group B Streptococci of Human and Bovine Origin—Physiological Characteristics and Pathogenicity

G. Hahn

Institut für Hygiene der Bundesanstalt für Milchforschung, Kiel, Federal Republic of Germany

Streptococcus agalactiae (group B streptococci) is one of the most important strains causing mastitis in cattle. The same species can be isolated more and more as aetiological agent of severe infectious diseases in humans, especially in newborns, responsible for lethal septicaemia and meningitis.

Therefore it is of predominant interest whether humans can be infected by cattle, for example, by consumption of raw milk, and vice versa. For this reason data are presented of the incidence of group B streptococci in humans and bovines and also epidemiological investigations in human carriers. To differentiate these strains, cultural (type of haemolysis and colony, morphology of chains), biochemical (fermentation of lactose and glycerol, resistance to methylene blue) and serological (distribution of types) characteristics are compared, also pathogenicity for cattle and mice. Current investigations deal with bactericidal tests by human and bovine leucocytes *in vitro* as an indicator for the specific pathogenicity.

The results evaluated until now show a typical accumulation of the mentioned characteristics in either human or bovine strains. But on the other hand there is a distinct coincidence which indicates an adaptation phenomenon so that, in our opinion, there do not exist different human and bovine species of group B streptococci.

Session II

Amino Acid and Peptide Utilization by *Streptococcus thermophilus* in Relation to Yoghurt Manufacture

P. A. Shankar and F. L. Davies

National Institute for Research in Dairying, Shinfield, Reading RG2 9AT, Berks, England

During fermentation of milk to yoghurt with mixed cultures of *Streptococcus thermophilus* and *Lactobacillus bulgaricus*, the latter organism hydrolyses protein to products which stimulate the growth of the streptococcus. The utilization of a range of such compounds by strains of *S. thermophilus* has, therefore, been investigated in relation to acid and flavour production.

Using a defined medium, glutamic acid, histidine and cysteine were found to be essential for the growth of *S. thermophilus* whereas in milk heated as for yoghurt manufacture (95°C for 10 min), glutamic acid, histidine, valine, methionine, leucine and tryptophan stimulated acid production and were all required to equal the stimulation elicited by *Lact. bulgaricus*. Stimulatory amino acids could be replaced by dipeptides (alanyl, valyl or leucyl but not glycyl), tri-, penta- or hexapeptides containing appropriate residues.

Appreciable dipeptidase (cell bound and extracellular) activity towards valyl, leucyl and alanyl (but not glycyl) peptides and aminopeptidase activity (cell bound) was demonstrated for *S. thermophilus*. Proteinase activity on the other hand was very low in contrast to the activity of this enzyme in *Lact. bulgaricus*.

The addition of certain amino acids influenced production of acetaldehyde (an important flavour compound of yoghurt) by *S. thermophilus* in milk. Its formation was stimulated by methionine and threonine (and the intermediates homocysteine and cystathionine) but completely inhibited by glycine and cysteine. The effects of methionine and glycine were also shown by oligopeptides containing these residues.

Growth and Acetoin Production by *Streptococcus lactis* subsp. *diacetylactis*
T. M. COGAN
Moorepark Research Centre, The Agricultural Institute,
Fermoy, Co. Cork, Ireland

Streptococcus lactis subsp. *diacetylactis* is a common component of the starter cultures used in Cheddar cheesemaking and differs from *S. lactis* in its ability to metabolize citrate to diacetyl, acetoin, acetate and CO_2 in the presence of an energy source.

Citrate utilization and lactate, diacetyl and acetoin production by eight strains of *S. lactis* at 21°C in MRS broth (de Man, J. C., Rogosa, M. & Sharpe, M. E. 1960 *Journal of Applied Bacteriology* **23**, 130) modified by the reduction of the glucose content to 1% and omission of sodium acetate and Tween 80, were measured. Citrate utilization and diacetyl and acetoin production began and continued at a steady rate as soon as growth was initiated. The strains could be divided into two groups (AR and NAR) depending on whether the acetoin level decreased after the utilization of citrate was complete.

The two acetoin-reducing (AR) strains reduced markedly the level of acetoin when almost all the citrate had disappeared from the medium while the six non-acetoin-reducing (NAR) strains ceased acetoin production coincident with the disappearance of citrate from the medium. The maximum amounts of acetoin produced averaged 110 μg ml^{-1} (range 105-115 μg ml^{-1}) for the AR strains and 264 μg ml^{-1} (174-310 μg ml^{-1} for the NAR strains. The maximum amounts of diacetyl produced were much lower in all strains averaging 9 μg ml^{-1} (5.8-16.7 μg ml^{-1}). All strains showed reduction of diacetyl after all the citrate was utilized. In the AR strains, the rate of lactate production was logarithmic in one strain only to 0.5 mg ml^{-1} (pH 6.0) compared to 1.9–2.3 mg ml^{-1} (pH 4.7–4.9) for the NAR strains. The other AR strain showed a continuously decreasing rate reaching a maximum of 1.05 mg ml^{-1} (pH 5.25).

Omission of Mn^{2+} or Mg^{2+} from the medium had no effect on either the rate of growth or acetoin reduction by the AR strains. When the AR strains were grown in skim milk containing 0.3% (w/v) peptone, no reduction of acetoin occurred once citrate was utilized and the rates of acetoin reduction and citrate utilization were lower than in MMRSC broth even though the rate of lactic acid production was somewhat faster. In the milk-peptone medium two rates of lactate production were evident, the faster rate ending at 0.4 mg ml^{-1} and the slower at 2.5 mg ml^{-1}.

The results suggest that acetoin reduction in the AR strains is a response to poor growth but this needs to be confirmed by further experimentation.

Isolation and Activity of Phospho-β-galactosidase Constitutive Mutants of *Streptococcus lactis* ML3

Y. A. MOUSTAFA AND R. DAVIES

National College of Food Technology, University of Reading, St. George's Avenue, Weybridge, Surrey, England

Continuous culture technique with lactose limitation was used to isolate mutants, derived from *S. lactis* ML3, capable of producing phospho-β-galactosidase constitutively in the presence of non-inducible substrates. The medium used for mutant selection and maintenance contained sodium lactobionate. In comparison with *S. lactis* ML3 parent cultures, the mutants grew faster in milk and consistently produced higher titratable acidity the lower the pH values in standard cheese-starter activity tests. Mutant stability and ability to outgrow parent cells has been demonstrated during repeated, daily milk-culture transfers. over an extended period.

Enterococci of Avian Caecum

S. J. JENKINS, J. R. WALTON AND R. M. GRIFFIN*

Department of Veterinary Preventive Medicine, University of Liverpool Veterinary Field Station, Neston, Wirral and *May & Baker Limited, Ongar Research Station, Fyfield Road, Ongar, Essex, England

Forty-four per cent of caecal enterococci from chickens could not be classified with any of the established species of faecal streptococci.

The Fimbriae of *Streptococcus salivarius*

P. S. HANDLEY AND A. HANNAN

*Department of Bacteriology and Virology, The Medical School,
University of Manchester, Oxford Road, Manchester, Lancs, England*

Streptococcus salivarius is known to adhere to the dorsal surface of the tongue and electronmicrographs of thin sections have shown it to have a 'fibrillar fuzzy coat', which appears to attach the bacteria to the oral epithelial cell membrane (Gibbons, R. J., van Houte, J. & Liljemark, W. F. 1972 *Journal of Dental Research* 51, 424-435).

Negatively stained preparations of six strains of *S. salivarius* revealed the presence of fimbriae on all cells. The fimbriae appeared to be unusual, being very short (< 0.2 μm) and very fine, completely covering the cell surface. They were a very constant feature, being unaffected by ageing, anaerobiosis, 1% (w/v) glucose in the medium, or growth on different media.

Trypsinization (1% in phosphate buffer) for 4 h did, however, completely remove all fimbriae—a 0.1% solution only reduced the numbers, The 'fuzz' seen in thin sections was also trypsin sensitive (Gibbons, R. J. *et al.* 1972 *Journal of Dental Research* 51, 424-435) and loss of this layer prevented the cells from adhering to the epithelial cell membrane. Since the fimbriae are both numerous and trypsin sensitive it seems likely that they comprise a major part of the fibrillar fuzzy layer and are intimately involved with attachment of *S. salivarius* to the tongue surface.

The Formation of Bacillary Forms by Streptococci

FIONA M. TOMLEY AND C. RUSSELL

Department of Bacteriology and Virology and Dental School, University of Manchester, Oxford Road, Manchester, Lancs, England

Streptococci with unusual shape have been reported from a variety of habitats but little work has been done on environmental factors influencing cell form. Three strains of *S. mitior*, Lancefield groups C, F and L, from the mouths of students, were studied. All produced Gram positive and negative cocci and rods. Using the light microscope the following were noted: coccus, ovoid coccus, short rod, long rod, swollen coccus, swollen or mis-shapened rod. Single cell isolates of cocci and rods showed that each gave rise to the other. No medium was found in which cells were all cocci or all rods; rich media, i.e. BHI blood agar and PYE agar encouraged pleomorphism and Gram negativity. Growth over 30-42°C did not change the proportions of rods to cocci or the appearance of the cells. Incubation anaerobically compared to aerobically, on solid or in liquid medium, had no effect. Change in pH had a marked effect, cells becoming more coccal, Gram positive and chained as pH was lowered, e.g. at pH 5.3, *S. mitior* group L produced entirely Gram positive chained cocci, whereas at pH 6.5 cells were Gram variable and many swollen forms appeared. The proportion of rods to cocci increased with length of incubation and number of subcultures, in both cases reaching 20% rods to 80% cocci. Electronmicroscopic examination was carried out, using thin sectioning and negative staining techniques, on organisms grown on blood agar and BHI blood agar. All rods were multiseptate, indicating malfunction of separation of cocci. The shapes and septum arrangements of cells grown on BHI blood agar suggested a morphological 'life cycle'. Cells were surrounded by a dense fibrillar layer.

Two streptococci, each isolated as the sole organism from a dental abscess, although microscopy indicated Gram positive and negative rods and cocci in the pus, were also examined. After 4-5 subcultures on BHI blood agar they exhibited bacillary forms. Thus, cell types seen in smears of pus but not on culture might sometimes be morphological variants of one streptococcal species.

Variation in Sensitivity to Penicillins and Cephalosporins of Streptococci of different Serotypes
M. J. BASKER AND B. SLOCOMBE
Beecham Pharmaceuticals Research Division, Brockham Park, Betchworth, Surrey RH3 7AF, England

A number of streptococci of different serotypes were tested for sensitivity to various penicillin and cephalosporin antibiotics. The streptococci examined were clinical isolates causing disease in either humans or animals and included representatives of Lancefield serogroups A, B, C, D, E, F, G, H, K and O, and a number of non-typable viridans streptococci. These strains were tested for sensitivity to benzylpenicillin (Penicillin G), phenoxymethylpenicillin (Penicillin V), ampicillin, amoxycillin, methicillin, cloxacillin, flucloxacillin, carbenicillin and ticarcillin and to cephalothin, cephaloridine, cephazolin and cephalexin.

Streptococcal serogroups other than B or D were very sensitive to all the test antibiotics. Penicillin G, penicillin V and amoxycillin were the most active of the penicillins with Minimum Inhibitory Concentrations of the order of 0.001-0.01 $\mu g\ ml^{-1}$. The descending order of activity of the other penicillins was ampicillin, flucloxacillin and cloxacillin, carbenicillin, ticarcillin and methicillin. Of the cephalosporins examined, cephalorindine was the most active and was often as effective as Penicillin G; cephalothin and cephazolin were rather less active, and cephalexin was considerably less active than cephaloridine.

Group B streptococci were slightly less sensitive to the penicillins than the other groups, although interestingly, this was not the case with the cephalosporins. Group D streptococci showed a bimodal distribution. Enterococci

(*S. faecalis* and *S. faecium*) were the least sensitive of the streptococci tested, particularly to the cephalosporins, whereas the non-enterococcal species (*S. bovis*) was almost as sensitive as other serotypes. These results emphasize the need for proper identification of streptococcal isolates to ensure optimal choice of therapy.

Subject Index

Acetaldehyde, in yoghurt, 271, 403
Acidosis, 221, 223, 224
Actinomyces spp., 182, 184, 186
Adherence, of streptococci to surfaces, 172, 173, 174, 175, 400, 401
Aerobic properties, of lactic acid bacteria, 52
Aerococci, classification of, 65, 66
Agglutinins, action of on cottage cheese, 291
Aldolases, FDP, of various streptococci, 13, 14, 15, 16, 19, 20, 37
Amines, toxic, formation of in cheese, 273, 274
Anaerobic streptococci, 31, 32, 33, 34, 245, 246, 247, 249, 250
Antibiotics, effect on growth of calves, 217
   function of to producer organism, 310, 311
   presence of in milk, 292, 293
   production of by streptococci, 274, 275, 297, 298
   resistance of streptococci to, 100, 101
   susceptibility of streptococci to, 190, 191, 192, 250, 254, 256, 407, 408
   use of in selective media, 394
Antibodies, to group B streptococci, 137, 138
Appendicitis, role of *Streptococcus milleri* in, 97
Arthritis, in pigs, 92, 150
ATP synthesis, by streptococci, 52, 59, 61, 63, 226

Bacteraemia, 78, 152
   transient, after dental extractions, 189, 190, 192
Bactericidal test, *in vitro*, 111, 112
Bacteriocins, production of by streptococci, 183, 184, 185
Bacteroides ochraceus, 181, 182
Bacteroides ruminicola, 220, 227
Bargen streptococcus, classification of the, 26
Bile-tolerant streptococci, 337, 338
Birds, intestinal flora of, 250, 254
Bloat, 212, 219, 220, 221, 233
Bloodstream, human, streptococci isolated from, 398
Bovine strains, of *Streptococcus agalactiae*, 128
Brain abscesses, *Streptococcus milleri* in, 192
Burns, streptococcal infection of, 90
Butyrivibrio fibrisolvens, 220

Caecum, ruminant, microflora of, 218, 219, 233
Calves, streptococci found in alimentary tract of, 210, 212, 216, 218
Canned foods, use of nisin for preservation of, 298, 299, 300
Capsulated organisms, 221
Carbohydrate metabolism, alternative pathways of in lactic streptococci, 270, 271
Caries, dental, streptococci as cause of, 185, 186, 187, 188
Carriage, of group B streptococci, 132, 133, 134, 138, 139
Catalase-like activity, in lactic acid bacteria, 57, 58
Cephalosporins, sensitivity of streptococci to, 407, 408
Cheese, Cheddar, 266, 272, 273, 288, 403
   cottage, 266, 271, 279, 281, 291
   DVI starters for, 284, 288, 291
   Edam, 266, 272
   Emmenthal, 272
   hard, 279, 283, 291
   incubation of cultures, 282
   mould-ripened, 272, 281
   off-flavours in, 273
   Roquefort, 272
   slow acid development in, 298
   Stilton, 281
   Streptococcus faecalis strains isolated from, 24
   Streptococcus thermophilus as starter culture for, 37, 266
   Swiss type, 298, 386
Chickens, intestinal flora of, 248, 249, 254, 255, 256, 257
Clostridia, spoilage of foods by, 298, 300
Cold, damage to streptococci caused by exposure to, 351, 352, 353, 354
Coliforms, as neonatal pathogens, 130
Colonies, of streptococci, 376, 377, 379, 380, 385, 387, 388, 389, 390, 391, 392, 393
Colonization, 134, 135
   rate of, 131, 132, 137
Combined cultures, use of lactic streptococci in, 266

Communicability, of streptococci, in man, 87, 88, 89, 90, 91
Corynebacterium pyogenes, 147
    as cause of bovine endocarditis, 148
    in pigs, 150
    in sheep, 153
Cows, acidosis in, 221, 222, 223, 224
    bacteria isolated from gastrointestinal tract of, 217, 218, 219
    bloat in, 219, 220, 221
    Corynebacterium pyogenes infection in, 147, 148
    pneumococcal infection in, 148
    rumen bacteria of, 213, 214
    streptococci isolated from, 4, 7, 13, 16, 19, 22, 26, 27, 28, 34, 36, 144, 145, 146, 147, 148
Cultures, starter, 263, 265, 266, 268, 271, 272, 273, 274, 275
    essential characteristics of, 281
    frozen, 282
    freeze-dried, 291
    mixed, 291, 293
Curd, 266, 271, 273
Cytochrome-like respiration, in streptococci, 58, 59, 61
Cytochromes, in lactic acid bacteria, 61

Deer, rumen organisms in, 215
Denmark, bovine mastitis in, 146, 147
Dental plaque, streptococci found in, 159, 161, 165, 171, 174, 180, 182, 184, 185, 191
Dextran, formation of by streptococci, 73, 74, 81, 82, 87, 161, 162, 228
Dextran sucrase activity, of *Streptococcus mutans*, 401
Diacetyl, in cheese, 271, 272
Diet, effect of on rumen bacterial populations, 214
Diseases, foodborne, 315, 316, 317, 318, 320, 325
Dogs, as source of bovine mastitis infections, 146
    streptococci isolated from, 15
DVI starters, for cheese production, 284, 288, 291

Emden-Meyerhof pathway, in *Streptococcus bovis*, 226
Endocarditis, 81, 82, 84, 86, 87, 93, 94, 97, 98, 100, 101, 127, 135
    after dental bacteraemia, 189, 190, 192
    bovine, 148
    in sheep, 152, 153
    porcine, 151

Endocarditis index, 81, 82
Enterococci, 245, 246, 247, 249, 250, 254, 321, 335, 336, 405
Enterococcus division, of streptococci, 5, 6, 22
Enterotoxin B, production of by *Staphylococcus aureus*, 311
Enzymes, involvement of in lactose utilization, 268, 269, 270, 272, 273
    secretion of by *Streptococcus bovis*, 227, 228
Epithelial cells, adhesion of group A streptococci to, 110, 111
Erysipelas, group A streptococcal infection as cause of, 83, 84, 109
    streptococci isolated from cases of, 3
Erysipelothrix rhusiopathiae, in pigs, 151
Erythrogenic toxins, production of by group A streptococci, 82, 83, 86
Escherichia coli, in water, 335, 336, 341, 342, 343, 345

Faecal coliform: faecal streptococci ratios, 340, 341, 342, 345, 346
Faecal streptococci, 22, 23, 24, 25, 26, 27, 28, 245, 249, 250, 335, 336, 338, 340, 341, 343, 345
Faeces, bovine, counts of streptococci in, 216, 217, 218
    flora of, 249, 250, 252
    ovine, strains of streptococci isolated from, 218
Fermentation tests, for the classification of streptococci, 4
Fish, group B streptococci isolated from, 14
Flavin systems, streptococci with, 61, 62
Fluoride, sensitivity of streptococci to, 176, 177
Food, spoilage of, 298, 300
    streptococci in, 392
    tonsillitis outbreaks spread by, 91, 92, 315
Fusobacterium fusiforme, 182

Gastroenteritis, foodborne, 316, 317
Genital tract, female, group B streptococci isolated from, 129, 132, 133, 134, 136
    male, detection of group B streptococci in, 133
Glandular fever, 325
Glomerulonephritis, see Nephritis
Glucose, uptake of by oral streptococci, 176
    utilization of by lactic acid bacteria, 53, 54, 56

SUBJECT INDEX

Glucosyl transferases, isolation of from oral streptococci, 179, 180
Gnotobiotic animals, infection of with streptococci, 180, 182, 185, 186, 248, 255
Grain diets, effect of on rumen bacterial populations, 214, 215
Group D antigen, in pediococci, 66
Growth depression, organisms involved in, 254, 255, 256, 258

Haemin, addition of to media for lactic acid bacteria, 57, 58, 59, 60, 63
Heart-reactive antibody, production of, 122
Heat, damage to streptococci caused by exposure to, 354, 355, 356
Hepatitis A virus, 317, 318
Horses, streptococci isolated from, 27, 28, 151, 152
 'strangles' in, 151, 152
Hydrogen peroxide production, by streptococci, 62, 63

Ileum, of ruminants, bacterial counts from, 216, 217, 218
Impetigo, streptococcal, 78, 87, 88, 89, 90, 94, 102, 109, 118, 119, 123
Incubation temperatures, for streptococci, 394
Index organisms, use of group D streptococci as, 317, 318, 319, 328, 394
Indicator organisms, 335, 336 (see also Index organisms)
Infants, human, faecal flora of, 249
 neonatal disease in, 13, 90, 93, 94, 95, 129, 130, 131, 134, 135, 136, 137, 138, 139
Intestine, streptococci in flora of, 384, 385
Invasiveness, of group A streptococci, 85, 86
Isoprenoid quinones, 400

Lactic acid, accumulation of in the rumen, 222, 223, 224
 production of by anaerobic streptococci, 31, 33
Lactic acid bacteria, 57, 58, 59, 60, 63, 400
Lactic acid utilizing bacteria, 221, 222, 224, 230
Lactic division, of streptococci, 5, 6, 263, 264, 265, 266, 267, 268, 269, 270, 271, 272, 273, 274, 275, 386, 387, 388, 389, 390, 391
Lactobacilli, in gastric epithelium of neonatal pigs, 398, 399

Lactobacillus brevis, utilization of glucose by, 54
Lactobacillus buchneri, utilization of glucose by, 54, 55, 56
Lactobacillus bulgaricus, 266, 268, 281, 386, 390, 402, 403
Lactobacillus plantarum, effect of hydrogen peroxide on, 63
Lactobacillus viridescens, 54, 55, 56
 effect of haemin on, 58
Lactose, utilization of by lactic streptococci, 268, 269, 270
Lancefield precipitin technique, 5
Lancefield serological grouping, 5, 7, 9, 17, 128, 129
Leuconostoc mesenteroides, growth of in presence of haemin, 58
 utilization of hexoses by, 53, 54
Leuconostocs, in the diary industry, 264, 265, 279, 391
Lewis protected system, for production of starter cultures, 282, 284
Lipoteichoic acid (LTA), 110
LIPS method, for controlling contamination of foods, 328, 329
Listeria monocytogenes, 394

M antigens, 108, 109, 110, 112, 113, 114, 116, 117, 121, 122, 123
M-associated protein, 113, 114, 117, 118, 122
M types, 108, 110, 111, 112, 114, 116, 117, 118, 119, 120, 121
Mannitol, fermentation of by streptococci, 26, 27
Mastitis, antibiotics used for treatment of, 292
 groups G and L streptococci as cause of, 146, 147
 pneumococcal infection in, 148
 streptococcal infection caused by milk of cow with, 91
 Streptococcus agalactiae as cause of, 13, 133, 144, 146, 147, 402
 Streptococcus dysgalactiae as cause of, 146, 147
 Streptococcus uberis as cause of, 34, 147, 148
 Streptococcus zooepidemicus as cause of, 146
Maternity hospitals, streptococcal infections in, 90
Meat, raw, streptococci in, 92

SUBJECT INDEX

Media, effect of on survival of bacteria after exposure to stress conditions, 349, 350, 354, 358, 361, 362, 363, 366
   for detection of streptococci in foods, 319, 320, 321, 325, 328
   for faecal streptococci, 337, 338
   isolation, for animal disease streptococci, 379
   for dairy streptococci, 386
   for food streptococci, 392
   for human disease streptococci, 375
   for intestinal streptococci, 384
   for oral streptococci, 380
   for water streptococci, 393
   for streptococci, 371
Megasphaera elsdenii, 215, 222, 230
Membrane filtration technique for water supply, 336, 337
Meningitis, in pigs, 92, 149
   neonatal, 90, 95, 97, 129, 130, 131, 134, 135, 136, 138, 399
   pneumococci as cause of, 80, 87, 95, 102
   streptococci isolated from outbreaks of, 16
   Streptococcus bovis as cause of, 93
Mice, adherence of streptococci to upper respiratory tract of, 400, 401
   outbreaks of cervical adenitis and pneumonia in, 108
Milk, causes of slow growth of lactic streptococci in, 274, 275
   group B streptococci infections caused by drinking of, 144
   presence of antibiotics in, 292, 293
   strains of streptococci isolated from, 14, 29, 31, 34, 36, 37, 38, 91
   utilization of proteins in by lactic streptococci, 267, 268
Monkeys, group A streptococci isolated from, 108, 112
   rhesus, bacterial fermentation in stomach of, 247
Mouse, organisms found in stomach of, 247, 249, 250, 257
Mycoplasmas, in the rumen, 231, 232

NADH, oxidation of, 58, 59, 61, 62
Neonatal disease, human, 90, 93, 94, 95, 131, 135, 136, 399
   strains of *Streptococcus agalactiae* responsible for, 13, 129, 130, 131, 134, 135, 136, 137, 138, 139, 402
Nephritis, 84, 91, 109, 111, 118, 120, 121, 123

Nisin, correlation between length of lag phase and cellular level of, 309, 310
   disappearance of, 307
   effect on growth of producer organism, 306, 307
   effect on spores, 300
   location of in producer organism, 305
   production of by strains of *Streptococcus lactis*, 29, 275
   resistance of *Clostridium botulinum* to, 300
   synthesis of, 301, 302, 303, 304, 305, 306
Nitrogen metabolism, of ruminants, 224, 225

Opacity factor, production of by group A streptococci, 114, 116, 117
Oral streptococci, 19, 20, 21, 76, 77, 318, 325, 380, 381
Oxidative phosphorylation, 58, 59

Pediococci, classification of 64, 66
Pediococcus pentosaceus, comparison of with strains of *Streptococcus faecalis* and *Streptococcus faecium*, 66
Penicillin, for treatment of bloat, 220, 221
   sensitivity of streptococci to, 407
Penicillium roqueforti, 281
Peptides, utilization of by streptococci, 268
Peptostreptococcus genus, 33, 246
Peroxidase system, sensitivity of cheese starters to, 292
   in *S. faecalis*, 63
pH, of growth medium, effect of on *Streptococcus lactis*, 358
Phage infections, of lactic streptococci, 275, 283, 291, 293
Phage-typing system, for group B streptococci, 399
Phage-unrelated 'rotations', use of in dairy industry, 291
Phages, in the rumen, 231, 232
Pigs, arthritis in, 92, 150, 151
   cervical lymphadenitis in, 150
   Corynebacterium pyogenes in, 150
   endocarditis in, 151
   group L streptococci in, 147, 151
   groups R and S streptococci in, 149, 150
   growth depression in, 258
   intestinal flora of, 249, 258
   meningitis in, 92, 149, 150
   neonatal, streptococci and lactobacilli in gastric epithelium of, 398

SUBJECT INDEX

strains of streptococci isolated from, 14, 15, 16, 92
Streptococcus suis in, 149, 150
Pneumococci, 73, 76, 78, 95
  antibiotic resistance in, 101
  classification of, 3, 4, 9, 17, 19
  infections in calves caused by, 148
  mastitis caused by, 148
  meningitis caused by, 80
  middle-ear infection caused by, 78, 101
  streptococcal pneumonia caused by, 78
  toxic effects of, 86
Pneumonia, pneumococcal, 95, 102
  streptococcal, 78, 87
Polysaccharides, extracellular, production of by oral streptococci, 177, 178, 179, 180
  intracellular, synthesis of, 180
Poultry, intestinal flora of, 252, 254, 258
  streptococcal diseases of, 153
Precipitation test, for M typing, 112
Pregnancy, carriage of group B streptococci in, 129, 132, 134, 137
Protozoa, in the rumen, 231, 232
Pyogenic division, of streptococci, 5, 6, 9, 10, 11, 12, 13, 14, 15, 16, 17, 72, 73, 81, 215

R antigens, 109
Rabbit, intestinal flora of, 248
Rat, streptococci in the mouth of, 172, 173, 180, 181, 182, 186
Recovery, of streptococci from cold-induced damage, 358, 359, 366
  from heat-induced damage, 359, 360, 361, 362, 365, 366
Resistance, of group D streptococci, 317
  of hepatitis A virus, 317, 318
Rheumatic fever, 84, 97, 102, 114, 118, 120, 121, 122, 123, 191
River water samples, faecal streptococci in, 337, 338, 340, 341, 342, 343, 345
Rodents, outbreaks of cervical adenitis and pneumonia in, 108
Rumen, isolation of *Streptococcus lactis* from, 264
  streptococci in the, 208, 209, 210, 211, 212, 213, 214, 215, 216, 219, 220, 221, 222, 224, 225, 226, 227, 228, 229, 230, 231, 232, 233

Sanitizers, effect on *Streptococcus faecalis*, 356

Scarlet fever, group A streptococcus as cause of, 82, 83, 84, 118, 123, 315
Selective agents, 372, 373
Selenomonas ruminantium, 224
Septic sore throat, 315, 320
Septicaemia, 77, 78, 79, 80, 95, 96, 129, 130, 131, 136, 138, 139
Serological grouping, of streptococci, 6, 7, 8, 9, 13, 14, 15, 16, 17, 18, 19, 20, 21, 38
Serotypes, distribution of among streptococcal isolates, 135, 136
Sewage samples, faecal streptococci in, 337, 338, 340, 341, 342, 343, 345, 346
Sheep, streptococcal diseases of, 152, 153
  streptococci isolated from rumen of, 213, 214, 215
Skin, streptococci isolated from, 112, 119, 120, 121
Slime, formation of by *Streptococcus bovis* on sucrose agar, 26, 27
  production of by rumen bacteria, 220, 221
Sodium azide, selective activity of, 372, 373
Soft agar culture, for lactic acid bacteria, 52, 53, 54, 56
Sows, group B streptococci isolated from vagina of, 144, 147
Starch hydrolysis, in *Streptococcus bovis* and *Streptococcus equinus*, 218, 227, 233
Starter cultures, for the dairy industry, 263, 265, 266, 268, 271, 272, 273, 274, 275, 386, 387, 391
  bulk, 283, 284
  concentrated, 284, 288
  essential characteristics of, 281
  frozen, 282, 283
  mixed, 291, 293
Streptococci, anaerobic, 31, 32, 33, 34, 245, 246, 247, 249, 250
  faecal, 22, 23, 24, 25, 26, 27, 28, 97, 245, 249, 250, 335, 336, 338, 340, 341, 343, 345
  FDP aldolases of, 13
  group A, 78, 82, 83, 84, 85, 86, 87, 88, 89, 90, 91, 92, 94, 95, 130, 315, 320, 328, 376, 377
  group B, 76, 78, 80, 86, 90, 92, 95, 96, 144, 252, 316, 320, 376, 377, 399, 402
  group C, 78, 80, 90, 95, 146, 151, 152, 376

SUBJECT INDEX

group D, 148, 149, 151, 153, 209, 210, 213, 215, 225, 228, 230, 246, 252, 257, 316, 317, 318, 321, 322, 324, 325, 328, 335, 392, 393
group E, 150, 153
group G, 78, 80, 90, 95, 146, 153, 376
group H, 170
group L, 146, 147, 150, 151, 153
group N, 215, 257, 264, 265, 266, 268, 269, 270, 271, 279
group R, 149
group S, 149
β-haemolytic, 375
human and animal strains of, 252
non-haemolytic, 74, 75, 76
oral, 19, 20, 21, 76, 77, 318, 325, 380, 381
Sherman's divisions of, 4, 5, 6, 7, 9, 22
Streptococcaceae family, genera included in, 1, 2, 33
Streptococcus acidominimus, 34, 36, 251
Streptococcus agalactiae, 4, 6, 7, 13, 73, 127, 128, 129, 130, 131, 132, 133, 134, 135, 136, 137, 138, 139
as cause of bovine mastitis, 144, 146, 147, 292
media for, 376, 377, 379, 394, 398, 402
responsible for neonatal disease, 14, 19
Streptococcus anginosus, 3, 17, 20
Streptococcus avium, 22, 25, 26, 28
in human faeces, 77, 251, 385
Streptococcus bovis, 4, 7, 13, 16, 19, 22, 26, 27, 28, 38, 74, 76, 77, 81, 93, 100, 210, 212, 213, 214, 217, 218, 219, 220, 221, 222, 224, 225, 226, 227, 228, 229, 230, 231, 232, 233, 247, 252, 254, 321, 335, 341, 381, 385, 394, 395, 399
Streptococcus constellatus, 33
Streptococcus cremoris, 4, 29, 31, 37, 264, 265, 266, 267, 268, 271, 283, 285, 297, 386, 387
Streptococcus diacetylactis, 29, 31, 62, 264, 265, 266, 271, 283, 297, 386, 387, 403
Streptococcus durans, 24, 232
Streptococcus dysgalactiae, 12, 13, 219, 292, 379
Streptococcus equi, 12, 13, 108, 151, 152
Streptococcus equinus, 3, 7, 19, 22, 26, 27, 28, 38, 210, 218, 227, 252, 321, 335, 341, 385
Streptococcus equisimilis, 12, 13, 73, 108, 150, 152, 398

Streptococcus faecalis, 3, 6, 7, 16, 22, 24, 25, 26, 28, 31, 34, 37, 38, 39, 73, 77, 78, 81, 91, 100, 153, 210, 212, 213, 216, 217, 218, 219, 226, 229, 231, 232, 248, 250, 252, 255, 257, 335, 337, 342, 398, 408
comparison of strains of with *Pediococcus pentosaceus*, 66
damage to from exposure to cold, 352, 354
damage to from exposure to heat, 354, 355, 356
differentiation of from *Streptococcus faecium*, 67
effect of haemin on, 58, 59, 60
effect of sanitizers on, 356
formation of catalase by, 57
grown on glucose, 53, 61
hydrogen peroxide destroying systems in, 63, 64
in the mouth, 172, 173, 190
media for, 373, 395
motile strains of, 24, 25
oxidase activity in strains of, 61, 62
recovery of from cold-induced damage, 358, 359
recovery of from heat-induced damage, 361, 362, 363, 364
use of in cheese starters, 274
utilization of polyols and myo-inositol by, 54
Streptococcus faecalis subsp. *faecalis*, 250, 252, 256
Streptococcus faecalis subsp. *liquefaciens*, 225, 226, 248, 255, 256, 257, 258
Streptococcus faecium, 4, 22, 24, 25, 26, 27, 28, 34, 38, 73, 77, 385, 395
comparison of with *Pediococcus pentosaceus*, 66
differentiation of from *Streptococcus faecalis*, 67
heat damage to, 354, 365
hydrogen peroxide production by, 62, 63
in intestinal flora, 250, 252
in ruminants, 210, 213, 216, 217, 218, 224, 229, 232, 233
motile strains of, 24, 25
serological types of, 254
utilization of polyols and myo-inositol by, 52, 53, 54
Streptococcus faecium subsp. *durans*, 255
Streptococcus glycerinaceus, 4, 24
Streptococcus hansenii, 33, 250
Streptococcus infrequens, 14, 150

SUBJECT INDEX

Streptococcus intermedius, 33, 250
Streptococcus inulinaceus, 4, 26, 27, 28, 254
Streptococcus lactis, 4, 13, 28, 29, 31, 39, 62, 215, 264, 265, 266, 267, 268, 283, 285, 288, 297, 304, 310, 358, 386, 387, 390, 403, 405
   cold damage to, 351, 352, 353
Streptococcus lentus, 14, 108
Streptococcus liquefaciens, 4
Streptococcus mastitidis, 4, 128
Streptococcus milleri, 17, 19, 20, 73, 74, 75, 76, 77, 80, 81, 87, 93, 94, 97, 100, 158, 162, 163, 171, 186, 188, 189, 190, 192, 318, 380, 381, 397, 398
Streptococcus mitior, 20, 73, 100, 158, 161, 162, 163, 164, 170, 173, 174, 188, 189, 190, 380, 381, 398, 406, 407
Streptococcus mitis, 3, 20, 73, 162, 174, 181, 182, 184, 318, 328
Streptococcus morbillorum, 33, 250
Streptococcus mutans, 20, 21, 74, 76, 81, 93, 100, 318, 380, 381, 382, 383, 401
   effect of haemin on, 58
   genospecies of, 21
   in the mouth, 157, 158, 159, 160, 163, 164, 167, 168, 169, 170, 175, 176, 177, 178, 179, 180, 181, 182, 183, 184, 185, 186, 187, 188, 189, 190, 191
   serotypes of, 168, 169, 170
Streptococcus pleomorphus, 33, 34, 251
Streptococcus pneumoniae, 17, 19
Streptococcus pyogenes, 3, 6, 9, 12, 13, 17, 19, 31, 73, 107, 108, 109, 110, 111, 112, 113, 114, 115, 116, 117, 118, 119, 120, 121, 122, 123, 136, 292, 398, 401
Streptococcus salivarius, 3, 19, 74, 76, 158, 162, 164, 171, 172, 173, 174, 175, 178, 181, 182, 183, 184, 186, 189, 190, 219, 318, 328, 380, 381, 398, 399, 401, 406
Streptococcus sanguis, 20, 21, 58, 73, 74, 81, 87, 100, 157, 158, 160, 161, 164, 167, 170, 172, 173, 174, 175, 178, 179, 181, 182, 183, 184, 186, 188, 190, 318, 328, 398
Streptococcus subacidus, 14
Streptococcus suis, 7, 16, 108, 149, 150
Streptococcus thermophilus, 4, 36, 37, 265, 266, 268, 270, 281, 386, 387, 388, 389, 390, 394, 402, 403

Streptococcus turkosis, 397
Streptococcus uberis, 34, 36, 215, 219, 292, 379
Streptococcus viridans, 20
Streptococcus zooepidemicus, 12, 13, 91, 92, 143, 146, 151, 152, 153
Sucrose, use of in experiments on dental caries, 185, 186

T antigens, 109, 116
Teeth, streptococci on the surface of, 171, 172, 174, 180
Tellurite resistance, recovery of, 361, 363, 364
Tests, for the classification of streptococci, 3, 4, 164, 165
   for the dairy industry, 283
Thallous acetate, selective activity of, 373
Throat, streptococci in the, 76, 89, 112, 119, 120, 121, 400
Tongue, streptococci on the, 171, 172
Tonsillitis, streptococcal, 88, 89, 90, 94, 111, 118, 119, 123
   foodborne outbreaks of, 91, 92, 315
Toxaemia, 79, 86
Toxins, erythrogenic, production of by group A streptococci, 82, 83, 86
Tropics, nephritis in, 84, 102
   rheumatic fever in, 84, 102

Uric acid, decomposition of by *Streptococcus faecalis*, 257
Urinary tract, group B streptococci found in infections of, 129

Vaccines, for dental caries, 188
Vagina, of sows, streptococci in the, 144, 147
   streptococci in the, 76, 77, 78, 90, 130, 134, 137, 139
Veillonella alcalescens, symbiotic relationship of with *Streptococcus mutans*, 182
Viridans division, of streptococci, 5, 6, 73, 74, 76, 80, 81, 93, 97, 98, 158, 163, 189, 209, 210, 246, 397

Water, streptococci in, 393
Wounds, streptococcal infection of, 90

Yeast extract, stimulatory effect of on lactic streptococci, 275, 284
Yoghurt, manufacture of, 266, 271, 279, 281, 282, 292, 293, 386, 388, 389, 390, 402, 403